Electric Power Applications

Electric Power Applications

Editors

Federico Barrero
Mario Bermúdez

MDPI

Basel • Beijing • Wuhan • Barcelona • Belgrade • Novi Sad • Cluj • Manchester

Editors
Federico Barrero
University of Seville
Seville, Spain

Mario Bermúdez
University of Seville
Seville, Spain

Editorial Office
MDPI
St. Alban-Anlage 66
4052 Basel, Switzerland

This is a reprint of articles from the Special Issue published online in the open access journal *Applied Sciences* (ISSN 2076-3417) (available at: https://www.mdpi.com/journal/applsci/special_issues/electric_power_applications).

For citation purposes, cite each article independently as indicated on the article page online and as indicated below:

Lastname, A.A.; Lastname, B.B. Article Title. *Journal Name* **Year**, *Volume Number*, Page Range.

ISBN 978-3-0365-9336-4 (Hbk)
ISBN 978-3-0365-9337-1 (PDF)
doi.org/10.3390/books978-3-0365-9337-1

Cover image courtesy of Mario Bermúdez Guzmán

Contents

About the Editors

Federico Barrero

Prof. Dr. Federico Barrero received an M.S. and a Ph.D. in electrical and electronic engineering from the University of Seville, Seville, Spain, in 1992 and 1998, respectively. In 1992, he joined the Department of Electronic Engineering at the University of Seville, where he currently serves as a Full Professor. He received the Best Paper Award from the IEEE Transactions on Industrial Electronics in 2009 and IET Electric Power Applications for 2010–2011. He is the coauthor of more than 300 studies.

Mario Bermúdez

Dr. Mario Bermúdez received a B.Eng. in industrial engineering from the University of Málaga, Málaga, Spain, in 2014 and a Ph.D. in electrical/electronic engineering jointly from Arts et Métiers ParisTech, Lille, France, and the University of Seville, Seville, Spain, in 2018. He is currently an Associate Professor at the Department of Electrical Engineering at the University of Seville. His research interests include the modeling and control of multiphase drives, digital-signal-processor-based systems, and electrical vehicles.

applied sciences

MDPI

Editorial

Special Issue on Electric Power Applications

Federico Barrero [1] and Mario Bermúdez [2,*]

[1] Department of Electronic Engineering, University of Seville, 41004 Seville, Spain; fbarrero@us.es
[2] Department of Electrical Engineering, University of Seville, 41004 Seville, Spain
* Correspondence: mbermudez4@us.es

The continuing trend toward greater electrification in consumer, commercial, industrial, and transportation applications promises a dynamic and increasingly important role for power electronics. The term 'power electronics' refers to electric and electronic circuits whose primary function is to process energy. The growing penetration of power electronics in energy systems requires attention, with its principal challenges being cost reduction and reliability. Power electronic systems are indispensable parts of modern engineering applications, covering a wide range of engineering branches.

This Special Issue of *Applied Sciences* entitled "Electric Power Applications" includes 16 published papers, confirming the scientific community's interest in this area. The topics of interest for the call included modern applications of electricity, power electronics converters, electric motor drives, renewable energy systems, power filters, distributed power, and power grid equipment. A summary of the contributions in this Special Issue can be found below.

In [1], a current closed-loop control strategy based on an improved QPIR (quasi-proportional integral resonant) controller is presented for a three-phase LCL grid-connected inverter. The proposed controller enables the independent control of active power and reactive power without coupling between the α axis and β axis in a two-phase static coordinate system. Compared with traditional controllers, the improved QPIR control strategy has higher grid-connected current control accuracy under the condition of stable and fluctuating grid frequencies while also showing a good dynamic response.

Since the number of distribution terminal units (DTUs) integrated in power systems is gradually increasing, it is necessary to reduce the fault incidence of DTU devices and improve the efficiency of fault elimination. This was the aim in [2], where the authors propose a DTU fault analysis method using an association-rule-mining algorithm. The practicality of the method is demonstrated by experiments using a realistic DTU fault database, allowing for the conclusion that DTU fault incidence can be reduced, and fault elimination ability can be enhanced.

The authors of [3] explored the simultaneous operation of the imbalance-netting process (INP) and the cross-border activation of regulating reserves (CBRR), proposing a function for correction power adjustment to prevent undesirable simultaneous activation. Extensive dynamic simulations were carried out in a testing system with three control areas to evaluate the impact on the frequency quality, load-frequency control, and performance. The results confirm that the simultaneous operation of the INP and CBRR reduces the balancing energy and decreases the unintended exchange of energy, thereby improving load frequency and performance.

The authors of [4] developed an adaptive and scalable protection coordination (ASPC) system for overcurrent relays (OCRs) in distributed-generator-integrated distribution networks consisting of two performance stages. The first stage determines the min–max confidence interval of the fault current for different fault types, while in the second stage, three common algorithms (Particle Swarm Optimization, the Gravitational Search Algorithm, and a genetic algorithm) are used to find the optimal setting parameters of the OCRs

Citation: Barrero, F.; Bermúdez, M. Special Issue on Electric Power Applications. *Appl. Sci.* **2023**, *13*, 11574. https://doi.org/10.3390/app132011574

Received: 12 October 2023
Accepted: 19 October 2023
Published: 23 October 2023

that satisfy all the coordination constraint conditions. A real 22 kV distributed-generator-integrated distribution network was used as a testbed to validate the relay coordination results obtained by the ASPC system.

The influence of an inverter-interfaced distributed generation (IIDG) unit's response during transmission network faults is analyzed in [5]. An improved power-based fast fault current (FFC) concept is proposed to adjust fault current injection according to the required active and reactive power. This proposal is compared with the current-based FFC concept in different 110 kV networks with loop and radial topologies and for different short-circuit capacities of the aggregated network supply.

In the field of the maintenance of electric power system devices and networks, time-based maintenance (TBM) is still widely used. Recently, many companies have gradually been introducing condition-based maintenance and its upgraded version, called reliability-centered maintenance (RCM), which considers the reliability of the entire system and not only individual elements. An RCM model is developed and tested for a transmission substation in [6], where the planning and actual performance of maintenance are carried out using an optimization algorithm. The novelty of this paper consists in its integrated treatment of all maintenance processes that are included in the pre-processing phase and used in the optimization process for reliability-centered maintenance. The proposed RCM optimization algorithm is tested and compared with a previous TBM in an existing Slovenian substation, resulting in the conclusion that RCM allows for the maintenance of the reliability and technical condition of the equipment and components while reducing maintenance costs.

The extension of the Direct Torque Control (DTC) strategy to symmetrical five-phase induction machines with distributed windings in normal and faulty operation is described in [7]. Virtual voltage vectors were used to nullify the stator voltage and current components in the x–y plane. The experimental results obtained under healthy and faulty conditions (one and two open phases) show that the speed, torque, and flux references were successfully tracked in all cases. This led the authors to conclude that DTC is viable and can be extended to the case of multiphase drives, particularly when the required control goals are robustness, low computational cost, and a natural fault-tolerant capacity.

In [8], the authors solve the issues of optimal supply and the purchase of reactive power in the IEEE 30-bus power system when considering voltage stability and reducing total generation and operational costs. A modified version of the artificial bee colony (MABC) algorithm is used to solve optimization problems. Additionally, its results are compared with those of an artificial bee colony algorithm, the particle swarm optimization algorithm, and a genetic algorithm. The optimization results prove that the proposed MABC algorithm has lower active power loss and reactive power cost and a better voltage profile than the other algorithms.

In [9], the impact of installing photovoltaic solar panels on diesel–electric trains is analyzed in order to improve energy efficiency and reduce greenhouse gas emissions. The energy produced by the solar panels is quantified and compared with that produced by an auxiliary diesel generator during six different journeys worldwide under various solar resource potentials and meteorological weather conditions. The results reveal that the minimum annual fuel reduction of auxiliary generators provided by the solar panels was above 50% in all cases, demonstrating the feasibility of the proposal.

A new type of converter for power factor correction applications, named Independent Double-Boost Interleaved Converter (IDBIC), is analyzed in [10]. It is based on three voltage levels at the output combined with interleaved operation at the input and high voltage gain. The proposed converter was tested through simulation and experimentation to highlight the overall impact of the solution, with the authors concluding that the total harmonic distortion and power factor improve near the converter-rated power. The authors also observed that the converter has better performance in terms of power quality at higher voltage gain, while its efficiency is lower at a low supply voltage. According to the authors, the proposed solution can be successfully applied to electric vehicles, high-power electrical traction, rapid-battery-charging applications, and chemical electrolysis processes.

The research in [11] provides a theoretical design for transformer winding resistance under short-circuit conditions. The damping, stiffness, and mass parameters between windings are comprehensively considered, and a classical "mass-spring-damping" axial vibration mathematical model of transformer windings is established, yielding a more detailed and refined mode shape structure compared with the previous V-type and M-type mode shapes. In addition, the authors analyze the displacement and acceleration distributions of the model when a transformer with different damping ratios is short circuited at different frequencies.

In [12], the authors propose a fault diagnosis method based on a gradient-boosting decision tree (GBDT) to improve the efficiency of the protection systems in an intelligent substation. This method presents a higher fault diagnosis accuracy compared with the existing methods based on recurrent neural networks and random forest algorithms. Additionally, the GBDT algorithm also performs very well when faced with inadequate data (e.g., false alarms of fault information and multiple faults), allowing the authors to conclude that it can play a better role in practical applications.

In [13], a novel method is introduced to facilitate the selection process of the generators in a ship power plant. This method uses different parameters related to the life-cycle costs of engines and the operating routines of ships along with other cost factors such as flat annual cost, maintenance, personnel, or the acquisition and installation costs. The results provide valuable insight into the total cost from every aspect and present the optimum generator for minimal expenditure and a maximum return of investment.

The state-of-the-art technology for energy harvesting (EH) used in wireless sensors that acquire real-time electrical data in transmission lines is reviewed and analyzed in [14]. Although EH systems for transmission lines are reviewed therein, many other applications could potentially benefit from introducing wireless sensors with an EH capacity, such as power transformers, distribution switches, or low- and medium-voltage power lines, among others. The authors conclude that, among the different existing energy harvesting methods, there is not a universal solution since the most suitable EH system depends on the characteristics of each application, such as the geographical location, the intensity of the current flowing in the line, or the line's voltage.

Since renewable generators are usually connected to an electrical power system through power electronic converters lacking natural responses to frequency variations, [15] focuses on proposing a solution based on advanced controls that allow such generators to participate in frequency control. Particularly, this paper studies the benefits of introducing power reserve control in photovoltaic generators and extended optimal power-point-tracking control in wind generators to provide frequency control in low-inertia power systems and for the interactions between them. The results show that these control systems help in stabilizing system frequency thanks to the cooperative action of both types of renewable generators.

In [16], a novel triple phase shift (TPS) closed-loop control scheme of a dual active bridge (DAB) LCC resonant DC/DC converter is presented to perform the unity power factor operation of wireless charging at an optimized rectifier AC load resistance. The primary-side inverter phase shift angle regulates the battery charging current/voltage, while the secondary side rectifier phase shift angle controls the rectifier AC load resistance to match its optimized settings. Finally, the inverter-to-rectifier phase shift angle is set to achieve the unity power factor operation of a DAB rectifier and inverter. Simulation analyses and experimental tests were carried out in a small power laboratory experimental setup to verify the proposed TPS closed-loop control scheme.

Author Contributions: Conceptualization, all authors; methodology, all authors; software, all authors; validation, all authors; formal analysis, all authors; investigation, all authors; resources, all authors; data curation, M.B.; writing—original draft preparation, all authors; writing—review and editing, all authors; visualization, all authors; supervision, F.B.; project administration, F.B.; funding acquisition, F.B. All authors have read and agreed to the published version of the manuscript.

Acknowledgments: This Special Issue would not be possible without the hard work of the authors, reviewers, and the editorial team of *Applied Sciences*. The Guest Editors would like to take this opportunity to thank the authors who responded to the call and made valuable contributions. We are also deeply indebted to the reviewers whose feedback, comments, and suggestions helped the authors to improve their papers. Finally, we would like to note our gratitude to the dedicated editorial team of *Applied Sciences* and give special thanks to the section managing editors for their excellent support during every stage of the development of this project.

Conflicts of Interest: The authors declare no conflict of interest.

References

1. Li, Y.; Zhang, J.; Hao, Z.; Tian, P. Improved PR Control Strategy for an LCL Three-Phase Grid-Connected Inverter Based on Active Damping. *Appl. Sci.* **2021**, *11*, 3170. [CrossRef]
2. Zhang, X.; Tang, Y.; Liu, Q.; Liu, G.; Ning, X.; Chen, J. A Fault Analysis Method Based on Association Rule Mining for Distribution Terminal Unit. *Appl. Sci.* **2021**, *11*, 5221. [CrossRef]
3. Topler, M.; Polajžer, B. Evaluation of the Simultaneous Operation of the Mechanisms for Cross-Border Interchange and Activation of the Regulating Reserves. *Appl. Sci.* **2021**, *11*, 8188. [CrossRef]
4. Bui, D.; Le, P.; Nguyen, T.; Nguyen, H. An Adaptive and Scalable Protection Coordination System of Overcurrent Relays in Distributed-Generator-Integrated Distribution Networks. *Appl. Sci.* **2021**, *11*, 8454. [CrossRef]
5. Polajžer, B.; Grčar, B.; Černelič, J.; Ritonja, J. Power-Based Concept for Current Injection by Inverter-Interfaced Distributed Generations during Transmission-Network Faults. *Appl. Sci.* **2021**, *11*, 10437. [CrossRef]
6. Kitak, P.; Belak, L.; Pihler, J.; Ribič, J. Maintenance Management of a Transmission Substation with Optimization. *Appl. Sci.* **2021**, *11*, 11806. [CrossRef]
7. Bermúdez, M.; Barrero, F.; Martín, C.; Perales, M. Performance Analysis of Direct Torque Controllers in Five-Phase Electrical Drives. *Appl. Sci.* **2021**, *11*, 11964. [CrossRef]
8. Shokouhandeh, H.; Latif, S.; Irshad, S.; Ahmadi Kamarposhti, M.; Colak, I.; Eguchi, K. Optimal Management of Reactive Power Considering Voltage and Location of Control Devices Using Artificial Bee Algorithm. *Appl. Sci.* **2022**, *12*, 27. [CrossRef]
9. Fayad, A.; Ibrahim, H.; Ilinca, A.; Sattarpanah Karganroudi, S.; Issa, M. Energy Efficiency Improvement of Diesel−Electric Trains Using Solar Energy: A Feasibility Study. *Appl. Sci.* **2022**, *12*, 5869. [CrossRef]
10. Szekely, N.; Salcu, S.; Suciu, V.; Pintilie, L.; Fasola, G.; Teodosescu, P. Power Factor Correction Application Based on Independent Double-Boost Interleaved Converter (IDBIC). *Appl. Sci.* **2022**, *12*, 7209. [CrossRef]
11. Gao, S.; Sun, L.; Tian, Y.; Liu, H. Research on the Distribution Characteristics of Transformer Axial Vibration under Short-Circuit Conditions Considering Damping Parameters. *Appl. Sci.* **2022**, *12*, 8443. [CrossRef]
12. Ding, W.; Chen, Q.; Dong, Y.; Shao, N. Fault Diagnosis Method of Intelligent Substation Protection System Based on Gradient Boosting Decision Tree. *Appl. Sci.* **2022**, *12*, 8989. [CrossRef]
13. Michalopoulos, P.; Tsekouras, G.; Kanellos, F.; Prousalidis, J. Optimal Selection of the Diesel Generators Supplying a Ship Electric Power System. *Appl. Sci.* **2022**, *12*, 10463. [CrossRef]
14. Riba, J.; Moreno-Eguilaz, M.; Bogarra, S. Energy Harvesting Methods for Transmission Lines: A Comprehensive Review. *Appl. Sci.* **2022**, *12*, 10699. [CrossRef]
15. Radaelli, L.; Martinez, S. Frequency Stability Analysis of a Low Inertia Power System with Interactions among Power Electronics Interfaced Generators with Frequency Response Capabilities. *Appl. Sci.* **2022**, *12*, 11126. [CrossRef]
16. Cheong, Y.; Cao, S.; Naayagi, R.; Lee, S. Triple Phase Shift Control of Wireless Charging DAB LCC Resonant Converter for Unity Power Factor Operation with Optimized Rectifier AC Load Resistance. *Appl. Sci.* **2022**, *12*, 11871. [CrossRef]

applied sciences

MDPI

Article

Improved PR Control Strategy for an LCL Three-Phase Grid-Connected Inverter Based on Active Damping

Yahui Li, Jing Zhang, Zhenghang Hao * and Peng Tian

College of Electrical Engineering, Guizhou University, Guiyang 550025, China; liyahui0505@163.com (Y.L.); zhangjing@gzu.edu.cn (J.Z.); gzgytianpeng@163.com (P.T.)
* Correspondence: zhshi@gzu.edu.cn; Tel.: +86-136-2851-1067

Featured Application: The controller of a three-phase grid-connected inverter is studied and improved, which provides support for the independent control of power and the design of the controller.

Abstract: Aiming at the problem of power coupling and complicated decoupling in the *d-q* coordinate system of a three-phase grid-connected inverter, a current closed-loop control strategy based on an improved QPIR (quasi-proportional integral resonant) controller in the α-β two-phase static coordinate system is proposed. Firstly, the mathematical model of an LCL three-phase grid-connected inverter is established, and its instantaneous power calculation equation is deduced. Secondly, the frequency method is applied to compare and analyze the proportional resonant, quasi-proportional resonant, and improved current controller, and the appropriate improved controller parameters are obtained according to the traditional proportional integral controller parameter design method and the weight coefficient. Finally, the improved controller is compared with the traditional controller in the simulation model of the LCL three-phase grid-connected inverter based on active damping. The results show that the proposed improved current control strategy has good dynamic response characteristics, can realize the non-static error control of grid-connected current, and realizes the decoupling control of active power and reactive power when the load jumps. At the same time, the results also prove the superiority of the proposed control strategy and verify its effectiveness.

Keywords: LCL filter; grid-connected inverter; active damping; decoupling control

Citation: Li, Y.; Zhang, J.; Hao, Z.; Tian, P. Improved PR Control Strategy for an LCL Three-Phase Grid-Connected Inverter Based on Active Damping. *Appl. Sci.* **2021**, *11*, 3170. https://doi.org/10.3390/app11073170

Academic Editor: Federico Barrero

Received: 28 February 2021
Accepted: 31 March 2021
Published: 2 April 2021

1. Introduction

High power factor and low grid-connected current total harmonic distortion are common requirements for grid-connected inverters [1–3]. Pulse-width modulation (PWM) technology has a wide range of applications in the field of inverters, but a grid-connected inverter using PWM control technology will produce many switching frequency sub-harmonics, which seriously threaten the power quality and safe operation of the power grid [4–6]. In order to reduce the harmonic content of the grid-connected current, an appropriate grid-connected current filter must be adopted [7–9]. At present, L-type and LCL-type filters are widely used in harmonic suppression of grid-connected inverter output current. Under the same capacity, an LCL filter has strong attenuation ability to high frequency interference and has a good prospect for engineering application. However, the inherent disadvantage of the low damping resonance of an LCL filter in inverter control technology is still under discussion [10,11]. Compared with the passive damping control method, active damping can provide an appropriate damping coefficient without increasing the extra loss of the system, thus achieving the purpose of weakening the resonant peak [12–14]. Common active damping control techniques include capacitor current feedback [15,16], capacitor voltage primary differential feedback [17], grid-connected current secondary differential feedback [18,19], and multi-state variable combined control [20,21], etc. Among them, the capacitance current proportional feedback damping effect is good, easy to realize, and the application is also the most common.

For the control method of a three-phase grid-connected inverter, the current common method is to convert it from a three-phase stationary coordinate system to a two-phase stationary coordinate system (α-β) or two-phase synchronous rotating coordinate system (d-q) [22–24]. In the d-q coordinate system, all control variables are DC flow, the classical PI controller can be used to realize the non-static error control of grid-connected current, and the implementation method is flexible and simple, so it has been widely used. However, there is coupling in the d-q coordinate system of the three-phase grid-connected inverter. If the active power and reactive power can be controlled independently, decoupling must be carried out. For L-type grid-connected inverters, there is only one group of coupling terms between the d axis and q axis, and the decoupling process is relatively simple. However, for LCL-type three-phase grid-connected inverters, there are three groups of coupling terms between the d axis and q axis, and the decoupling process is very complicated [25]. In practical application, due to factors such as system parameter error and interference, it is almost impossible for the inverter to achieve complete power decoupling. In the α-β coordinate system, there is no coupling term between the α axis and β axis, which creates favorable conditions for independent control of active power and reactive power. However, in the α-β coordinate system, all control quantities are AC quantities. In order to realize static error-free control, unless the gain of the controller is infinite, it is particularly important to find a controller with high gain [26–28].

As for the control technology of grid current in a three-phase grid-connected inverter, the commonly used control methods include proportional-integral (PI) control, proportional-resonant (PR) control, and quasi-proportional-resonant (QPR) control. The PI control structure is simple and easy to implement and is suitable for DC flow control, but the AC PI controller cannot realize non-static error control [29]. Increasing the feedforward link of the grid voltage ratio can reduce the steady-state error of grid-connected current, but cannot effectively suppress the harmonics [30]. Theoretically, the PR controller has infinite gain at the specific frequency, but the PR controller has poor adaptability to grid frequency fluctuations, so quasi-proportional resonant controllers are widely used [31,32]. The QPR controller has strong adaptability to grid frequency fluctuations, but the gain at the fundamental frequency is limited, so it can only approximately realize error-free tracking [33,34]. In addition, there are hysteresis control, repetitive control, and H_∞ control, but their control performance is not significantly better than that of PI, PR, and QPR controllers [28]. For three-phase LCL grid-connected inverters, few studies consider the steady-state error of grid-connected current and the power grid frequency fluctuation at the same time, and relevant control technologies need further research.

This paper studies the controller of the three-phase LCL grid-connected inverter in the α-β coordinate system. A current closed-loop control strategy based on an improved QPIR controller is proposed while considering the steady-state error of grid-connected current, power decoupling, and grid frequency fluctuations. Theoretical analysis and case simulation show that the QPIR controller can track the grid-connected current without static error and has good output waveform quality, which proves the correctness and effectiveness of the controller.

This paper consists of the following parts: In Section 2, the mathematical model of the LCL three-phase grid-connected inverter is established, and the advantages of independent power control in the α-β coordinate system are pointed out. In Section 3, the control method of active damping is given, and compared with the advantages and disadvantages of a traditional PR controller and QPR controller, an improved QPIR controller strategy is proposed. In Section 4, the simulation results and some discussion under different conditions are given through an LCL three-phase grid-connected inverter simulation model. The conclusion is given in Section 5.

2. Mathematical Model and Control Method

2.1. Mathematical Model of the Three-Phase LCL-Type Grid-Connected Inverter

The main circuit topology of the three-phase grid-connected inverter with the LCL filter is shown in Figure 1, wherein L_1 is the inductance on the inverter side; L_2 is the inductance on the grid side; C is a filter capacitor. The inverter supplies power to the power grid through the LCL filter. U_{dc} is the DC bus voltage, u_a, u_b, and u_c are the midpoint voltages of each bridge arm of the inverter, i_{1a}, i_{1b}, and i_{1c} are the three-phase inductor currents on the inverter side, i_{2a}, i_{2b}, and i_{2c} are the three-phase inductor currents on the grid side, i_{Ca}, i_{Cb}, and i_{Cc} are the three-phase currents of the filter capacitor, u_{Ca}, u_{Cb}, and u_{Cc} are the three-phase voltages of the filter capacitor, u_{ga}, u_{gb}, and u_{gc} are the three-phase grid voltages, N is the neutral point of the LCL filter, N' is the neutral point of the grid, and Q_1-Q_6 represent the six IGBT switch tubes of the inverter.

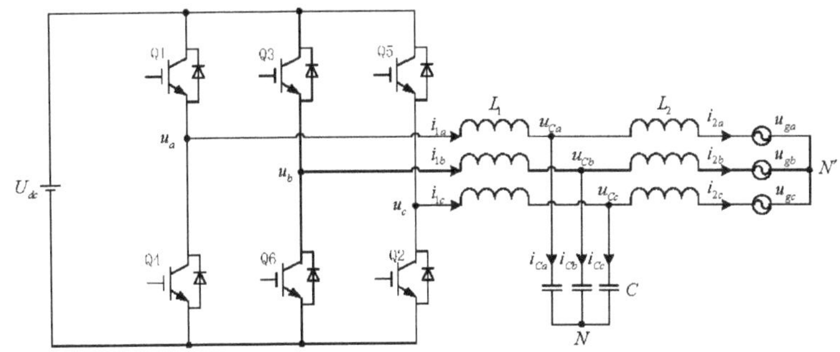

Figure 1. LCL-type three-phase grid-connected inverter topology.

Assume that the three-phase grid voltage is balanced. According to Figure 1, select the inverter side inductor currents i_{1a}, i_{1b}, i_{1c}, the grid side inductor currents i_{2a}, i_{2b}, i_{2c}, and the filter capacitor voltages u_{Ca}, u_{Cb}, u_{Cc} as state variables; the state equation in the three-phase static coordinate system can be obtained, as shown in Equation (1).

$$\begin{cases} L_1\frac{di_{1k}}{dt} = u_k - u_{Ck} \\ L_2\frac{di_{2k}}{dt} = u_{Ck} - u_{gk} \quad k = a,b,c \\ C\frac{du_{Ck}}{dt} - i_{1k} - i_{2k} \end{cases} \tag{1}$$

2.2. Mathematical Model of the α-β Coordinate System

Edith Clark transformation was carried out on Equation (1) to obtain the state equation of the three-phase inverter in the two-phase stationary coordinate system, and then Laplace transformation was carried out to obtain the transfer function in the two-phase stationary coordinate system, as shown in Equations (2) and (3).

$$\begin{cases} \begin{bmatrix} u_\alpha \\ u_\beta \end{bmatrix} = \begin{bmatrix} u_{C\alpha} \\ u_{C\beta} \end{bmatrix} + L_1 p \begin{bmatrix} i_{1\alpha} \\ i_{1\beta} \end{bmatrix} \\ \begin{bmatrix} u_{C\alpha} \\ u_{C\beta} \end{bmatrix} = \begin{bmatrix} u_{g\alpha} \\ u_{g\beta} \end{bmatrix} + L_2 p \begin{bmatrix} i_{2\alpha} \\ i_{2\beta} \end{bmatrix} \\ \begin{bmatrix} i_{1\alpha} \\ i_{1\beta} \end{bmatrix} = \begin{bmatrix} i_{2\alpha} \\ i_{2\beta} \end{bmatrix} + Cp \begin{bmatrix} u_{C\alpha} \\ u_{C\beta} \end{bmatrix} \end{cases} \tag{2}$$

$p = \frac{\mathrm{d}}{\mathrm{d}t}$, and Laplace transformation was carried out on Equation (2) to obtain the mathematical model in the s domain, as shown in Equation (3):

$$\begin{cases} \begin{bmatrix} u_\alpha \\ u_\beta \end{bmatrix} = \begin{bmatrix} u_{C\alpha} \\ u_{C\beta} \end{bmatrix} + L_1 s \begin{bmatrix} i_{1\alpha} \\ i_{1\beta} \end{bmatrix} \\ \begin{bmatrix} u_{C\alpha} \\ u_{C\beta} \end{bmatrix} = \begin{bmatrix} u_{g\alpha} \\ u_{g\beta} \end{bmatrix} + L_2 s \begin{bmatrix} i_{2\alpha} \\ i_{2\beta} \end{bmatrix} \\ \begin{bmatrix} i_{1\alpha} \\ i_{1\beta} \end{bmatrix} = \begin{bmatrix} i_{2\alpha} \\ i_{2\beta} \end{bmatrix} + Cs \begin{bmatrix} u_{C\alpha} \\ u_{C\beta} \end{bmatrix} \end{cases} \tag{3}$$

As can be seen from Equation (3), there is no coupling in the two-phase static coordinate system three-phase LCL grid-connected inverter. This advantage can be fully utilized to decouple the active power and reactive power. The LCL type three-phase grid-connected inverter will be directly controlled in a two-phase stationary coordinate system.

2.3. Instantaneous Power Calculation

The calculation equation of instantaneous active power in the three-phase static coordinate system is as follows:

$$p = u_a i_a + u_b i_b + u_c i_c \tag{4}$$

Equivalent transformation is carried out on Equation (4):

$$p = [u_a\ u_b\ u_c] \begin{bmatrix} i_a \\ i_b \\ i_c \end{bmatrix} = [u_a\ u_b\ u_c] C^{-1} C \begin{bmatrix} i_a \\ i_b \\ i_c \end{bmatrix} = \left[(C^{-1})^T \begin{bmatrix} u_a \\ u_b \\ u_c \end{bmatrix} \right]^T \cdot C \begin{bmatrix} i_a \\ i_b \\ i_c \end{bmatrix} = \left[(C^{-1})^T C^{-1} C \begin{bmatrix} u_a \\ u_b \\ u_c \end{bmatrix} \right]^T \cdot C \begin{bmatrix} i_a \\ i_b \\ i_c \end{bmatrix} \tag{5}$$

From the Clark transformation equation, it can be seen that:

$$\begin{bmatrix} u_\alpha \\ u_\beta \\ u_0 \end{bmatrix} = C \begin{bmatrix} u_a \\ u_b \\ u_c \end{bmatrix}, \begin{bmatrix} i_\alpha \\ i_\beta \\ i_0 \end{bmatrix} = C \begin{bmatrix} i_a \\ i_b \\ i_c \end{bmatrix} \tag{6}$$

where the transformation matrix is:

$$C = \frac{2}{3} \begin{bmatrix} 1 & -\frac{1}{2} & -\frac{1}{2} \\ 0 & \frac{\sqrt{3}}{2} & -\frac{\sqrt{3}}{2} \\ \frac{1}{2} & \frac{1}{2} & \frac{1}{2} \end{bmatrix}, C^{-1} = \begin{bmatrix} 1 & 0 & 1 \\ -\frac{1}{2} & \frac{\sqrt{3}}{2} & 1 \\ -\frac{1}{2} & -\frac{\sqrt{3}}{2} & 1 \end{bmatrix}$$

By inserting Equation (5) and Equation (6) into Equation (4), we can obtain:

$$p = [u_\alpha\ u_\beta\ u_0] \left(C^{-1} \right)^T C^{-1} \begin{bmatrix} i_\alpha \\ i_\beta \\ i_0 \end{bmatrix} \tag{7}$$

Through further calculation, the expression of instantaneous active power in two-phase stationary coordinate system is as follows:

$$p = \frac{3}{2}(u_\alpha i_\alpha + u_\beta i_\beta + 2u_0 i_0) \tag{8}$$

Similarly, according to the above calculation method of active power, the expression of instantaneous reactive power can be obtained as follows:

$$q = \frac{3}{2}(u_\beta i_\alpha - u_\alpha i_\beta) \tag{9}$$

With the above instantaneous power expression, the grid-connected inverter can be controlled by the power outer loop.

3. Analysis of the Control Strategy

3.1. LCL-Type Three-Phase Grid-Connected Inverter Control Structure

The control structure diagram of the LCL-type grid-connected inverter is shown in Figure 2. After collecting the voltage and current of the three-phase power grid from the external large power grid, through Edith Clark transformation, u_α, u_β and i_α, i_β are obtained. According to the instantaneous power calculation principle, the current reference values $i_{2\alpha}^*$ and $i_{2\beta}^*$ of the current loop can be obtained and then compared with the grid-connected feedback currents $i_{2\alpha}$ and $i_{2\beta}$. After the error signal is adjusted by the improved $G_{QPIR}(s)$ controller, the PWM controller controls the on and off status of the switch tube.

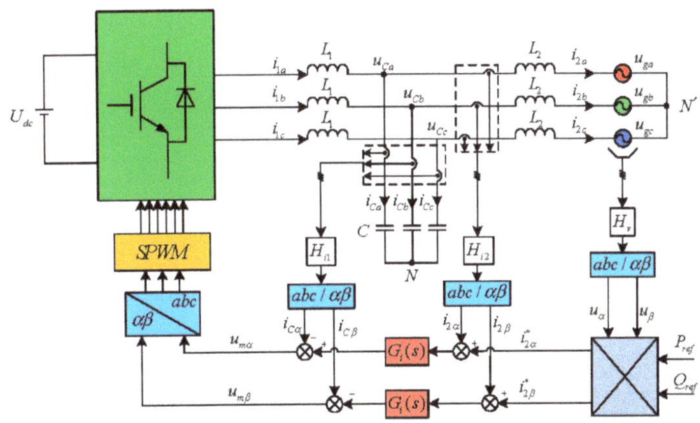

Figure 2. Threephase LCL-type grid-connected inverter with active damping in the α-β coordinate system.

According to Equation (3) and considering the control system structure given in Figure 2, the s domain model of the three-phase LCL grid-connected inverter controlled in the α-β coordinate system can be obtained, as shown in Figure 3. K_{PWM} is the transfer function of the voltage-source three-phase inverter bridge modulated by PWM, and H_{i1} and H_{i2} are the capacitance current feedback coefficient and grid-connected current feedback coefficient, respectively.

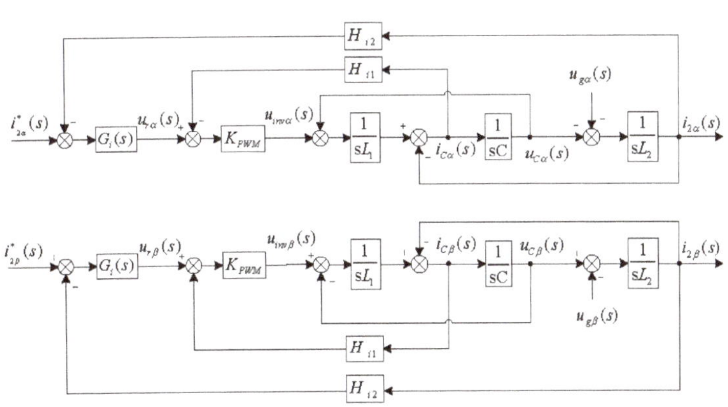

Figure 3. Control block diagram of the grid-connected inverter with LCL filter in the α-β coordinate system.

3.2. Capacitor Current Feedback Active Damping

Capacitor current feedback active damping will not reduce the low-frequency gain and high-frequency harmonic attenuation capability of the LCL filter and will not increase the additional power loss of the system, thus having very good practical value [14].

See Appendix A for the capacitor current proportional feedback active damping parameters. Figure 4 is a Bode diagram of loop gain $T(s)$ after active damping with capacitor current feedback, where f_0 is the fundamental frequency, f_c is the cutoff frequency, and f_r is the resonant frequency. As can be seen from Figure 4, the introduction of capacitor current feedback can weaken the resonant peak of the LCL filter. With the increase in damping coefficient H_{i1}, the damping effect of the resonant peak will be better, and at the same time it will not affect the low frequency and high frequency characteristics of the system.

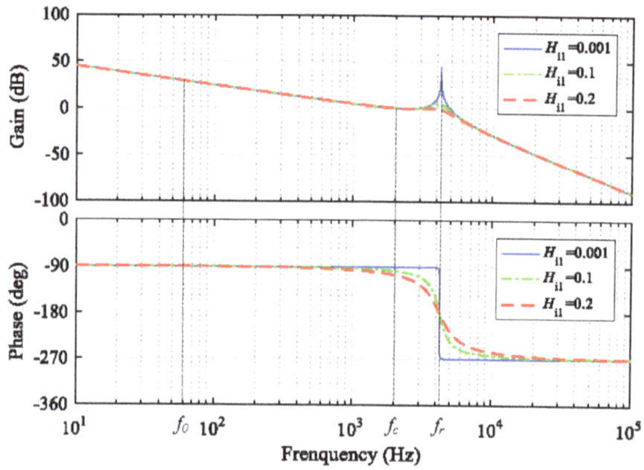

Figure 4. Bode diagram of $T(s)$ with capacitor current proportional feedback.

3.3. Performance Comparison of the PR and QPR Controller

In the current control of the grid-connected inverter in a α-β coordinate system, the PR controller can obtain higher gain at the fundamental wave of the power grid or a specific frequency and can eliminate the steady-state error of the grid-connected current or inhibit the influence of a specific subharmonic on the grid-connected current. The QPR controller has good robustness and adaptability to power grid frequency fluctuation. The following is a comparative analysis of these two controllers.

The transfer function of the PR controller is:

$$G_{PR}(s) = K_1 + \frac{K_2 s}{s^2 + \omega_0{}^2} \tag{10}$$

The transfer function of the QPR controller is:

$$G_{QPR}(s) = K_3 + \frac{K_4 \omega_i s}{s^2 + 2\omega_i s + \omega_0{}^2} \tag{11}$$

K_1 and K_2 in Equations (10) and (11) are the respective proportional coefficients and resonant coefficients of the PR controller; K_3 and K_4 are the respective proportional coefficient and quasi-resonance coefficient of the QPR controller; $\omega_0 = 2\pi f_0$ is the fundamental angular frequency; ω_i is the bandwidth of the resonance term considering the -3 dB requirement, i.e., the gain of the resonance term is $0.707K_4$ at $\omega_0 \pm \omega_i$.

Figure 5 is the Bode diagrams of $G_{PR}(s)$ and $G_{QPR}(s)$. For the LCL three-phase grid-connected inverter, the gain of the PR controller can obtain infinite gain at fundamental frequency, thus realizing non-static error control of the fundamental component of the grid-connected current. However, due to load changes and various interferences, the actual power grid frequency will fluctuate. When the grid frequency deviates from the fundamental frequency f_0 set by the PR controller, the gain of the PR controller will decrease rapidly, which will lead to the rapid increase in the steady-state error of the grid-connected current. The QPR controller adds parameter ω_i to the PR controller, so that it can obtain higher gain than the PR controller in a wider frequency band and can adapt to the power grid frequency fluctuation in a certain range at the same time. However, the gain at the fundamental frequency f_0 is much smaller than the gain of the PR controller, which indicates that the accuracy of the QPR controller is lower than that of the PR controller and will produce steady-state errors. Increasing the quasi-resonance coefficient K_4 can improve the gain at the fundamental frequency of the QPR controller, but due to the limitation of system stability, the value of K_4 cannot be too large.

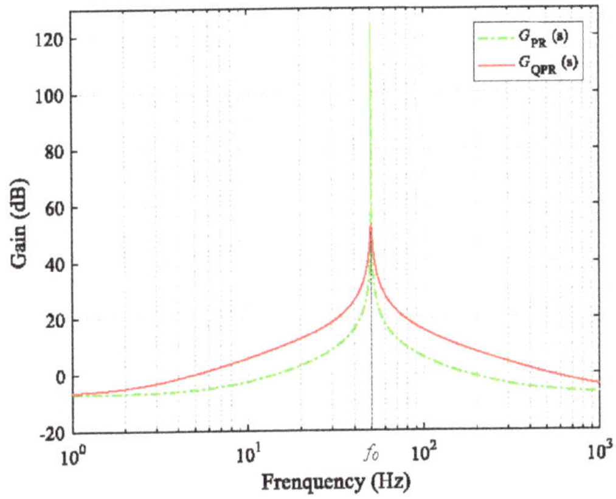

Figure 5. Bode diagram of PR and QPR controllers.

3.4. Improved Controller QPIR

As can be seen from the previous analysis, the LCL-type three-phase grid-connected inverter has no coupling in the two-phase stationary coordinate system, which creates favorable conditions for independent control of active power and reactive power. However, the control quantity in the two-phase stationary coordinate system is AC quantity, and the traditional PR and QPR controllers have their own defects, which decreases the control effect of the inverter. In order to solve this problem, this paper proposes an improved controller with proportional, integral, and quasi-resonant hybrid structures. Its transfer function is shown in Equation (12):

$$G_{QPIR} = K_P + \frac{K_{i0}}{s} + \frac{K_{r1}\omega_{i1}s}{s^2 + 2\omega_{i1}s + \omega_{01}^2} + \frac{K_{r2}\omega_{i2}s}{s^2 + 2\omega_{i4}s + \omega_{02}^2} + \frac{K_{r3}\omega_{i3}s}{s^2 + 2\omega_{i3}s + \omega_{03}^2} \tag{12}$$

$$K_i = K_{i0} + K_{r1}\omega_{i1} + K_{r2}\omega_{i2} + K_{r3}\omega_{i3} = \alpha K_i + \beta K_i + \gamma K_i + \eta K_i \tag{13}$$

In Equations (12) and (13), K_P is the proportional coefficient, K_{i0} is the integral coefficient, K_{r1}, K_{r2}, K_{r3} are quasi-resonant coefficients, K_i is the total integral resonant coefficient, α, β, γ, η are the weight coefficients of K_{i0}, K_{r1}, K_{r2}, K_{r3}, respectively, $\omega_{o1} = 2\pi(f_0 - \Delta f)$, $\omega_{o2} = 2\pi f_0$, $\omega_{o3} = 2\pi(f_0 + \Delta f)$ are the fundamental angular frequencies considering the frequency fluctuation of the power grid, ω_{i1}, ω_{i2}, ω_{i3}, ω_{i4} represent the resonant term bandwidth, and Δf is the frequency fluctuation range.

By inserting Equation (13) into Equation (12), another form can be obtained:

$$G_{QPIR} = K_P + K_i\left(\frac{\alpha}{s} + \frac{\beta s}{s^2 + 2\omega_{i1}s + \omega_{01}^2} + \frac{\gamma s}{s^2 + 2\omega_{i4}s + \omega_{02}^2} + \frac{\eta s}{s^2 + 2\omega_{i3}s + \omega_{03}^2}\right) \tag{14}$$

The Bode diagram of the improved controller $G_{QPIR}(s)$ is shown in Figure 6. In Figure 6, the total integrated resonance coefficients of the three controllers $G_{PR}(s)$, $G_{QPR}(s)$, and $G_{QPIR}(s)$ are all the same, but the improved controller $G_{QPIR}(s)$ has different weight coefficients and resonance bandwidths. As can be seen from the Bode diagram of Figure 6, the improved QPIR controller is basically consistent with the Bode diagram of the QPR controller in other frequency bands except near the fundamental frequency. It can be seen from the local amplified Bode diagram that the Bode diagram (red) of the improved QPIR controller has a wider frequency range than the traditional PR controller (blue), and at the same time has a higher gain near the fundamental frequency compared with the traditional QPR controller (green), which can not only improve the steady-state accuracy of the grid-connected current but also has certain robustness to the fluctuation of the grid frequency.

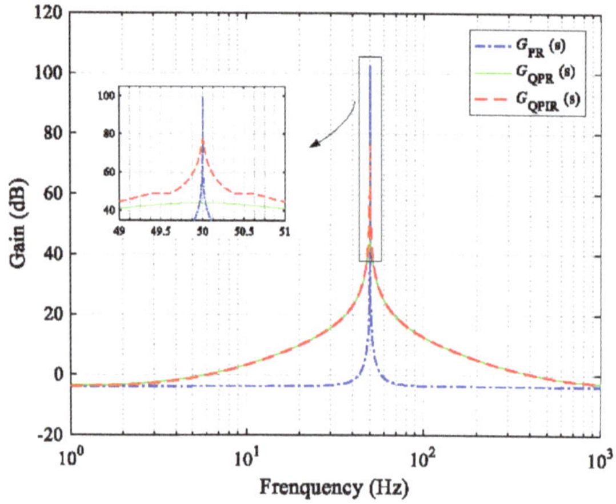

Figure 6. Bode diagram of PR, QPR, and QPIR controllers.

3.5. Implementation of the QPIR Controller

In order to facilitate the realization of the QPIR controller, suppose that parameter ω_{i2} is k times ω_{i4} (i.e., $\omega_{i2} = k\omega_{i4}$) and the value of k is determined by the actual values of ω_{i2} and ω_{i4}. According to Equation (12), the equivalent control block diagram can be obtained, as shown in Figure 7. The input signals of the controller pass through the proportional, integral, and quasi-proportional resonant modules, and the output signals are obtained after accumulation. For each basic module in Figure 7, we can use C language programming to achieve the output. In this way, the control strategy proposed in this paper can be implemented in the actual converter.

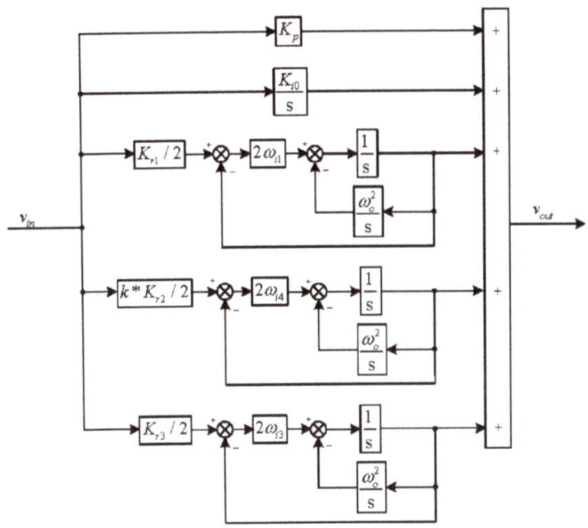

Figure 7. Implementation block diagram of the QPIR controller.

4. Case Analysis

In order to verify the correctness and effectiveness of the improved QPIR algorithm proposed in this paper, a system simulation model is built in MATLAB/Simulink simulation software. PR, QPR, and QPIR current controllers combined with capacitor current feedback active damping are used to simulate and compare the three-phase LCL grid-connected inverter. The system simulation parameters are shown in Table 1.

Table 1. Parameters of the LCL-type three-phase grid-connected inverter system.

Project	Parameter
DC bus voltage U_{dc}	750 V
Power grid line voltage u_g (RMS)	380 V,
Rated power P_N	40 kW
Inverter side inductance L_1	700 µH
Network side inductance L_2	110 µH
Filter capacitor C	15 µF
Fundamental frequency f_0	50 Hz
Switching frequency f_s	15 kHz
Capacitance current feedback coefficient H_{i1}	0.12
Power network current feedback coefficient H_{i2}	0.14
Power network voltage feedback coefficient H_v	1.00

It can be seen from reference [35] that when the controller parameters are the same, they also have similar performance. Therefore, in order to facilitate comparison, the QPIR controller and PR and QPR controllers are set to have similar parameters, that is, the QPIR controller parameters satisfy the following relationship:

$$K_P = K_1 = K_3 \quad (15)$$

$$K_i = K_2 = K_4 \quad (16)$$

For small photovoltaic power stations, the allowable range of grid frequency fluctuation is 49.5–50.5 Hz [36], i.e., the frequency deviation is $\Delta f = 0.5$ Hz. The system phase margin PM > 45° is selected to ensure a good dynamic response. The gain margin is GM > 3 dB for sufficient robustness. PR, QPR, and QPIR controllers are designed to meet the above conditions.

Considering the dynamic response and the attenuation of high-frequency noise, the system crossover frequency f_c usually has strict limits. Generally, $f_0 << f_c << f_s$, where f_0 is the fundamental frequency and f_s is the switching frequency. It can be seen from Figure 6 that when the total parameters of the controller are the same, the PR, QPR, and QPIR controllers have similar amplitude-frequency characteristic curves in the high-frequency part. In other words, for a given converter, irrespective of the controller used, the system crossover frequency is the same, so according to the system loop design requirements, set $K_p = K_1 = K_4 = 0.65$, $K_i = K_2\omega_i = K_4\omega_i = 2001$; for the calculation process, refer to the literature [37]. For the QPR controller, considering the fluctuation of the fundamental frequency, take the resonance bandwidth parameter $\omega_i = 2\pi\Delta f = \pi$ and further obtain $K_6 = K_i/\omega_i = 636.94$. For the PR controller, further calculation can be used to obtain $K_2 = 636.94$.

For the QPIR controller, considering the fluctuation of the fundamental frequency, set three different fundamental angular frequencies, namely $\omega_{o1} = 2\pi(f_0\text{-}\Delta f) = 311.02$, $\omega_{o2} = 2\pi f_0 = 314.16$, $\omega_{o3} = 2\pi(f_0 + \Delta f) = 317.3$, making the system gain as large as possible within the frequency fluctuation range. Considering that the fluctuation of the fundamental frequency obeys a normal distribution, in order to make the controller have certain robustness, it is hoped that the bandwidth around the angular frequency ω_{o2} should be larger, and the bandwidth around ω_{o1} and ω_{o3} should be smaller, so set the resonance bandwidth parameter $\omega_{i1} = \omega_{i3} = 0.4\pi$, $\omega_{i2} = 2\pi$. In order to ensure that the controller has a large gain near the fundamental frequency f_0 (49.8~50.2 Hz), it is necessary to satisfy $\omega_{i4} < \omega_{i2}$, so that its characteristic curve is similar to that of the PR controller, and $\omega_{i4} = 0.1$ can be obtained by the estimation method. At the same time, based on the idea that the fundamental frequency is dominant and the frequency fluctuation obeys the normal distribution, the quasi-resonance coefficient of the QPIR controller is mainly allocated reasonably to make full use of its respective advantages. Take $\alpha = 1/2001$, $\beta = 300/2001$, $\gamma = 1400/2001$, $\eta = 300/2001$.

Obviously, there is no coupling between the α axis and β axis in Figure 3, and there is duality. The following is an analysis of the system taking only the α axis as an example. According to the control block diagram of Figure 3, the open-loop transfer function of the system can be obtained as follows:

$$T(s) = \frac{H_{i2}K_{PWM}G_i(s)}{s^3L_1L_2C + s^2L_2CH_{i1}K_{PWM} + s(L_1 + L_2)} \tag{17}$$

Figure 8 is a Bode diagram of the loop gain T(s) of different controller systems plotted according to the parameters in Table 1. In the figure, T_{PR}(s), T_{QPR}(s), and T_{QPIR}(s) represent the loop gain using the PR, QPR, and QPIR controllers, respectively. With the same parameters, the phase margin of the PR controllers is 67.5°, and that of the QPR and QPIR controllers is 56.1° > 45°; their amplitude margin is 4.2 dB > 3 dB, which meets the design requirements. The cutoff frequency f_c is 1756 Hz, and the resonance frequency f_r is 4125 Hz. The gain (104 dB) at the fundamental frequency f_0 with the QPIR controller is much higher than that with the QPR (76.2 dB) controller. At the same time, the QPIR controller can also meet the requirements of adapting to the power grid frequency fluctuation in the range of $\Delta f = 0.5$ Hz, similar to the QPR controller.

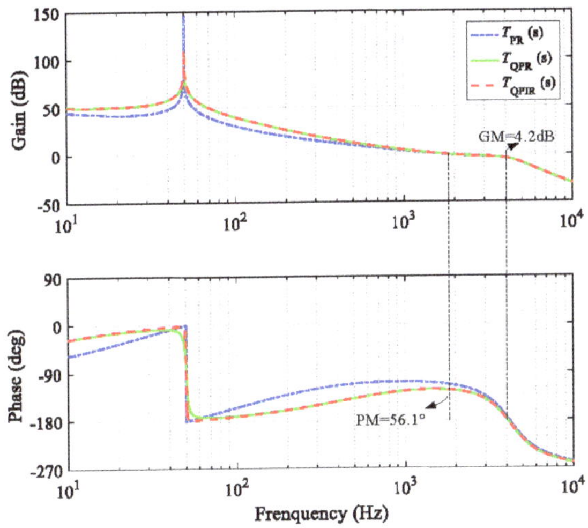

Figure 8. Bode diagram of $T(s)$ with PR, QPR, and QPIR controllers.

4.1. Analysis of Voltage and Current Based on the Improved QPIR Controller

The voltage and current waveforms, active power waveforms, and reactive power waveforms of the PR controller, QPR controller, and QPIR controller at a grid frequency of 49.5 Hz are shown in Figures 9–11, respectively.

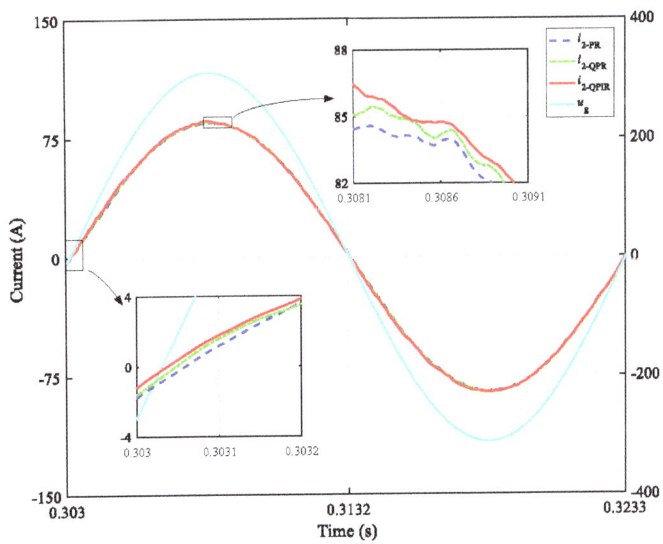

Figure 9. Steady-state waveforms of grid-connected current and voltage using PR, QPR, and QPIR controllers (at 49.5 Hz).

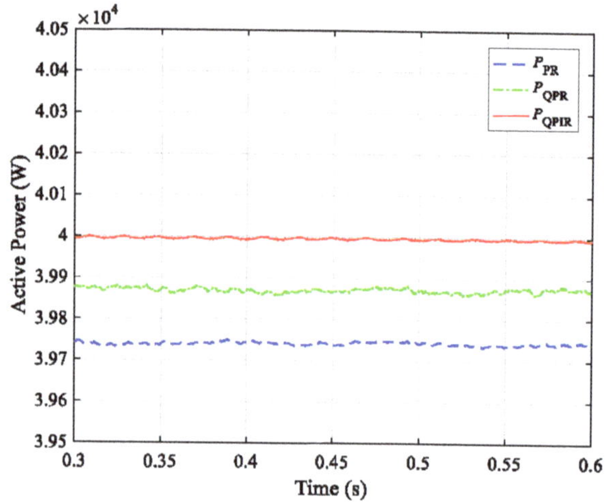

Figure 10. Active power waveforms controlled by PR, QPR, and QPIR controllers (at 49.5 Hz).

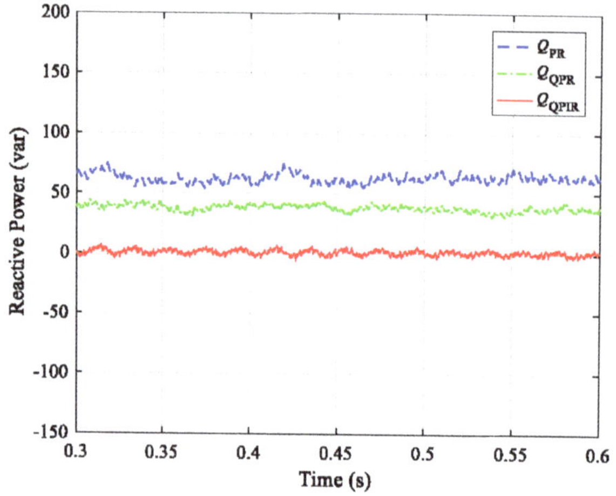

Figure 11. Reactive power waveforms controlled by PR, QPR, and QPIR controllers (at 49.5 Hz).

In Figure 9, when the PR controller is used, the phase error between grid-connected current and voltage is 0.09°, and the effective value of the grid-connected current is 60.26 A, which is less than the current loop reference value (60.606 A). In Figures 10 and 11, P represents active power and Q represents reactive power. When the PR controller is used, the active power is 39,740 W, which is less than the given value 40 kW, and the reactive power reaches 60 var. This shows that the PR controller has errors in controlling amplitude and phase.

When the QPR controller is used, the phase error between the grid-connected current and voltage is 0.06°, and the effective value of the grid-connected current is 60.44 A, which is slightly less than the reference value. In Figures 10 and 11, when the QPR controller is used, the active power is 39,870 W and the reactive power is 40 var. It is verified that the gain of the QPR controller at 49.5 Hz is not large enough to realize static error-free adjustment.

When the QPIR controller is used, the grid-connected current basically has no phase error. The effective value of the grid-connected current is 60.60 A, and the error between the grid-connected current and the reference value is basically 0. As shown in

Figures 10 and 11, when the QPIR controller is used, the active power approaches 40 kW and the reactive power approaches 0 var. Compared with the QPR controller, the fluctuation range of active power and reactive power is smaller. It is verified that the QPIR controller can eliminate the static error of the system, achieve the operating condition of the unit power factor, and also improve the utilization rate of DC side voltage.

The voltage and current waveforms, active power waveforms, and reactive power waveforms of the PR controller, QPR controller, and QPIR controller at a grid frequency of 50.5 Hz are shown in Figures 12–14, respectively.

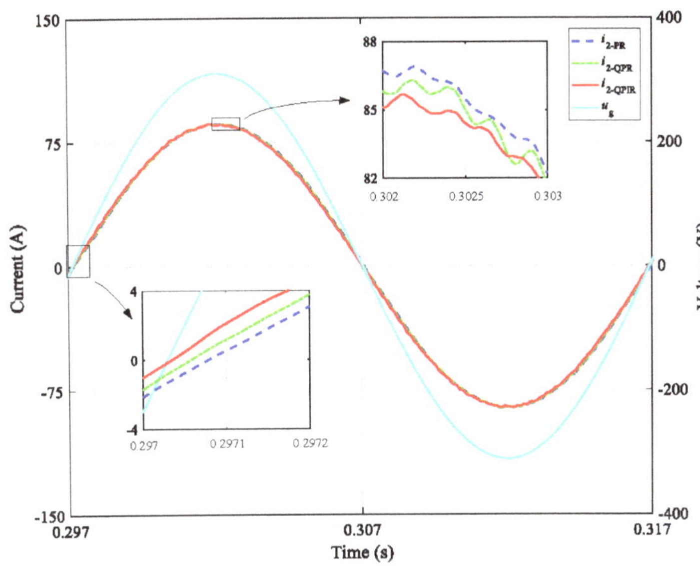

Figure 12. Steady-state waveforms of grid-connected current and voltage using PR, QPR, and QPIR controllers (at 50.5 Hz).

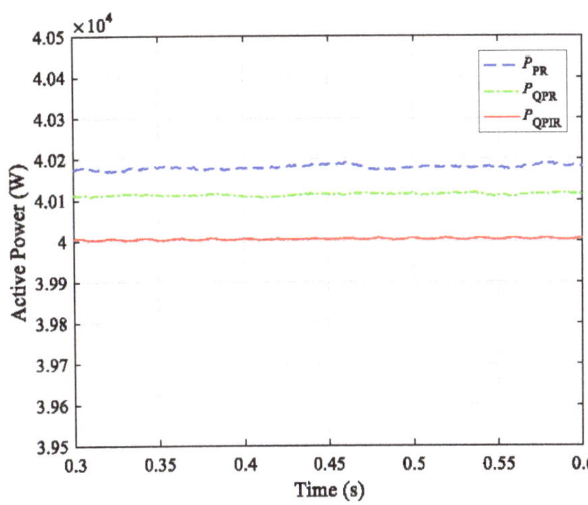

Figure 13. Active power waveforms controlled by PR, QPR, and QPIR controllers (at 50.5 Hz).

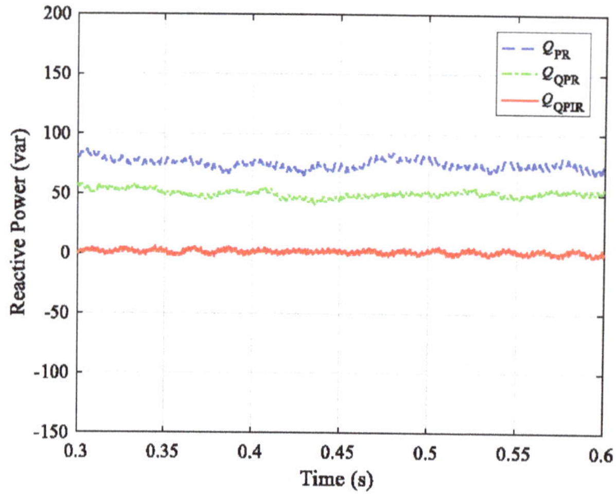

Figure 14. Reactive power waveforms controlled by PR, QPR, and QPIR controllers (at 50.5 Hz).

In Figure 12, when the PR controller is used, the phase error between the grid-connected current and voltage is 0.11°, and the effective value of the grid-connected current is 60.91 A, which is greater than the current loop reference value (60.606 A). In Figures 13 and 14, P represents active power and Q represents reactive power. When the PR controller is used, the active power is 40,180 W, which is greater than the given value 40 kW, and the reactive power reaches 80 var. This shows that the PR controller has errors in controlling amplitude and phase.

When the QPR controller is used, the phase error between grid-connected current and voltage is 0.07°, and the effective value of grid-connected current is 60.8 A, which is slightly higher than the reference value. In Figures 13 and 14, when the QPR controller is used, the active power is 40,110 W and the reactive power is 50 var. It is verified that the gain of the QPR controller at 50.5 Hz is not large enough to realize static error-free adjustment.

When the QPIR controller is used, the grid-connected current basically has no phase error. The effective value of the grid-connected current is 60.61 A, and the error between the grid-connected current and the reference value is basically 0. As shown in Figures 13 and 14, when the QPIR controller is used, the active power approaches 40 kW and the reactive power approaches 0 var. It is verified that the QPIR controller can eliminate the static error of the system and achieve the operating condition of unit power factor.

4.2. Step Response Analysis Based on the Improved QPIR Controller

In order to verify the coupling relationship between active power and reactive power in the two-phase stationary coordinate system, the active power and reactive power jumped at 0.2 s and 0.3 s, and the power waveform is shown in Figure 15.

In Figure 15, P represents active power and Q represents reactive power. At 0.2 s, the active power suddenly jumps from 0 to full load (40 kW). It can be seen that when the active power jumps, the reactive power is hardly affected. At 0.3 s, the reactive power suddenly jumps from 0 to 40 kvar, and when the reactive power suddenly changes, the active power is hardly affected.

Figure 16 is a waveform of active power and reactive power following the reference power. At 0.2 s, when the active power reference quantity P_{ref} jumps, the active power output by the inverter can quickly follow the target value without overshoot. At 0.3 s, when the reactive power reference value Q_{ref} jumps, the reactive power output by the inverter can quickly follow the target value without overshoot.

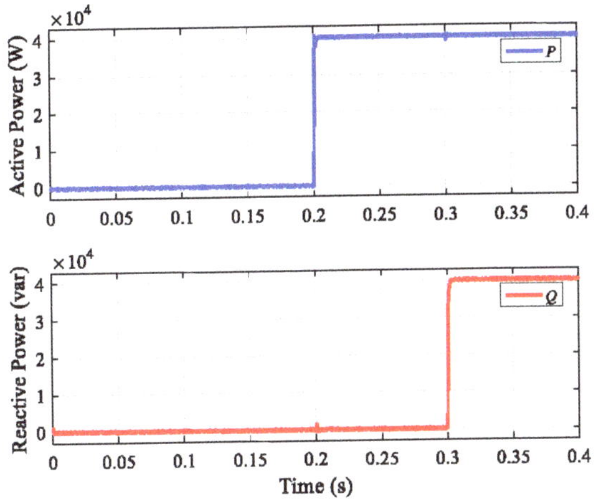

Figure 15. Step waveforms of active and reactive power.

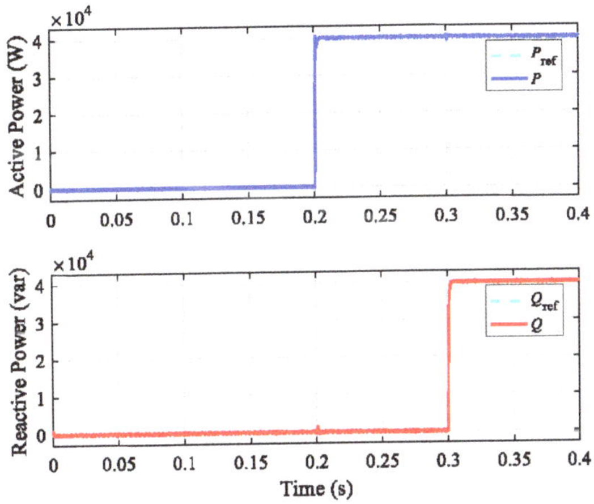

Figure 16. Step tracking waveforms of active and reactive power.

Through the comparison between Figures 15 and 16, it can be seen that there is no coupling relationship between active power and reactive power in the two-phase static coordinate system, and the improved QPIR controller can quickly follow the target value, which verifies the correctness of the previous analysis.

4.3. Frequency Fluctuation Analysis Based on the Improved QPIR Controller

Figure 17 is a chart of the power grid frequency fluctuation. A sawtooth wave is used to simulate the power grid frequency fluctuation, and the frequency fluctuation range is 49.5–50.5 Hz. Figure 18 shows power waveforms of the system using the QPIR controller and PR controller under the condition of grid frequency fluctuation according to Figure 17. Figure 19 shows power waveforms of the system using the QPIR controller and QPR controller under the condition of grid frequency fluctuation according to Figure 17.

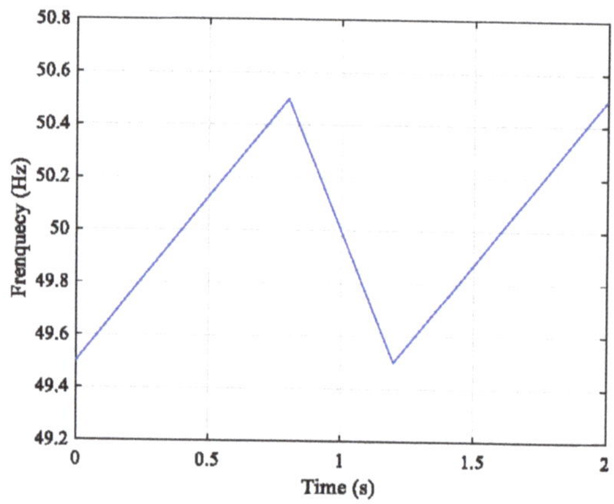

Figure 17. Power grid frequency fluctuation diagram.

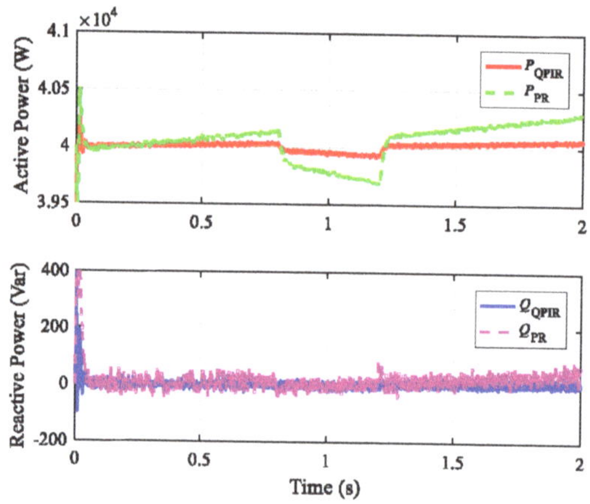

Figure 18. Power waveforms controlled by the QPIR and PR controllers when the grid frequency fluctuates.

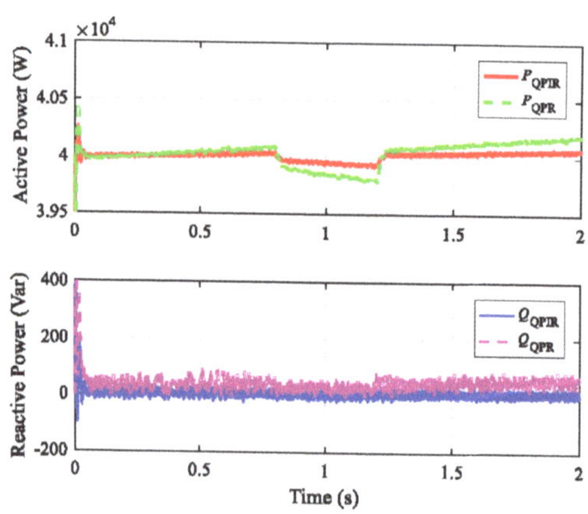

Figure 19. Power waveforms controlled by the QPIR and QPR controllers when the grid frequency fluctuates.

As can be seen from the power waveform in Figure 18, when the power grid frequency fluctuates, both the QPIR controller and PR controller systems can maintain stable operation. When the power grid frequency fluctuates, the active power changes with the change in frequency, but the active power fluctuation of the QPIR controller is obviously smaller than that of the PR controller, and the reactive power is almost not affected by the frequency change. For the PR controller, the maximum deviation of active power is 300 W (at 1.2 s) and the deviation of reactive power is 45 var. For the improved QPIR controller, the maximum deviation of active power is only 75 W (at 1.2 s), and the reactive power approaches to 0. This shows that the control accuracy of the QPIR controller is significantly higher than that of the PR controller when the power grid frequency fluctuates.

As can be seen from the power waveform in Figure 19, when the power grid frequency fluctuates, the active power fluctuation of the QPIR controller is obviously smaller than that of the QPR controller, and the reactive power is almost not affected by the frequency change. For the QPR controller, the maximum deviation of active power is 200 W (at 1.2 s) and the deviation of reactive power is 50 var. For the improved QPIR controller, the maximum deviation of active power is only 75 W (at 1.2 s), and the reactive power approaches 0. This shows that the control accuracy of the QPIR controller is significantly higher than that of the QPR controller when the power grid frequency fluctuates, which verifies that the QPIR controller can adapt to the fluctuation of power grid frequency within the normal range.

The simulation results show that the steady-state control accuracy of the QPIR controller is higher than that of the PR and QPR controllers under the condition of stable power grid frequency of 49.5 Hz and 50.5 Hz. Under the condition of power grid frequency fluctuation and load variation, the steady-state control accuracy of the QPIR controller is also higher than that of the PR and QPR controllers, which has stronger adaptability to power grid frequency fluctuation and verifies the correctness and effectiveness of the theoretical analysis.

5. Conclusions

In this paper, an improved current control strategy (QPIR) for a three-phase LCL grid-connected inverter based on active damping is proposed, and the simulation and example analysis were carried out using MATLAB/Simulink software. The following conclusions were obtained:

1. The LCL grid-connected inverter based on active damping can realize independent control of active power and reactive power without coupling between the α axis and β axis in a two-phase static coordinate system.
2. Compared with the traditional PR and QPR controllers, the improved QPIR controller has higher grid-connected current control accuracy under the condition of stable and fluctuating grid frequency.
3. The improved QPIR controller can realize the fast follow-up of active power and reactive power by the inverter, and when the active power jumps, the reactive power is not affected. When the reactive power jumps, the active power is not affected.

To sum up, the improved QPIR current controller proposed in this paper realizes the independent control of active power and reactive power of the three-phase grid-connected inverter in the two-phase static coordinate system. It also ensures the control accuracy and rapidity of the grid-connected current and has certain practical significance for high-power factor applications and improving the utilization rate of the DC side.

Author Contributions: Z.H. contributed to the project idea and the discussion of the results. Y.L. contributed to the specific strategy, theoretical analysis, simulation experiment design, data analysis, discussion of the results, and conclusions. J.Z. reviewed the final manuscript. P.T. contributed to data analysis and the simulation experiment. All authors have read and agreed to the published version of the manuscript.

Funding: This research was funded by Guizhou Province Science and Technology Innovation Talent Team Project ([2018] 5615).

Institutional Review Board Statement: Not applicable.

Informed Consent Statement: Not applicable.

Data Availability Statement: Not applicable

Conflicts of Interest: The authors declare no conflict of interest.

Nomenclature

PWM	pulse width modulation
PI	proportional integral
PR	proportional resonant
QPR	quasi proportional resonant
QPIR	quasi proportional integral resonant
k = a, b, c	three-phase
u_k	midpoint voltage of bridge arm
i_{1k}	inductor current on inverter side
i_{2k}	grid side current
i_{Ck}	filter capacitance current
u_{Ck}	filter capacitance voltage
U_{dc}	DC-bus voltage
u_{gk}	grid voltage

Appendix A

Table A1. System parameters for capacitive current feedback active damping.

Symbol	Quantity	Parameter
U_{dc}	DC bus voltage	750 V
u_g	Power grid line voltage (RMS)	380 V
L_1	Inverter side inductance	700 μH
L_2	Network side inductance	110 μH
C	Filter capacitor	15 μF
V_{tri}	Amplitude of triangle carrier wave	4.58 V
K_{pwm}	Transfer function of modulated wave to inverter bridge	81.87
H_{i1}	Capacitance current feedback coefficient	0.001, 0.1, 0.2
H_{i2}	Power network current feedback coefficient	0.20
H_v	Power network voltage feedback coefficient	1.00

References

1. Hassaine, L.; OLias, E.; Quintero, J.; Salas, V. Overview of power inverter topologies and control structures for grid connected photovoltaic systems. *Renew. Sustain. Energy Rev.* **2014**, *30*, 796–807. [CrossRef]
2. Blaabjerg, F.; Teodorescu, R.; Liserre, M.; Timbus, A.V. Overview of Control and Grid Synchronization for Distributed Power Generation Systems. *IEEE Trans. Ind. Electron.* **2006**, *53*, 1398–1409. [CrossRef]
3. Gomes, C.C.; Cupertino, A.F.; Pereir, H.A. Damping techniques for grid-connected voltage source converters based on LCL filter: An overview. *Renew. Sustain. Energy Rev.* **2018**, *81*, 116–135. [CrossRef]
4. Nardi, C.; Stein, C.M.O.; Carati, E.G.; Costa, J.P.; Cardoso, R. A methodology of LCL filter design for grid-tied power converters. *IEEE Trans. Ind. Electron.* **2006**, *53*, 1398–1409.
5. Zhang, J.; Xiong, G.; Meng, K.; Yu, P.; Yao, G.; Dong, Z. An improved probabilistic load flow simulation method considering correlated stochastic variables. *Int. J. Electr. Power Energy Syst.* **2019**, *111*, 260–268. [CrossRef]
6. Zhang, J.; Fan, L.; Zhang, Y. A Probabilistic Assessment Method for Voltage Stability Considering Large Scale Correlated Stochastic Variables. *IEEE Access* **2020**, *8*, 5407–5415. [CrossRef]
7. Sahoo, A.K.; Shahani, A.; Basu, K.; Mohan, N. LCL filter design for grid-connected inverters by analytical estimation of PWM ripple voltage. In Proceedings of the IEEE Applied Power Electronics Conference and Exposition, Fort Worth, TX, USA, 16–20 March 2014; pp. 1281–1286.
8. Zhang, H.; Ruan, X.; Lin, Z.; Wu, L.; Ding, Y.; Guo, Y. Capacitor Voltage Full Feedback Scheme for LCL-Type Grid-Connected Inverter to Suppress Current Distortion Due to Grid Voltage Harmonics. *IEEE Trans. Power Electron.* **2021**, *36*, 2996–3006. [CrossRef]

9. Elsaharty, M.A.; Ashour, H.A. Passive L and LCL filter design method for grid-connected inverters. In Proceedings of the IEEE Conference in Innovative Smart Grid Technologies, Kuala Lumpur, Malaysia, 20–23 May 2014; pp. 13–18.
10. Li, N.; Wang, Y.; Niu, R.; Guo, W.; Lei, W.; Wang, Z. A novel LCL filter parameter design method basing on resonant frequency optimization method of three-level NPC grid connected inverter. In Proceedings of the IEEE Conference on Power Electronics, Hiroshima, Japan, 18–21 May 2014; pp. 160–165.
11. Reznik, A.; Simões, M.G.; Durra, A.A.; Muyeen, S.M. LCL filter design and performance analysis for grid interconnected systems. *IEEE Trans. Ind. Appl.* **2014**, *50*, 1225–1232. [CrossRef]
12. Alzola, R.P.; Liserre, M.; Blaabjerg, F.; Sebastián, R.; Dannehl, J.; Fuchs, F.W. Analysis of the passive damping losses in LCL-filter-based grid converters. *IEEE Trans. Power Electron.* **2013**, *28*, 2642–2646. [CrossRef]
13. Ahn, H.M.; Oh, C.Y.; Sung, W.Y.; Ahn, J.H.; Lee, B.K. Analysis and Design of LCL Filter with Passive Damping Circuits for Three-phase Grid-connected Inverters. In Proceedings of the IEEE 2015 9th International Conference on Power Electronics and ECCE Asia (ICPE-ECCE Asia), Seoul, Korea, 1–5 June 2015; pp. 1–5.
14. Pan, D.; Ruan, X.; Bao, C.; Li, W.; Wang, X. Capacitor-current feedback active damping with reduced computation delay for improving robustness of LCL-type grid-connected inverter. *IEEE Trans. Power Electron.* **2014**, *29*, 3414–3427. [CrossRef]
15. Jia, Y.; Zhao, J.; Fu, X. Direct Grid Current Control of LCL-Filtered Grid-Connected Inverter Mitigating Grid Voltage Disturbance. *IEEE Trans. Power Electron.* **2014**, *29*, 1532–1541.
16. Zou, Z.; Wang, Z.; Cheng, M. Modeling, Analysis, and Design of Multifunction Grid-Interfaced Inverters with Output LCL Filter. *IEEE Trans. Power Electron.* **2014**, *29*, 3830–3839. [CrossRef]
17. Dannehl, J.; Fuchs, F.W.; Hansen, S.; Paul, B.T. Investigation of active damping approaches for PI-based current control of gridconnected pulse width modulation converters with LCL filters. *IEEE Trans. Ind. Appl.* **2010**, *46*, 1509–1517. [CrossRef]
18. Xu, J.; Xie, S.; Tang, T. Active damping-based control for grid connected LCL-filtered inverter with injected grid current only. *IEEE Trans. Ind. Electron.* **2014**, *61*, 4746–4758. [CrossRef]
19. Hanif, M.; Khadkikar, V.; Xiao, W.; Kirtley, J.L. Two Degrees of Freedom Active Damping Technique for LCL Filter-Based Grid Connected PV Systems. *IEEE Trans. Ind. Electron.* **2014**, *61*, 2795–2830. [CrossRef]
20. Dannehl, J.; Fuchs, F.W.; Thøgersen, P.B. PI State Space Current Control of Grid-Connected PWM Converters With LCL Filters. *IEEE Trans. Power Electron.* **2010**, *25*, 2320–2330. [CrossRef]
21. Busada, C.A.; Jorge, S.G.; Solsona, J.A. Full-state feedback equivalent controller for active damping in LCL filtered grid connected inverters using a reduced number of sensors. *IEEE Trans. Ind. Electron.* **2015**, *62*, 5993–6002. [CrossRef]
22. Nishida, K.; Abded, T.; Nakaoka, M. A Novel Finite-Time Settling Control Algorithm Designed for Grid-Connected Three-Phase Inverter With an LCL-Type Filter. *IEEE Trans. Ind. Appl.* **2014**, *50*, 2005–2020. [CrossRef]
23. Yu, Y.; Li, H.; Li, Z.; Zhao, Z. Modeling and Analysis of Resonance in LCL-Type Grid-Connected Inverters under Different Control Schemes. *Energies* **2017**, *10*, 104. [CrossRef]
24. Dang, C.; Tong, X.; Song, W. Sliding-mode control in dq-frame for a three-phase grid-connected inverter with LCL-filter. *ELSEVIER* **2020**, *357*, 10159–10174. [CrossRef]
25. Zhang, H.; Xian, J.; Shi, J.; Wu, S.; Ma, Z. High Performance Decoupling Current Control by Linear Extended State Observer for Three-Phase Grid-Connected Inverter With an LCL Filter. *IEEE Access* **2020**, *8*, 13119–13127. [CrossRef]
26. Firuzi, M.F.; Roosta, A.; Gitizadeh, M. Stability analysis and decentralized control of inverter-based ac microgrid. *Protection and Control of Modern Power Systems.* **2019**, *4*, 65–86.
27. Badal, F.R.; Das, P.; Sarker, S.K.; Das, S.K. A survey on control issues in renewable energy integration and microgrid. *Protection and Control of Modern Power Systems.* **2019**, *4*, 87–113. [CrossRef]
28. Liserre, M.; Sauter, T.; Hung, J.Y. Future energy systems: Integrating renewable energy sources into the smart power grid through industrial electronics. *IEEE Trans. Ind. Electron.* **2010**, *4*, 18–37. [CrossRef]
29. Annamraju, A.; Nandiraju, S. Coordinated control of conventional power sources and PHEVs using jaya algorithm optimized PID controller for frequency control of a renewable penetrated power system. *Prot. Control Mod. Power Syst.* **2019**, *4*, 343–355. [CrossRef]
30. Tian, P.; Li, Z.; Hao, Z. Doubly-Fed Induction Generator Coordination Control Strategy Compatible with Feeder Automation. *Electronics* **2020**, *9*, 18. [CrossRef]
31. Komurcugil, H.; Altin, N.; Ozdemir, S.; Sefa, I. Lyapunov-function and proportional resonant based control strategy for single-phase grid-connected VSI with LCL filter. *IEEE Trans. Ind. Electron.* **2016**, *63*, 2838–2849. [CrossRef]
32. Alemi, P.; Jeong, S.; Jeong, S.Y.; Lee, D.C. Active Damping of LLCL Filters Using PR Control for Grid-Connected Three-Level T-Type Converters. *Korean Institude Power Electron.* **2015**, *15*, 786–795. [CrossRef]
33. Tian, P.; Li, Z.; Hao, Z. A Doubly-Fed Induction Generator Adaptive Control Strategy and Coordination Technology Compatible with Feeder Automation. *Energies* **2019**, *12*, 4463. [CrossRef]
34. Gong, Z.; Zheng, X.; Zhang, H.; Wu, X.; Li, M. A QPR-Based Low-Complexity Input Current Control Strategy for the Indirect Matrix Converters with Unity Grid Power Factor. *IEEE Trans. Access.* **2019**, *99*, 38766–38777. [CrossRef]
35. Tian, P.; Li, Z.; Hao, Z. Improved Controller of LCL-Type Grid-connected Inverter System based on Active-Damping. In Proceedings of the 2018 10th International Conference on Modelling, Identification and Control (ICMIC), Guiyang, China, 2–4 July 2018; pp. 1–6.

36. Craciun, B.I.; Kerekes, T.; Sera, D.; Teodorescu, R. Overview of recent grid codes for PV power integration. In Proceedings of the IEEE International Conference on Optimization of Electrical and Electronic Equipment, Brasov, Romania, 24–26 May 2012; pp. 959–965.
37. Bao, C.; Ruan, X.; Wang, X.; Li, W.; Pan, D.; Weng, K. Step-by-Step Controller Design for LCL-Type Grid-Connected Inverter with Capacitor Current-Feedback Active-Damping. *IEEE Trans. Power Electron.* **2014**, *29*, 1239–1253.

applied sciences

MDPI

Article

A Fault Analysis Method Based on Association Rule Mining for Distribution Terminal Unit

Xuecen Zhang [1], Yi Tang [1,2,*], Qiang Liu [3], Guofeng Liu [3], Xin Ning [4] and Jiankun Chen [3]

1 School of Cyber Science and Engineering, Southeast University, Nanjing 210096, China; XZHANG067@e.ntu.edu.sg
2 School of Electrical Engineering, Southeast University, Nanjing 210096, China
3 State Grid Nanjing Power Supply Company, Nanjing 210000, China; liuqiang63025@gmail.com (Q.L.); mitsushima123@gmail.com (G.L.); chenjiankun355@gmail.com (J.C.)
4 State Grid Corporation of China, Beijing 100032, China; Xin-ning@sgcc.com.cn
* Correspondence: tangyi@seu.edu.cn

Abstract: With the development of distribution networks, large amounts of distribution terminal units (DTU) are gradually integrated into the power system. However, limited numbers of maintenance engineers can hardly cope with the pressure brought about by the substantial increase of DTU devices. As DTU fault would pose a threat to the stable and safe operation of power systems; thus, it is rather significant to reduce the fault incidence of DTU devices and improve the efficiency of fault elimination. In this paper, a DTU fault analysis method using an association rule mining algorithm was proposed. Key factors of DTU fault were analyzed at first. Then, the main concept of the Eclat algorithm was illustrated, and its performance was compared with FP-growth and Apriori algorithms using DTU fault databases of different sizes. Afterwards, a DTU fault analysis method based on the Eclat algorithm was proposed. The practicality of this method was proven by experiment using a realistic DTU fault database. Finally, the application of this method was presented to demonstrate its effectiveness.

Keywords: association rule mining; distribution terminal unit; Eclat algorithm; fault analysis

Citation: Zhang, X.; Tang, Y.; Liu, Q.; Liu, G.; Ning, X.; Chen, J. A Fault Analysis Method Based on Association Rule Mining for Distribution Terminal Unit. *Appl. Sci.* **2021**, *11*, 5221. https://doi.org/10.3390/app11115221

Academic Editors: Federico Barrero and Mario Bermúdez

Received: 30 April 2021
Accepted: 2 June 2021
Published: 4 June 2021

1. Introduction

The distribution terminal unit (DTU) collects real-time operation data of distribution networks and uploads them to the distribution automation system (DAS) or the supervisory control and data acquisition (SCADA) system. With DTU devices, DAS/SCADA can monitor the operating state of distribution networks and control secondary-side devices remotely, which shortens fault elimination time and reduces the maintenance cost to some extent. Nevertheless, with the development of distribution networks, large-scale integration of DTU also imposes pressure on maintenance work. DTU faults happen frequently while the number of maintenance workers is limited [1–5]. Moreover, DTU faults may lead to device failure, which would have severe impacts on the distribution network. Therefore, lowering fault incidence and the enhancing fault elimination ability are vital for guaranteeing the safe and stable operation of distribution networks and power systems. During DTU operation, significant fault data is generated and recorded which, however, is not fully utilized.

Data mining techniques are effective approaches for analyzing the fault of devices in power systems. The studied methods are mainly based on expert systems [6–8], state estimation (e.g., the hierarchical clustering method [9], state evaluation [10,11], and analytical hierarchy process [12–15]), machine learning (e.g., neural networks [16–19] and SVM [20–23]), and association rule mining algorithms [24–30]. Methods based on expert systems and state estimation involve evaluating devices based on human experience, which is relatively subjective. Machine learning methods and association rule mining algorithms analyze faults based on realistic operation data of devices. Therefore, they are more objective compared to the former two methods. Machine learning methods are innovative and effective in detecting and identifying faults. However, they usually require large amounts of high-quality data to achieve good performance. Compared with machine learning, methods based on association rule mining algorithms are not strictly restricted to the amount of data. Many methods based on association rule mining algorithms are designed for power transformers [24–26,29,30]. They are used to find relationships between transformer faults and operation data such as the gas component, voltage, and current. However, these methods are not feasible for DTU because it possesses a specific configuration that is completely different from the transformer. Studies have also utilized association rule mining algorithms to find possible factors that lead to faults of the distribution system [27] and secondary-side devices [28]. These approaches are relatively general rather than specific. Very few studies have analyzed the fault of DTU devices in a comprehensive and objective way. However, large-scale integration of DTU devices makes the investigation of DTU fault analysis necessary and urgent.

In order to improve the maintenance efficiency and facilitate the integration of DTU devices, a DTU fault analysis method based on an association rule mining algorithm was proposed. Firstly, the physical configuration of DTU was constructed, and factors leading to DTU fault and fault types were analyzed. Then, three different association rule mining algorithms were evaluated using DTU fault data; the Eclat algorithm, which has the best performance, was selected for this purpose. Afterwards, the DTU fault analysis method was put forward and applied to realistic DTU fault databases. Finally, application examples of this method were illustrated and recommendations for future work were suggested. The unique contributions of this paper are as follows:

1. This paper studied different aspects of DTU faults (i.e., factors which lead to faults, different fault types, and fault modules of devices); the proposed method can realize fault cause analysis and fault diagnosis, which analyzes DTU faults in a comprehensive way.
2. The proposed data-driven fault analysis method only relied on operation data of DTU devices instead of human experience. Thus, it was objective rather than subjective.
3. The fault analysis method was oriented towards DTU devices, which is more specific than other general methods for the whole system. To the best of our knowledge, this was the first method that investigated the fault of DTU devices.
4. A series of simulations and applications demonstrated that the proposed method can effectively reduce DTU fault incidence and shorten fault elimination time.

The rest of this paper is organized as follows. Section 2 discusses the fault analysis of DTU devices. Section 3 illustrates the Eclat algorithm and evaluates three different association rule mining algorithms. Section 4 proposes a DTU fault analysis method based on Eclat algorithm. Section 5 describes the implementation of this method on realistic DTU fault databases and analyzes the outcome. Section 6 is the application of this proposed fault analysis method. Section 7 is the conclusion.

2. DTU Fault Analysis

DTU is an indispensable device in power systems that monitors the operating state of distribution networks [31,32], transmits telemetry data to DAS or SCADA systems, and controls connected objects in the distribution network by utilizing signals from DAS/SCADA (Figure 1).

2.1. Configuration of DTU

A picture and simplified configuration of a DTU are shown in Figure 2. It comprises 5 modules which are, respectively, the CPU module, the communication module, the sampling module, the control circuit module, and the power supply module. Among these, the CPU functions as the centralized controller, the communication module controls information interaction between the DTU and the main station, the control circuit module sends signals to breakers, and the power supply module provides electricity to other modules.

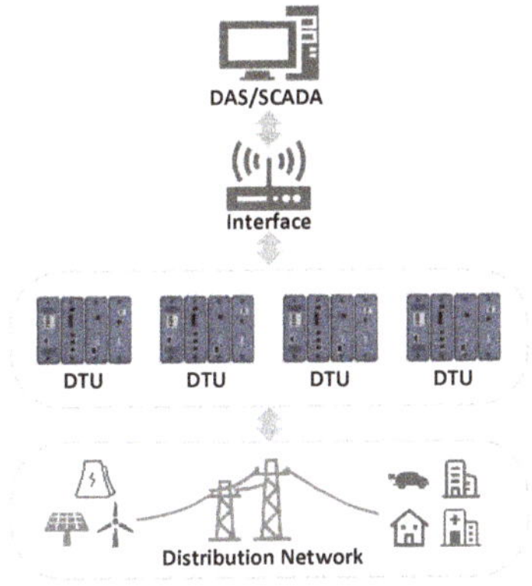

Figure 1. DTU in the power system.

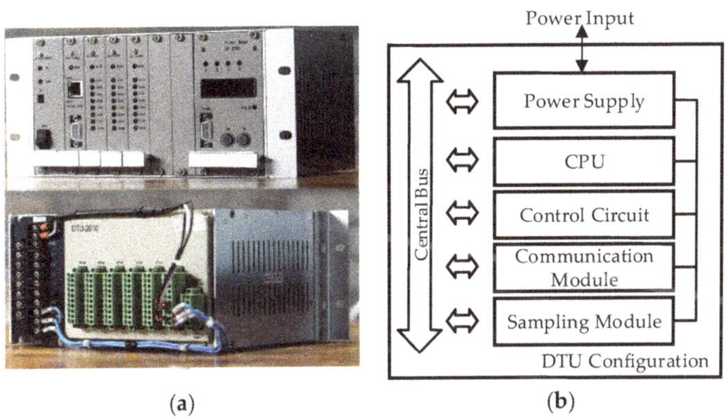

Figure 2. (**a**) Picture of DTU; (**b**) configuration of DTU.

2.2. DTU Fault Types

The DTU has several types of faults as summarized in Table 1. Functional faults can be divided into telemetering faults, remote signaling faults, and remote control faults [33]. A telemetering fault means the DTU is unable to measure the analog signals of the distribution network (e.g., voltage, current, and power). A remote signaling fault means that the DTU fails to measure signals of breakers and protection devices. A remote control fault means the DTU cannot send signals to control breakers and switches [34]. Faults that cause the DTU to stop working are called device faults. Among them, DTU offline means that the DTU is disconnected from the main station, and frequent online-offline means that the DTU is not working in a stable state.

Table 1. DTU fault types.

DTU Fault Types			
Functional Faults	Telemetering Fault	**Device Faults**	DTU Offline
	Remote Signaling Fault		Frequent Online-Offline
	Remote Control Fault		Other Faults

2.3. Factors Leading to DTU Fault

Many factors may give rise to DTU faults, which are summarized in Table 2.

Table 2. Factors leading to DTU fault.

Factors Leading to DTU Fault				
External Factors	Humidity	**Internal Factors**	Operating State	DTU Upgradation
	Wireless Signal Quality			Hardware Operating State
	Power Failure			Software Operating State
	Effects of Primary-Side Equipment		Historical State	Long Service Time
	Mis-operation			Family Defect

There are mainly 10 factors that can lead to DTU fault. Among external factors, humidity means that the moist environment causes fault of devices, effects of primary-side equipment refer to the malfunction of primary-side devices (e.g., breakers and transformers) [35], and mis-operation refers to engineers' failed operation of control devices. In terms of internal factors, operating state refers to the hardware and software operating state as well as the upgrading state of DTU devices, long service time means that device reaches or exceeds its lifespan, and family defect means that products of certain manufacturers may have certain types of faults.

2.4. Data Analysis of DTU Fault

DTU fault data usually have the same format as Table 3.

Table 3. DTU fault data format.

Time	Device ID	Brand	Fault Type	Fault Elimination Approach
2020/XX/XX	XXXX	B	T	F/M

Each data item contains the time, device ID, device brand B, fault type T, and fault elimination measure F/M. In which, fault type T refers to six types as shown in Table 1. Fault elimination approach contains information pertaining to the DTU fault module M (Figure 2) or factors F (Table 2). For instance, fault elimination approaches can be 'wireless signal issues and solved by restarting the communication module'; in this case, factor F is 'wireless signal quality' and the fault module is 'communication module'.

According to DTU fault data, brand B, fault type T, factors F, and fault module M are key elements. Therefore, item set FA containing these four elements was constructed as (1) to facilitate fault analysis:

$$FA = \{B,\ T,\ F,\ M\} \tag{1}$$

in which, there are p brand B, q types of fault T, i factors F, and j fault modules as shown in Equations (2)–(5).

$$B = \{B_1, B_2 \ldots, B_p\} \tag{2}$$

$$T = \{T_1, T_2 \ldots, T_q\} \tag{3}$$

$$F = \{F_1, F_2 \ldots, F_i\} \tag{4}$$

$$M = \{M_1, M_2 \ldots, M_j\} \tag{5}$$

3. Association Rule Mining Algorithm

Association rule mining was proposed to find the regularity of products bought by customers [36]. As for DTU fault analysis, we studied three frequently used association rule mining algorithms, which were Eclat [37], FP-growth [38], and Apriori [39]. The basic concept of the association rule, the main idea of Eclat algorithm, and the comparison of these three algorithms are illustrated in this section.

3.1. Indices of Association Rule

Taking item set $FA_\alpha = \left\{B_b, T_t, F_f, M_m\right\}$ as an example (in which $B_b \subseteq B$, $T_t \subseteq T$, $F_f \subseteq F$, $M_m \subseteq M$), the association rule can be expressed as $F_f \Rightarrow T_t$. This means if factor F_f happened, there is a possibility that fault T_t would also happen. Two indices used to evaluate the association rule are as follows:

3.1.1. Support

Support is a probability index for items. Taking item F_f as an example, support of one item can be expressed as Equation (6):

$$support(F_f) = \frac{count(F_f)}{count(itemset)} \tag{6}$$

where *count*(F_f) is the time that F_f happens in the database, *count*(*itemset*) is the number of item sets.

Support of two or more items can be calculated through Equation (7):

$$support(T_t, F_f, M_m) = \frac{count(T_t \cap F_f \cap M_m)}{count(itemset)} \tag{7}$$

where *count*($T_t \cap F_f \cap M_m$) is the time that T_t, F_f, and M_m happen in the same item set.

3.1.2. Confidence

Confidence is an index of reliability for the association rule. Taking the association rule $F_f \Rightarrow T_t$ as an example, the confidence of this rule can be calculated through Equation (8):

$$confidence(F_f \Rightarrow T_t) = \frac{count(F_f \cap T_t)}{count(F_f)} \tag{8}$$

where *count*($F_f \cap T_t$) is the number of F_f and T_t happen in the same item set.

When analyzing the relationship between different item sets, the minimum support threshold and minimum confidence threshold are often set in advance. Item sets satisfying the minimum support threshold are called frequent item sets. Rules satisfying the minimum confidence threshold are called strong rules.

3.2. Eclat Algorithm

Eclat algorithm [37] works in a vertical manner as shown in Figure 3.

In the first step, the database was scanned and all single items were listed, as well as their belonging item sets. The support value of each item equaled to the number of their belonging item sets. Next, the support value of each item was compared with the threshold, and those items meeting requirement constituted frequent 1-item sets. Then, every two items paired together, the intersection of their belonging item sets was listed. Afterwards, 2-item sets sharing one item combined with each other to form 3-item sets. This recursive process was continued until no item sets could be combined. The Eclat algorithm only scanned the whole database once, in the first step.

3.3. Generation of Strong Rules

After obtaining frequent item sets by association rule mining algorithms, the next step was to find strong rules. Taking the maximal frequent item set $\{B_1, T_2, F_3, M_4\}$ as an example, all candidate association rules $X \Rightarrow Y$ were listed, in which $X \subseteq \{B_1, T_2, F_3, M_4\}$, $Y \subseteq \{B_1, T_2, F_3, M_4\} - X$. For instance, $X = \{B_1, T_2\}$ and $Y = \{M_4\}$ or $X = \{B_1, T_2, F_3\}$ and

$Y = \{M_4\}$. Then, the confidence of each rule was compared with the minimum confidence threshold, and those rules meeting the requirement were considered as strong rules.

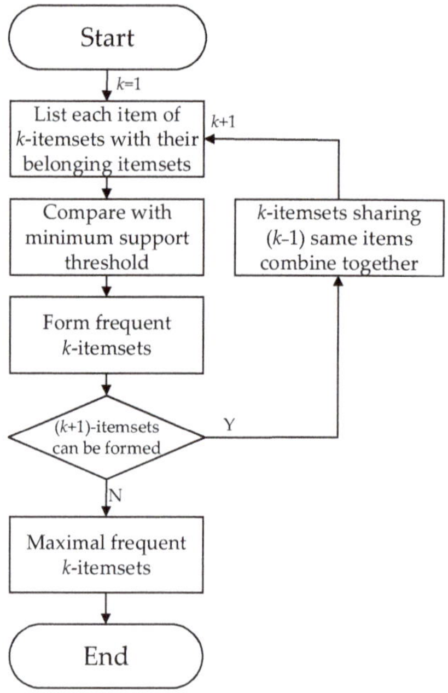

Figure 3. Procedure of the Eclat algorithm.

3.4. Comparison of Three Algorithms

Theoretically, Eclat is the best among all three algorithms as it only scans the whole database once. FP-growth is the second best since it scans database twice [38], while Apriori takes the longest time since it scans the database when generating frequent item sets [39]. To evaluate the performance of the three algorithms, five DTU fault databases of different sizes (100, 500, 1000, 5000, 10,000) were constructed, and simulation was conducted in Python 3.7 on a PC with an Intel Core i5-9400F CPU running at 2.90 GHZ and with 8.0 GB of RAM, the figure was plotted in MATLAB, version R2020b (The MathWorks, Inc., Natick, MA, USA). Time usage of three algorithms is shown in Figure 4.

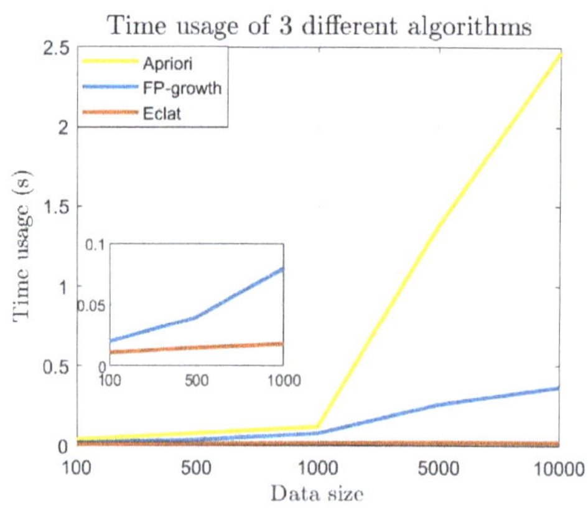

Figure 4. Time usage of the three algorithms in terms of different data sizes.

As can be seen from Figure 4, Eclat had better performance than FP-growth and Apriori. When the number of data is lower than 1000, the time usage difference of the three algorithms was nearly negligible. However, as the data size increased, the difference became much more obvious. When the number of data reached 10,000, the time usage of FP-growth was 15 times longer than Eclat, while Apriori was 103 times longer than Eclat. This outcome conformed to the theoretical analysis of three algorithms: Eclat takes the least time as it only scans the database once, while FP-growth and Apriori take longer due to additional scanning.

The Apriori algorithm is the easiest and most direct approach to mining frequent item sets, but it needs to generate a large number of candidate item sets and repeatedly scan the database, so it is rather costly. FP-growth deals with this problem by introducing a frequent pattern tree; by this means, it only scans the database twice. However, it also has disadvantages such as complexity and large memory occupation. Compared with Apriori and FP-growth, Eclat requires less time and is thus more efficient. However, as Eclat needs to repeatedly save item sets, it needs more memory space.

Strengths and weaknesses of three different algorithms are summarized in Table 4.

Table 4. Strengths and weaknesses of three algorithms.

Algorithm	Apriori	FP-Growth	Eclat
Time Usage	Most	Medium	Least
Scalability	Small and medium database	Large database	Large database
Complexity	Easy	Complex	Easy
Memory occupation	Small	Large	Medium

4. DTU Fault Analysis Method

DTU fault databases contain significant amounts of data; moreover, they are updated in real time. Therefore, the Eclat algorithm was utilized due to its best performance among the three algorithms.

The proposed DTU fault analysis method is shown in Figure 5.

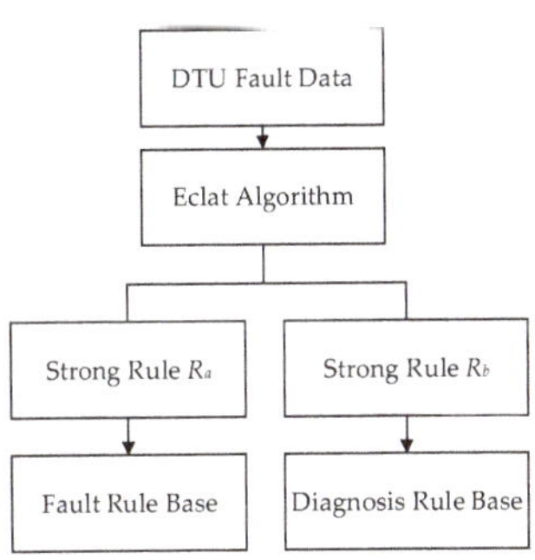

Figure 5. DTU fault analysis method.

Firstly, DTU fault data was sampled and recorded. By scanning the database, item sets *FA* were constructed. Then, Based on the Eclat algorithm, two types of strong rules were obtained:

$$R_a : F_f(B_b) \Rightarrow T_t \tag{9}$$

$$R_b : T_t \Rightarrow M_m \tag{10}$$

R_a represents the relationship between factor F and fault type T. It indicates that certain factor F_f may result in certain type of fault T_t, which reveals the rule of the fault. Brand B_b is one kind of factor F (family defect), but it is separated as an individual item in order to show if products of certain manufacturers B_b have a certain type of fault T_t. R_b includes fault type T and fault module M. It indicates when a certain type of fault T_t happens; there is a high probability that fault exists in corresponding module M_m. When fault T_t happens, diagnosing module M_m, before other modules, can shorten fault elimination time. Based on strong rules R_a and R_b, fault rule bases and diagnosis rule bases can be constructed. These two rule bases can be utilized to arrange planned maintenance and facilitate fault repair.

4.1. Fault Rule Base

Fault rule bases can be used to find the factors that lead to specific faults. Therefore, when one type of fault happens frequently and regularly, engineers can take measures to avoid corresponding factors in planned maintenance so as to lower DTU fault incidence. On the occasion when brand information is also included in the obtained strong rule $B_b \Rightarrow T_t$, manufacturer B_b can be informed that their products may have defects that can lead to a certain type of fault T_t. They should improve their products according to this information.

4.2. Diagnosis Rule Base

Diagnosis rule base reveals the relationship between fault types and fault modules. Higher confidence means that the rule is more reliable and is more likely to happen. Under circumstances wherein several rules exist, the rule with higher confidence should be considered before others, which means the corresponding fault module in that rule should be investigated at the first step. Other modules should be checked according to the rank of rule confidence. By these means, fault repair time can be shortened and efficiency of fault elimination can be improved.

5. Experiment and Results

The proposed fault analysis method for DTU was implemented on a realistic DTU fault database released by the State Grid Corporation of China.

5.1. Data Information

DTU fault data from Mar 2020 to Jun 2020 in one district of Nanjing, China was collected. After data preprocessing, a DTU fault database containing 12,320 item sets was obtained. By scanning the database, *FA* was constructed. Fault types T had five elements, which were the remote control fault, remote signaling fault, DTU offline, frequent online-offline, and other faults. Factors F have 10 kinds, including mis-operation, humidity, wireless signal quality, power failure, effects of primary-side equipment, hardware operating state, software operating state, DTU upgradation, family defects, and long-time service. Fault modules M had five elements, which were the CPU module, communication module, sampling module, power supply module, and control circuit module. Brands B had 15 elements.

5.2. Strong Rules of DTU Fault Data

Setting a reasonable support threshold and confidence threshold are of great importance. A higher support threshold may lead to fewer association rules obtained, and a lower confidence threshold may give rise to more strong rules obtained, which, however, are less reliable. Therefore, the support threshold and confidence threshold of the Eclat algorithm were set to be 0.03 and 0.8 respectively.

After applying the Eclat algorithm, 19 frequent 1-item sets, 40 frequent 2-item sets, 22 frequent 3-item sets, and 3 frequent 4-item sets were obtained. In total, 17 strong rules were mined from the database. Strong rules obtained from the DTU database are shown in Table 5.

Table 5. Strong rules obtained from DTU fault database.

No.	Strong Rule Type	Strong Rule	Support (%)	Confidence (%)	Lift
1	R_a	DTU Upgradation+Brand 4 ⇒ Remote Signaling Fault	8.26	100	5.01
2	R_a	Humidity ⇒ Remote Signaling Fault	3.60	98.1	3.97
3	R_a	Brand 10 ⇒ Remote Signaling Fault	6.71	97.9	1.97
4	R_a	Mis-operation ⇒ Remote Control Fault	4.78	96.9	3.70
5	R_a	Brand 2 + DTU Upgradation ⇒ Remote Signaling Fault	4.31	96.8	1.95
6	R_a	DTU Upgradation ⇒ Remote Signaling Fault	16.81	96.4	1.94
7	R_a	Brand2 + Wireless Signal Quality ⇒ DTU Offline	7.24	95.2	4.41
8	R_a	Wireless Signal Quality ⇒DTU Offline	8.94	93.6	4.34
9	R_a	Brand2 + Software Operating State ⇒ DTU Offline	4.17	92.2	2.24
10	R_b	Communication Module ⇒DTU Offline	10.38	81.7	3.79
11	$R_a + R_b$	Hardware Operating State ⇒ Remote Signaling Fault ⇒ Control Circuit	5.65	98.8	5.90
12	$R_a + R_b$	Hardware Operating State+Brand 2 ⇒ Remote Signaling Fault ⇒ Control Circuit	3.81	98.1	5.86
13	$R_a + R_b$	Wireless Signal Quality ⇒ Remote Signaling Fault ⇒ Communication Module	5.67	95.6	2.16
14	$R_a + R_b$	Wireless Signal Quality+Brand 2 ⇒ DTU Offline ⇒ Communication Module	4.56	95.2	4.42
15	$R_a + R_b$	Brand 2 ⇒ DTU Offline ⇒ CPU	6.79	93.1	2.36
16	$R_a + R_b$	Software Operating State+Brand 2 ⇒ DTU Offline ⇒ Control Circuit	7.16	90.4	2.29
17	$R_a + R_b$	Brand 2 ⇒ DTU Offline ⇒ Communication Module	8.96	82.6	3.83

R_a strong rule can be used to form the fault rule base. R_b strong rule can be utilized to form the diagnosis rule base. $R_a + R_b$ strong rule includes the relationship between fault type T, $F(B)$, and M, and it can be used to form both the fault rule base and diagnosis rule base.

The formed two rule bases are shown in Figure 6. The fault rule base comprised factors F, fault type T, and the confidence values of rules R_a (between F and T). The diagnosis rule base included fault type T, fault module M, and the confidence values of rules R_b.

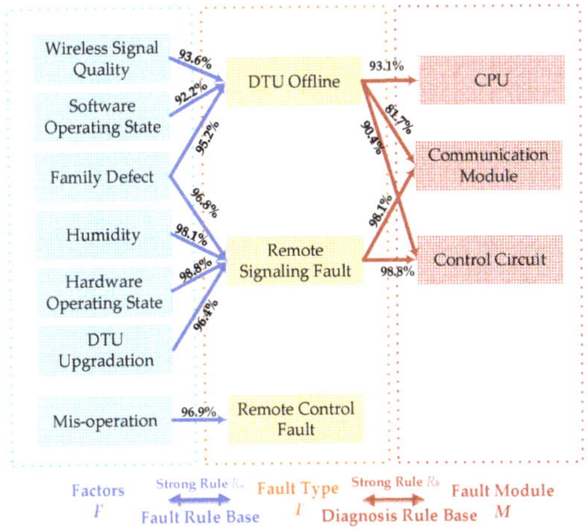

Figure 6. The fault rule base and diagnosis rule base.

The fault rule base and diagnosis rule base in Figure 6 can be used to support planned maintenance and fault repairing, respectively.

5.2.1. Planned Maintenance

In some cases, wireless signal quality, software operating states, and family defects cause DTU offline. Family defects, humidity, hardware operating states, and DTU upgradation may lead to remote signaling faults. Mis-operation is likely to cause remote control faults. These eight rules form the initial fault rule base. In planned maintenance, when a certain type of fault happens, corresponding factors in the fault rule base should be of great concern. Taking measures to avoid those factors can lower DTU fault incidence in the future.

5.2.2. Fault Repairing

When DTU offline happens, a fault usually exists in the CPU module, communication module, or control circuit module. Based on the rank of rule confidence, the CPU module is investigated at first; then, the communication module and control circuit module are checked. When a remote signaling fault happens, a fault usually exists in the control circuit or communication module. These five rules form the initial diagnosis rule base. During fault repairing, modules in the diagnosis rule base should be inspected first. By these means, fault elimination time can be shortened.

Strong rules involving brand information were not included in the above two rule bases, but they could be used to inform manufacturers about family defects of their products. These rules and their confidence values are shown in Figure 7.

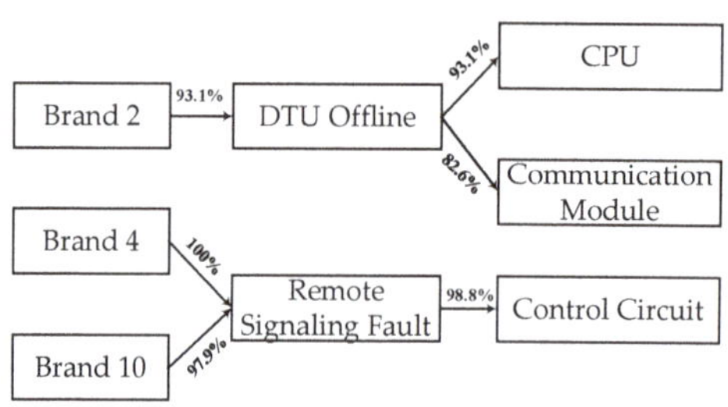

Figure 7. Strong rules including brand information.

For instance, brand 2 should pay attention to the CPU and communication modules of its products as well as the DTU offline issue. Brand 4 and brand 10 should improve control circuit modules of their products and deal with remote signaling faults.

6. Application Example

A fault analysis model was constructed and inserted into a DTU device. This model was based on the proposed method. It contained rule bases obtained from a historical DTU fault database. Rule bases were updated regularly according to the newly generated DTU fault data. When a fault happened, information of the fault type was sent into this model. The outputs were the analysis results of possible factors and corresponding fault modules in the rule bases. Information of factors was used to guide planned maintenance and information of fault modules was used for fault repairing.

For comparison, this fault analysis model was incorporated into 1000 DTUs in a certain area of Nanjing in China, while the other 4600 DTUs in the same area remained unchanged. Initial rule bases were formed as depicted in Figure 6. Based on the newly generated fault data, rule bases were updated every week. Additionally, planned maintenance was also scheduled on a weekly basis.

DTU with a fault analysis model is represented as DTU_1, the other type without the model is marked as DTU_2. From September 2020 to December 2020, fault times and fault rate are listed in Table 6. Noted that fault times *FT* (Equation (11)) are the average fault times of all devices in each group for 3 months, while fault rate *FR* (Equation (12)) refers to fault times per device per hour.

$$FT = \frac{1}{N}\sum_{n=1}^{N} FT_n \tag{11}$$

$$FR = \frac{1}{Nt}\sum_{n=1}^{N} FT_n \tag{12}$$

where N is the number of devices, FT_n is the fault times of nth device throughout 3 months, and t represents the total counted hours.

Table 6. Fault times and fault rate of two groups of DTUs.

DTU_1		DTU_2	
Fault times	Fault rate	Fault times	Fault rate
34.44	0.00159	79.45	0.03678

As can be seen from Table 6, the fault times and fault rate of DTU_1 decreased to more than half of DTU_2. This was due to the prevention of risk factors in planned maintenance.

Mean fault elimination time in terms of different fault types is shown in Figure 8. Fault elimination time is defined as the duration from the fault happening to the fault being eliminated.

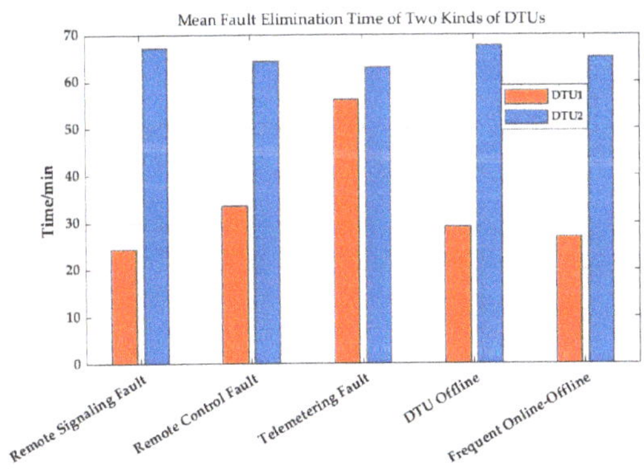

Figure 8. Mean fault elimination time of two groups of DTUs.

As shown in Figure 8, the mean fault elimination time of DTU_1 reduced to less than half of DTU_2 in terms of all fault types except the telemetering fault. Time reduction was due to the model's guidance for engineers during fault repairing. In some cases, faults could be eliminated by restarting certain modules remotely; engineers no longer needed to be on the scene. This kind of saved time was also reflected in the reduction of mean fault elimination time.

As for the telemetering fault, the saved time of DTU_1 was not as remarkable as other fault types. After analyzing the DTU fault database, it was found that the telemetering fault is not common in this area. The support value of the telemetering fault was 2.12%, while the support values of other common faults were 26.20% (remote control fault), 49.79% (remote signaling fault), and 21.54% (DTU offline). Therefore, the insufficient case of the telemetering fault limited the improvement of this model.

7. Discussion

The proposed method was intended for enhancing the efficiency of DTU maintenance work. Constructing a fault rule base and diagnosis rule base can support planned maintenance and fault repairing. This method showed a good performance in terms of lowering fault incidence and shortening fault elimination time. These improvements were due to the prevention of potential fault causes and guidance for fault repairing. Nonetheless, the proposed method also had some limitations:

1. Its performance was affected by the amount of available data.

As shown in Figure 8, insufficient data limited the improvement of this method. It is reasonable since this is a data-driven method which required a certain amount of data to achieve satisfactory results.

2. Results were not always reliable.

Since fault data was updated according to the real-time operation of DTU devices, newly generated rules were probably not in the initial rule bases and some of them might have been unreliable. Therefore, these new rules must be verified before they are incorporated into rule bases. Additionally, strong rules founded by association rule mining algorithms were sometimes redundant and repetitive; in this case, manual inspection is indispensable.

8. Conclusions and Future Work

This paper proposed a fault analysis method based on an association rule mining algorithm for DTU so as to improve the efficiency of fault elimination. The selection of algorithm for this method was proven to be reasonable through simulation, and the practicality of this method was proven by a realistic DTU fault database. This method can form fault rule bases and diagnosis rule bases, which can be used for supporting planned maintenance and fault repairing. In this way, DTU fault incidence can be reduced and fault elimination ability can be enhanced. Comparative experiment in the application example demonstrated the effectiveness of this method.

However, the performance of this method was not satisfying when not enough data were available. Moreover, since this method only relied on objective data, the results were not always reliable. A possible solution is to combine expertise to verify the generated rules as well as remove the redundant and invalid rules.

It is recommended that, with the successful application of this fault analysis and diagnosis method, more research that combines this objective data-driven method with subjective expertise should be conducted to improve the reliability of fault analysis. Additionally, machine-learning techniques can be incorporated into this method to analyze fault and estimate the state of devices so as to improve the accuracy of fault diagnosis. With the integration of more DTU devices, lifecycle management of devices should be studied to facilitate the application of this method and enhance its positive impacts. Moreover, in order to reduce maintenance pressure, future work can be conducted on eliminating manual processes and realizing intelligent maintenance.

Author Contributions: Conceptualization, X.Z and Y.T.; methodology, X.Z.; investigation, G.L.; resources, J.C.; writing—original draft preparation, X.Z.; writing—review and editing, Y.T.; supervision, Q.L.; project administration, X.N.; funding acquisition, Q.L. All authors have read and agreed to the published version of the manuscript.

Funding: This research was funded by State Grid Corporation of China. Program name: Research on automatic configuration and intelligent operation and maintenance technology of large-scale plug and play power distribution terminal. Grant number: SGJSNJ00KHJS2000556.

Institutional Review Board Statement: Not applicable.

Informed Consent Statement: Not applicable.

Data Availability Statement: Restrictions apply to the availability of these data. Data was obtained from State Grid Corporation of China. Data is available on request from the corresponding author.

Conflicts of Interest: The authors declare no conflict of interest.

References

1. Gu, J.C.; Huang, Z.J.; Wang, J.M.; Hsu, L.C.; Yang, M.T. High Impedance Fault Detection in Overhead Distribution Feeders Using a DSP-Based Feeder Terminal Unit. *IEEE Trans. Ind. Appl.* **2021**, *57*, 179–186. [CrossRef]
2. Xu, B.; Li, T.; Xue, Y. Smart distribution grid and distribution automation. *Autom. Electr. Power Syst.* **2009**, *33*, 38–41.
3. Han, G.; Xu, B. Modeling of intelligent distribution terminal according to IEC61850. *Electr. Power Autom. Equip.* **2011**, *31*, 104–107.
4. Han, G.; Xu, B.; Suonan, J.; Chen, Y. Implementation of automatic discovery technology for distribution terminal. *Autom. Electr. Power Syst.* **2012**, *36*, 82–85.
5. Zhao, J.; Dong, Z.; Wen, F. Data science for energy systems: Theory, techniques and prospect. *Autom. Electr. Power Syst.* **2017**, *41*, 1–11, 19.
6. Lin, C.E.; Ling, J.M.; Huang, C.L. An expert system for transformer fault diagnosis using dissolved gas analysis. *IEEE Trans. Power Deliv.* **1993**, *8*, 231–238. [CrossRef]
7. Yongli, Z.; Yang, Y.H.; Hogg, B.W. An expert system for power systems fault analysis. *IEEE Trans. Power Syst.* **1994**, *9*, 503–509. [CrossRef]
8. Minakawa, T.; Ichikawa, Y.; Kunugi, M. Development and implementation of a power system fault diagnosis expert system. *IEEE Trans. Power Syst.* **1995**, *10*, 932–940. [CrossRef]
9. Rao, W.; Wang, F.; Ding, J. Family defect assessment of electric equipment based on improved hierarchical clustering method. *Zhejiang Electr. Power* **2013**, *32*, 9–13.
10. Liu, M.; Wang, K.; Feng, S. Study of comprehensive state evaluation method for distribution terminal device. *Mech. Electr. Inf.* **2016**, *36*, 95–96.
11. Zhu, J.; Leng, H.; Wang, Y. Research on state evaluation for distribution automation terminals. *Distrib. Util.* **2017**, *34*, 59–63.
12. Cai, J.; Yang, C. Study on distribution automation terminal state evaluation method based on analytic hierarchy process. *Guangxi Electr. Power* **2018**, *41*, 6–9.
13. Li, M.; Han, X.; Yang, M. Basic concept and theoretical study of condition-based maintenance for power transmission system. In Proceedings of the 2011 Communications in Computer and Information Science, Wuhan, China, 21 August 2011; pp. 43–52.
14. Yang, Z.; Shen, Y.; Zhang, J. Online monitoring and state diagnosis method for distribution terminal battery. *High. Volt. Eng.* **2016**, *42*, 121–126.
15. Zhang, P.; Su, Y.; Han, Y. Judgment on failure distribution type and estimation of failure rate for distribution terminals. *Proc. CSU-EPSA.* **2018**, *30*, 61–65.
16. Bernieri, A.; Betta, G.; Pietrosanto, A. A neural network approach to instrument fault detection and isolation. In Proceedings of the 10th Anniversary IMTC/94 Advanced Technologies in I & M. 1994 IEEE Instrumentation and Measurement Technology Conference (Cat. No. 94CH3424-9), Hamamatsu, Japan, 10–12 May 1994; pp. 139–144.
17. Zhang, Y.; Ding, X.; Liu, Y. An artificial neural network approach to transformer fault diagnosis. *IEEE Trans. Power Deliv.* **1996**, *11*, 1836–1841. [CrossRef]
18. Meng, K.; Dong, Z.; Wang, D.H. A self-adaptive RBF neural network classifier for transformer fault analysis. *IEEE Trans. Power Syst.* **2010**, *25*, 1350–1360. [CrossRef]
19. Yang, Q.; Li, J.; Le Blond, S. Artificial neural network based fault detection and fault location in the DC microgrid. *Energy Procedia* **2016**, *103*, 129–134. [CrossRef]
20. Ganyun, L.V.; Haozhong, C.; Haibao, Z. Fault diagnosis of power transformer based on multi-layer SVM classifier. *Electr. Power Syst. Res.* **2005**, *74*, 1–7. [CrossRef]
21. Bacha, K.; Souahlia, S.; Gossa, M. Power transformer fault diagnosis based on dissolved gas analysis by support vector machine. *Electr. Power Syst. Res.* **2012**, *83*, 73–79. [CrossRef]
22. Bhattacharya, B.; Sinha, A. Intelligent fault analysis in electrical power grids. In Proceedings of the 2017 IEEE 29th International Conference on Tools with Artificial Intelligence (ICTAI), Boston, MA, USA, 6–8 November 2017; pp. 985–990.
23. Chen, M.J.; Lan, S.; Chen, D.Y. Machine learning based one-terminal fault areas detection in HVDC transmission system. In Proceedings of the 2018 8th International Conference on Power and Energy Systems (ICPES), Colombo, Sri Lanka, 21–22 December 2018; pp. 278–282.
24. Lee, L.; Cheng, Y.; Xie, L. An integrated method of set pair analysis and association rule for fault diagnosis of power transformers. *IEEE Trans. Dielectr. Electr. Insul.* **2015**, *22*, 2368–2378.
25. Sheng, G.; Hou, H.; Jiang, X. A novel association rule mining method of big data for power transformers state parameters based on probabilistic graph model. *IEEE Trans. Smart Grid* **2016**, *9*, 695–702. [CrossRef]
26. Chen, Y.; Du, X.; Zhou, L. Transformer defect correlation analysis based on apriori algorithm. In Proceedings of the 2016 IEEE International Conference on High Voltage Engineering and Application, Chengdu, China, 19 September 2016.
27. Doostan, M.; Chowdhury, B.H. Power distribution system fault cause analysis by using association rule mining. *Electr. Power Syst. Res.* **2017**, *152*, 140–147. [CrossRef]

28. Zhang, Y.; Hu, C.; Huang, S. Apriori algorithm based data mining and analysis method for secondary device defects. *Autom. Electr. Power Syst.* **2017**, *41*, 147–151, 163.
29. Li, H.; Wang, Y.; Liang, X. Fault prediction of power transformer by association rules and markov. In Proceedings of the 2018 IEEE International Conference on High Voltage Engineering and Application (ICHVE), Athens, Greece, 10 September 2018.
30. Guo, C.; Wang, B.; Wu, Z. Transformer failure diagnosis using fuzzy association rule mining combined with case-based reasoning. *IET Gener. Trans. Distrib.* **2020**, *14*, 2202–2208. [CrossRef]
31. Lin, D.; Liu, Q.; Li, F. Optimal Placement of Multiple Feeder Terminal Units Using Intelligent Algorithms. *Appl. Sci.* **2020**, *10*, 299. [CrossRef]
32. Sodin, D.; Rudež, U.; Mihelin, M. Advanced Edge-Cloud Computing Framework for Automated PMU-Based Fault Localization in Distribution Networks. *Appl. Sci.* **2021**, *11*, 3100. [CrossRef]
33. Tao, W.; Ma, M.; Fang, C. Design and Application of a Distribution Network Phasor Data Concentrator. *Appl. Sci.* **2020**, *10*, 2942. [CrossRef]
34. Orozco-Henao, C.; Suman Bretas, A.; Marín-Quintero, J. Adaptive impedance-based fault location algorithm for active distribution networks. *Appl. Sci.* **2018**, *8*, 1563. [CrossRef]
35. Li, G.; Chen, Q.; Zhang, J. Novel Faulted Section Location Method for Distribution Network Based on Status Information of Fault Indicating Equipment. *Appl. Sci.* **2020**, *10*, 5910. [CrossRef]
36. Agrawal, R.; Imieliński, T.; Swami, A. Mining association rules between sets of items in large databases. In Proceedings of the 1993 ACM SIGMOD International Conference on Management of Data, Washington, DC, USA, 26–28 May 1993; pp. 207–216.
37. Zaki, M.J. Scalable algorithms for association mining. *IEEE Trans. Knowl. Data Eng.* **2000**, *12*, 372–390. [CrossRef]
38. Han, J.; Pei, J.; Yin, Y. Mining frequent patterns without candidate generation. *ACM Sigmod Rec.* **2000**, *29*, 1–12. [CrossRef]
39. Agrawal, R.; Srikant, R. Fast algorithms for mining association rules. In Proceedings of the 20th International Conference on Very Large Data Bases, Santiago, Chile, 12–15 September 1994; pp. 487–499.

applied sciences

MDPI

Article

Evaluation of the Simultaneous Operation of the Mechanisms for Cross-Border Interchange and Activation of the Regulating Reserves

Marcel Topler * and Boštjan Polajžer

Faculty of Electrical Engineering and Computer Science, University of Maribor, SI-2000 Maribor, Slovenia; bostjan.polajzer@um.si
* Correspondence: marcel.topler@um.si

Abstract: This article examines the mechanisms for cross-border interchange of the regulating reserves (RRs), i.e., the imbalance-netting process (INP) and the cross-border activation of the RRs (CBRR). Both mechanisms are an additional service of frequency restoration reserves in the power system and connect different control areas (CAs) via virtual tie-lines to release RRs and reduce balancing energy. The primary objective of the INP is to net the demand for RRs between the cooperating CAs with different signs of interchange power variation. In contrast, the primary objective of the CBRR is to activate the RRs in the cooperating CAs with matching signs of interchange power variation. In this way, the ancillary services market and the European balancing system should be improved. However, both the INP and CBRR include a frequency term and thus impact the frequency response of the cooperating CAs. Therefore, the impact of the simultaneous operation of the INP and CBRR on the load-frequency control (LFC) and performance is comprehensively evaluated with dynamic simulations of a three-CA testing system, which no previous studies investigated before. In addition, a function for correction power adjustment is proposed to prevent the undesirable simultaneous activation of the INP and CBRR. In this way, area control error (ACE) and scheduled control power are decreased since undesired correction is prevented. The dynamic simulations confirmed that the simultaneous operation of the INP and CBRR reduced the balancing energy and decreased the unintended exchange of energy. Consequently, the LFC and performance were improved in this way. However, the impact of the INP and CBRR on the frequency quality has no unambiguous conclusions.

Keywords: frequency quality; load-frequency control; regulating reserves; cross-border interchange; cross-border activation

Citation: Topler, M.; Polajžer, B. Evaluation of the Simultaneous Operation of the Mechanisms for Cross-Border Interchange and Activation of the Regulating Reserves. *Appl. Sci.* **2021**, *11*, 8188. https://doi.org/10.3390/app11178188

Academic Editors: Federico Barrero and Mario Bermúdez

Received: 11 August 2021
Accepted: 1 September 2021
Published: 3 September 2021

1. Introduction

1.1. Motivation and Literature Review

The mechanisms for cross-border interchange and activation of the regulating reserves (RRs) are evolving due to the expensive balancing energy, and are included in the European Union's current regulations [1,2]. They are put in operation in continental Europe by the members of the European network of transmission system operators for electricity (ENTSO-E) [3]. Since the first implementation of the cross-border interchange of the RRs, i.e., the imbalance-netting process (INP), in 2008, the cumulative value of all the netted imbalances amounted to more than €600 million by the third quarter of 2020 [4]. The total monthly volume of netted imbalances for September 2020 was 698.69 GWh, which amounts to €13.38 million. Moreover, the monthly avoided positive and negative RRs activations amount to a minimum of 10% to as much as 85%. The further development of the INP with the functionality of the cross-border activation of the RRs (CBRR), will additionally reduce the activation of the RRs and increase the associated savings.

To avert the activation of the RRs with different signs in the cooperating CAs, and thus reduce the RRs, a grid-control cooperation (GCC) platform was implemented in Germany, where four transmission system operators (TSOs), i.e., 50 Hertz, Amprion, TenneT, and TransnetBW have been collaborating since 2008 [5]. In the following years, many continental European countries joined and the GCC platform developed into the international GCC (IGCC) platform, with the aim to further reduce RRs and increase the reliability of the power system's operation [6–9]. Hence, the INP was developed and put in operation, where the cooperating CAs with different signs of power variations can exchange the balancing energy and thus compensate the power variations [10,11]. Therefore, CAs with a surplus of power can exchange with CAs having power shortages [12]. A comparable approach, i.e., area control error (ACE) diversity interchange, was implemented in North America in 1993, but does not consist of actual responses from the control units [13–15].

A further reduction of power-system operating costs and increasingly stringent requirements for the quality of the Load-Frequency Control (LFC) defined by the new network codes require further development of the INP with a functionality that will enable CBRR [16]. Therefore, in the first quarter of 2020, the development of the CBRR started that will be put in operation in continental Europe in 2022 [3]. The aim of the development and operation of the CBRR is to improve the ancillary services market and the European balancing system [17]. Similar to the INP, the same control–demand approach and implementation in the control structure will be used for the CBRR. However, the primary objective of the CBRR will be the activation of the RRs in the cooperating CAs and importing into its own CA, thereby reducing the balancing energy [18]. CBRR will only be achievable if the cooperating CAs have matching signs of power deviations. Consequently, CAs with a surplus of power can activate the RRs only in CAs with a surplus of power. Both mechanisms, i.e., INP and CBRR, reduce the balancing energy, while releasing the RRs and, therefore, reducing the associated economic costs. This increases the economic benefits, as the energy exchanged by the INP and activated by the CBRR is additionally financially compensated [19].

A basic schematic diagram of the LFC, INP, and CBRR is given in Figure 1 (left), where the order of the operation is clearly seen. Note here that the correction power is the output of the INP | CBRR block. The main distinction among the INP and the CBRR is in the requirements to compensate for the imbalances among the cooperating CAs. The aim of the INP is to avert the simultaneous activation of RRs with different signs in cooperating CAs, i.e., to net the demand for balancing energy between CAs with different signs of demand power. In contrast, the aim of CBRR is to activate the demand for balancing energy in cooperating CAs with matching signs of demand power. The INP and the CBRR link all the CAs to a joint portal of virtual tie-lines where the INP or CBRR optimization is performed. Note that a virtual tie-line means an additional input of the controllers of the cooperating CAs that has the same effect as a measuring value of a physical interconnector and allows exchange of electric energy between the cooperating CAs [1]. The main objectives of the INP optimization are given in [20,21], whereas the main objectives of the CBRR optimization are given in [22,23].

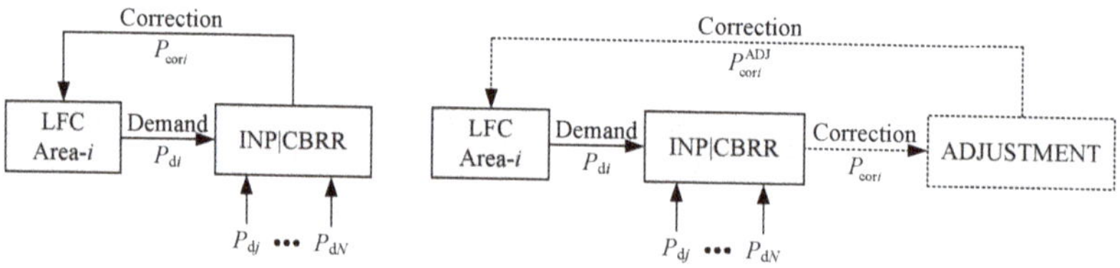

Figure 1. Schematic diagram of the LFC, INP, CBRR (**left**) and the function for correction power adjustment (**right**).

There has been a surge in the application of machine learning and statistical framework to solve similar problems focused in this paper. The authors in [24] explore the influencing factors of consumer purchase intention of cross-border e-commerce based on a wireless network and machine learning in order to provide decision support for the operation of e-commerce and to promote the better development of cross-border e-commerce. Several model-based experimental design techniques have been developed for design in domains with partial available data about the underlying process. The authors in [25] focus on a powerful class of model-based experimental design called the mean objective cost of uncertainty. To achieve a scalable objective-based experimental design, a graph-based mean objective cost of uncertainty with Bayesian optimization framework is proposed. A thorough review of the issues of data localization and data residency is given in [26], in addition to clarifying cross-border data flow restrictions and the impact of cross-border data flows in Asia.

1.2. Contribution and Structure of the Paper

Generally, the INP and the CBRR should have a positive impact on the LFC and performance. However, the quality of the frequency is continually decreasing [27]. Therefore, the impact of the INP and the CBRR on the frequency quality, the LFC, and performance in a three-CA test system was examined separately in [20,22] with dynamic simulations. In [21], the impact of INP on power-system dynamics is shown and an eigenvalue analysis of a two-CA system is conducted. The impact of CBRR on the power-system dynamics is shown in [23], and a modified implementation of the CBRR is proposed that has no impact on the system's eigendynamics. This article extends these earlier results with an in-depth evaluation of the simultaneous operation of the mechanisms for cross-border interchange and activation of the RRs, which was not studied before. Dynamic simulations are performed for all the cases where the simultaneous operation of the INP and the CBRR is possible. In addition, a function for correction-power adjustment is proposed as one of the contributions of this paper, since small delays in demand power sign change could cause undesired simultaneous activation of the INP and the CBRR. In this way, ACE and scheduled control power are decreased, since udesired correction is prevented. A basic schematic diagram of the LFC, INP, and CBRR and the function for correction power adjustment is given in Figure 1 (right), where the order of operation is clearly seen. Note here that the correction power is the output of the adjustment block. As far as we know, no researchers have examined the impact of the simultaneous operation of the INP and CBRR on the LFC and performance.

This article consists of the following parts: In Section 2, the elemental concepts of the LFC, the INP and the CBRR are described. Simultaneous operation of the INP and the CBRR is also described. Additionally, a function for correction power adjustment is proposed as one of the main contributions of this article, which prevents undesired correction. Section 3 describes indicators for evaluation of the frequency quality, LFC and performance, rate of change of frequency (RoCoF), balancing energy, unintended exchange of energy and energy exchange. In Section 4, a three-CA test system with the INP and the CBRR is described. Two types of test cases were performed with the dynamic simulations, i.e., step change of the load and the random load variation. The primary contribution of this article is given in Section 5, where the impact is given of the simultaneous operation of the INP and the CBRR on the frequency quality, the LFC, and performance. Lastly, Section 6 outlines the main conclusions and outlines future work.

2. LFC, INP, and CBRR

2.1. LFC

An interconnected power system consists of a large number of CAs that are connected via tie-lines. Each individual TSO maintains the frequency of each CA within predefined standard limits and the tie-line power flows with neighboring CAs within prespecified tolerances, which are the main objectives of the LFC [28]. This is generally accomplished by reducing the ACE, which is, for the CA i, defined as

$$ACE_i = \Delta P_i + B_i \Delta f_i, \quad (1)$$

where $\Delta P_i = (P_{ai} - P_{si})$ and $\Delta f_i = (f_{ai} - f_{si})$ denote interchange power variation and frequency deviation, respectively. Note that P_{ai} and f_{ai} denote actual values, while P_{si} and f_{si} denote scheduled values. Furthermore, B_i is the frequency-bias coefficient [1]. Note that $ACE_i > 0$ denotes that the generation is higher than the load; hence, the CA is denoted as "long". Similarly, a CA is denoted as "short" when $ACE_i < 0$.

The basic LFC framework of the CA i is shown in Figure 2, where LPF denotes a low pass filter and SH a sample and hold, with the value of a sampling time $T_s = 2$ s. A negative control-feedback is characterized as -1 gain and PI denotes a proportional-integral controller. Scheduled control power ΔP_{sci} denotes the output of LFC, which is appointed to the participating control units that change the electrical control power ΔP_{ei} appropriately. Neglecting the losses, then ΔP_{ei} is, for the CA i, defined as

$$\Delta P_{ei} = \Delta P_{Li} + \Delta P_i, \quad (2)$$

where ΔP_{Li} denotes the load-power variation. Note that ΔP_{ei} is well-known as the balancing energy, whereas, instead of LFC reserve, the term RR is generally used.

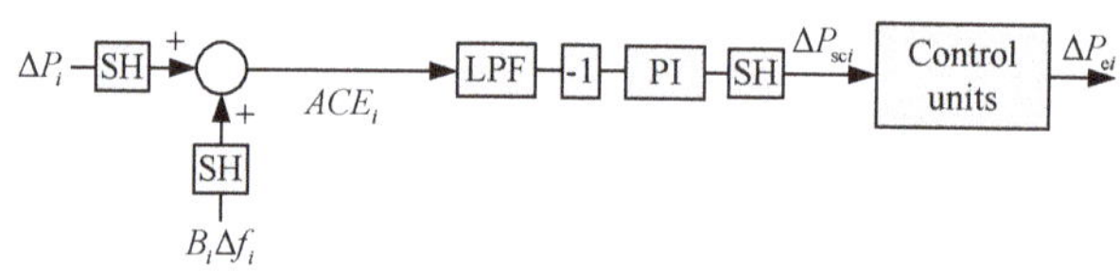

Figure 2. Schematic block diagram representation of the LFC in the CA i.

2.2. INP

The basic principle of the INP operation and a steady-state correction value calculation with the INP for three CAs is given in [20]. A control–demand concept is used for the INP and a schematic block diagram representation is shown in Figure 3. Commonly, N CAs can be connected via virtual tie-lines, i.e., they can all activate the INP via the interchange factors $K_i^{\text{INP}}, K_j^{\text{INP}}, \ldots, K_N^{\text{INP}}$ marked with the green rectangle. Note that the factor K_j^{INP} represents the size of the INP interchange of the j-th CA in the i-th CA, where $K_j^{\text{INP}} = 0$ determines that the possible INP interchange is equal to 0%, whereas $K_j^{\text{INP}} = 1$ determines that the possible INP interchange is equal to 100%.

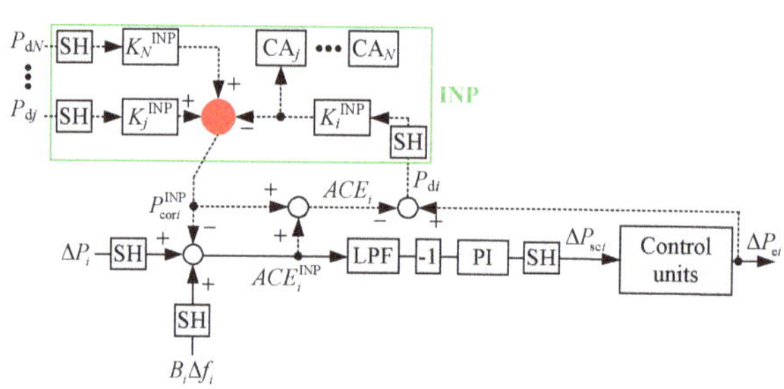

Figure 3. Schematic block diagram representation of the LFC in the CA i (solid line) with the INP (solid and dotted line).

Moreover, the values of $K_i^{\text{INP}}, K_j^{\text{INP}}, \ldots, K_N^{\text{INP}}$ can be different. The cooperating CAs are connected to the "red" summator, forming virtual tie-lines, as shown in Figure 3. The input variables to the "red" summator are the demand powers of the cooperating CAs, i.e., P_{di},

$P_{dj}, \ldots, P_{dN}$. The demand power of the CA i characterizes the maximum interchange power for the CA i among the cooperating CAs and is defined as

$$P_{di} = \Delta P_{ei} - ACE_i \tag{3}$$

according to [3,29].

Introducing (1) and (2) in (3) gives the following relation:

$$P_{di} = \Delta P_{Li} - B_i \Delta f_i. \tag{4}$$

The power imbalance between generation and load in addition to $K_i^{\text{INP}} P_{di}$ from the CA i, $K_j^{\text{INP}} P_{dj}$ from CA $j, \ldots, K_N^{\text{INP}} P_{dN}$ from CA N is, for the CA i, defined as

$$ACE_i^{\text{INP}} = B_i \Delta f_i + \Delta P_i + K_i^{\text{INP}} P_{di} - (K_j^{\text{INP}} P_{dj} +, \ldots, + K_N^{\text{INP}} P_{dN}). \tag{5}$$

The output variables of the "red" summator are the correction powers of the cooperating CAs, i.e., $P_{\text{cor}i}^{\text{INP}}, P_{\text{cor}j}^{\text{INP}}, \ldots, P_{\text{cor}N}^{\text{INP}}$, determined with a delay of T_s due to the SH. The correction power of the CA i characterizes the maximum interchange power for the CA i among the cooperating CAs with a different sign of ACE_i, and is included as

$$ACE_i^{\text{INP}} = (B_i \Delta f_i + \Delta P_i) - P_{\text{cor}i}^{\text{INP}}, \tag{6}$$

where the terms in brackets denote ACE_i.

Moreover, only CAs with different signs of demand power, i.e., $\text{sign}(P_{di}) \neq \text{sign}(P_{dj})$, can net the demand for balancing energy. If two or more cooperating CAs are "short", then CBRR is used instead of the INP and vice versa [23]. Therefore, the cooperating CAs must be "short" and "long" in order to net the demand power through the INP. Hence, the balancing energy in CAs that net the balancing energy from the cooperating CAs can be reduced, and simultaneously the RR is released. The $P_{\text{cor}i}^{\text{INP}}, P_{\text{cor}j}^{\text{INP}}, \ldots, P_{\text{cor}N}^{\text{INP}}$ is determined by the INP optimization module, considering numerous target functions, as given in [20].

Considering N CAs, then the $P_{\text{cor}i}^{\text{INP}}$ is, for the CA i, expressed as

$$P_{\text{cor}i}^{\text{INP}} = -P_{di} K_i^{\text{INP}} + P_{dj} K_j^{\text{INP}} +, \ldots, + P_{dN} K_N^{\text{INP}}. \tag{7}$$

Considering (4), then the $P_{\text{cor}i}^{\text{INP}}$ between N CAs is, for the CA i, expressed as

$$P_{\text{cor}i}^{\text{INP}} = -(\Delta P_{Li} - B_i \Delta f_i) K_i^{\text{INP}} + (\Delta P_{Lj} - B_j \Delta f_j) K_j^{\text{INP}} +, \ldots, + (\Delta P_{LN} - B_N \Delta f_N) K_N^{\text{INP}}. \tag{8}$$

In this way, the correction power of the CA i compensates the load variation that is varied by the frequency variation of the cooperating CAs. From a system point of view, this corresponds to additional frequency-based feedback and cross-couplings with cooperating CAs, which inseparably changes the eigendynamics of the CA i [21].

2.3. CBRR

The basic principle of the CBRR operation and a steady-state correction-value calculation with the CBRR for three CAs is given in [22]. The same control–demand concept is used for the CBRR as is currently used for the INP, and a schematic block diagram representation is shown in Figure 4 [3]. Similar to INP, N CAs can be connected via the virtual tie-lines, i.e., they can all activate the CBRR via the activation factors $K_i^{\text{CBRR}}, K_j^{\text{CBRR}}, \ldots, K_N^{\text{CBRR}}$ marked with the green rectangle. Note that the factor K_j^{CBRR} represents the size of the CBRR activation for the CA j in the CA i, where $K_j^{\text{CBRR}} = 0$ determines that the possible CBRR activation is equal to 0%, whereas $K_j^{\text{CBRR}} = 1$ determines that the possible CBRR activation is equal to 100%. Moreover, the values of $K_i^{\text{CBRR}}, K_j^{\text{CBRR}}, \ldots, K_N^{\text{CBRR}}$ can be different. The cooperating CAs are connected to the "red" summator, forming virtual tie-lines, as shown in Figure 4.

The input variables to the "red" summator are the demand powers of the cooperating CAs, i.e., $P_{\mathrm{d}i}$, $P_{\mathrm{d}j}$, ..., $P_{\mathrm{d}N}$. The demand power of the CA i characterizes the maximum activation power for the CA i among the cooperating CAs, defined as (3), according to [3,29].

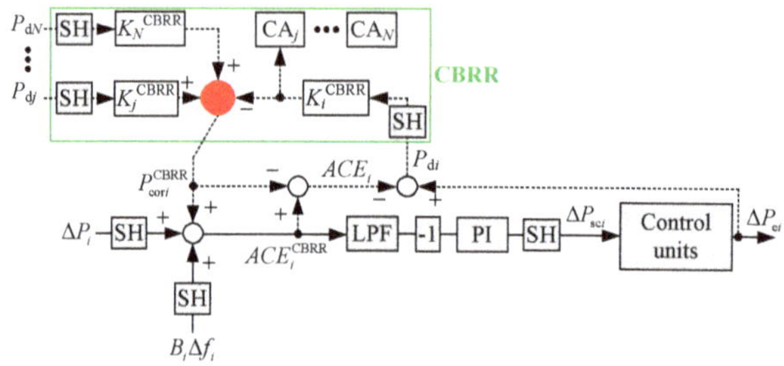

Figure 4. Schematic block diagram representation of the LFC in the CA i (solid line) with the CBRR (solid and dotted line).

Similar to INP, by introducing (1) and (2) in (3), (4) is obtained.

The power imbalance between generation and load in addition to $K_i^{\mathrm{CBRR}}P_{\mathrm{d}i}$ from the CA i, $K_j^{\mathrm{CBRR}}P_{\mathrm{d}j}$ from the CA j, ..., $K_N^{\mathrm{CBRR}}P_{\mathrm{d}N}$ from the CA N is, for the CA i, defined as

$$ACE_i^{\mathrm{CBRR}} = B_i\Delta f_i + \Delta P_i - K_i^{\mathrm{CBRR}}P_{\mathrm{d}i} + (K_j^{\mathrm{CBRR}}P_{\mathrm{d}j}+,\ldots,+K_N^{\mathrm{CBRR}}P_{\mathrm{d}N}). \tag{9}$$

The output variables of the "red" summator are the correction powers of the cooperating CAs, i.e., $P_{\mathrm{cor}i}^{\mathrm{CBRR}}$, $P_{\mathrm{cor}j}^{\mathrm{CBRR}}$, ..., $P_{\mathrm{cor}N}^{\mathrm{CBRR}}$, determined with a delay of T_s due to the SH. The correction power of the CA i characterizes the maximum activation power for the CA i among the cooperating CAs with matching sign of ACE_i, and is included as

$$ACE_i^{\mathrm{CBRR}} = (B_i\Delta f_i + \Delta P_i) + P_{\mathrm{cor}i}^{\mathrm{CBRR}}, \tag{10}$$

where the terms in brackets denote ACE_i.

Moreover, only CAs with matching sign of demand power, i.e., $\mathrm{sign}(P_{\mathrm{d}i}) = \mathrm{sign}(P_{\mathrm{d}j})$, can activate the demand for balancing energy. If any of the cooperating CAs are "long" and the others are "short", then INP is used instead of the CBRR and vice versa [21]. Therefore, the cooperating CAs must be either "short" or "long", depending on whether a positive or negative CBRR is activated. Hence, the balancing energy in CAs that activates the balancing energy in the cooperating CAs can be reduced, and simultaneously the RR is released. The $P_{\mathrm{cor}i}^{\mathrm{CBRR}}$, $P_{\mathrm{cor}j}^{\mathrm{CBRR}}$, ..., $P_{\mathrm{cor}N}^{\mathrm{CBRR}}$ is determined by the CBRR optimization module, considering numerous target functions, as given in [22].

Considering N CAs, then the $P_{\mathrm{cor}i}^{\mathrm{CBRR}}$ is, for the CA i, expressed as

$$P_{\mathrm{cor}i}^{\mathrm{CBRR}} = -P_{\mathrm{d}i}K_i^{\mathrm{CBRR}} + P_{\mathrm{d}j}K_j^{\mathrm{CBRR}}+,\ldots,+P_{\mathrm{d}N}K_N^{\mathrm{CBRR}}. \tag{11}$$

Considering (4), then the $P_{\mathrm{cor}i}^{\mathrm{CBRR}}$ between N CAs is, for the CA i, expressed as

$$P_{\mathrm{cor}i}^{\mathrm{CBRR}} = -(\Delta P_{\mathrm{L}i} - B_i\Delta f_i)K_i^{\mathrm{CBRR}} + (\Delta P_{\mathrm{L}j} - B_j\Delta f_j)K_j^{\mathrm{CBRR}}+,\ldots,+(\Delta P_{\mathrm{L}N} - B_N\Delta f_N)K_N^{\mathrm{CBRR}}. \tag{12}$$

Similar to the INP, the correction power of the CA i compensates the load variation that is varied by the frequency variation of the cooperating CAs. From a system point of view, this corresponds to an additional frequency-based feedback and cross-couplings with cooperating CAs, which inseparably changes the eigendynamics of the CA i [23].

2.4. Simultaneous Operation of the INP and the CBRR

In the cooperating CAs, ΔP_{Li} changes randomly and continuously. Consequently, cases of ΔP_{di} sign changes can occur, resulting in a continuous switching between the INP and the CBRR, which causes undesirable ΔP_{cori} sign changes. Such a situation might occur when the signs of ΔP_{di} and ΔP_{dj} are changed with a short time delay. Therefore, a function for P_{cori} adjustment is proposed as one of the contributions of this article.

A schematic block diagram representation for P_{cori} adjustment in relation to P_{di} of the CA i is shown in Figure 5. The signs of the two successive samples, i.e., $P_{di,k-1}$ and $P_{di,k}$, are compared with the relational operator, whose output is connected to a switch, marked with "c". Two states are possible, i.e.,

- State 1: $\text{sign}(P_{di,k-1}) = \text{sign}(P_{di,k})$ and
- State 2: $\text{sign}(P_{di,k-1}) \neq \text{sign}(P_{di,k})$.

For State 1, the switch position is "1" and the output variable is $P_{cori}^{ADJ} = P_{cori}$. For State 2, the switch position is "2" and the output variable is $P_{cori}^{ADJ} = 0$.

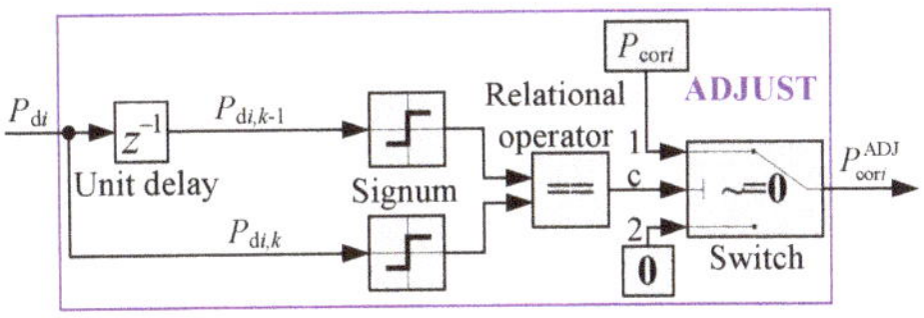

Figure 5. Schematic block diagram representation for P_{cori} adjustment in the CA i.

The implementation of the P_{cori} adjustment in the INP and the CBRR framework is shown in Figure 6. The simultaneous operation of the INP and the CBRR, considering the P_{cori} adjustment, can be described using the pseudo-code, as shown in Alogorithm 1.

Algorithm 1: The code of simultaneous operation of the INP and the CBRR.

```
if sign(P_di) = sign(P_dj), i ≠ j then
    activate CBRR & adjust P_cori, P_corj
else if sign(P_di) ≠ sign(P_di), i ≠ j then
    activate INP & adjust P_cori, P_corj
end if.
```

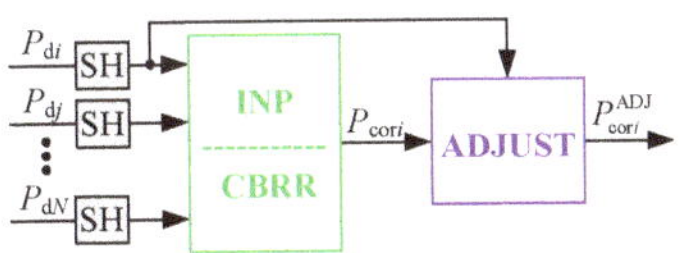

Figure 6. Schematic block diagram representation for P_{cori} adjustment with the INP and the CBRR in the CA i.

An example of the simultaneous operation of the INP and the CBRR, with and without the P_{cori} adjustment, is shown in Figure 7. The values of the loads were set in such a way that the simultaneous operation of the INP and the CBRR was possible. In the time interval of 80–100 s, the INP operated between CA_1–CA_3 and CA_2–CA_3, whereas the CBRR operated between CA_1–CA_2. During this time interval, P_{cori} due to the INP was possible in all three CAs, whereas P_{cori} due to CBRR was possible only in CA_1 and CA_2. At $t = 100$ s, a simultaneous step change of the load in CA_1 and CA_3 was applied, whereas, in CA_2, it was applied with a delay, i.e., at $t = 100.05$ s, as seen at $t = 102$ s due to SH with $T_s = 2$ s. Time responses of P_{di} and P_{cori} at $t = 100$ s with one step-size activation of the INP in CA_1 and CA_3, and the CBRR in CA_3 is clearly seen in Figure 7a (without P_{cori} adjustment). Note that P_{cori} due to CBRR should be zero in CA_3. Additionally, at $t = 102$ s, one step-size activation of the CBRR in CA_2 is also seen. However, in Figure 7b (with P_{cori} adjustment), at $t = 100$ s and $t = 102$ s, the value of P_{cori} was zero in all CAs. In this way, delayed P_{di} sign changes have no impact on the switching between the INP and the CBRR. Moreover, the P_{cori} variation is significantly reduced using the proposed adjustment.

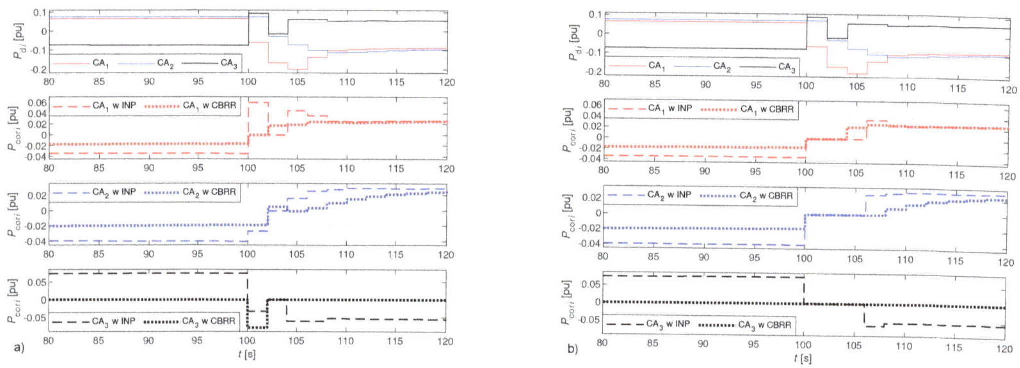

Figure 7. Time responses of P_{di} and P_{cori} for a three-CA testing system without P_{cori} adjustment (**a**) and with P_{cori} adjustment (**b**), where "w INP" is with the INP and "w CBRR" is with the CBRR.

3. Indicators for Evaluation of LFC, INP, and CBRR Provision

The impact of the simultaneous operation of the INP and CBRR on the frequency quality, the LFC, and performance is evaluated using 15-min averages [1].

3.1. Performance Indicators

Frequency quality is evaluated with the standard deviation of Δf_i, denoted as $\sigma_{\Delta fi}$. In addition, the LFC and performance is evaluated with the standard deviation of ACE_i, denoted as σ_{ACEi} [1,30].

3.2. RoCoF

RoCoF is the time derivative of the power system's frequency, i.e., $\frac{df_i}{dt}$ [31]. The mean value of RoCoF_i, denoted as μ_{RoCoF_i}, is evaluated individually for positive and negative values, denoted as $\mu_{\text{RoCoF}_{i+}}$ and $\mu_{\text{RoCoF}_{i-}}$.

3.3. Standard Deviation and Mean Value of RRs

RRs assist in active power balance to correct the imbalance in the transmission grid and lead the power system frequency to the normal frequency range [27]. The standard deviation and mean value of ΔP_{sci} are calculated individually for positive and negative values, denoted as $\sigma_{\Delta P_{sci+}}$, $\sigma_{\Delta P_{sci-}}$ and $\mu_{\Delta P_{sci+}}$, $\mu_{\Delta P_{sci-}}$.

3.4. Balancing Energy

The balancing energy enables TSOs to cost-effectively compensate for power and voltage variation in the transmission grid [2]. It is the actual electrical control power that is, for a particular period of time, calculated as $\Delta W_{ei} = \int_0^t \Delta P_{ei} dt$. Individual positive and negative values, denoted as ΔW_{ei+} and ΔW_{ei-}, are calculated.

3.5. Unintended Exchange of Energy

The unintended exchange of energy is determined by the difference between interchange power variation and correction power [2] that is, for a particular period of time, calculated as $\Delta W_{uni} = \int_0^t (\Delta P_i - P_{cori}) dt$. Individual positive and negative values, denoted as ΔW_{uni+} and ΔW_{uni-}, are calculated.

3.6. Energy Exchange

The energy exchange through the INP and CBRR is defined as the actual interchanged or activated power between cooperating CAs [32], that is, for a particular period of time, calculated as $W_{\mathrm{cor}i} = \int_0^t P_{\mathrm{cor}i} dt$. Additionally, positive and negative values are calculated, individually for the INP, denoted as $W_{\mathrm{cor}i+}^{\mathrm{INP}}$, $W_{\mathrm{cor}i-}^{\mathrm{INP}}$, and individually for the CBRR, denoted as $W_{\mathrm{cor}i+}^{\mathrm{CBRR}}$, $W_{\mathrm{cor}i-}^{\mathrm{CBRR}}$.

4. Dynamic Simulations

A three-CA test system was used for the dynamic simulations, where CA_1–CA_2 and CA_2–CA_3 were connected by physical tie-lines, whereas CA_1–CA_3 were not connected with a tie-line. Moreover, all three CAs were connected with virtual tie-lines due to the INP and the CBRR. A Matlab/SIMULINK model was used, where the dynamic simulations were performed with a 50 ms step size.

4.1. Dynamic Model

4.1.1. Structure

The basic schematic block diagram representation of a single CA, characterized with a linearized, low-order, time-invariant model, is shown in Figures 8 and 9 [29,33]. Note that the INP and CBRR implementation is not shown. The generator-load dynamic is described by the rotor inertia H_i and the damping D_i. Moreover, three different types of the turbine-governor systems were considered, i.e., hydraulic, steam reheat, and steam non-reheat. A constant droop characteristic R_{ni} was assumed. In addition, the ramping rate and the participation factors α_{ni} of the control units were also taken into account. The tie-line between the connected CAs is described by the synchronizing coefficient T_{ij} [34]. Furthermore, a 1st-order LPF is modeled by a time constant $T_{\mathrm{LPF}i}$, while the PI controller is modeled by a gain K_{ri} and a time constant T_{ri}.

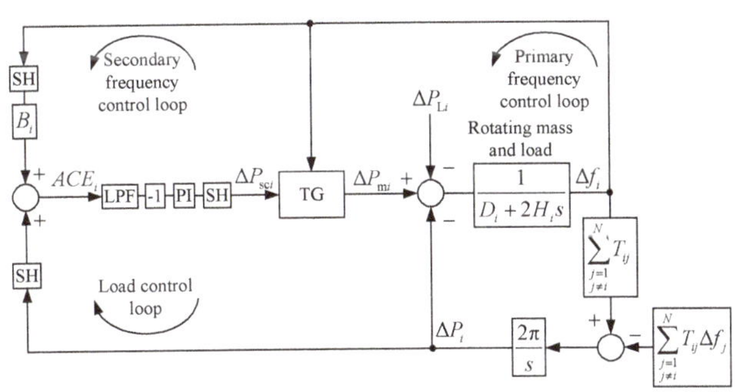

Figure 8. Schematic block diagram representation of the CA i without the INP and the CBRR.

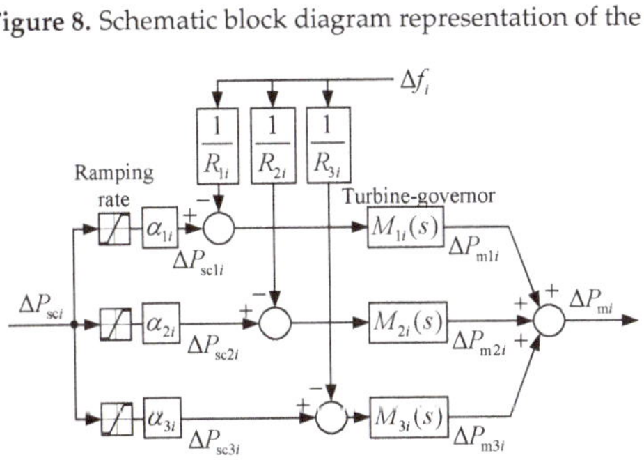

Figure 9. Schematic block diagram representation of the turbine-governor system with primary frequency control of the CA i.

4.1.2. Parameters

Common parameter values were set for a three-CA test system as given in Table 1 [28,29]. Note that the frequency-bias coefficient was determined as a constant, i.e., $B_i = 1/R_{1i} + 1/R_{2i} + 1/R_{3i} + D_i$. The model parameters were set equally for all three CAs. One cycle of the LFC, INP, and CBRR was incorporated with $T_s = 2$ s.

Table 1. Parameter values for a three-CA test system.

Parameter	Value	Parameter	Value
H_i	0.1 pu s	α_{ni}	1/3
D_i	0.01 pu/Hz	K_{ri}	0.3
T_{ij}	1/30 pu/Hz	R_{ni}	3 Hz/pu
T_{LPFi}	0.3 s	T_{ri}	30 s
ramp rate	value	ramp rate	value
hydraulic	±100 puMW/min	rehat	±10 puMW/min
non-reheat	±20 puMW/min	–	–

4.2. Test Cases

The maximum possible compensation with the INP and CBRR was considered. Dynamic simulations were performed so that the loads of individual CAs were altered during the simulation. In addition, two types of test cases were performed, i.e., step change of the load and the random load variation.

Moreover, the inertia time constant H_i, the tie-line parameter T_{ij}, and the droop characteristic R_i have a considerable impact on frequency quality according to [35,36]. Therefore, different values of H_i, T_{ij}, and R_i were used to show the impact of the simultaneous operation of the INP and CBRR on the indicators for evaluation of LFC, INP, and CBRR provision.

4.2.1. Step Change of Load

The values of the loads were set in such a way that the simultaneous operation of the INP and the CBRR was possible. At $t = 0$ s, a simultaneous step change of the loads was applied and the magnitudes were set as $\Delta P_{L1} = 0.06$ pu, $\Delta P_{L2} = 0.07$ pu and $\Delta P_{L3} = -0.08$ pu. In addition, at $t = 100$ s, the magnitudes were set as $\Delta P_{L1} = -0.06$ pu and $\Delta P_{L3} = 0.08$ pu, whereas, at $t = 100.05$ s, the magnitude was set as $\Delta P_{L2} = -0.07$ pu. The resulting load is shown in Figure 10a. Consequently, the INP operated between CA_1–CA_3 and CA_2–CA_3, while the CBRR operated only between CA_1–CA_2. Note that this case is used in Section 2.4.

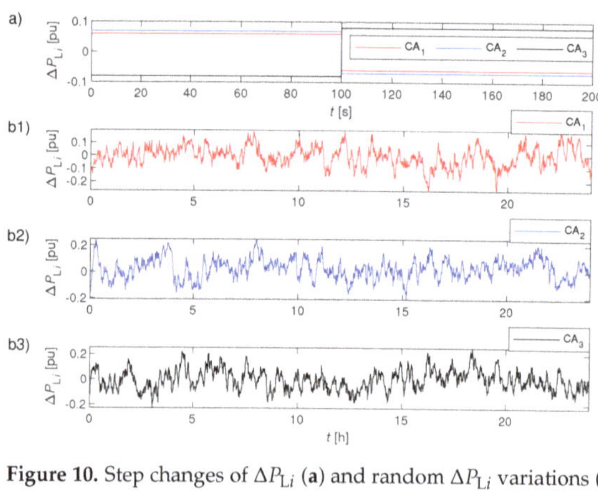

Figure 10. Step changes of ΔP_{Li} (**a**) and random ΔP_{Li} variations (**b1**–**b3**) for a three CA test system.

Appl. Sci. **2021**, *11*, 8188

4.2.2. Random Load Variation

The random load was modeled as a linear, stochastic, time-invariant, first-order system with two components [37]. A low-frequency component captures the trend changes with a quasi-period of 10–30 min, whereas the residual component captures fluctuations with a quasi-period of several minutes. Measurements of an open-loop ACE in an undisclosed CA were used to determine the model parameters. The resulting normalized load for a three-CA test system with different signs is shown in Figure 10(b1–b3), where the random load was changed every 60 s for 24 h. The statistical parameters of the random loads for all three CAs are given in Table 2, where μ_{Li} and σ_{Li} denote the meanvalue and the standard deviation of the *i*-th load, while the correlation between the *i*-th and *j*-th loads is denoted as ρ_{Lij}. The correlation ρ_{Lij} is extremely small, which allows both the INP and CBRR to operate simultaneously.

Table 2. Statistical parameters of random loads.

μ_{L1} [pu·10^3]	μ_{L2} [pu·10^3]	μ_{L3} [pu·10^3]
−2.501	29.526	−0.693
σ_{L1} [pu·10^3]	σ_{L2} [pu·10^3]	σ_{L3} [pu·10^3]
74.891	76.890	78.827
ρ_{L12}	ρ_{L23}	ρ_{L31}
−0.056	0.059	0.004

5. Results

Dynamic simulations with and without the INP and CBRR were performed for a three-CA test system. The impact of the simultaneous operation of the INP and CBRR on the frequency quality, the LFC, and performance was evaluated with the obtained results. Note that the results shown in this section refer to a three-CA test system, whereas the basic principle is applicable to *N* CAs as shown in Section 2. In addition, the results cannot be generalized to the dynamics of the INP and CBRR.

5.1. *Step Change of Load*

The time responses to the step change of ΔP_{Li} are shown in Figures 11–13. In Figure 11 (left), it is clear that the frequency deviations Δf_i in all three CAs appeared following a step change of ΔP_{Li} that was applied. After the first step change, Δf_1 and Δf_2 were negative due to the positive step change of ΔP_{Li}, whereas Δf_3 was positive due to the negative step change of ΔP_{Li}. Note that after the second step change, the signs were opposite. The primary frequency control decreased $|\Delta f_i|$ in about 15–25 s after the step change of ΔP_{Li}; then, LFC decreased $|\Delta f_i|$ slowly. The results show that the impact of the INP and CBRR on Δf_i is not significant.

The impact of the INP and CBRR is shown more obviously in Figures 11 (right) and 12 (left). In all three CAs, the values of ACE_i, ΔP_{sci} and ΔP_{ei} were decreased with the INP and CBRR. Furthermore, the INP and CBRR clearly increased ΔP_i, due to the increased tie-line power flow between the CAs.

The signs of P_{di} and P_{cori} are opposite, as shown in Figure 12 (right). A 2 s time delay is seen, due to SH with $T_s = 2$ s. Note that the time responses of P_{di} and P_{cori} were already described in Section 2.4 and Figure 7b, where the same case was performed.

The time responses ACE_i and ΔP_{sci} with and without the P_{cori} adjustment function are shown in Figure 13. Without the P_{cori} adjustment, ACE_i was increased and ΔP_{sci} was undesirably increased, due to switching between the INP and the CBRR, which caused undesirable ΔP_{cori} sign changes. Clearly, in all three CAs, the values of ACE_i and ΔP_{sci} were decreased with the P_{cori} adjustment. Note that, in Figure 13, the same example is performed as described in Section 4.2.1 and shown in Figures 11 and 12. The difference can be seen because, in Figures 11 and 12, the comparison without and with the INP and CBRR is shown, whereas, in Figure 13, the comparison without and with the P_{cori} adjustment is shown.

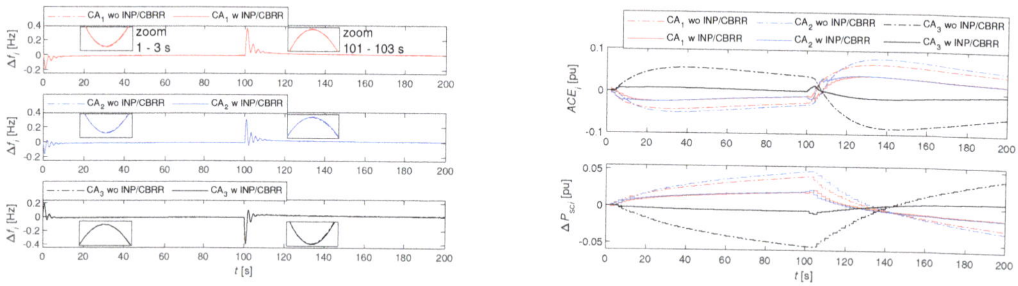

Figure 11. Time responses of Δf_i (**left**) and time responses of ACE_i and ΔP_{sci} (**right**) for a three-CA test system, where "wo INP/CBRR" is without and "w INP/CBRR" is with the INP and CBRR.

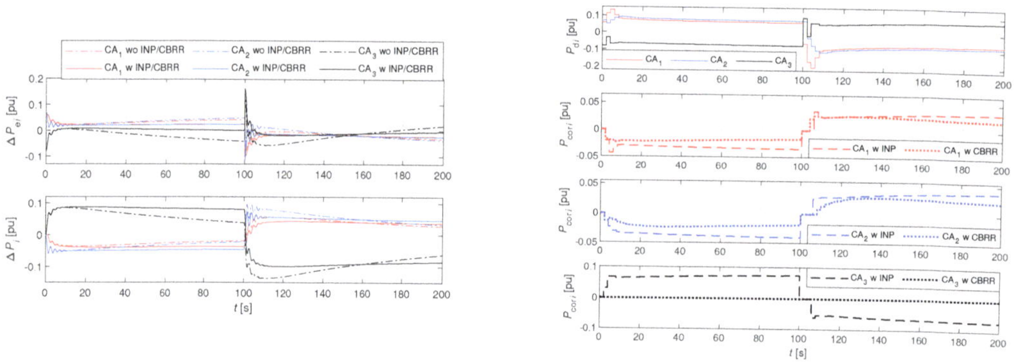

Figure 12. Time responses of ΔP_{ei} and ΔP_i (**left**) and time responses of P_{di} and P_{cori} (**right**) for a three-CA test system, where "wo INP/CBRR" is without and "w INP/CBRR" is with the INP and CBRR, whereas "w INP" is with the INP and "w CBRR" is with the CBRR.

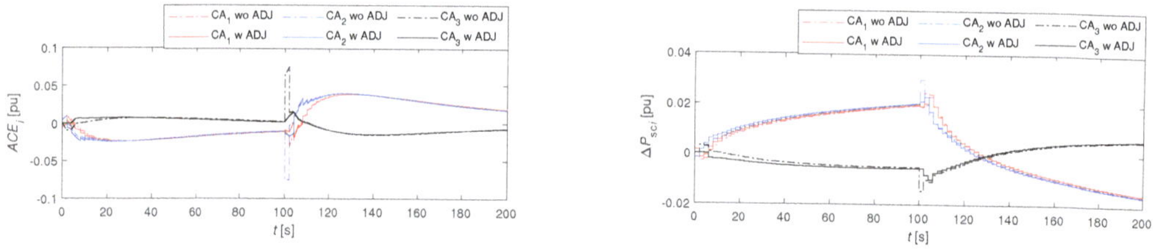

Figure 13. Time responses of ACE_i (**left**) and time responses of ΔP_{sci} (**right**) for a three-CA test system, where "wo ADJ" is without and "w ADJ" is with the P_{cori} adjustment function.

5.2. Random Load Variation

Simulations of the simultaneous operation of the INP and CBRR were performed to show the impact of an individual mechanism. The simulations were also performed separately, i.e., operation of only the INP or only the CBRR. The results are given in Tables 3–5.

There are no unambiguous conclusions about the impact of the mechanisms on $\sigma_{\Delta fi}$, and the differences are extremely small, as shown in Table 3. However, both mechanisms reduce σ_{ACEi}, INP slightly more than CBRR. The reduction is most pronounced when both mechanisms operate simultaneously.

Table 3. Performance indicators, RoCoF, mean value, and standard deviation of RRs.

Indicator	$\sigma_{\Delta fi}$ [mHz]				σ_{ACEi} [pu·10^3]			
	wo	INP/CBRR	INP	CBRR	wo	INP/CBRR	INP	CBRR
CA_1	5.263	5.256	5.251	5.272	8.873	6.310	6.654	8.649
CA_2	5.252	5.256	5.250	5.260	9.612	5.844	7.033	8.185
CA_3	5.265	5.267	5.273	5.260	8.823	6.063	6.439	8.281
indicator	$\mu_{\text{RoCoF}i+}$ [mHz/s]				$\mu_{\text{RoCoF}i-}$ [mHz/s]			
	wo	INP/CBRR	INP	CBRR	wo	INP/CBRR	INP	CBRR
CA_1	0.682	0.606	0.611	0.677	−0.724	−0.645	−0.674	−0.692
CA_2	0.879	0.776	0.832	0.818	−1.233	−1.039	−1.072	−1.193
CA_3	0.752	0.690	0.722	0.711	−0.841	−0.740	−0.743	−0.830
indicator	$\mu_{P\text{sc}i+}$ [pu·10^3]				$\mu_{P\text{sc}i-}$ [pu·10^3]			
	wo	INP/CBRR	INP	CBRR	wo	INP/CBRR	INP	CBRR
CA_1	46.17	36.60	31.20	46.38	−47.08	−28.87	−29.69	−45.97
CA_2	58.62	32.95	43.77	49.68	−41.10	−23.06	−22.40	−31.63
CA_3	53.09	35.58	39.46	47.99	−50.11	−29.81	−31.01	−44.21
indicator	$\sigma_{P\text{sc}i+}$ [pu·10^3]				$\sigma_{P\text{sc}i-}$ [pu·10^3]			
	wo	INP/CBRR	INP	CBRR	wo	INP/CBRR	INP	CBRR
CA_1	36.68	26.35	28.83	32.17	40.43	18.55	23.49	34.44
CA_2	45.13	25.70	39.43	31.00	29.59	18.25	25.95	24.32
CA_3	39.36	26.98	32.78	31.03	31.67	20.14	24.18	26.12

Legend: wo—without the INP/CBRR, INP/CBRR—simultaneous operation, INP—separate operation, CBRR—separate operation.

Both mechanisms reduce the $|\mu_{\text{RoCoF}i+}|$ and $|\mu_{\text{RoCoF}i-}|$, with the reduction being most pronounced when both mechanisms operate simultaneously according to Table 3. However, there are no unambiguous conclusions as to which mechanism reduces $|\mu_{\text{RoCoF}i+}|$ and $|\mu_{\text{RoCoF}i-}|$ more, and, in most cases, it is the INP.

When both mechanisms operate simultaneously, $|\mu_{P\text{sc}i+}|$ and $|\mu_{P\text{sc}i-}|$ are greatly reduced according to Table 3. However, the results of the separate operation of the mechanisms show that the impact of INP is greater than the impact of CBRR. This is expected, as the CBRR only activates the RRs in the cooperating CAs. Moreover, the reduction of $|\sigma_{P\text{sc}i+}|$ and $|\sigma_{P\text{sc}i-}|$ is most noticeable when both mechanisms operate simultaneously. However, the results of the separate operation of the mechanisms show that the impact of the INP is greater than the impact of the CBRR.

The conclusions for $|\Delta W_{\text{e}i+}|$ and $|\Delta W_{\text{e}i-}|$ are similar to $|\mu_{P\text{sc}i+}|$ and $|\mu_{P\text{sc}i-}|$ according to Table 4, which is expected, as this indicator describes the response of the control units.

When both mechanisms operate simultaneously, $|\Delta W_{\text{un}i+}|$ and $|\Delta W_{\text{un}i-}|$ are reduced according to Table 4, except in one case where only $|\Delta W_{\text{un}i+}|$ was increased, while $|\Delta W_{\text{un}i-}|$ was reduced considerably. Furthermore, the results of the separate operation of the mechanisms show that the INP almost completely eliminates unintentional deviations, while the impact of the CBRR is not very pronounced.

When both mechanisms operate simultaneously, $|W^{\text{INP}}_{\text{cor}i+}|$, $|W^{\text{INP}}_{\text{cor}i-}|$, $|W^{\text{CBRR}}_{\text{cor}i+}|$ and $|W^{\text{CBRR}}_{\text{cor}i-}|$ are slightly reduced compared to separate operation of the INP and CBRR according to Table 5. This is due to the $P_{\text{cor}i}$ adjustment mechanism, which is only required in the case of simultaneous operation of the INP and CBRR.

Moreover, simulations of the simultaneous operation of the INP and CBRR were performed for different values of H_i, T_{ij}, and R_i. The results are given in Figures 14–16.

There are no unambiguous conclusions about the impact of the mechanisms on $\sigma_{\Delta fi}$, and the differences are extremely small, as shown in Figure 14 (left). In addition, the impact of H_i and T_{ij} is not clear, whereas the increase of R_i results in an increase of $\sigma_{\Delta fi}$. However, both mechanisms significantly reduce σ_{ACEi}, as shown in Figure 14 (right), whereas H_i, T_{ij} and R_i have no impact on σ_{ACEi}.

Table 4. Balancing energy and unintended exchange of energy.

Indicator	**ΔW_{ei+} [pu h]**				**ΔW_{ei-} [pu h]**			
	wo	**INP/CBRR**	**INP**	**CBRR**	**wo**	**INP/CBRR**	**INP**	**CBRR**
CA_1	8.533	7.465	6.478	9.127	−9.469	−5.216	−5.229	−8.699
CA_2	16.033	8.481	10.565	12.690	−4.713	−3.014	−3.306	−3.983
CA_3	9.785	7.567	7.603	9.224	−10.059	−5.172	−6.000	−8.518
indicator	ΔW_{uni+} [pu h]				ΔW_{uni-} [pu h]			
	wo	INP/CBRR	INP	CBRR	wo	INP/CBRR	INP	CBRR
CA_1	7.750	5.937	0.628	6.769	−4.531	−3.551	−0.607	−4.023
CA_2	3.916	5.431	0.721	6.358	−9.803	−2.745	−0.733	−2.857
CA_3	7.462	5.771	0.611	6.302	−4.794	−3.218	−0.620	−3.444

Legend: wo—without the INP/CBRR, INP/CBRR—simultaneous operation, INP—separate operation, CBRR—separate operation.

Table 5. Energy exchange.

Indicator	**W^{INP}_{cori+} [pu h]**			**W^{INP}_{cori-} [pu h]**		
	INP/CBRR	**INP**	**CBRR**	**INP/CBRR**	**INP**	**CBRR**
CA_1	5.147	5.165	0	−2.965	−2.976	0
CA_2	2.012	2.016	0	−6.068	−6.083	0
CA_3	4.824	4.840	0	−2.950	−2.961	0
indicator	W^{CBRR}_{cori+} [pu h]			W^{CBRR}_{cori-} [pu h]		
	INP/CBRR	INP	CBRR	INP/CBRR	INP	CBRR
CA_1	3.472	0	4.047	−4.831	0	−5.405
CA_2	2.783	0	2.884	−7.285	0	−8.745
CA_3	3.844	0	4.590	−5.610	0	−6.475

Legend: INP/CBRR—simultaneous operation, INP—separate operation, CBRR—separate operation.

Generally, both mechanisms reduce the $|\mu_{RoCoFi+}|$ and $|\mu_{RoCoFi-}|$, as shown in Figure 15. In addition, an increase of H_i and R_i results in a decrease of $|\mu_{RoCoFi+}|$ and $|\mu_{RoCoFi-}|$, whereas an increase of T_{ij} results in an increase of $|\mu_{RoCoFi+}|$ and $|\mu_{RoCoFi-}|$. Note that, when $T_{ij} = 1/15$ pu/Hz, $|\mu_{RoCoFi+}|$ and $|\mu_{RoCoFi-}|$ is increased with the mechanisms.

When both mechanisms operate simultaneously, $|\mu_{Psci+}|$ and $|\mu_{Psci-}|$ are greatly reduced, as shown in Figure 16. However, H_i, T_{ij} and R_i have no impact on $|\mu_{Psci+}|$ and $|\mu_{Psci-}|$.

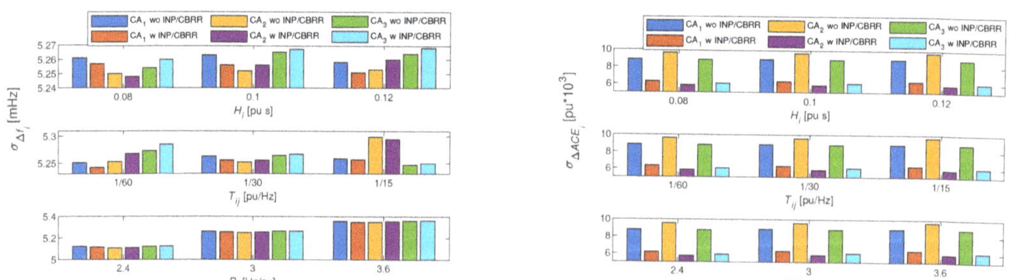

Figure 14. Average values of $\sigma_{\Delta f_i}$ (**left**) and average values of σ_{ACE_i} (**right**) for different values of H_i, T_{ij} and R_i, where "wo INP/CBRR" is without and "w INP/CBRR" is with the INP and CBRR.

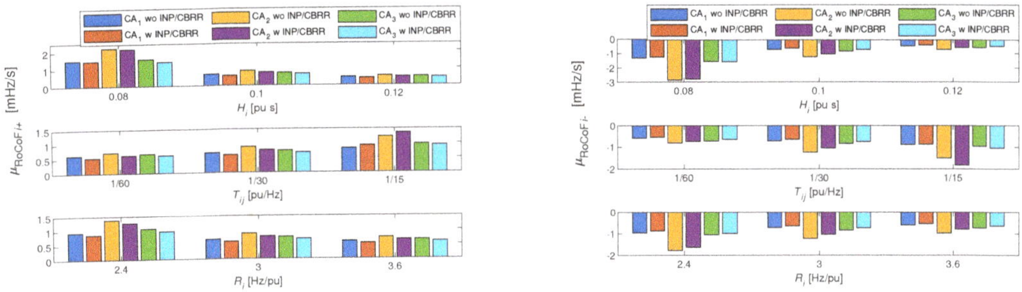

Figure 15. Average values of $\mu_{\mathrm{RoCoF}_{i+}}$ **(left)** and average values of $\mu_{\mathrm{RoCoF}_{i-}}$ **(right)** for different values of H_i, T_{ij} and R_i, where "wo INP/CBRR" is without and "w INP/CBRR" is with the INP and CBRR.

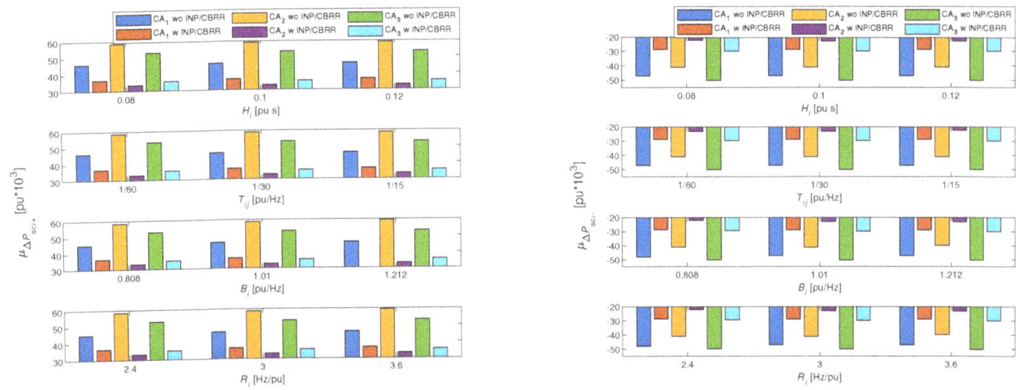

Figure 16. Average values of $\mu_{\Delta P_{\mathrm{sci}+}}$ **(left)** and average values of $\mu_{\Delta P_{\mathrm{sci}-}}$ **(right)** for different values of H_i, T_{ij} and R_i, where "wo INP/CBRR" is without and "w INP/CBRR" is with the INP and CBRR.

6. Conclusions

This article discusses the simultaneous operation of the INP and CBRR, which no previous studies have investigated. Extensive dynamic simulations of a three-CA test system with the simultaneous operation of the INP and CBRR were performed to evaluate their impact on the frequency quality, the LFC, and performance.

The results confirmed the conclusions in [20–23], where the INP and CBRR were analyzed separately. The results of the step change of load and the random load variation confirmed that the impact of the INP and CBRR on frequency deviations has no unambiguous conclusions. In addition, the impact of inertia and synchronizing coefficient is not clear, whereas the increase of droop characteristic results in an increase of frequency deviations. However, both mechanisms reduce the ACE deviations—the INP slightly more than CBRR. The reduction is most pronounced when both mechanisms operate simultaneously. Moreover, the function for correction power adjustment additionally decreased ACE and scheduled control power, which prevents undesirable switching between the INP and CBRR. Both mechanisms also reduce the RoCoF, and the reduction is most pronounced when both mechanisms operate simultaneously. However, there are no unambiguous conclusions as to which mechanism reduces the RoCoF more, and, in most cases, it is the INP. Increase of inertia and droop characteristic results in a decrease of RoCoF, whereas increase of synchronizing coefficients results in an increase of RoCoF. When both mechanisms operate simultaneously, the scheduled control power is greatly reduced and the impact of the INP is greater than the impact of the CBRR. Similarly, the balancing energy as well as the unintended exchange of energy are greatly reduced when both mechanisms operate simultaneously. However, there is no impact of inertia, synchronizing coefficient, and droop characteristic on scheduled control power. Due to the function for correction power adjustment, which prevents undesirable activation of the INP and CBRR, energy exchange was slightly reduced, as expected. Because of the reduced unintended exchange of energy, beneficial economic consequences can be anticipated when the INP and CBRR operate simultaneously.

One of the tough challenges for all researchers in this domain is the dynamic dimensioning of RRs, considering the INP and CBRR. This article clearly shows that the simultaneous operation of the INP and CBRR reduces the activation of the RRs, which is currently not considered in the reserve dimensioning process.

Author Contributions: Supervision, B.P.; Writing—original draft, M.T.; Writing—review and editing, B.P. All authors have read and agreed to the published version of the manuscript.

Funding: This work was supported by ARRS under Projects P2-0115 and L2-7556.

Institutional Review Board Statement: Not applicable.

Informed Consent Statement: Not applicable.

Data Availability Statement: This study did not report any data.

Conflicts of Interest: The authors declare no conflict of interest.

Abbreviations

The following abbreviations are used in this manuscript:

RR	Regulating Reserve
INP	Imbalance Netting Process
CBRR	Cross-Border Activation of the Regulating Reserve
CA	Control Area
LFC	Load–Frequency Control
ACE	Area Control Error
ENTSO-E	European Network of Transmission System Operators for Electricity
GCC	Grid Control Cooperation
TSO	Transmission System Operator
IGCC	International Grid Control Cooperation
RoCoF	Rate of Change of Frequency
LPF	Low Pass Filter
SH	Sample and Hold
PI	Proportional-Integral

References

1. EUR-Lex. Establishing a Guideline on Electricity Transmission System Operation. Commission Regulation (EU) 2017/1485. Available online: https://eur-lex.europa.eu/eli/reg/2017/1485/oj (accessed on 15 July 2021).
2. EUR-Lex. Establishing a Guideline on Electricity Balancing. Commission Regulation (EU) 2017/2195. Available online: https://eur-lex.europa.eu/eli/reg/2017/2195/oj (accessed on 15 July 2021).
3. ENTSO-E. *Proposal for the Implementation Framework for a European Platform for the Exchange of Balancing Energy from Frequency Restoration Reserves*; ENTSO-E: Brussels, Belgium, 2020.
4. ENTSO-E. *IGCC Regular Report on Social Welfare*; ENTSO-E: Brussels, Belgium, 2020.
5. Zolotarev, P.; Gökeler, M.; Kuring, M.; Neumann, H.; Kurscheid, E.M. Grid Control Cooperation—A Framework for Technical and Economical Crossborder Optimization for Load-Frequency Control. In Proceedings of the 44th International Conference on Large High Voltage Electric Systems (Cigre'12), Paris, France, 26–30 August 2012.
6. Pawalek, A.; Hofmann, L. Utilization of HVDC-Systems in the International Grid Control Cooperation. In Proceedings of the IEEE Milan PowerTech 2019, Milan, Italy, 23–27 June 2019.
7. Pawalek, A.; Hofmann, L. Comparison of Methods for the Simulation of Dynamic Power Flows in the International Grid Control Cooperation. In Proceedings of the IEEE Electronic Power Grid (eGrid) 2018, Charleston, SC, USA, 12–14 November 2018.
8. Cronenberg, A; Sager, N.; Willemsen, S. Integration of France Into the International Grid Control Cooperation. In Proceedings of the IEEE International Energy Conference 2016 (ENERGYCON'16), Leuven, Belgium, 4–8 April 2016.
9. ENTSO-E. *Stakeholder Document for the Principles of IGCC*; ENTSO-E: Brussels, Belgium, 2020.
10. ENTSO-E. *Supporting Document for the Network Code on Load-Frequency Control and Reserves*; ENTSO-E: Brussels, Belgium, 2013.

11. Zolotarev, P. Social Welfare of Balancing Markets. In Proceedings of the 14th International Conference on the European Energy Market (EEM) 2017, Dresden, Germany, 6–9 June 2017.
12. Avramiotis, F.I.; Margelou, S.; Zima, M. Investigations on a Fair TSO-TSO Settlement for the Imbalance Netting Process in European Power System. In Proceedings of the 15th International Conference on the European Energy Market (EEM) 2018, Lodz, Poland, 27–29 June 2018.
13. Oneal, A.R. A Simple Method for Improving Control Area Performance: Area Control Error (ACE) Diversity Interchange ADI. *IEEE Trans. Power Syst.* **1995**, *10*, 1071–1076. [CrossRef]
14. Zhou, N.; Etingov, P.V.; Makarov, J.V.; Guttromson, R.T.; McManus, B. Improving Area Control Error Diversity Interchange (ADI) Program by Incorporating Congestion Constraints. In Proceedings of the IEEE PES T&D 2010, New Orleans, LA, USA, 19–22 April 2010.
15. Apostolopoulou, D.; Sauer, P.W.; Dominguez-Garcia, A.D. Balancing Authority Area Coordination With Limited Exchange of Information. In Proceedings of the IEEE Power and Energy Society General Meeting 2015, Denver, CO, USA, 26–30 July 2015.
16. ENTSO-E. *Anual Report 2019 Edition*; ENTSO-E: Brussels, Belgium, 2020.
17. EXPLORE. Target Model For Exchange of Frequency Restoration Reserves. Available online: https://www.tennet.eu/news/detail/explore-target-model-for-exchange-of-frequency-restoration-reserves/ (accessed on 15 July 2021).
18. ENTSO-E. *Consultation on the Design of the Platform for Automatic Frequency Restoration Reserve (AFRR) of PICASSO Region*; ENTSO-E: Brussels, Belgium, 2012.
19. Gebrekiros, T.Y. *Analysis of Integrated Balancing Markets in Northern Europe under Different Market Design Options*; NTNU: Trondheim, Norway, 2015.
20. Topler, M.; Polajžer, B. Impact of Imbalance Netting Cooperation on Frequency Quality and Provision of Load-Frequency Control. In Proceedings of the IEEE International Conference on Environment and Electrical Engineering 2019 and IEEE Industrial and Commercial Power Systems Europe (EEEIC/I&CPS Europe) 2019, Genoa, Italy, 11–14 June 2019.
21. Topler, M.; Ritonja, J.; Polajžer, B. The Impact of the Imbalance Netting Process on Power System Dynamics. *Energies* 2019, *12*, 1–18. [CrossRef]
22. Topler, M.; Polajžer, B. Evaluation of the Cross-Border Activation of the Regulating Reserve with Dynamic Simulations. In Proceedings of the IEEE International Conference on Environment and Electrical Engineering 2020 and IEEE Industrial and Commercial Power Systems Europe (EEEIC/I&CPS Europe) 2020, Madrid, Spain, 9–12 June 2020.
23. Topler, M.; Ritonja, J.; Polajžer, B. Improving the Cross-Border Activation of the Regulating Reserve to Enhance the Provision of Load-Frequency Control. *IEEE Access* 2020, *8*, 170696–170712. [CrossRef]
24. Lu, C.W.; Lin, G.H.; Wu, T.J.; Hu, I.H.; Chang, Y.C. Influencing Factors of Cross-Border E-Commerce Consumer Purchase Intention Based on Wireless Network and Machine Learning. *Secur. Commun. Netw.* 2021, *2021*, 9984213. [CrossRef]
25. Imani, M.; Ghoreishi, S.F. Graph-Based Bayesian Optimization for Large-Scale Objective-Based Experimental Design. *IEEE Trans. Neural Netw. Learn. Syst.* 2021, 1–13. [CrossRef]
26. Meltzer, J.P.; Lovelock, P. Regulating for a Digital Economy—Understanding the Importance of Cross-Border Data Flows in Asia. *Glob. Econ. Dev. Work. Pap.* 2018, *113*, 1–51.
27. ENTSO-E. *Operational Reserve Ad Hoc Team Report*; ENTSO-E: Brussels, Belgium, 2012.
28. Bevrani, H. *Robust Power System Frequency Control*; Springer: New York, NY, USA, 2009.
29. Kundur, P. *Power System Stability and Control*; McGraw-Hill: New York, NY, USA, 1994.
30. NERC. *Frequency Response Standard, Background Document*; NERC: Atlanta, GA, USA, 2012.
31. ENTSO-E. *Rate of Change of Frequency (ROCOF) Withstand Capability*; ENTSO-E: Brussels, Belgium, 2017.
32. ACER. *Benefits from Balancing Market Integration (Imbalance Netting and Exchange of Balancing Energy)*; ACER: Ljubljana, Slovenia, 2018.
33. Anderson, P.M.; Mirheydar, M. A Low-Order System Frequency Response Model. *IEEE Trans. Power Syst.* 1990, *5*, 720–729. [CrossRef]
34. Elgerd, O.I.; Fosha, C.E. Optimum Mmegawatt-Frequency Control of Multiarea Electric Energy Systems. *IEEE Trans. Power App. Syst.* 1970, *89*, 556–563. [CrossRef]
35. Vorobev, P.; Greenwood, D.M.; Bell, J.H; Bialek, J.W.; Taylor, P.C.; Turitsyn, K. Deadbands, Droop, and Inertia Impact on Power System Frequency Distribution. *IEEE Trans. Power Syst.* 2019, *34*, 3098–3108. [CrossRef]
36. Kundur, P. Definition and Classification of Power System Stability IEEE/CIGRE Joint Task Force on Stability Terms and Definitions. *IEEE Trans. Power Syst.* 2004, *19*, 1387–1401.
37. Janecek, E.; Cerny, V.; Fialova, A.; Fantík, J. A New Approach to Modelling of Electricity Transmission System Operation. In Proceedings of the IEEE PES Power Systems Conference and Exposition 2006, Atlanta, GA, USA, 29 October–1 November 2006.

applied sciences

MDPI

Article

An Adaptive and Scalable Protection Coordination System of Overcurrent Relays in Distributed-Generator-Integrated Distribution Networks

Duong Minh Bui [1,*], Phuc Duy Le [2,3], Thanh Phuong Nguyen [3] and Hung Nguyen [3]

[1] Department of Electrical and Computer Engineering (ECE), Faculty of Engineering, Vietnamese-German University (VGU), Thu Dau Mot City 75000, Vietnam
[2] Department of Administration, Ho Chi Minh City Power Corporation, Ho Chi Minh City 70000, Vietnam; phucld91@gmail.com
[3] HUTECH Institute of Engineering, Ho Chi Minh City University of Technology (HUTECH), Ho Chi Minh City 70000, Vietnam; nt.phuong@hutech.edu.vn (T.P.N.); n.hung@hutech.edu.vn (H.N.)
* Correspondence: duong.bm@vgu.edu.vn; Tel.: +84-918-163-356

Abstract: Integration of distributed generators (DGs) into a distribution network (DN) can cause coordination challenges of overcurrent relays (OCRs) because of different fault-current contributions of DGs as well as the directional change in fault currents. Therefore, the OCRs should be properly coordinated to maintain their adaptability and scalability to protect the DG-integrated distribution network. In this study, an adaptive and scalable protection coordination (ASPC) approach has been developed for the OCRs in a DG-contained distribution network based on two implementation stages. At the first stage, the reliability improvement of fault-current calculation results is performed by determining the min-max confidence interval of fault current for each different fault type, which is the basis for properly selecting tripping and pick-up thresholds of definite-time and inverse-time OC functions in the same OCR. At the second stage, optimization algorithms are used for calculating protection-curve coefficients and Time-Dial Setting (TDS) multiplier for the inverse-time OC functions in the OCR. A real 22 kV DG-integrated distribution network which is simulated by ETAP software is considered a reliable test-bed to validate the proposed ASPC system of OCRs in the multiple-DG-contained distribution network. In addition, the coordination results of OCRs can be obtained by three common optimization algorithms, Particle Swarm Optimization (PSO), Gravitational Search Algorithm (GSA), and Genetic Algorithm (GA). These relay coordination results have shown an effective protection combination of the definite-time OC functions (50P and 50G) and the inverse-time OC functions (51P and 51G) in the same OCR to get the adaptable and scalable DN protection system.

Keywords: distributed generator; distribution network; overcurrent relay; fault protection; relay coordination; meta-heuristic algorithms

Citation: Bui, D.M.; Le, P.D.; Nguyen, T.P.; Nguyen, H. An Adaptive and Scalable Protection Coordination System of Overcurrent Relays in Distributed-Generator-Integrated Distribution Networks. *Appl. Sci.* **2021**, *11*, 8454. https://doi.org/10.3390/app11188454

Academic Editors: Federico Barrero and Mario Bermúdez

Received: 26 August 2021
Accepted: 9 September 2021
Published: 12 September 2021

1. Introduction

1.1. Motivation and Assumption

Distributed generators have been recently developed to utilize sustainable and clean energy sources for electrical energy conversion. DGs can operate as ancillary services to improve power-supply reliability indices, SAIFI (System Average Interruption Frequency Index) and SAIDI (System Average Interruption Duration Index), to a distribution network. However, the operation characteristics of DGs could lead to certain difficulties in an overcurrent protection system of DN because of the remarkable change in fault-current values observed by overcurrent relays (OCRs). Specifically, fault-current values depend on DG types, locations, and technologies. For example, rotating-based distributed generators (RBDGs) can contribute high fault currents whereas inverter-based distributed generators (IBDGs) can only inject the limited fault currents about 1.2~2.0 p.u. to a distribution network [1–3]. Moreover, high penetration of DGs may lead to a magnitude decrease in fault-currents contributed from the utility, which can reduce the sensitivity and selectivity of OCRs on feeders of the distribution network [4–8]. Malfunction issues of OCRs in the DG-integrated DN could be over-reach or under-reach of OC relays, loss of sensitivity/selectivity, blinding of overcurrent protection, false tripping or sympathetic tripping of overcurrent protection [9].

Several studies are performed on protection coordination of overcurrent relays in a DG-integrated DN. In [10], risks of protection miscoordination of OCRs or reclosers are indicated in each fault isolation and service restoration (FISR) plan when the DN needs to change its topology due to faulted events. In [11], the authors point out that the allowable fault-current contribution of IBDGs can significantly impact a DN's overcurrent protection system. In [12–15], the threshold setting of OCRs is only valid for typical DG capacities. In [16], the authors have introduced a hybrid PSO-based protection coordination method to adjust TDS parameters of directional overcurrent relays. The paper [17] proposed an optimization function to calculate tripping and pick-up thresholds of OCRs in a range of pre-defined values. In [18], an adaptive overcurrent protection system is developed for fault detection and isolation in a DG-penetrated distribution network. The research paper [19] has proposed an ANN-based method to divide a distribution network into multiple zones, and then an overcurrent protection scheme for each zone is developed. In [20], an effective overcurrent protection system is developed to protect a distribution network when considering two on-grid and off-grid operation modes of DGs. From the above references, it could be generally concluded that the accurate calculation of fault-current values can strongly impact the protection coordination results of OCRs on a DN having DGs. That assumption could be more considered in this study by contributing a novel proposal to determine the min-max confidence interval of fault currents according to each different fault type. The estimated confidence interval of fault currents will then be the basis for properly selecting tripping and pick-up thresholds of the OCRs in the DN.

The DG-integrated distribution networks mostly use digital relays with overcurrent and sequence-component-based protection functions. Accordingly, overcurrent-based protection functions, e.g., 50, 51, 67, or sequence-component-based protection functions, e.g., 46, 47, 50REF, 46BC, $3I_0$, $3V_0$, can be properly re-adjusted when the topology of DN is changed under fault events. As presented in [18], a low-voltage microgrid (LVMG) protection system using directional overcurrent protection functions in digital relays is developed along with its coordination strategies. The paper [21] presents a protection coordination solution based on symmetrical and differential current components to detect faults in a grid-connected microgrid. In [22], the authors propose a definite-time overcurrent protection model regarding the placement of DGs in a distribution network. In [23], a novel protection coordination algorithm is developed to improve the selectivity of the protection system through using a three-level communication network, directional overcurrent protection functions, and under/overvoltage-based protection functions, which are available in digital relays. On the other hand, references [24,25] have considered sequence-component-based protection functions, e.g., positive-, negative-, zero-sequence components, along with a GOOSE (Generic Object Oriented System Event) solution for detecting, classifying and isolating symmetrical and asymmetric faults in microgrids or DG-penetrated distribution networks. Generally, embedding several different protection functions into a digital relay could be a feasible solution to improve the adaptability and flexibility of a DN protection system, which has been applied for this study.

1.2. Literature Review

Protection coordination of OCRs in a distribution network is severely affected by the presence of DGs [9]. According to different DG penetration levels, DG sites, and fault types and locations in the DN, determining the optimal setting values of pick-up current, TPS multiplier, and inverse-time curve coefficients of overcurrent relay is very necessary to ensure the adaptability and reliability of a DN protection system. Many coordination methods of overcurrent relays can be classified by trial-and-error methods [26], topological analysis methods [27], and optimization methods [28–30]. The latter is commonly used in recent years because the operating time of OC relays can be optimized against various coordination constraints as well as relay characteristic curves. The optimal coordination of overcurrent relays consists of an objective function (OF), parameters, constraints and a selected optimization method to solve the defined problem. A literature survey for the objective functions with parameters and constraints, type of OC relays, and optimization algorithms has been shown as the following.

- Objective function and setting parameters: In [29–36], objective functions are based on the overall operation time of primary and back-up overcurrent relays, where time-dial setting (TDS) multipliers are considered as the optimized parameters. In [37–50], the optimized parameters of OFs are pick-up currents or both the pick-up currents and TDS multipliers. In [51,52], a multi-objective function (MOF) with the overall operation time of relays and protection coordination constraints has been developed. The optimization parameters of the MOF are TDS multipliers, pick-up currents and coefficients of time-current characteristic curves. In [52], the authors have explained why adaptive coordination schemes are paid more attention to an overcurrent protection system of the DG-integrated DN when compared to others such as: immediate disconnection of DGs under a fault event; capacity limitation of installed DGs; protection system improvement by using more circuit breakers for sectionalization, distance relays, or directional OC relays (DOCRs); the use of fault current limiters to preserve the original settings of OC relays; and the use of fault ride through (FRT) strategies of inverter-based DGs. Based on the references [15,19,42,43], an adaptive protection coordination (APC) scheme must first update the data from the latest change of DG-integrated distribution network, e.g., the operation status of DGs, the opened/closed status of circuit breakers (CBs), and then calculate power flows and perform the fault analysis to obtain the input data for optimal coordination algorithms. The APC scheme could require a data center to send/receive the data to/from digital relays before or after the sudden changes/disturbances of the DG-penetrated DN. A supervisory control and data acquisition (SCADA) system is needed to perform this adaptive coordination scheme. Moreover, digital relays can be remotely controlled through communication channels. According to the above references, the combination of digital relays, the SCADA system, and an appropriate optimization algorithm will be necessary for the adaptive coordination of OCRs in the DG-integrated DN. The optimal settings of both TDS multipliers and protection-curve coefficients for the inverse-time OC function could be easily implemented in a digital relay. Furthermore, several previous studies have not considered both the reliability of fault-current calculation results and the proper selection of tripping and pick-up thresholds of OCRs in the DN for the optimization of the objective function. Last but not least, it is needed to consider the operation characteristics of DG units, e.g., plug-and-play, peer-to-peer characteristics [53–55], for parameter settings of the adaptable relay coordination system in the DG-contained DN.
- Type of OC relays: In [29,39,40], the optimal coordination of non-directional OC relays has been considered. In [12,56], different operation characteristic curves of non-directional OC relays have been analyzed in detail. The OC relays can be classified into three main types: instantaneous, definite-time, and inverse-time OC relays. If the measured current exceeds a tripping-current threshold, the instantaneous OC relay sends a tripping signal immediately to circuit breakers, while the definite-time overcurrent relay will send a trip signal after a pre-defined time-delay. The inverse-time OC relay operates with a typically mathematical function with certain parameters to form a protection curve as followed by IEC 60255-3 or IEEE C37.112-2018 standards [57,58]. The optimal coordination algorithm of OC relays usually focuses on the inverse-time function. In [59,60], the admittance-based inverse-time OC relay is used to improve the sensitivity of fault detection in the DN and microgrids. In [61–64], the voltage-based inverse-time OC relay is used to be more sensitive to different

fault types and reduce the total operation time of relays. In [65], the single and dual settings of OC relays are proposed to increase the selectivity of the protection system under the high penetration of DGs into the DN. In general, the above-mentioned works only consider the inverse-time OC function as an objective function to be optimized, however, it can be proposed to combine the definite-time OC function with the inverse-time OC function to protect a DG-based distribution network. The reference [66] has solved this research gap, but how to demonstrate the reliability of calculated fault-current values in the DG-integrated distribution network has not been paying attention to. In fact, minimum and maximum fault-current values of each fault type are the basis to properly set-up the inverse-time and define-time OC functions in the protection system.

- Coordination algorithms of OC relays: In [39,54], two PSO (Particle Swarm Optimization) and GS (Gravitational Search) algorithms are used to calculate TDS multipliers and pick-up currents for the OC relays in a DG-contained DN. In [53], a microgenetic algorithm is used to calculate overcurrent protection settings under any change in the DN configuration. In [66], a Firefly Algorithm (FA) is applied to coordinate the definite-time OC functions and the inverse-time OC functions in the DN protection system. In [67], a continuous genetic algorithm (CGA) is used for the optimal coordination of OC relays in a ring-type distribution system. In [68], the Firefly and Chaotic Firefly algorithms have been applied to solve the coordination problem of OC relays. In [69], a modified PSO algorithm is proposed to calculate the optimal relay settings. Generally, the main objective of coordination algorithms is to achieve the possible minimum tripping times through the optimal parameter settings of each OC relay. Typically, this study proposes to use three very common techniques such as GA, GSA, and hybrid PSO-GSA (Gravitational Search Algorithm) to calculate the TDS multiplier and inverse-time curve coefficients for each OC relay in a DG-contained distribution network. To explain that, the GA is a well-known optimization algorithm, whereas the GSA and the hybrid PSO-GSA are recently developed and also applied for protection coordination of OC relays in the DG-contained DN because of high convergence to the global optimal solution with several different constraints [39]. Moreover, relay coordination results of the GA can be used as the standard results to compare with the results of the GSA and the hybrid PSO-GSA.
- Directional OC (DOC) relays and coordination algorithms: In [17,24,44–46,48–50], the optimal coordination of DOC relays in the distribution network has been performed. Reference [17] uses a simplex algorithm to solve the protection coordination of DOC relays as a linear programming problem. References [24,42] use a differential evolution (DE) algorithm for the optimal coordination of bi-directional OC relays in the closed-loop distribution networks. In [31], a non-dominated sorting genetic algorithm-II (NSGA-II) is proposed for the coordination of DOC relays to minimize the total operating time of primary and backup relays with a multi-objective function. More recently, heuristic techniques, such as Cuckoo optimization algorithm [38], Electromagnetic Field Optimization (EFO) algorithm [41], the Hybrid GA-NLP Approach [44], evolutionary algorithm and linear programming [45], a biogeography-based optimization (BBO) algorithm [46], Symbiotic Organism Search Optimization technique [48], seeker algorithm [49], ant colony optimization (ACO) [43], bee colony optimization (BCO) [70], an Imperialistic Competition Algorithm [71], teaching learning-based optimization (TLBO) [72], harmony search algorithm [73], or firefly algorithm [74] are also used as powerful tools to solve the optimization coordination problem of DOC relays in the distribution network. In general, it can be concluded that many recent studies on coordination algorithms are performed for DOC relays in a DG-integrated distribution network because the presence of DGs leads to the directional change of fault currents. The operation characteristic of DOC relays is similar to that of OC relays excepting for the directional-change detection of fault currents [75–80]. In conclusion, it should be noted that the purpose of this survey is to provide a more comprehensive

view of many recent protection coordination algorithms for DOC relays in the DN. The directional-change detection functions of fault currents are out of the research scope so that this study will only focus on the development of an adaptive and scalable protection coordination system of overcurrent relays in a DG-integrated DN.

1.3. Contributions and the Paper Structure

According to an aforementioned literature survey, it is needed to develop an adaptive and scalable protection coordination (ASPC) approach for OCRs in the DG-integrated DN to improve the reliability of protection system. The ASPC approach for OCRs has two main performance stages: (i) Stage I–increasing the reliability of fault-current calculation results and selecting tripping and pick-up current thresholds for the OCRs; and (ii) Stage II–applying the optimization algorithms to calculate the coefficients and Time-Dial-Setting (TDS) multiplier for the inverse-time OC functions in the OCR. More clearly, Stage I finds the min-max confidence interval of fault currents for each different fault type (e.g., 3ph-G fault, 2ph-G fault, 1ph-G fault, and ph-ph fault) which is the basis for properly selecting tripping and pick-up thresholds of definite-time and inverse-time OC functions in the same OCR. Stage II uses three common optimization algorithms, Particle Swarm Optimization (PSO), Gravitational Search Algorithm (GSA), and Genetic Algorithm (GA) to determine the coefficients and TDS multiplier for the inverse-time OC functions. The objective function of the ASPC approach for the OCRs is formulated by three certain constraints consisting of Coordination Time Interval (CTI), relay operating time and TDS multiplier. Novel contributions and assumptions related to the ASPC system of the OC relays are briefly shown as follows:

- Increasing the reliability of fault-current calculation results: According to the literature survey, either nominal bus-voltage values or voltage values calculated right after some pre-determined cycles are commonly considered as the reference voltages for the power-flow analysis and fault-current calculation. However, the use of nominal bus-voltage values could lead to inaccurate calculations of power flows and fault currents in the distribution network, which can result in the protection miscoordination of OCRs. Therefore, this study will propose a novel statistical data-filtering method to determine a min-max confidence interval of load power $[P_{\text{load\_min}}, P_{\text{load\_max}}]$ and load current $[I_{\text{load\_min}}, I_{\text{load\_max}}]$ at each load bus in the DN, as referred in Sections of Forecasting a Min-Max Confidence Interval of Load Power at Each Load Bus and Calculating a Min-Max Confidence Interval of Load Current at Each Load Bus, respectively. Next, a current-injection-based power-flow (CIBPF) analysis method is used to calculate the min-max confidence interval of bus voltage at any i-th bus on the DN, $[V_{i\_\min}, V_{i\_\max}]$ instead of using either the nominal bus-voltage values or the voltage values calculated right after some pre-determined cycles from the power-flow analysis, as referred to Section of The Current-Injection-Based Power Flow Analysis. Based on the confidence intervals of bus-voltages in the distribution network, an adaptable fault analysis technique depicted in Section of An Adaptable Fault Analysis Technique is proposed to calculate the min-max confidence intervals of fault currents contributed by different DG units and fault current contributed by the utility which are used for effectively selecting tripping and pick-up thresholds of definite-time and inverse-time OC functions in the same OCR. In other words, instead of using a fixed fault-current value for each fault type with the unconsidered reliability, this paper has contributed an adaptable fault analysis technique to calculate the min-max confidence interval of fault current. The maximum and minimum confidence thresholds of fault current are then used to select tripping and pick-up thresholds of definite-time and inverse-time OC functions, respectively.
- Propose to use both definite-time and inverse-time protection functions in the same OCR to protect a DG-based distribution network.

- Propose to use the SCADA system to update the operation status of DGs and the opened/closed status of circuit breakers (CBs), and to remotely control digital relays for the ASPC approach.
- The ASPC system of OCRs can be effectively operated regarding different fault scenarios in the DG-based distribution system as well as the 'on-grid' or 'off-grid' operation modes for DG units in the DN.
- A real 22 kV DG-integrated distribution network which is simulated by ETAP software considered to be a reliable test-bed to validate the proposed ASPC system of OCRs.
- Coordination results of the OCRs are based on three optimization algorithms, Particle Swarm Optimization (PSO), Gravitational Search Algorithm (GSA) and Genetic Algorithm (GA), as referred in Section 2.4.

The remaining sections of the paper are organized as follows. Section 2 presents the proposed ASPC approach for OCRs in the DG-based DN. This section presents an objective function of OCR protection coordination, coordination constraints of the OCRs in the DN, the reliable and accurate calculation method of minimum and maximum fault currents for each fault type in the DN having DGs, selection of pick-up and tripping currents for OCRs, and three common optimization algorithms, GSA, hybrid PSO-GSA and GA. Section 3 describes a case study of the practical 22 kV DG-integrated distribution network. Section 4 indicates the optimal protection coordination results of OCRs in the practical 22 kV DN, analysis and discussion on these relay coordination results. Last but not least, Section 5 contains the conclusions of the study.

2. A Proposed ASPC Approach for OCRs in a DG-Integrated Distribution Network

2.1. Modelling of OC Relay Characteristics

In this study, the definite-time OC functions (e.g., 50P for the phase-to-phase fault protection; 50G for the ground fault protection) are deployed as primary protection, while the inverse-time OC functions (e.g., 51P for phase-to-phase fault protection and 51G for ground fault protection) are used for both primary and backup protection. Parameter settings of definite-time and inverse-time OC functions are separate; however, it always allows that the primary protection is activated before the backup protection. The IEEE C37.112™-2018 standard characteristics are considered for the relay coordination problem [57,58]. The operating time equation of the standard inverse-time OC function is given by Equation (1):

$$T_{F51\_ik} = TDS_i \left[\frac{A}{\left(I_{f\_ik} / I_{pu\_i} \right)^B - 1} + C \right] \quad (1)$$

where T_{F51_ik} is the operation time of the i-th relay (R_i) for a k fault location in the protected line/zone of the distribution network; TDS_i and I_{pu_i} are the time dial setting (TDS_i) and pick-up current (I_{pu_i}) parameters of the i-th relay, respectively; I_{f_ik} is the fault current seen by the i-th relay for the fault at k location; and A, B, and C are the coefficients of the inverse-time protection curve. Figure 1 illustrates the protection characteristic curves of definite-time (50) and inverse-time (51) OC functions.

2.2. Objective Function of Overcurrent Protection Coordination

The objective is to find the optimal value of TDS_i and three A, B, and C coefficients of each OCR, while I_{pu_i} and I_{f_ik} are properly selected from fault analysis results of the DG-based distribution network. The coordination problem of OCRs is to minimize the total operating time of all relays working as the primary protection but still satisfying certain constraints and maintaining the reliability of the protection system. The objective function is expressed by Equation (2) as follows:

$$\min Z = \sum_{i=1}^{n} T_{F51\_ik,pri} \quad (2)$$

where n is the total number of relays; and $T_{F51\_ik,pri}$ is the operating time of the i-th OC relay at the k faulted location working as the primary protection. This study proposes to only consider one faulted point/location for each protected line/zone, normally at the remote end of the protected line/zone where the minimum fault current can be calculated as the basis to select a pick-up current threshold for the primary OC relay of this protected line/zone in the distribution network, as referred to Section 2.3.4.

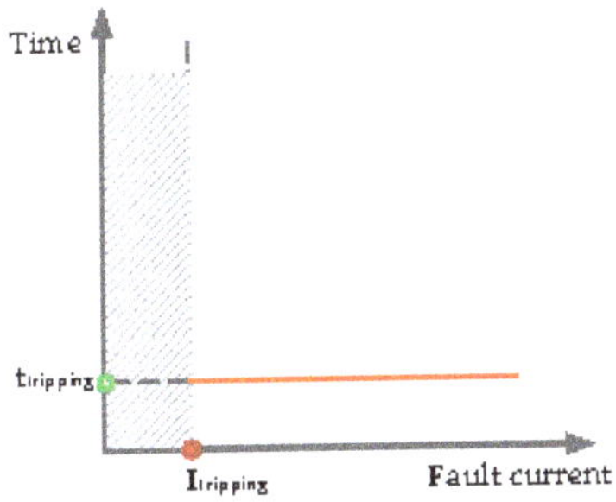

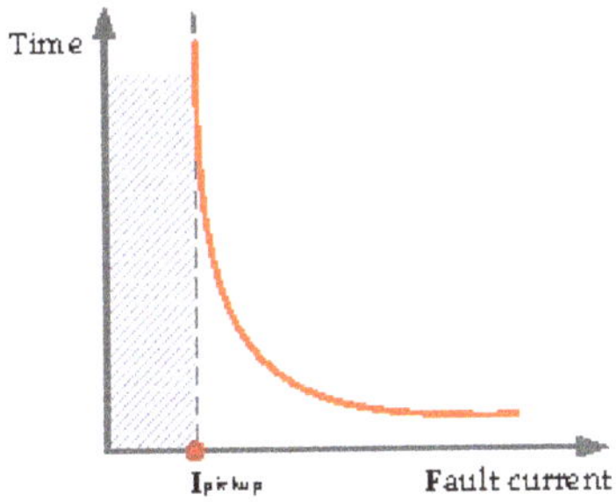

Figure 1. Illustration of protection characteristic curves (**a**) 50P/G–definite-time overcurrent function and (**b**) 51P/G–inverse-time overcurrent function.

On the other hand, it is most common to use the OCR pairs, e.g., one primary relay and one backup relay for each OCR pair, to detect and clear a given fault in the protected line/zone. However, to improve the scalability and selectivity of an overcurrent relay coordination system in the DN having DGs, this study proposes to use the OCR triples, i.e., one for the primary relay and two other consecutive ones for the backup relays. The total tripping time of the OCR triple is formulated as follows:

- For the upward direction of fault current to the relay R_i:

$$2*\text{CTI}^{\min} \le t_{F51,(upward-upward-i)k,backup} + t_{F51,(upward-i)k,backup} + t_{F51,ik,pri} \le 2*\text{CTI}^{\max} \quad (3)$$

- For the backward direction of fault current to the relay R_i:

$$2*\text{CTI}^{\min} \le t_{F51,(backward-backward-i)k,backup} + t_{F51,(backward-i)k,backup} + t_{F51,ik,pri} \le 2*\text{CTI}^{\max} \quad (4)$$

where CTI is a coordination time interval between two adjacent relays, commonly CTI $\in \left\{\text{CTI}^{\min},\text{CTI}^{\max}\right\} = \{0.2, 0.5\}$ [57,58]; $t_{F51,ik,pri}$ is the operation time of the i-th relay R_i when it works as primary protection for a k faulted location in the protected line/zone; $t_{F51,(upward-i)k,backup}$ is the backup operating time of the upward adjacent relay of the relay R_i; $t_{F51,(upward-upward-i)k,backup}$ is the backup operating time of the last upward adjacent relay of the relay R_i in the OCR triple; $t_{F51,(backward-i)k,backup}$ is the backup operating time of the backward adjacent relay of the relay R_i; and $t_{F51,(backward-backward-i)k,backup}$ is the backup operating time of the last backward adjacent relay of the relay R_i in the OCR triple.

In general, the optimization of the objective function is subjected to the following constraint conditions, (i) relay characteristic constraints; and (ii) protection coordination constraints. The former contains the relay operating time, TDS multiplier, coefficients of the inverse-time curve, and the pick-up current of OC relays. The latter considers the operation time difference of each pair of primary and backup OCRs as well as the total tripping time of each OCR triple.

2.3. Coordination Constraints of OCRs

2.3.1. Coordination Time Interval

According to IEEE 242-2001 standards, a CTI between primary and backup OCRs is usually selected in a range of 0.2 s to 0.5 s [57,58]. The CTI considers main factors such as the operating time of CBs, current transformer (CT) errors, the signal sending time of OCRs, and a safety margin [75]. If the primary relay fails to clear a fault given at the location k, then the backup relay will be activated to send the tripping signals to the related CBs after a certain CTI plus the failed operating time of the primary relay. The operating time T_{F51_jk} of the j-th backup relay (R_j) for a k faulted location in the network is expressed by Equation (5).

$$T_{F51\_jk} - T_{F51\_ik} \geq \text{CTI} \tag{5}$$

where T_{F51_jk} is calculated by Equation (6) as follows.

$$T_{F51\_jk} = TDS_j \left[\frac{A}{\left(I_{f\_jk} / I_{pu\_j} \right)^B - 1} + C \right] \tag{6}$$

where TDS_j and I_{pu_j} are the time-dial setting and pick-up current parameters of the j-th back-up relay, respectively; and I_{f_jk} is the fault current seen by the j-th back-up relay (R_j) regarding a k fault location in the distribution network. It is noted that all OC relays on a feeder of distribution network will have the same pick-up current threshold I_{pu}. The selection of pick-up current threshold for the OC relays is specifically presented in Section 2.3.4.

2.3.2. Boundary on Relay Operating Time

The operating time of OC relays has a practical boundary by two limits, $T_{ik}^{\min}$ and $T_{ik}^{\max}$, which can be formulated as the following:

$$T_{ik}^{\min} \leq T_{ik} \leq T_{ik}^{\max}, \forall i = 1, 2, \ldots n \tag{7}$$

where $T_{ik}^{\min}$ and $T_{ik}^{\max}$ are the minimum and maximum operation times of the i-th relay R_i when it operates as primary protection for a faulted location k in the DN, respectively. The value of $T_{ik}^{\min}$ depends on relay manufacturers, usually being higher than 0.01 s, while $T_{ik}^{\max}$ is the critical fault clearing time to avoid the damage of equipment as well as maintain the stability of power system, usually being lower than one second [32,33].

The Time-Dial Setting (TDS) is bounded by an available range of settings supplied by the relay manufacturer, which can be expressed as follows:

$$TDS_i^{\min} \leq TDS_i \leq TDS_i^{\max}, \forall i = 1, 2, \ldots n \tag{8}$$

where $TDS_i^{\min}$ and $TDS_i^{\max}$ are the minimum and maximum TDS values of the i-th relay R_i, respectively. The minimum and maximum available TDS values are usually in a range of 0.025 to 1.2 [32,33]. This study selects an available range of 0.01 to 3.2, respectively, for $TDS_i^{\min}$ and $TDS_i^{\max}$.

Similarly, the boundary of A, B, and C coefficients of inverse-time protection curve can be formulated by:

$$\begin{cases} A_i^{\min} \leq A_i \leq A_i^{\max} \\ B_i^{\min} \leq B_i \leq B_i^{\max}, \forall i = 1, 2, \ldots n \\ C_i^{\min} \leq C_i \leq C_i^{\max} \end{cases} \tag{9}$$

where $A_i^{\min}$ and $A_i^{\max}$ are 0.009 and 150, respectively; $B_i^{\min}$ and $B_i^{\max}$ are 0.02 and 2.5, respectively; $C_i^{\min}$ and $C_i^{\max}$ are zero and 1.5, respectively, in this study. The number of intervals in a pre-determined range of A_i, B_i, and C_i is 50.

2.3.3. Calculating the Min-Max Confidence Interval of Fault Currents in the DN Having DGs

To increase the reliability of fault-current calculation results in the DN having DGs, this section will present a novel statistical data-filtering method to determine a min-max confidence interval of load power $[P_{load\_min}, P_{load\_max}]$ and load current $[I_{load\_min}, I_{load\_max}]$ at each load bus in the DN, as seen in Sections of Forecasting a Min-Max Confidence Interval of Load Power at Each Load Bus and Calculating a Min-Max Confidence Interval of Load Current at Each Load Bus, respectively. Next, a current-injection-based power-flow (CIBPF) analysis method is used to calculate the min-max confidence interval of bus voltage at any i-th bus on the DN,$[V_{i\_min}, V_{i\_max}]$, instead of using either nominal bus-voltage values or voltage values calculated right after some pre-determined cycles from the power-flow analysis, as referred to Section of The Current-Injection-Based Power Flow Analysis. Based on the confidence intervals of bus-voltages in the distribution network, an adaptable fault analysis technique depicted in Section of An Adaptable Fault Analysis Technique is proposed to calculate the min-max confidence intervals of fault currents contributed by different DG units and fault current contributed by the utility which are used for effectively selecting tripping and pick-up thresholds of definite-time and inverse-time OC functions in the same OCR. In general, this adaptable fault analysis method contains four main steps: (i) forecasting a min-max confidence interval of load power at each load bus on a feeder of distribution network [81,82]; (ii) calculating a min-max confidence interval of load current at each load bus on the feeder; (iii) doing the current-injection-based power flow analysis to find a min-max confidence interval of voltage at each bus where has already installed the OC relay; and (iv) performing an adaptable fault analysis method to determine a min-max confidence interval of fault currents for each fault type in the DG-contained distribution network.

(i) Forecasting a Min-Max Confidence Interval of Load Power at Each Load Bus

Figure 2 shows a flow diagram of forecasting a min-max confidence interval of load power at each load bus in the DN. According to the work [82], there are seven main steps to forecast a min-max confidence interval of load power at each load bus on a feeder of distribution network as the following.

- Step 1—Input the historical load profiles at each load bus and perform the wrangling of the input data.
- Step 2—Calculate Probability Density Function (PDF) of load data and check whether the load data are in a Gaussian distribution or not. If the input load data have a Gaussian distribution, it is continued to Step 6.
- Step 3—If the input load data are not normally distributed, then the "differencing" method is used to eliminate the data trend by creating a new differential load data series in the one-day-ahead basis and the PDF of new differential load data series will be next calculated. When its PDF is in a Gaussian distribution, Step 6 will be implemented.
- Step 4—If the new differential data series is still not normally distributed, a Principal Components Analysis (PCA) method is conducted; and then the PDF is re-calculated for new differential load sub-datasets. If the new differential load sub-datasets are normally distributed, then it is continued with Step 6; otherwise, it goes to Step 5.
- Step 5—A 'dendrogram' method is conducted over the whole new differential load data series to explore the relationship among 15-min, 30-min, or hourly load data points, and then cluster them into many smaller differential load sub-datasets. Subsequently, it will go to Step 6.
- Step 6—The input load data at Step 2, the new differential load data series at Step 3, the new differential load sub-datasets at Step 4, or the smaller differential load sub-datasets at Step 5 are filtered-out with a possible confidence range of 13 levels, specifically from 90% to 99% with an increasing interval of 1%, 4.5-sigma (~99.73%), 5.5-sigma (~99.9937%), and 6-sigma (~99.99966%). Then, it is shown that the best con-

fidence interval of load data will be determined by an artificial neural network (ANN) forecasting model with the lowest Mean Absolute Percentage Error (MAPE) metric.

- Step 7—Applying the best confidence interval at Step 6 to eliminate unexpected outliers/noises of the original input load data; and then a min-max confidence interval of load power, $[P_{load\_min}, P_{load\_max}]$, at each load bus will be calculated by Equation (10).

$$\begin{cases} P_{\text{load\_max}} = \mu_P + \frac{Z}{\sqrt{N}}\sigma_P \\ P_{\text{load\_min}} = \mu_P - \frac{Z}{\sqrt{N}}\sigma_P \end{cases} \quad (10)$$

where a Z factor is taken from a normal/Gaussian distribution table corresponding to the best confidence level; μ_P is the mean value of load power at each load bus; N is the number of observed load data points; and σ_P is the standard deviation of the analyzed load data.

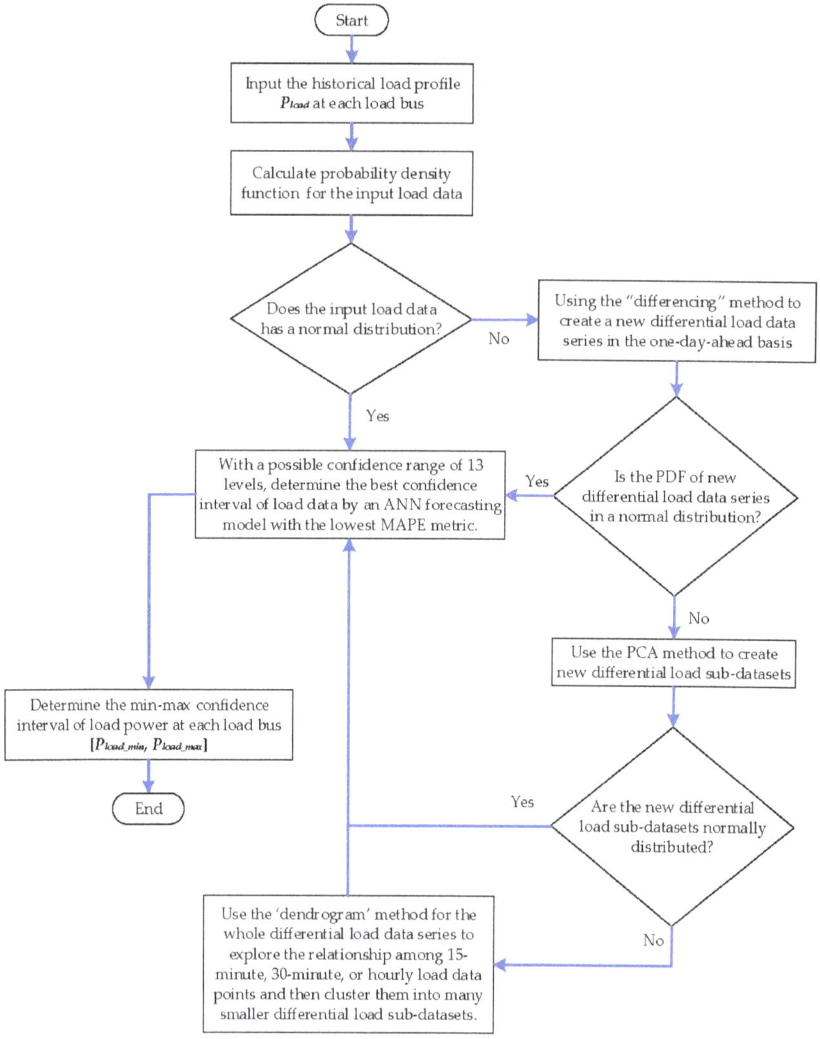

Figure 2. A flow diagram of forecasting a min-max confidence interval of load power at each load bus in the DN.

(ii) Calculating a Min-Max Confidence Interval of Load Current at Each Load Bus

Based on the min-max confidence interval of load power at each load bus in the DN and the nominal bus-voltage value, a min-max confidence interval of load current $[I_{\text{load\_min}}, I_{\text{load\_max}}]$ at each load bus can be calculated by:

$$\begin{cases} I_{\text{load\_min}} = \frac{P_{\text{load\_min}}}{V_{\text{nom}}} \\ I_{\text{load\_max}} = \frac{P_{\text{load\_max}}}{V_{\text{nom}}} \end{cases} \quad (11)$$

where V_{nom} is the nominal bus-voltage value. The minimum and maximum load current values, $I_{\text{load\_min}}$ and $I_{\text{load\_max}}$, at each bus are used for the current injection-based power flow analysis in the distribution network with DGs.

(iii) The Current-Injection-Based Power Flow Analysis

After the min-max confidence thresholds of load current at each load bus have been appropriately determined, they will be used for calculating voltage-fluctuating thresholds at all buses of the DG-based distribution network through the current-injection-based power-flow (CIBPF) analysis [83–86]. Then, the calculated voltage-fluctuating thresholds will be input to an adaptable fault analysis technique as mentioned in Section of An Adaptable Fault Analysis Technique. The modified (CIBPF) analysis method uses two matrices which are (i) branch currents (BC) matrix and (ii) bus voltages (BV) matrix. By considering the i-th bus in the distribution network, the power injected into this i-th bus can be expressed by Equation (12).

$$S_i = (P_i + jQ_i) = (P_{G,i} - P_{L,i}) + j(Q_{G,i} - Q_{L,i}), \forall i = 1, 2, \ldots, N \quad (12)$$

where $P_{G,i}$ and $Q_{G,i}$ are the active and reactive power of a generation source at the i-th bus, respectively; and $P_{L,i}$ and $Q_{L,i}$ are the active and reactive power of a load connected to the i-th bus, respectively. It is noted that a distribution network is assumed to have N buses.

The injection of equivalent current to the i-th bus at the k-th iteration of the power flow analysis is calculated by Equation (13) as below:

$$I_i^k = I_i^{real}\left(V_i^k\right) + jI_i^{imag}\left(V_i^k\right) = \left(\frac{P_i + jQ_i}{V_i^k}\right)^*, \forall i = 1, 2, \ldots, N \quad (13)$$

where V_i^k and I_i^k are the bus voltage and the equivalent current injected to the i-th bus at the k-th iteration, respectively; I_i^{real} and I_i^{imag} are real and imaginary components of the equivalent current injected to the i-th bus, which are considered as a function of V_i^k.

By considering a DG-integrated distribution network as in Figure 3, the injected-to-bus power can be properly converted to the injected-to-bus current by conducting Equation (13). According to [84,86], the relationship between the $[BC]$ matrix and the $[BV]$ matrix can be obtained through Kirchhoff laws. The [BC] matrix is determined by the injected-to-bus currents, as shown in Equation (14).

$$\begin{bmatrix} B_1 \\ B_2 \\ B_3 \\ B_4 \\ B_5 \end{bmatrix} = \begin{bmatrix} 1\,1\,1\,1\,1 \\ 0\,1\,1\,1\,1 \\ 0\,0\,1\,1\,0 \\ 0\,0\,0\,1\,0 \\ 0\,0\,0\,0\,1 \end{bmatrix} \begin{bmatrix} I_2 \\ I_3 \\ I_4 \\ I_5 \\ I_6 \end{bmatrix} \quad (14)$$

$$\begin{cases} I_6 = I_{load6} - I_{DG2} \\ I_5 = I_{load5} - I_{DG1} \\ I_4 = I_{load4} \\ I_3 = I_{load3} \\ I_2 = I_{load2} \end{cases} \quad (15)$$

$$I_{load\_l} \in \left\{ \frac{P_{load\_l\_min}}{V_{nom}}; \frac{P_{load\_l\_max}}{V_{nom}} \right\};\ l = 1 \ldots N_{load} \tag{16}$$

$$I_{DG\_d} = \left(\frac{P_{DG\_d} + jQ_{DG\_d}}{V_i} \right)^*;\ d = 1 \ldots N_{DG} \tag{17}$$

which B_1, B_2, B_3, B_4 and B_5 are the branch currents on the distribution network; I_2, I_3, I_4, I_5 and I_6 are the equivalent currents injected into the bus; I_{load_l} is the load current at a l-th load bus on the DN, with $l = 1, \ldots, N_{load}$; N_{load} is the total number of load buses; I_{DG_d} is the source current of the d-th DG injected to the bus, with $d = 1, \ldots, N_{DG}$; N_{DG} is the total number of distributed generators penetrated to the DN; $P_{load\_l\_min}$ and $P_{load\_l\_max}$ are the minimum and maximum thresholds of load power at the l-th load bus, respectively, as calculated in Section of Forecasting a Min-Max Confidence Interval of Load Power at Each Load Bus; V_{nom} is the nominal bus-voltage value of the distribution network; P_{DG_d} and Q_{DG_d} represent the active and reactive power of the d-th DG, respectively; and V_i is the i-th bus voltage on the distribution system.

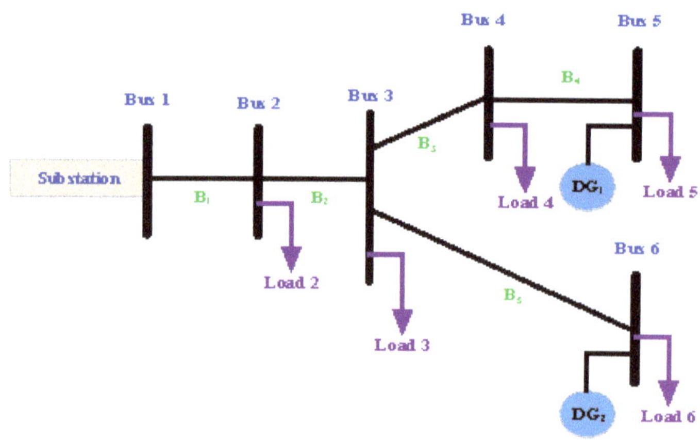

(a) A distribution system contains six buses and two DGs

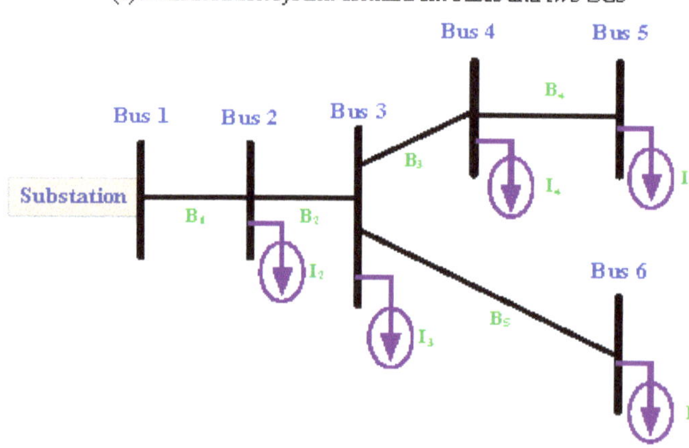

(b) The equivalent currents are injected into the buses on a single-line diagram of DN

Figure 3. A typical single-line diagram of DN with the integration of distributed generators.

In general, Equation (14) can be re-written as the following:

$$[BC] = [C][I] \tag{18}$$

where a $[C]$ matrix is an upper triangular matrix filled by "0" or "1".

Next, the relationship of branch-currents matrix $[BC]$ and bus-voltages matrix $[BV]$ is shown in Equation (19):

$$[\Delta V] = [BV] = \begin{bmatrix} V_1 \\ V_1 \\ V_1 \\ V_1 \\ V_1 \end{bmatrix} - \begin{bmatrix} V_2 \\ V_3 \\ V_4 \\ V_5 \\ V_6 \end{bmatrix} = \begin{bmatrix} Z_{12} & 0 & 0 & 0 & 0 \\ Z_{12} & Z_{23} & 0 & 0 & 0 \\ Z_{12} & Z_{23} & Z_{34} & 0 & 0 \\ Z_{12} & Z_{23} & Z_{34} & Z_{45} & 0 \\ Z_{12} & Z_{23} & 0 & 0 & Z_{36} \end{bmatrix} \begin{bmatrix} B_1 \\ B_2 \\ B_3 \\ B_4 \\ B_5 \end{bmatrix} \quad (19)$$

$$V_j = V_i - B_i Z_{\text{ij}} \rightarrow B_i Z_{\text{ij}} = V_i - V_j, \ \forall i,j = 1,2,\ldots,N; \ i \neq j \quad (20)$$

which V_i is the i-th bus voltage; V_j is the j-th bus-voltage; and Z_{ij} is the line impedance between bus i and bus j. Equation (19) can be generally re-written as follows.

$$[\Delta V] = [BV] = [Z][BC] \quad (21)$$

where $[\Delta V]$, or called $[BV]$, is the matrix of voltage drop between the bus i and the bus j on the DN; and $[Z]$ is a lower triangular matrix of line impedances in the system.

In summary, when the min-max confidence interval of load power at the l-th load bus, $[P_{load\_l\_min}, P_{load\_l\_max}]$, has been already calculated/forecasted, the min-max confidence interval of load current, $[I_{load\_l\_min}, I_{load\_l\_max}]$, will be then appropriately determined. Next, the min-max confidence interval of equivalent current injected into the i-th bus, is also promptly defined. After that, the $[BC_{min}]$ and $[BC_{max}]$ matrices are calculated by Equation (18), with $[BC_{\text{min}}] = [B_{1\_\text{min}} B_{2\_\text{min}} \cdots B_{\text{i\_min}}]^T$ and $[BC_{\text{max}}] = [B_{1\_\text{max}} B_{2\_\text{max}} \cdots B_{\text{i\_max}}]^T$. Finally, the $[\Delta V_{min}]$ and $[\Delta V_{max}]$ voltage-drop matrices will be determined by Equation (21). As a result, the min-max confidence interval of bus voltage at any i-th bus on the DN, $[V_{i\_\text{min}}, V_{i\_\text{max}}]$, can be effectively calculated, specifically for $[V_1] - [B_{i\_\text{min}}] = [V_{i\_\text{max}}]$ and $[V_1] - [B_{i\_\text{max}}] = [V_{i\_\text{min}}]$ in this study. It is worth noting that V_1 is the nominal voltage value of bus 1 (also called the slack bus) in the distribution network.

- Solving the power flow problem in case of a radial distribution network

  To obtain the $[\Delta V]$ matrix, Equation (18) is substituted to Equation (21) as the following:

$$[\Delta V] = [Z][C][I] = [PF][I] \quad (22)$$

The $[PF]$ matrix in Equation (22) can be determined by a Lower-Upper (LU) factorization solution. The $[PF]$ matrix can be factorized by two $[Z]$ and $[C]$ matrices, where $[C]$ is the upper triangular matrix and $[Z]$ is the lower triangular matrix. The use of LU-decomposition and forward/backward algorithms allows skipping the calculation of Jacobian matrices or admittance matrices. Therefore, the computation time of power flows can be significantly reduced. As a result, the LU-decomposition-based power-flow analysis can be adaptable to the online power-flow analysis for a DG-integrated distribution system, as referred to Equation (23).

$$\begin{cases} I_i^k = I_i^{real}\left(V_i^k\right) + jI_i^{imag}\left(V_i^k\right) = \left(\frac{P_i + jQ_i}{V_i^k}\right)^* \\ \left[\Delta V_i^{k+1}\right] = [PF]\left[I_i^k\right] \\ [PF] = [Z][C] \\ \left[V_i^{k+1}\right] = [V_1] - \left[\Delta V_i^{k+1}\right] \end{cases} \quad (23)$$

where k is the iteration number of the current-injection-based power-flow analysis method.

- Solving the power flow problem in case of a ring-type distribution network

Figure 4 shows a single-line diagram of meshed distribution network. A new branch is connected from bus 5 to bus 6, hence, the equivalent currents at all buses are calculated as the following. Equation (19) can be re-formulated by:

$$[\Delta V] = [BV] = \begin{bmatrix} V_1 \\ V_1 \\ V_1 \\ V_1 \\ V_1 \\ 0 \end{bmatrix} - \begin{bmatrix} V_2 \\ V_3 \\ V_4 \\ V_5 \\ V_6 \\ 0 \end{bmatrix} = \begin{bmatrix} Z_{12} & 0 & 0 & 0 & 0 & 0 \\ Z_{12} & Z_{23} & 0 & 0 & 0 & 0 \\ Z_{12} & Z_{23} & Z_{34} & 0 & 0 & 0 \\ Z_{12} & Z_{23} & Z_{34} & Z_{45} & 0 & 0 \\ Z_{12} & Z_{23} & 0 & 0 & Z_{36} & 0 \\ 0 & 0 & Z_{34} & Z_{45} & -Z_{36} & Z_{56} \end{bmatrix} \begin{bmatrix} B_1 \\ B_2 \\ B_3 \\ B_4 \\ B_5 \\ B_6 \end{bmatrix} \tag{24}$$

$$\begin{bmatrix} \Delta V \\ 0 \end{bmatrix} = [Z_{new}] \begin{bmatrix} BC \\ B_{new} \end{bmatrix} \tag{25}$$

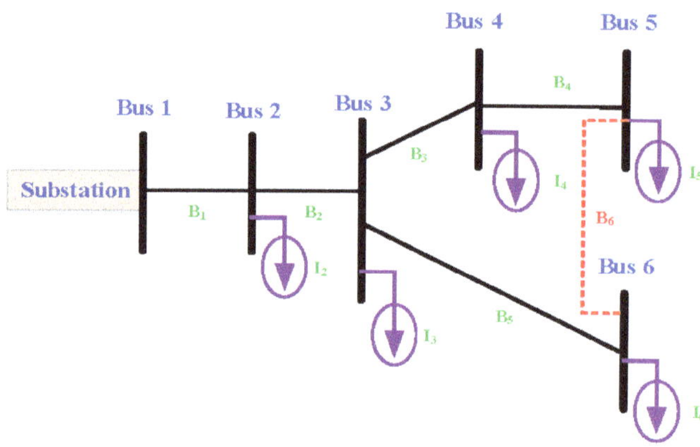

Figure 4. A typical single-line diagram of meshed distribution network.

Generally, a new branch B_k is built on a distribution network to create a ring topology; specifically, the new branch is connected from the bus i to the bus j. The initial [Z] matrix can be re-calculated by the following steps: (i) adding a new row to the initial [Z] matrix and each element of the added new row is calculated by subtracting the j-th row from the i-th row; (ii) an element of the new row that belongs to a diagonal of [Z] matrix is also an intersectional position of the new row and the new column of the [Z] matrix; and it will be an impedance of the new branch B_k.

In addition, Equation (14) is also re-written as follows:

$$\begin{bmatrix} B_1 \\ B_2 \\ B_3 \\ B_4 \\ B_5 \\ B_6 \end{bmatrix} = \begin{bmatrix} 1 & 1 & 1 & 1 & 1 & 0 \\ 0 & 1 & 1 & 1 & 1 & 0 \\ 0 & 0 & 1 & 1 & 0 & 1 \\ 0 & 0 & 0 & 1 & 0 & 1 \\ 0 & 0 & 0 & 0 & 1 & -1 \\ 0 & 0 & 0 & 0 & 0 & 1 \end{bmatrix} \begin{bmatrix} I_2 \\ I_3 \\ I_4 \\ I_5 \\ I_6 \\ B_6 \end{bmatrix} \tag{26}$$

$$\begin{bmatrix} BC \\ B_{new} \end{bmatrix} = [C_{new}] \begin{bmatrix} I \\ B_{new} \end{bmatrix} \tag{27}$$

A new branch B_k is built in a meshed distribution network, which is connected from the bus i to the bus j. The [C] matrix is then re-calculated by the following steps: (i) copying the elements of the column related to the i-th bus to the k-th column (called a new column) and subtracting the elements of the column related to the j-th bus from this new column; and (ii) adding the value (+1) to the k-th row and k-th column in the [C] matrix.

Substituting two Equations (27) and (25) into Equation (22), it shows that:

$$\begin{bmatrix} \Delta V_{new} \\ 0 \end{bmatrix} = [Z_{new}][C_{new}]\begin{bmatrix} I \\ B_{new} \end{bmatrix} = \begin{bmatrix} AM^T \\ MN \end{bmatrix}\begin{bmatrix} I \\ B_{new} \end{bmatrix} \tag{28}$$

Then, the Kron reduction method is applied to Equation (28), it becomes:

$$[\Delta V_{new}] = \left[A - M^T N^{-1} M\right][I] = [PF_{new}][I] \tag{29}$$

As a result, Equation (29) with the $[PF_{new}]$ matrix can be solved by the LU factorization method as mentioned in Equation (23).

(iv) An Adaptable Fault Analysis Technique

According to the work in [6], a simplified and automated fault-current calculation approach can be appropriately modified for this study. The total short-circuit current observed by the r-th OCR for a faulted location in the DG-contained DN is calculated by Equation (30). The configuration of DG-based distribution network is changeable due to two plug-and-play and peer-to-peer characteristics of DGs, so a real-time management system (RTMS) using the SCADA function is deployed to observe the operating states of DGs as well as the working status of protective devices like OCRs and other devices such as CBs, tie switches, load break switches, or reclosers in the DN.

$$I_{f,r} = I_{f,grid-r} + \sum_{i=1}^{N_{DG}} \left(k_{ri} * I_{f,DGi} * T_{DGi}\right), \forall I_{f,r} \in \left(I_{f,r}^{\min}, I_{f,r}^{\max}\right) \tag{30}$$

where $I_{f,r}$ is the total short-circuit current observed by the r-th OCR for a faulted location in the DN; $I_{f,grid-r}$ is the grid fault-current component seen by the r-th OCR for the considered fault location; N_{DG} is the total number of IBDGs and RBDGs integrated to the DN; k_{ri} is a fault-current division coefficient of the i-th DG impacting the operation of the r-th OCR; $I_{f,DGi}$ is the fault current at terminals of the i-th DG; and T_{DGi} is the operating status of the i-th DG, specifically, $T_{DGi} = 1$ for the "on-grid" mode and $T_{DGi} = 0$ for the "off-grid" mode.

Generally, if a distribution system contains the N_{DG} total number of DGs and the M_{OCPR} total number of OCRs, the fault-current division coefficient of each DG unit impacting the operation of each OCR can be properly calculated by Equation (30). Symmetrical impedances are used for calculating the fault-current division coefficients of DGs and the grid sources impacting all OCRs in the DN. Due to the SCADA function of RTMS, the symmetrical impedance matrices can be continuously updated corresponding to different DN topologies as well as the operating status of DGs such that the fault-current division coefficients of DGs will be quickly determined. Equation (30) can be re-written by:

$$\begin{cases} I_{f,r}^{\min} = I_{f,grid-r}^{\min} + \sum_{i=1}^{N_{DG}} \left(k_{ri} * I_{f,DGi}^{\min} * T_{DGi}\right) \\ I_{f,r}^{\max} = I_{f,grid-r}^{\max} + \sum_{i=1}^{N_{DG}} \left(k_{ri} * I_{f,DGi}^{\max} * T_{DGi}\right) \end{cases} \tag{31}$$

where $I_{f,r}^{\min}$ and $I_{f,r}^{\max}$ are the min-max confidence thresholds of total short-circuit current observed by the r-th OCR for a faulted location in the DN; $I_{f,grid-r}^{\min}$ and $I_{f,grid-r}^{\max}$ are the min-max confidence thresholds of grid fault-current for a given fault location; $I_{f,DGi}^{\min}$ and $I_{f,DGi}^{\max}$ are the minimum and maximum fault-current values of the i-th IBDG/RBDG at their terminals which are pre-defined, specifically, 5 pu~10 pu for RDBGs [87,88] and 1.0~2.0 pu for IBDGs [89,90].

- Determining the min-max confidence thresholds of fault current from the grid

To calculate $I_{f,grid-r}^{\min}$ and $I_{f,grid-r}^{\max}$, the min-max confidence interval of bus voltages, $[V_{i\_\min}, V_{i\_\max}]$, must be determined as shown in Section of The Current-Injection-Based Power Flow Analysis. Then, the values $V_{i\_\min}$ and $V_{i\_\max}$ are respectively substituted to

$V_i(0)$ as depicted in Equation (32). It is assumed that the total number of OCRs is equal to that of buses in the DN, i.e., $M_{OCPR} = N$.

$$\begin{cases} \begin{bmatrix} V_{f,1,k} \\ \cdots \\ V_{f,i,k} \\ \cdots \\ V_{f,N,k} \end{bmatrix} = \begin{bmatrix} V_1 \\ \cdots \\ V_i(0) \\ \cdots \\ V_N(0) \end{bmatrix} - \left([Z_{\text{bus}}]_{N\times N} + \left[Z_{f,k}\right]_{N\times N} \right) \begin{bmatrix} I^{\min}_{f,grid-1,k} \\ \cdots \\ I^{\min}_{f,grid-i,k} \\ \cdots \\ I^{\min}_{f,grid-N,k} \end{bmatrix}, V_i(0) = V_{i\_\min}, \forall i \in N \\ \begin{bmatrix} V_{f,1,k} \\ \cdots \\ V_{f,i,k} \\ \cdots \\ V_{f,N,k} \end{bmatrix} = \begin{bmatrix} V_1 \\ \cdots \\ V_i(0) \\ \cdots \\ V_N(0) \end{bmatrix} - \left([Z_{\text{bus}}]_{N\times N} + \left[Z_{f,k}\right]_{N\times N} \right) \begin{bmatrix} I^{\max}_{f,grid-1,k} \\ \cdots \\ I^{\max}_{f,grid-i,k} \\ \cdots \\ I^{\max}_{f,grid-N,k} \end{bmatrix}, V_i(0) = V_{i\_\max}, \forall i \in N \end{cases} \quad (32)$$

where $V_{f,i,k}$ is the fault voltage of the i-th bus for a k faulted location in the DN; $V_i(0)$ is the min/max pre-fault bus-voltage at the i-th bus; the values $I^{\min}_{f,grid-r}$ and $I^{\max}_{f,grid-r}$ are equal the values $I^{\min}_{f,grid-i}$ and $I^{\max}_{f,grid-i}$, respectively, because each bus i has already installed the r-th OCR in the system; $[Z_{\text{bus}}]_{N\times N}$ is the Thevenin impedance matrix with a size of $N \times N$; and $\left[Z_{f,k}\right]_{N\times N}$ is the fault impedance at the k faulted location; normally, the faulted location is considered at the buses of a distribution system. For the asymmetrical faults, the positive- negative-, and zero-sequence Thevenin impedance matrices $\left[Z^{012}_{bus}\right]_{N\times N}$ need to be calculated. The subscripts 0, 1, 2 are represented as zero-, positive-, and negative-sequence components, respectively. Hence, Equation (32) can be re-written as the following.

$$\begin{cases} V^{012}_{f,i,k} = V^{012}_i(0) - \left(\left[Z^{012}_{bus}\right]_{N\times N} + \left[Z_{f,k}\right]_{N\times N} \right) \begin{bmatrix} I^{012,\min}_{f,grid-1,k} \\ \cdots \\ I^{012,\min}_{f,grid-i,k} \\ \cdots \\ I^{012,\min}_{f,grid-N,k} \end{bmatrix}, V^{012}_i(0) = V^{012}_{i\_\min}, \forall i \in N \\ V^{012}_{f,i,k} = V^{012}_i(0) - \left(\left[Z^{012}_{bus}\right]_{N\times N} + \left[Z_{f,k}\right]_{N\times N} \right) \begin{bmatrix} I^{012,\max}_{f,grid-1,k} \\ \cdots \\ I^{012,\max}_{f,grid-i,k} \\ \cdots \\ I^{012,\max}_{f,grid-N,k} \end{bmatrix}, V^{012}_i(0) = V^{012}_{i\_\max}, \forall i \in N \end{cases} \quad (33)$$

- Determining fault-current division coefficients of DGs in the DN

In general, the min-max confidence thresholds of grid fault current seen by the r-th relay for a given fault, $I^{\min}_{f,grid-r}$ and $I^{\max}_{f,grid-r}$, can be used as the basis for parameter settings of the r-th relay in the DN. A fault-current division coefficient of the i-th RBDG impacting the operation of the r-th relay is expressed by:

$$\begin{cases} k^{\min}_{ri} = \frac{I^{\min}_{f,DGi-r}}{I^{\max}_{f,DGi}} = \left(\frac{V_{i\_\min}}{Z_{ir}}\right) / I^{\max}_{f,DGi}, \forall i \neq r \\ k^{\max}_{ri} = \frac{I^{\max}_{f,DGi-r}}{I^{\min}_{f,DGi}} = \left(\frac{V_{i\_\max}}{Z_{ir}}\right) / I^{\min}_{f,DGi}, \forall i \neq r \\ k_{ri} = 1 \text{ for } i = r \end{cases} \quad (34)$$

where $k^{\min}_{ri}$ and $k^{\max}_{ri}$ are the minimum and maximum fault-current division coefficients of the i-th RBDG with respect to the r-th relay, respectively; it is assumed that the i-th RBDG is connected to the i-th bus in the DN, so Z_{ir} is the impedance from the bus i to the bus r that is determined from the $[Z_{\text{bus}}]_{N\times N}$ matrix. In the case of $i = r$, the k_{ri} coefficient is unit; and the value k_{ri} is in an available range of [0~1].

On the other hand, the fault-current division coefficient k_{ri} of the i-th IBDG impacting the operation of the r-th relay can be expressed by Equation (35):

$$\begin{cases} \left. \begin{array}{l} k_{i-downstream\_r} = \frac{Z_{i-upstream\_r}}{Z_{i-downstream\_r}+Z_{i-upstream\_r}} \\ k_{i-upstream\_r} = \frac{Z_{i-downstream\_r}}{Z_{i-downstream\_r}+Z_{i-upstream\_r}} \end{array} \right. \text{for a ring topology} \\ k_{ri} = 1 \text{ for a radial topology} \end{cases} \quad (35)$$

where the subscripts '*downstream_r*' and '*upstream_r*' represent the fault current direction of the i-th IBDG flowing into the downstream and upstream sites of the r-th relay respectively, with regard to a given fault location in the DN. When a fault occurs in the distribution system, the IBDG will limit its fault current contribution and operate as a fault current source. Therefore, the coefficient k_{ri} equals to 1 for a radial topology of DN. For the ring-type/meshed DN, it is necessary to calculate the downstream and upstream fault-current division coefficients $k_{i-downstream\_r}$ and $k_{i-upstream\_r}$ of the i-th DG regarding to the r-th relay, as referred to Equation (35). The $k_{i-downstream\_r}$ coefficient is selected in case the r-th relay is used to detect the fault at its upstream direction, whereas the $k_{i-upstream\_r}$ coefficient is selected only if the r-th relay is designed to detect the fault at its downstream direction.

In general, a $[K]$ coefficient matrix can be determined to show the fault-current contribution of N_{DG} DGs to M_{OCPR} relays (noted that the number of OCRs equals to the number of buses in the DN) as expressed in Equation (36).

$$[K] = \begin{bmatrix} k_{11} & k_{1i} & k_{1N_{DG}} \\ \cdots & \cdots & \cdots \\ k_{r1} & k_{ri} & k_{rN_{DG}} \\ \cdots & \cdots & \cdots \\ k_{M_{OCPR}1} & k_{M_{OCPR}i} & k_{M_{OCPR}N_{DG}} \end{bmatrix} \text{ with } \begin{cases} i = 1 \ldots N_{DG} \\ r = 1 \ldots M_{OCPR} = 1 \ldots N \end{cases} \quad (36)$$

where the k_{ri} element of $[K]$ matrix can be k_{ri}^{min}, k_{ri}^{max}, 1, $k_{i-downstream-r}$, or $k_{i-upstream-r}$ according to the type of DGs (e.g., IBDG or RBDG), the location of DGs, the location of OCRs, and the topology of DN (e.g., radial topology or ring topology). In addition, the $[K]$ matrix can also be applied to the unbalanced faults in the distribution system.

2.3.4. Selection of Pick-Up Currents for OC Relays

The pick-up current I_{pu} of OCR should be selected to be higher than the maximum possible load current and lower than the minimum fault current with a reliable security margin. The measurement error of current transformers (CTs) should also be considered for the pick-up current selection. The boundary of pick-up current of the r-th OCR, $I_{pu,r}^{\min}$ and $I_{pu,r}^{\max}$, is determined as below:

$$\begin{cases} I_{pu,r}^{\min} = \left(OLF + K_L \frac{e_{CT\%}}{100}\right) \frac{I_{load,r}^{\max}}{CTR} \\ I_{pu,r}^{\max} = \left(\frac{1}{3} - K_{SC} \frac{e_{CT\%}}{100}\right) \frac{I_{f,r}^{\min}}{CTR} \\ I_{pu,r}^{selected} = \frac{I_{pu,r}^{\min} + I_{pu,r}^{\max}}{2} \end{cases} \quad (37)$$

where OLF is the overload factor selected to 1.15 [91]; K_L and K_{SC} are the security factors to be higher than 1; $e_{CT\%}$ is the measurement error of current transformers selected to 10%; $I_{load,r}^{\max}$ is the maximum load current seen by the r-th relay based on the power-flow analysis results in Section of The Current-Injection-Based Power Flow Analysis; CTR is the current transformer ratio; $I_{f,r}^{\min}$ is the minimum total fault current observed by the r-th relay (normally considering a phase-to-phase (ph-ph) fault located at a remote end of the protected line) as presented in Section of An Adaptable Fault Analysis Technique; a

1/3 factor in Equation (37) is referred by the references [92,93]. In this study, the selected pick-up current $I_{pu,r}^{selected}$ of the r-th OCR will be the average of $I_{pu,r}^{\min}$ and $I_{pu,r}^{\max}$.

2.3.5. Selection of Tripping Currents for the OC Relays

Table 1 shows the selection of tripping currents of the OCRs in the DG-based distribution network. The fault-current analysis is performed at the remote end of the protected zones/lines in the DN, with four main types including (i) three phase-to-ground (3ph-G) fault, (ii) double-phase to-ground (2ph-G) fault, (iii) phase-to-phase (ph-ph) fault and iv) single-phase to-ground (1ph-G) fault. As a novel contribution of the paper, the protection coordination of the OCRs including 50P, 50G, 5P, and 51G functions is established for the maximum fault-current value of each fault type.

Table 1. Fault-current selection for the 50 and 51 OC functions.

Different OC Functions	Selection of Fault/Tripping Currents, I_{f_rk}	Descriptions
50P: Definite-time phase OC protection function	$I_{f\_rk\_F50P} = I_{f,r,ph-ph}^{\max}$ with $r = 1 \ldots N$	k is a given fault location at the remote end of the protected line/zone; and $I_{f,r,ph-ph}^{\max}$ is the maximum ph-ph fault current.
51P: Inverse-time phase OC protection function	$I_{f\_rk\_F51P} = I_{f,r,ph-ph}^{\max}$ with $r = 1 \ldots N$	
50G: Definite-time ground OC protection function	$I_{f\_rk\_F50G} = \min\left(I_{f,r,3ph-G}^{\max}, I_{f,r,2ph-G}^{\max}, I_{f,r,1ph-G}^{\max}\right)$ with $r = 1 \ldots N$	$I_{f,r,3ph-G}^{\max}, I_{f,r,2ph-G}^{\max}$ and $I_{f,r,1ph-G}^{\max}$ are the maximum 3ph-G, 2ph-G, and 1ph-G fault currents for a given fault location, respectively.
51G: Inverse-time ground OC protection function	$I_{f\_rk\_F51G} = \min\left(I_{f,r,3ph-G}^{\max}, I_{f,r,2ph-G}^{\max}, I_{f,r,1ph-G}^{\max}\right)$ with $r = 1 \ldots N$	

2.3.6. Summary of the Proposed ASPC Approach for OCRs in a DG-Based Distribution Network

Figure 5 shows a flow-chart of the ASPC approach for OCRs in the DN with DGs. There are five basic steps as the following:

- Step 1: Input the maximum fault-current values detected by each OCR corresponding to four different fault types (e.g., ph-ph, 1ph-G, 2ph-G, and 3ph-G) that occurred at the remote end of the protective zones/lines.
- Step 2: Select the pick-up and tripping/fault currents for the OC relays and define the tripping time of 50P and 50G functions, as shown in Table 1.
- Step 3: Based on the objective function Z in Equation (2), determine A, B, and C factors and TDS multiplier of 51P and 51G functions in each OCR by using three meta-heuristic algorithms, GSA, hybrid PSO&GSA and GA.
- Step 4: Check the tripping time of 50P/G and 51P/G functions and the constraint conditions of CTI between the primary and backup relays for the OCR pairs and the OCR triples as referred to Equations (3)–(5).
- Step 5: Select the appropriate protection coordination results for OCRs which are satisfied with the constraint conditions; and update the setting parameters to the related OCRs in the DG-based DN.

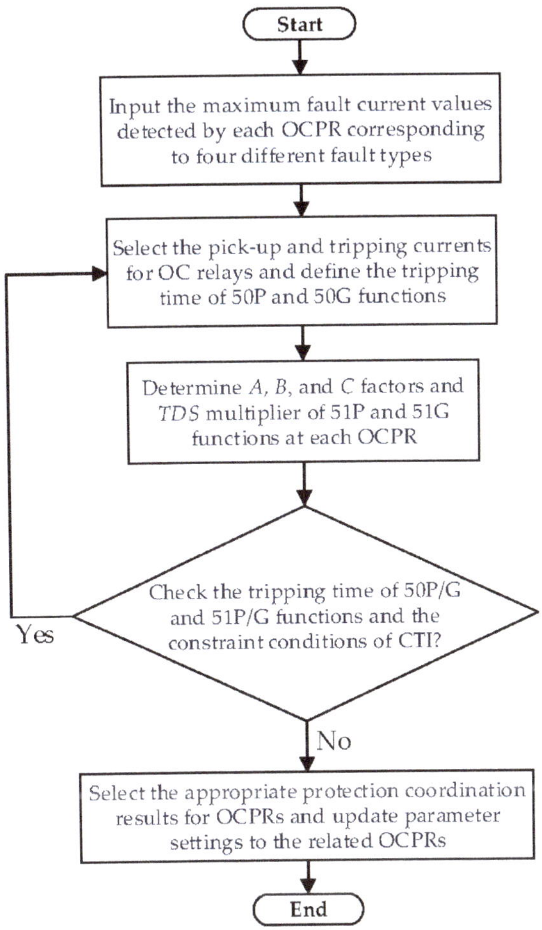

Figure 5. A general flow-chart of the proposed ASPC approach for OCRs in a DG-based distribution network.

2.4. Protection Coordination Algorithms

2.4.1. Gravitational Search Algorithm (GSA)

A Gravitational Search Algorithm (GSA) is based on the gravitational force and Newton's law of motion [94]. By considering a system with N agents (or N masses) with the d-th dimension in a certain search space, the gravitational force from the j-th agent to the i-th agent at the t-th time, $F_{ij}^d(t)$, can be expressed by Equation (38):

$$F_{ij}^d(t) = G(t)\frac{M_{pi}(t) \times M_{aj}(t)}{R_{ij}(t) + \in}\left(x_j^d(t) - x_i^d(t)\right) \quad (38)$$

where, $M_{aj}(t)$ is the active gravitational mass regarding to the j-th agent in the time t; $M_{pi}(t)$ is the passive gravitational mass regarding to the i-th agent; $\in$ is a small constant; $R_{ij}(t)$ represents the Euclidian distance between the agents i and j; x_i^d and x_j^d are the position of the i-th agent and the j-th agent at the d-th dimension, respectively; and $G(t)$ is a gravitational constant at the time t as referred to Equation (39).

$$G(t) = G_0 \times \exp(-\alpha * iter/max\_iter) \quad (39)$$

where G_0 and α are the initial value of gravitational constant and the descending coefficient, respectively, normally selected to 20 and 100; '*iter*' is the current number of iterations; and '*max_iter*' is the maximum number of iterations. In other words, $G(t)$ is a function of the initial value G_0 and is gradually decreased with respect to the time to control the search accuracy of the algorithm.

The Euclidian distance, $R_{ij}(t)$, between two agents i and j is calculated as the following.

$$R_{ij}(t) = ||X_i(t),\ X_j(t)||_2 \tag{40}$$

where the position of the i-th agent and the j-th agent with n-dimensions, X_i and X_j, can be defined as below:

$$\begin{aligned} X_i &= \left(x_i^1, \ldots, x_i^d, \ldots, x_i^n\right), i = 1, 2, \ldots, N \\ X_j &= \left(x_j^1, \ldots, x_j^d, \ldots, x_j^n\right), j = 1, 2, \ldots, N \\ X &= \{X_1, \ldots, X_i \ldots, X_j, \ldots, X_N\} \end{aligned} \tag{41}$$

For the stochastic search algorithm, the total gravity force impacting the i-th agent at the d-th dimension, $F_i^d(t)$, is the sum of d-th components of the forces generated from other agents in a set of N agents with the random weighting factor $rand_j$ in an available interval [0, 1].

$$F_i^d(t) = \sum_{j=1, j\neq i}^{N} rand_j F_{ij}^d(t) \tag{42}$$

The acceleration of the i-th agent at the time t with the d-th dimension, $A_i^d(t)$, can be calculated by:

$$A_i^d(t) = \frac{F_i^d(t)}{M_{ii}(t)} \tag{43}$$

where $M_{ii}(t)$ is the inertial mass of the i-th agent at the time t.

For each iteration step, the velocity and position of the i-th agent are updated by the following equations.

$$Vel_i^d(t+1) = rand_i \times Vel_i^d(t) + A_i^d(t) \tag{44}$$

$$X_i^d(t+1) = X_i^d(t) + Vel_i^d(t+1) \tag{45}$$

where the next velocity of the i-th agent, $Vel_i^d(t+1)$, is the sum of a fraction of its current velocity $rand_i \times Vel_i^d(t)$ and its acceleration $A_i^d(t)$; '$rand_i$' is a random variable in the interval [0, 1]; and the next position of the i-th agent $X_i^d(t+1)$ is the sum of its current position, $X_i^d(t)$, and its next velocity, $Vel_i^d(t+1)$. It is noted that the use of a random variable is to give a randomized search characteristic of the algorithm.

The gravitational and inertia masses of the i-th agent can be calculated by the fitness evaluation which is referred to Equation (2) in Section 2.2. The greater mass is, the more efficient agent is. The better agents will have the higher attractive forces and the slower movement. The gravitational and inertia masses of the agent are assumed to be equal. The value of masses can be updated as follows.

$$M_{ai} = M_{pi} = M_{ii} = M_i,\ \forall i = 1, 2, \ldots, N \tag{46}$$

$$m_i(t) = \frac{fit_i(t) - worst(t)}{best(t) - worst(t)} \tag{47}$$

$$M_i(t) = \frac{m_i(t)}{\sum_{j=1}^{N} m_j(t)} \tag{48}$$

where $fit_i(t)$ is the fitness value of the i-th agent at the time t; $best(t)$ is the optimal solution, whereas $worst(t)$ is the worst solution of the algorithm. The $best(t)$ and $worst(t)$ parameters can be determined depending on two following problems.

- For the minimization value problem:

$$best(t) = \min\{fit_j(t)\}_{j\in\{1,\dots,N\}} \tag{49}$$

$$worst(t) = \max\{fit_j(t)\}_{j\in\{1,\dots,N\}} \tag{50}$$

- For the maximization value problem:

$$best(t) = \max\{fit_j(t)\}_{j\in\{1,\dots,N\}} \tag{51}$$

$$worst(t) = \min\{fit_j(t)\}_{j\in\{1,\dots,N\}} \tag{52}$$

All the procedures are involved in the GSA to find the optimal parameters of the OCRs referred to Figure 6.

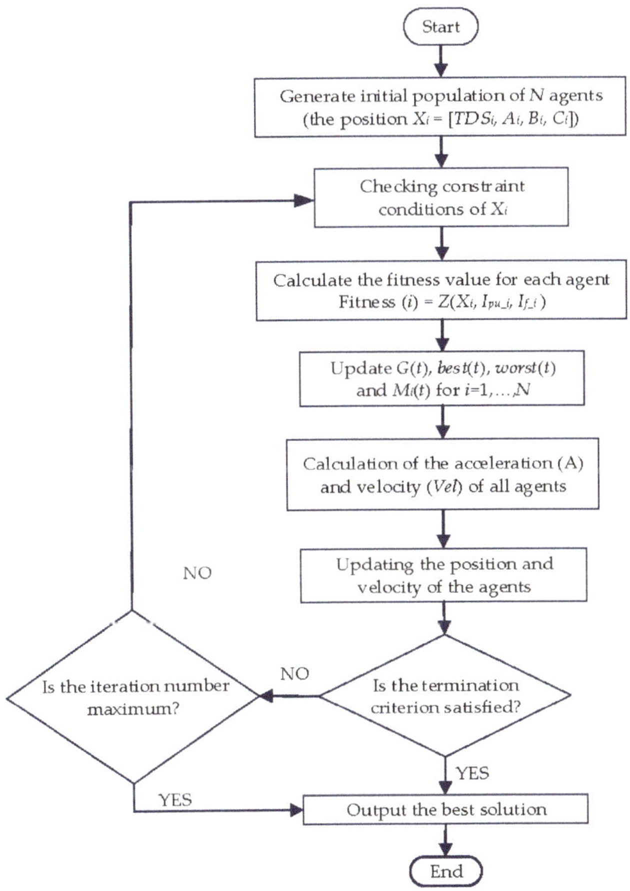

Figure 6. A flow chart of GSA applied to the ASPC approach of OCRs.

2.4.2. Hybrid Particle Swarm Optimization and Gravitational Search Algorithm (Hybrid PSO&GSA)

In the particle swarm optimization algorithm, the feasible solutions are called the particles/agents that fly in the d-dimensions search space by following the current optimal particles [95]. Let s_i and v_i denote the position and velocity of the i-th particle, respectively;

at each iteration, the velocity of each particle is updated by using its current position, velocity and its distance from the *pbest* to the *gbest* as shown in Equation (53).

$$\begin{aligned}
v_{i,d}^{k+1} &= \omega * v_{i,d}^{k} + C_1 * rand() * \frac{(pbest^k{}_i - s_{i,d}^k)}{\Delta t} + C_2 * rand() * \frac{(gbest^k - s_{i,d}^k)}{\Delta t} \\
v_{i,d}^{k} &\in [-v_{\max}, v_{\max}], v_{\max} = ks_{\max}, 0.1 \le k \le 1 \\
s_{i,d}^{k} &\in [-s_{\max}, s_{\max}] \\
v_i^k &= \left(v_{i,1}^k, v_{i,2}^k, \ldots, v_{i,d}^k\right) \\
s_i^k &= \left(s_{i,1}^k, s_{i,2}^k, \ldots, s_{i,d}^k\right) \\
pbest^k{}_i &= \left(pbest^k{}_{i,1} pbest^k{}_{i,2}, \ldots, pbest^k{}_{i,d}\right) \\
gbest^k &= \left(pbest^k{}_1, pbest^k{}_2, \ldots, pbest^k{}_N\right)
\end{aligned} \tag{53}$$

where $v_{i,d}^k$ is the current velocity of the i-th particle at the k-the iteration with the d-th dimension; $rand()$ is a random variable in the interval [0, 1]; $s_{i,d}^k$ is the current position of the i-th particle at the k-the iteration; C_1 and C_2 are the acceleration coefficients; $pbest^k{}_i$ is the personal best position of the i-th particle in the d-dimensions; $gbest^k$ is the global best position of all the particles in the d-dimensions; and Δt is the time step of the algorithm. Moreover, the weighting function ω can be expressed by Equation (54).

$$\omega = \omega_{\max} - \left(\frac{\omega_{\max} - \omega_{\min}}{iter_{\max}}\right) * iter \tag{54}$$

where $\omega_{\max}$ and $\omega_{\min}$ are the maximum and minimum weighting values, respectively, which are constant; $iter$ is the current iteration number; and $iter_{\max}$ is the maximum number of iterations. The algorithm starts with the $\omega_{\max}$ value and decreases to the $\omega_{\min}$ value when the number of iterations increases.

In addition, each particle can update its position at the next time step (iteration) by using its current velocity to explore the search space for a better solution as follows:

$$s_{i,d}^{k+1} = s_{i,d}^{k} + v_{i,d}^{k+1} * \Delta t \tag{55}$$

where Δt is commonly set to one in this study. For the minimization value problem, the personal best position, *pbest*, is updated after the k-th iteration according to Equation (56).

$$pbest^{k+1}{}_i = \begin{cases} pbest^k{}_i & \text{if} \quad f\left(s_i^{k+1}\right) \ge f\left(pbest^k{}_i\right) \\ s_i^{k+1} & \text{if} \quad f\left(s_i^{k+1}\right) < f\left(pbest^k{}_i\right) \end{cases} \tag{56}$$

where f is the fitness/objective function which is referred to Equation (2) in Section 2.2.

The agents in the GSA do not share the population information with each other agent, so the PSO algorithm can be integrated with the GSA to improve the global optimal searching capability. In other words, the GSA is to use for the local searching ability while the PSO algorithm is to use for the global searching ability. The velocity of each agent can be updated by the PSO and the acceleration of the agent is calculated by the GSA, therefore, this combination is called a hybrid PSO-GSA. The velocity and the position of the i-th agent are updated by the following two equations [96,97]:

$$v_{i,d}^{k+1} = \omega * v_{i,d}^{k} + C_1' * rand() * A_{i,d}^{k} + C_2' * rand() * \left(gbest^k - s_{i,d}^k\right) \tag{57}$$

$$s_{i,d}^{k+1} = s_{i,d}^{k} + v_{i,d}^{k+1} \tag{58}$$

where ω is the inertia weighting factor, $v_{i,d}^k$, $s_{i,d}^k$ and $A_{i,d}^k$ are the current velocity, current position, and current acceleration of the i-th particle at the k-th iteration with the d-th dimension, respectively; and C_1' and C_2' are constant acceleration coefficients, respectively.

The acceleration coefficients C_1' and C_2' can be considered as the exponential functions defined by:

$$C_i' = C_{start} * \left(\frac{C_{end}}{C_{start}}\right)^{1/(1+iter/iter_{\max})} \quad (59)$$

where C_{start} is the initial value; C_{end} is the final value; $iter_{\max}$ is the maximum iteration number; and $iter$ is the current iteration number.

In conclusion, Figure 7 shows the basic steps in the hybrid PSO-GSA to find the optimal setting parameters of the OCRs in the DN.

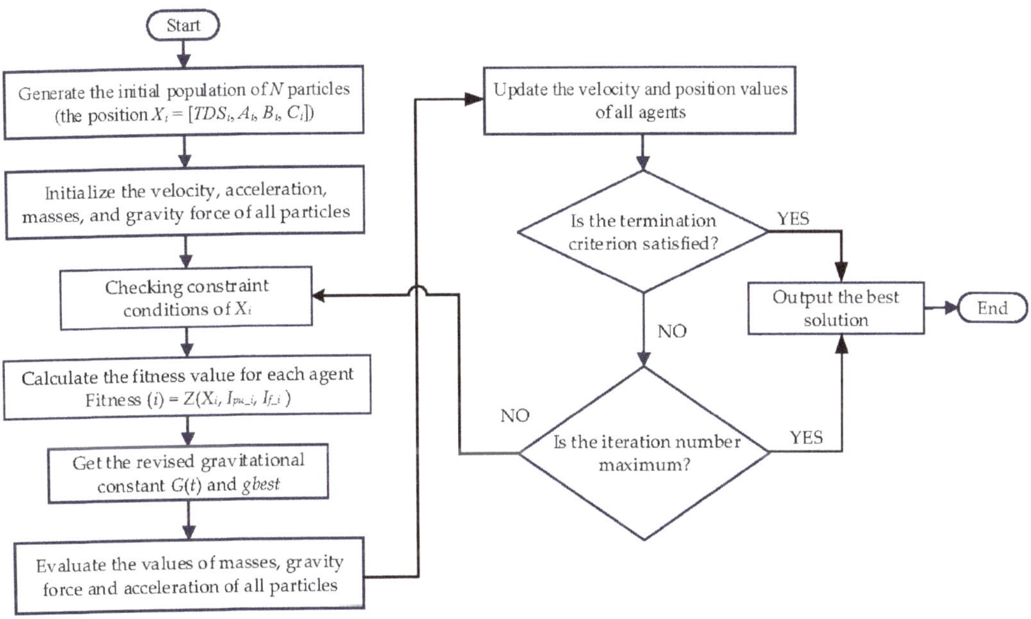

Figure 7. A flow chart of hybrid PSO-GSA applied to the proposed ASPC approach of OCRs.

2.4.3. Genetic Algorithm

The application of genetic algorithm (GA) to the ASPC approach of OCRs in the DG-based DN is shown in Figure 8. The algorithm selects the optimal TDS setting and A, B and C coefficients of protection characteristic curve. In general, the GA works in the process of biological selection, crossover, mutation and combination [98,99]. Survival of the fittest among all feasible solutions in the binary form of genes (or string structures) is considered in the GA. For each generation, a new set of potential solutions is created by the fittest of the previous set of solutions. The control variables are represented in the string structures. The GA randomly initializes the solutions, then, the fitness evaluation of each solution is performed. The fitness function is the same as the objective function Z in Equation (2) of Section 2.2. The solution has the higher fitness which can be probably selected for a new generation; this procedure is also called the biological selection or reproduction. Crossover and mutation steps are applied to get an available range of feasible solutions. When the new generation is obtained, the algorithm will repeatedly work until the termination condition is satisfied. In Figure 8, the counter of generation, G, is initially set to 1. The parameter I is defined as the counter of population, which will stop until getting the population size.

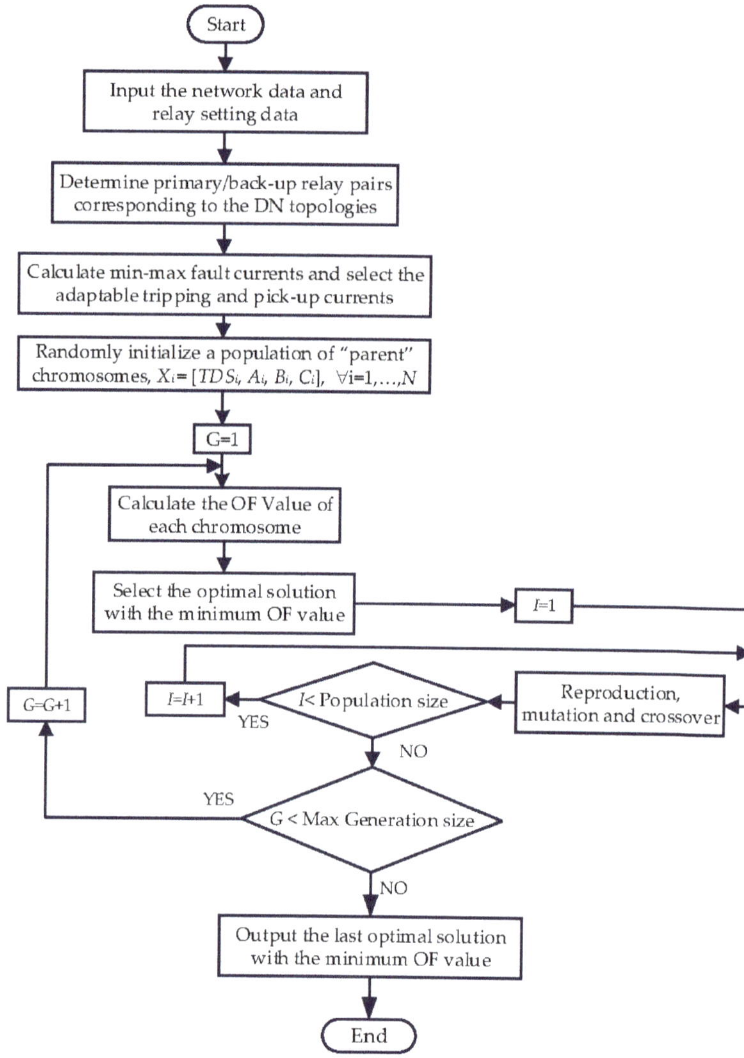

Figure 8. A flow chart of GA applied to the proposed ASPC approach of OCRs in the DN.

3. A Real 22 kV DG-Integrated Distribution Network: A Case Study

A real 22 kV DG-contained distribution network of Ho Chi Minh City, Vietnam is simulated by the ETAP software for evaluating the reliable fault analysis results and validating the proposed ASPC approach of OCRs, as seen in Figure 9. It is worth noting that the real 22 kV DG-integrated distribution network is used for this study instead of standard IEEE distribution systems. To explain that, the use of real-time load data of this DN is completely appropriate to determine the min-max confidence thresholds of load power at each load bus, the min-max confidence thresholds of load current, the min-max confidence interval of bus voltages, and the min-max confidence interval of fault currents contributed by the grid and the DG units as presented in Section 2.3.3. In the steady-state (normal) operation mode, the ASPC approach is performed to give the optimal solution of TDS multiplier and *A*, *B*, and *C* coefficients for inverse-time OC functions and to select tripping and pick-up current thresholds respectively for the definite-time and inverse-time

OC functions in the OCRs of a feeder supplied by the 110/22 kV Can-Gio power substation such as a 471 Hao-Vo OC relay, a REC Hao-Vo OC relay, and a REC Can-Thanh 163 OC relay. Moreover, the relay coordination in the real distribution network is discussed only in radial topologies. As seen in Figure 9, the feeder supplied by the 110/22 kV Can-Gio power substation is separated from the feeder supplied by the 110/22 kV An-Nghia power substation through opening a LBS (load breaker switch) Can-Thanh 97 to be operated in the radial topology. In Vietnam, the radial distribution systems are used for the normal (long-term) operation mode, while the ring-type distribution systems are only established for ancillary services or post-fault service restoration scenarios (the short-term operation mode). Therefore, the protection coordination solutions of OCRs are mostly analyzed for the radial systems, while the coordination solutions of DOC relays in the ring-type systems are limited in this case study.

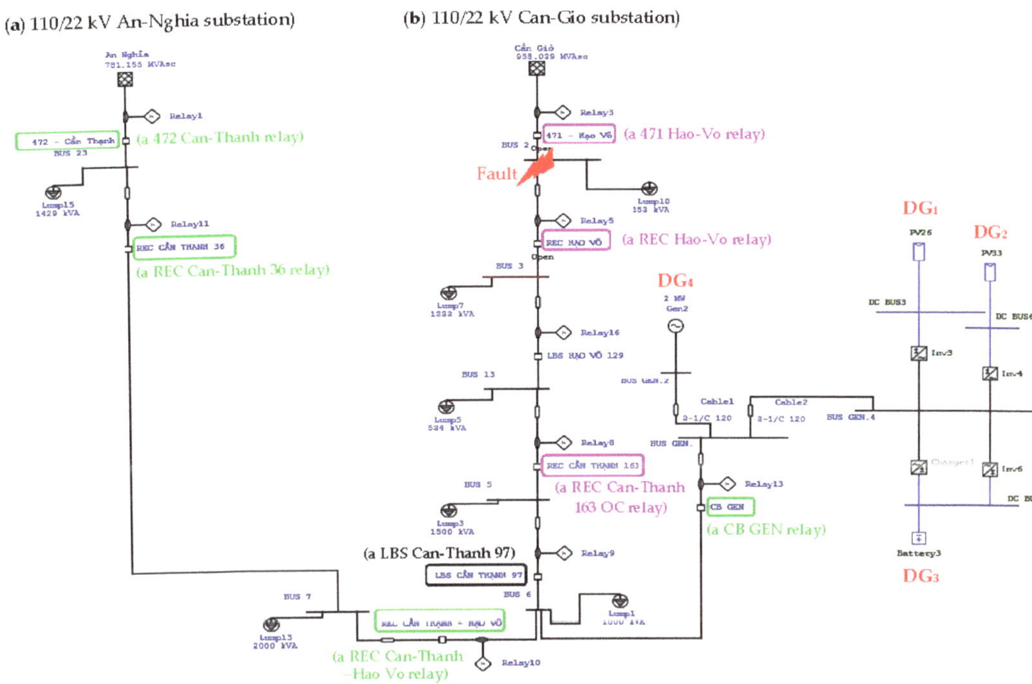

Figure 9. A real 22 kV DG-based distribution network is supplied by two power substations, namely, the 110/22 kV Can-Gio substation and the 110/22 kV An-Nghia substation, three IBDGs (DG_1, DG_2, and DG_3), and one RBDG (DG_4).

Figure 10 shows a simplified single-line diagram of the real 22 kV DG-integrated DN so that pre-fault and post-fault DN topologies can be indicated more clearly. Noted that a 3ph-G fault is given at Bus 2 of the DN. More clearly, at the pre-fault operation mode, a LBS (load breaker switch) Can-Thanh 97 is normally opened, the real 22 kV DN operates with two separate feeders. One feeder is supplied by the 110/22 kV Can-Gio power substation; and a power flow of this feeder is indicated by a dashed line (a). The other feeder is supplied by the 110/22 kV An-Nghia substation; and its power flow direction is shown by a dashed line (b). When the 3ph-G fault occurs at Bus 2, a 471 Hao-Vo relay and a REC Hao-Vo relay will give the tripping signals to the related circuit breakers, namely a 471 Hao-Vo circuit breaker and a REC Hao-Vo recloser, to isolate the fault at Bus 2. After clearing the fault, according to the proposed FISR plans [10,11], the protection coordination results of the related OCRs will be updated by the ASPC system through the SCADA function. By specifically considering the given fault at Bus 2, there are three feasible FISR plans which will be the basis to find the optimal parameter settings of the related OCRs. The first FISR plan is to use an adjacent distribution feeder to restore the power for interrupted customers through closing the LBS Can-Thanh 97 device and opening a circuit breaker controlled by a CB GEN relay. The power flow direction of the first FISR plan is indicated by a dashed line (c). The second FISR plan is to use three IBDGs and one RBDG in the DN as ancillary services. DG_1 and DG_2 are the 0.5 MWp photovoltaic (PV) generation systems. DG_3 is a 7.3 MWh and 1 MW battery energy storage system (BESS). DG_4 is a 2 MW diesel generator. The second FISR plan is performed by closing the LBS Can-Thanh 97 device and the CB of DG sources while opening the REC Can-Thanh Hao-Vo recloser. The power flow direction of the second FISR plan is represented by a dashed line (d). The last FISR plan is to use both the neighboring distribution feeder and all DG sources. The power flow direction of the last plan is followed by the two dashed lines (c) and (d). As a result, the ASPC approach should give at least three optimal protection coordination solutions for the related OCRs corresponding to the three above FISR plans in the case of the 3ph-G fault that occurred at Bus 2 of the real 22 kV DN.

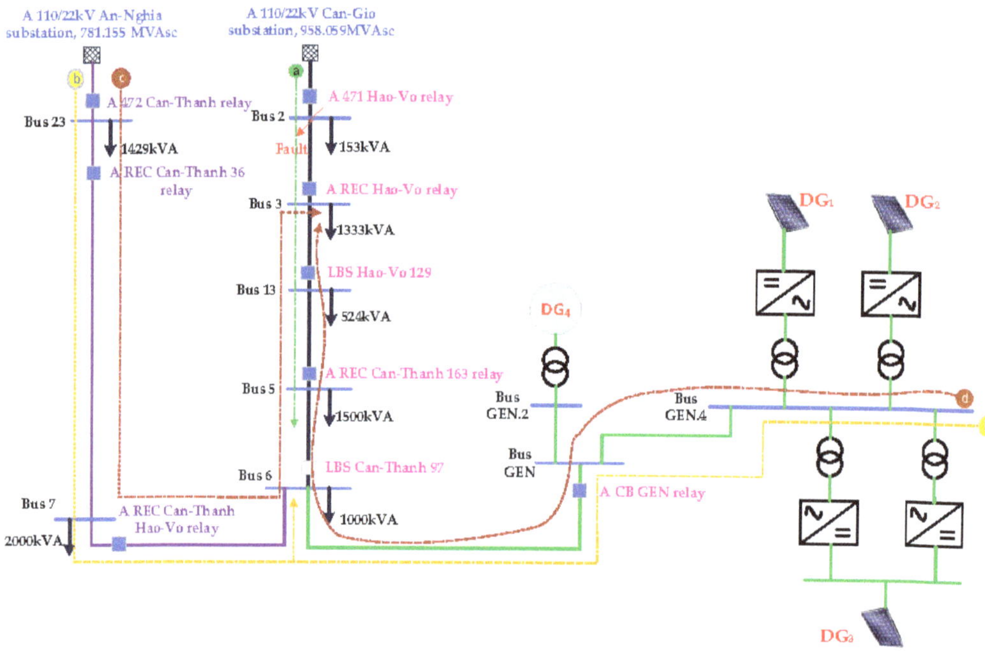

Figure 10. A simplified single-line diagram of the real 22 kV DG-integrated DN.

4. Overcurrent Protection Coordination Results, Analysis and Discussion

4.1. Reliable Fault-Current Calculation Results

As summarized from Section of An Adaptable Fault Analysis Technique, instead of using a fixed fault-current value for each fault type with the unconsidered reliability, this paper has contributed an adaptable fault analysis technique to calculate the min-max confidence interval of fault current for each different fault type. The maximum and minimum confidence thresholds of fault current are used to select tripping and pick-up thresholds of definite-time and inverse-time OC functions in the OCR of the DG-integrated DN, respectively. By considering a feeder only supplied by the 110/22 kV Can-Gio power substation from a real 22 kV DG-based distribution network, Table 2 shows the fault analysis results of this feeder. It is noted that Bus 5 is the end bus of the feeder. Three OC relays are used to protect this feeder including a 471 Hao-Vo OCR, a REC Hao-Vo OCR, and a REC Can-Thanh 163 OCR. The maximum fault currents can be seen by the OCRs in the

DN when the various fault types occur at their protected zones/lines. Specifically, the REC Can-Thanh 163 OCR operates as the primary protection for faults at Bus 5. The maximum fault-current value of each fault type such as 3ph-G, 2ph-G, ph-ph and 1ph-G faults at Bus 5 is seen by this REC Can-Thanh 163 OCR, which is also shown in Table 2. Similarly, the REC Hao-Vo OCR operates as the primary protection for faults at Bus 13. The maximum fault-current value of each fault type such as 3ph-G, 2ph-G, ph-ph and 1ph-G faults at Bus 13 is seen by this REC Hao-Vo OCR. In addition, the 471 Hao-Vo OCR operates as the primary protection for faults at Bus 2. The maximum fault-current value of each fault type such as 3ph-G, 2ph-G, ph-ph and 1ph-G faults at Bus 2 is seen by this 471 Hao-Vo OCR.

Table 2. Numerical fault-current calculation results are seen by the OCRs in a considered feeder of the DG-based DN.

Fault Types	**Maximum Fault Currents Detected by the OCRs**		
	Maximum Fault Currents at Bus 2 Are Detected by a 471 Hao-Vo OCR	**Maximum Fault Currents at Bus 13 Are Detected by a REC Hao-Vo OCR**	**Maximum Fault Currents at Bus 5 Are Detected by a REC Can-Thanh 163 OCR**
3ph-G fault	25,796 A	11,911 A	4902 A
2ph-G fault	24,347 A	10,742 A	4426 A
1ph-G fault	22,340 A	10,316 A	4246 A
ph-ph fault	20,469 A	6488 A	2340 A

As mentioned in Section 3, after a 3ph-G fault at Bus 2 of the feeder is cleared, three feasible FISR plans are proposed. Table 3 presents the maximum fault-current calculation results observed by the OCRs corresponding to four different operating scenarios of the DN, including:

- Scenario 1–the DN topology before the fault occurs at Bus 2, i.e., the feeder supplied by the 110/22 kV Can-Gio power substation, as referred to the dashed line (a) in Figure 10.
- Scenario 2–the DN topology for the first FISR plan, as referred to the dashed line (c) in Figure 10.
- Scenario 3–the DN topology for the second FISR plan, as referred to the dashed line (d) in Figure 10.
- Scenario 4–the DN topology for the third FISR plan, as referred to the two dashed lines (c) and (d) in Figure 10.

Moreover, the OCR triple for each topology including one primary relay and two backup relays is also shown in Table 3. It is noted that some faulted locations can be only detected and cleared by the OCR pairs. The maximum fault-current values which are calculated from four different fault types at the buses of the real 22 kV distribution network are indicated in this table by an order-based format, specifically [$the\ maximum\ 3ph - G\ fault\ current$;$the\ maximum\ 2ph - G\ fault\ current$;$the\ maximum\ 1ph - G\ fault\ current$; $the\ maximum\ ph - ph\ fault\ current$].

Table 3. Numerical fault-current calculation results are detected by the OCRs corresponding to four different operation scenarios of the real 22 kV distribution network.

(a) Scenario 1–A DN topology before the fault occurs at Bus 2:
The LBS Can-Thanh 97 is normally opened; and the feeder is only supplied by the 110/22 kV Can-Gio power substation.

Faults at Bus 2	Faults at Bus 13	Faults at Bus 5
Primary protection: the 471 Hao-Vo OCR	*Primary protection*: the REC Hao Vo OCR	*Primary protection*: the REC Can-Thanh 163 OCR
Backup protection: None	*Backup protection*: the 471 Hao-Vo OCR	*Backup protection*: the REC Hao-Vo OCR and the 471 Hao-Vo OCR
Maximum fault currents at Bus 2 are detected by the 471 Hao-Vo OCR: [25.8; 24.3; 22.3; 20.4] kA	Maximum fault currents at Bus 13 are detected by the REC Hao-Vo OCR: [11.9; 10.7; 10.3; 6.5] kA	Maximum fault currents at Bus 5 are detected by the REC Can-Thanh 163 OCR: [4.9; 4.4; 4.2; 2.3] kA

(b) Scenario 2–A DN topology for the first FISR plan:
As referred to the dashed line (c) in Figure 10; the REC Hao-Vo recloser is opened to isolate the fault at Bus 2; then the LBS Can-Thanh 97 is closed; and a CB GEN of DG sources is still opened.

Faults at Bus 7	Faults at Bus 5	Faults at Bus 3
Primary protection: the REC Can-Thanh 36 OCR	*Primary protection*: the REC Can Thanh–Hao Vo OCR	*Primary protection*: the REC Can-Thanh 163 OCR
Backup protection: the 472 Can-Thanh OCR	*Backup protection*: the REC Can Thanh 36 OCR and the 472 Can-Thanh OCR	*Backup protection*: the REC Can Thanh–Hao Vo OCR and the REC Can Thanh 36 OCR
Maximum fault currents at Bus 7 are detected by the REC Can-Thanh 36 OCR: [21.5; 21.4; 20.1; 18.6] kA	Maximum fault currents at Bus 5 are detected by the REC Can Thanh–Hao Vo OCR: [3.2; 2.8; 2.7; 1.3] kA	Maximum fault currents at Bus 3 are detected by the REC Can-Thanh 163 OCR: [2.6; 2.4; 2.2; 1.15] kA

(c) Scenario 3–A DN topology for the second FISR plan:
As referred to the dashed line (d) in Figure 10; the REC Hao-Vo recloser is opened to isolate the fault at Bus 2; the REC Can Thanh–Hao Vo recloser is opened; and then a CB GEN of DG sources is closed.

Faults at Bus 5	Faults at Bus 3
Primary protection: the OCR at the CB GEN	*Primary protection*: the REC Can-Thanh 163 OCR
Backup protection: None	*Backup protection*: the OCR at the CB GEN
Maximum fault currents at Bus 5 are detected by the OCR at the CB GEN: [1.1; 1.0; 0.9; 0.8] kA	Maximum fault currents at Bus 3 are detected by the REC Can-Thanh 163 OCR: [0.85; 0.80; 0.75; 0.70] kA

(d) Scenario 4–A DN topology for the third FISR plan:
As referred to the two dashed lines (c) and (d) in Figure 10; the REC Hao-Vo recloser is opened to isolate the fault at Bus 2; and then both the REC Can Thanh–Hao Vo recloser and the CB GEN are simultaneously closed.

Faults at Bus 7	Faults at Bus 5	Faults at Bus 3
Primary protection: the REC Can-Thanh 36 OCR	*Primary protection*: the REC Can Thanh–Hao Vo OCR and the CB GEN OCR	*Primary protection*: the REC Can-Thanh 163 OCR
Backup protection: the 472 Can-Thanh OCR	*Backup protection*: the REC Can-Thanh 36 OCR and the 472 Can-Thanh OCR	*Backup protection*: the OCR at CB GEN, the REC Can Thanh–Hao Vo OCR, and the REC Can-Thanh 36 OCR
Maximum fault currents at Bus 7 are detected by the REC Can-Thanh 36 OCR: [21.7; 21.6; 20.4; 18.8] kA	Maximum fault currents at Bus 5 are detected by the REC Can Thanh–Hao Vo OCR: [3.4; 3.1; 3.0; 1.8] kA and by the OCR at the CB GEN: [0.85; 0.80; 0.75; 0.70] kA	Maximum fault currents at Bus 3 are detected by the REC Can-Thanh 163 OCR: [2.9; 2.7; 2.6; 1.7] kA

4.2. Protection Coordination Results in Scenario 1

Optimal protection coordination results of the OCRs in Scenario 1 which are calculated by the GA, GSA and hybrid PSO-GSA are indicated in Table 4. In this table, I_f is the tripping-current threshold selected for the 50P and 50G functions of the OCRs, as referred to Section 2.3.5. $t_{trip}^{defined}$ is the pre-defined tripping time of the 50P and 50G functions of the OCRs. Moreover, the TDS multiplier and *A*, *B*, and *C* coefficients of the 51P and 51G OC functions of the OCRs are shown in the table. It is noted that the CTI is selected to 0.30~0.35 s. For example, when a 3ph-G fault is at Bus 2, the 471 Hao-Vo relay is the primary relay while there is not any backup relay. The tripping time of the 471 Hao-Vo relay is 0.01 s to clear the fault at Bus 2. On the other hand, when a typical 3ph-G fault is given at the end point of the feeder (at Bus 5) as referred to a dashed line (a) in Figure 10, the REC Can-Thanh 163 OCR is the primary relay while the REC Hao-Vo relay and the 471 Hao-Vo relay are the backup relays. Based on the optimal setting parameters obtained by the ASPC system for these OCRs, their tripping time t_{trip} to clear the fault at Bus 5 are also indicated in Table 4. The results have demonstrated that the CTI constraints from Equations (3)–(5) are completely satisfied.

Table 4. Optimal protection coordination results of the OCRs in Scenario 1 and their tripping time when the fault is at Bus 5.

The OCRs	**50P**		**50G**		**51P**				**51G**			
						GA	**GSA**	**PSO-GSA**		**GA**	**GSA**	**PSO-GSA**
The 471 Hao-Vo OCR	I_f	20,469 A	I_f	22,340 A	A	13.06	57.94	68.34	A	50.75	89.14	93.49
					B	2.14	2.44	2.49	B	2.12	2.31	2.50
					C	1.11	0.38	1.43	C	0.71	0.83	1.48
	$t_{trip}^{defined}$	0.01 s	$t_{trip}^{defined}$	0.01 s	TDS	0.41	0.33	0.21	TDS	0.067	0.05	0.05
					The tripping time of the relay for the typical fault at Bus 5:							
					t_{trip}	0.62 s	0.61 s	0.62 s	t_{trip}	0.62 s	0.61 s	0.62 s
The REC Hao-Vo OCR	I_f	6488 A	I_f	10316 A	A	36.23	72.58	5.81	A	78.07	74.84	13.74
					B	1.32	2.42	2.50	B	2.36	2.32	2.37
					C	1.07	0.22	0.00	C	0.25	0.53	0.00
	$t_{trip}^{defined}$	0.01 s	$t_{trip}^{defined}$	0.01 s	TDS	0.05	0.15	2.24	TDS	0.03	0.03	0.17
					The tripping time of the relay for the typical fault at Bus 5:							
					t_{trip}	0.32 s	0.31 s	0.31 s	t_{trip}	0.32 s	0.31 s	0.31 s
The REC Can-Thanh 163 OCR	I_f	2340 A	I_f	4246 A	A	0.036	59.36	0.01	A	3.95	59.97	0.01
					B	2.14	2.31	0.48	B	2.39	2.29	1.72
					C	1.00	0.01	0.00	C	0.3	0.72	0.01
	$t_{trip}^{defined}$	0.01 s	$t_{trip}^{defined}$	0.01 s	TDS	0.01	0.01	1.17	TDS	0.01	0.01	0.01
					The tripping time of the relay for the typical fault at Bus 5:							
					t_{trip}	0.01 s	0.01 s	0.01 s	t_{trip}	0.01 s	0.01 s	0.01 s

Figure 11 shows the optimal protection characteristic curves of the OC relays in Scenario 1 of the DN. Regarding the obtained coordination results of 51P/G and 50P/G functions, Relay 1 represents the OC protection function embedded into the REC Can-Thanh 163 recloser; Relay 2 represents the OC protection function integrated into the REC Hao-Vo recloser; and Relay 3 indicates the OC protection function used to control the 471 Hao-Vo circuit breaker. When the 3ph-G fault occurs at Bus 2, the maximum 3ph-G fault current is 25.8 kA as shown in Table 3, so the 471 Hao-Vo OCR only takes 0.01 s to send a tripping signal to the 471 Hao-Vo CB. When the 3ph-G fault occurs at Bus 5, the maximum 3ph-G fault current is 4.9 kA as indicated in Table 3, so the primary operation time of the REC Can-Thanh 163 OCR is 0.01 s, the backup operation times of the REC Hao-Vo OCR and the 471 Hao-Vo OCR are 0.31 s~0.32 s and 0.61 s~0.62 s, respectively, according to results of three different algorithms as seen at the table and the left-side of Figure 11. Additionally, I_{trip} represents the tripping-current threshold of 50P/G functions of the REC Can-Thanh 163 OCR. It can be observed that the characteristic curves of 51P/G functions in the OCR are certainly followed by the exponential function. However, there is a proposed protection combination of 51P/G and 50P/G functions, the inverse-time characteristic curves of 51P/G functions are terminated at the tripping-current threshold of

50P/G functions. In other words, in the same OCR, the 51P/G functions will be neglected when the fault current exceeds the tripping-current threshold of 50P/G functions. That could be an interesting contribution of this study regarding the coordination combination of 51P/G and 50P/G functions. For example, when considering the REC Hao-Vo OCR, the tripping current I_f of its 50P function is set to 6488 A as shown in Table 4. Therefore, this tripping current is considered as a terminated point of the inverse-time characteristic curve of the 51P function in the same REC Hao-Vo OCR. That can be clearly illustrated by the curve of Relay 2, i.e., a red curve of the plots at the right-side of Figure 11.

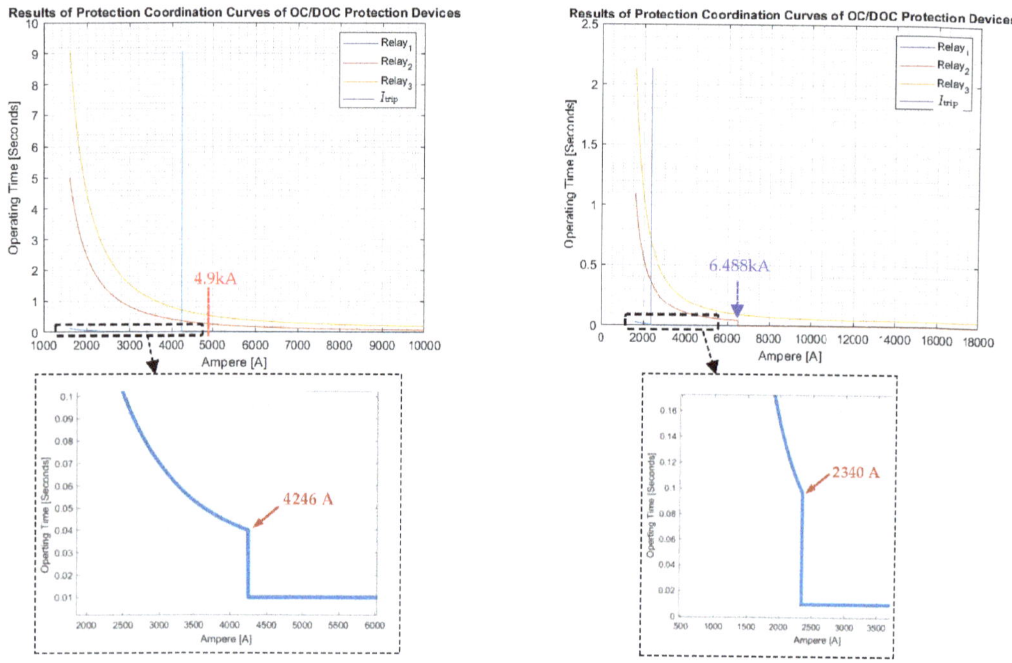

(**a**) The OC protection coordination results of 50G and 51G functions (at the left-side) and 50P and 51P functions (at the right-side) are obtained by the GSA.

Figure 11. *Cont.*

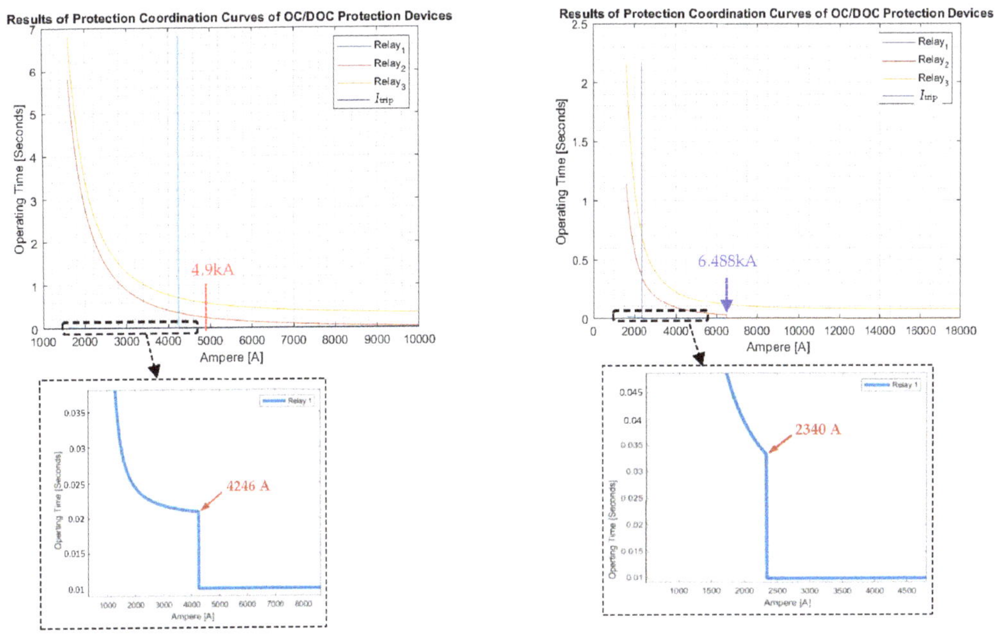

(b) The OC protection coordination results of 50G and 51G functions (at the left-side) and 50P and 51P functions (at the right-side) are obtained by the hybrid PSO-GSA.

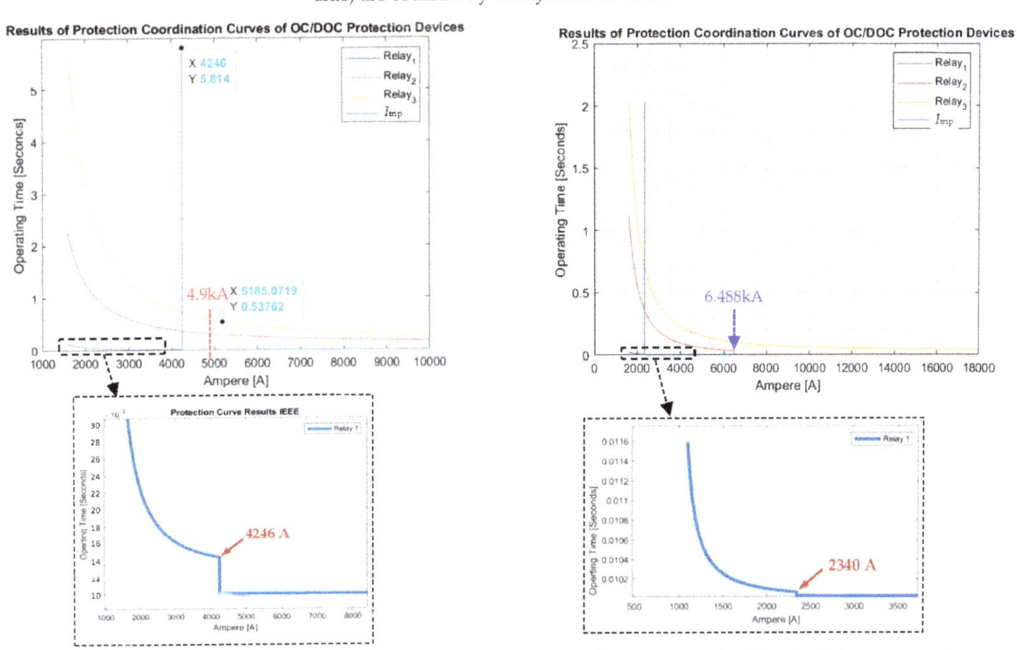

(c) The OC protection coordination results of 50G and 51G functions (at the left-side) and 50P and 51P functions (at the right-side) are obtained by the GA.

Figure 11. Optimal protection characteristic curves of the OC relays in Scenario 1 of the 22 kV distribution system.

4.3. Protection Coordination Results in Scenario 2

Scenario 2 is a distribution network topology of the first FISR plan after the 3ph-G fault at Bus 2 is cleared. As referred to the dashed line (c) in Figure 10, the REC Hao-Vo recloser is opened to isolate the fault at Bus 2. Then the LBS Can-Thanh 97 is closed to restore the power of interrupted customers while a CB GEN circuit breaker of DG sources is still opened. The optimal protection coordination results of the OCRs in Scenario 2 which are calculated by the GA, GSA and hybrid PSO-GSA are indicated in Table 5. In this table, I_f is the tripping-current threshold selected for the 50P and 50G functions of the OCRs including the REC Can-Thanh 36 OCR, the REC Can-Thanh Hao-Vo OCR, and the REC Can-Thanh 163 OCR. $t_{trip}^{defined}$ is the pre-defined tripping time of the 50P and 50G functions of these OCRs. Moreover, the TDS multiplier and A, B, and C coefficients of the 51P and 51G OC functions of the OCRs are shown in the table. The CTI is still selected to 0.30~0.35 s. It is worth noting that the parameter settings of the REC Can-Thanh 163 OCR must be simultaneously updated to be adaptable to Scenario 2 instead of Scenario 1. To validate the optimal setting parameters obtained by the ASPC system for the above OCRs, a 3ph-G fault is given at Bus 3. The REC Can-Thanh 163 OCR is the primary relay to detect and clear this fault while the REC Can-Thanh 36 OCR and the REC Can-Thanh Hao-Vo OCR operate as the backup relays. Their tripping time t_{trip} to clear the fault at Bus 3 are also indicated in Table 5. The achieved results have demonstrated that the CTI constraints from Equations (3)–(5) are completely satisfied.

Table 5. Optimal protection coordination results of the OCRs in Scenario 2 and their tripping time when the fault is at Bus 3.

The OCRs	50P		50G		51P				51G			
						GA	GSA	PSO-GSA		GA	GSA	PSO-GSA
The REC Can-Thanh 36 OCR					A	6.71	52.45	126.51	A	9.23	33.59	73.50
	I_f	18,595 A	I_f	20,076 A	B	2.31	2.31	2.07	B	1.74	1.82	2.50
					C	0.09	0.93	0.02	C	0.11	0.98	1.50
					TDS	0.73	0.09	0.03	TDS	0.024	0.01	0.01
	$t_{trip}^{defined}$	0.01 s	$t_{trip}^{defined}$	0.01 s	The tripping time of the relay for the typical fault at Bus 3:							
					t_{trip}	0.65 s	0.62 s	0.65 s	t_{trip}	0.65 s	0.62 s	0.65 s
The REC Can-Thanh Hao-Vo OCR					A	7.77	44.37	0.60	A	0.01	30.55	0.01
	I_f	1286 A	I_f	2755 A	B	2.46	1.79	2.28	B	0.60	2.19	1.75
					C	0.47	0.28	0.02	C	0.89	0.79	0.74
					TDS	0.26	0.03	1.26	TDS	0.37	0.01	0.46
	$t_{trip}^{defined}$	0.01 s	$t_{trip}^{defined}$	0.01 s	The tripping time of the relay for the typical fault at Bus 3:							
					t_{trip}	0.33 s	0.32 s	0.33 s	t_{trip}	0.33 s	0.32 s	0.33 s
The REC Can-Thanh 163 OCR					A	0.09	60.60	0.01	A	0.01	0.74	0.01
	I_f	1167 A	I_f	2501 A	B	0.44	2.33	1.39	B	1.51	0.89	2.06
					C	0.30	0.46	0.00	C	0.97	0.87	0.01
					TDS	0.03	0.01	3.00	TDS	0.01	0.01	0.43
	$t_{trip}^{defined}$	0.01 s	$t_{trip}^{defined}$	0.01 s	The tripping time of the relay for the typical fault at Bus 3:							
					t_{trip}	0.01 s	0.01 s	0.01 s	t_{trip}	0.01 s	0.01 s	0.01 s

Figure 12 shows the optimal protection characteristic curves of the OC relays in Scenario 2 of the DN. Regarding the obtained coordination results of 51P/G and 50P/G functions, Relay 1 represents the OC protection function embedded into the REC Can-Thanh 163 recloser; Relay 2 represents the OC protection function integrated into the REC Can-Thanh Hao-Vo recloser; and Relay 3 indicates the OC protection function used to control the REC Can-Thanh 36 recloser. When the 3ph-G fault occurs at Bus 3, the maximum 3ph-G fault current is 2.6 kA as indicated in Table 3, so the primary operation time of the REC Can-Thanh 163 OCR is 0.01 s. The backup operation times of the REC Can-Thanh Hao-Vo OCR and the REC Can-Thanh 36 OCR are 0.32~0.33 s and 0.62~0.65 s, respectively, according to results of three different algorithms as seen at the table and the left-side of Figure 12. Additionally, I_{trip} represents the tripping-current threshold of 50P/G protection functions of the REC Can-Thanh 163 OCR. By further considering the REC Can-Thanh 163 OCR, the tripping current I_f of its 50G function is set to 2501 A as shown in Table 5. Thus, this tripping current is considered as a terminated point of the inverse-time characteristic curve of the 51G function in the same REC Can-Thanh 163 OCR. Similarly, the tripping current I_f of the 50G function of the REC Can-Thanh Hao-Vo OCR which is 2755 A is considered as a finishing point of the inverse-time characteristic curve of its 51G function, as clearly illustrated by the curve of Relay 2, i.e., a red curve of the plots at the left-side of Figure 12.

4.4. Protection Coordination Results in Scenario 3

Scenario 3 is a distribution network topology of the second FISR plan after the 3ph-G fault at Bus 2 is cleared. As referred to the dashed line (d) in Figure 10, the REC Hao-Vo recloser is opened to isolate the fault at Bus 2. Then the LBS Can-Thanh 97 is closed to restore the power of interrupted customers while the REC Can-Thanh–Hao-Vo recloser is opened and a CB GEN circuit breaker of DG sources is closed. The optimal protection coordination results of the OCRs in Scenario 3 which are calculated by the GA, GSA and hybrid PSO-GSA are indicated in Table 6. In this table, I_f is the tripping-current threshold selected for the 50P and 50G functions of the OCRs including the CB GEN OCR and the REC Can-Thanh 163 OCR. $t_{trip}^{defined}$ is the pre-defined tripping time of the 50P and 50G functions of these OCRs. Moreover, the TDS multiplier and A, B, and C coefficients of the 51P and 51G OC functions of the above OCRs are shown in the table. The CTI is still selected to 0.30~0.35 s. It is worth noting that the parameter settings of the REC Can-Thanh 163 OCR must be simultaneously updated to be adaptable to Scenario 3 instead of Scenario 1 or Scenario 2. To validate the optimal setting parameters obtained by the ASPC system for the above OCRs, a 3ph-G fault is given at Bus 3. The REC Can-Thanh 163 OCR operates as the primary protection to detect and clear this fault while the CB GEN OCR operates as the backup protection. Their tripping time t_{trip} to clear the fault at Bus 3 are also indicated in Table 6. The achieved results have demonstrated that the CTI constraints from Equations (3)–(5) are completely satisfied.

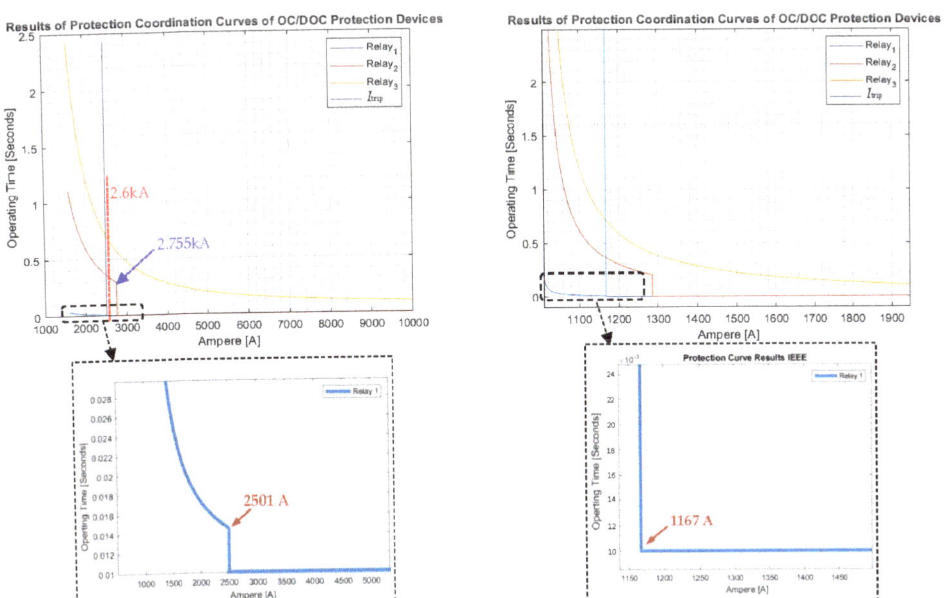

(a) The OC protection coordination results of 50G and 51G functions (at the left-side) and 50P and 51P functions (at the right-side) are obtained by the GSA.

Figure 12. *Cont.*

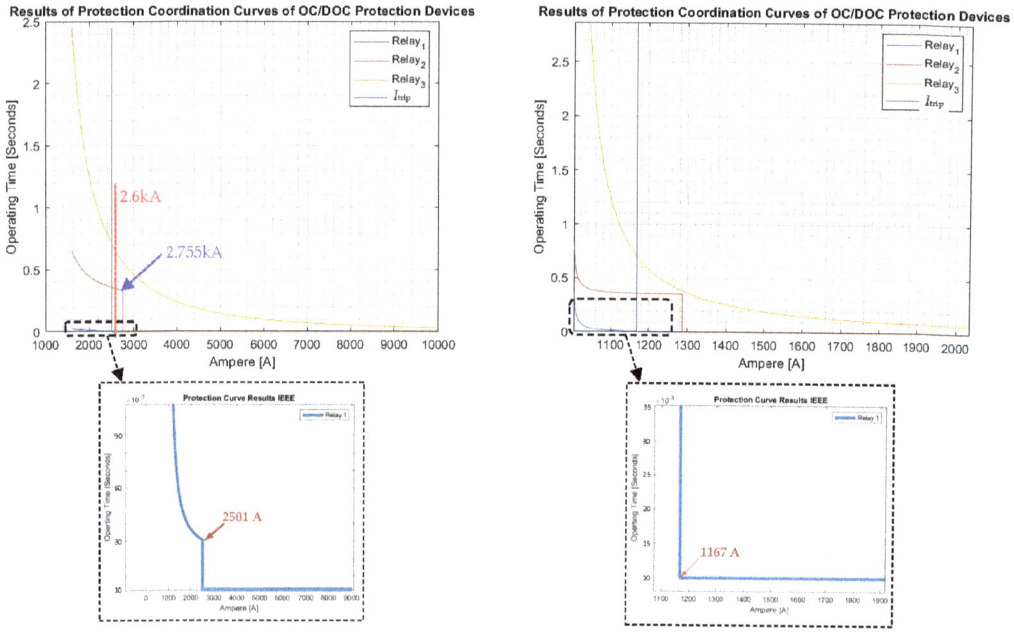

(**b**) The OC protection coordination results of 50G and 51G functions (at the left-side) and 50P and 51P functions (at the right-side) are obtained by the hybrid PSO-GSA.

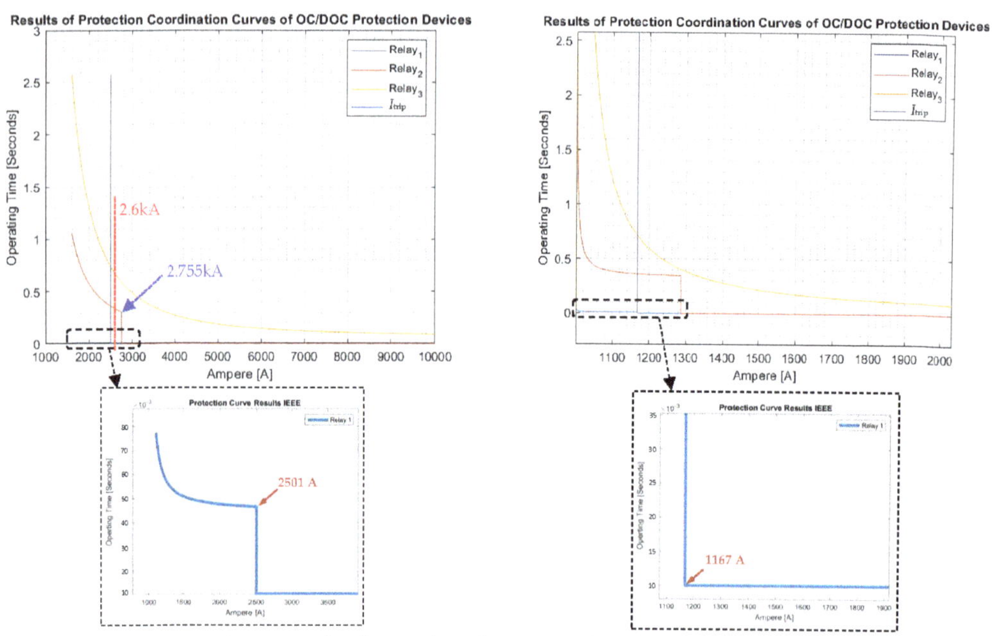

(**c**) The OC protection coordination results of 50G and 51G functions (at the left-side) and 50P and 51P functions (at the right-side) are obtained by the GA.

Figure 12. Optimal protection characteristic curves of the OC relays in Scenario 2 of the 22 kV distribution system.

Table 6. Optimal protection coordination results of the OCRs in Scenario 3 and their tripping time when the fault is at Bus 3.

The OCRs	50P		50G		51P				51G			
						GA	GSA	PSO-GSA		GA	GSA	PSO-GSA
A CB GEN OCR	I_f	816 A	I_f	910 A	A	4.39	57.46	0.01	A	0.23	67.95	0.01
					B	2.41	2.17	2.50	B	2.45	2.24	2.25
					C	1.18	0.59	0.05	C	0.76	0.80	0.80
					TDS	0.09	0.01	7.16	TDS	0.35	0.01	0.45
	$t_{trip}^{defined}$	0.01 s	$t_{trip}^{defined}$	0.01 s	The tripping time of the relay for the typical fault at Bus 3:							
					t_{trip}	0.35 s	0.32 s	0.35 s	t_{trip}	0.35 s	0.32 s	0.35 s
The REC Can-Thanh 163 OCR	I_f	700 A	I_f	750 A	A	1.11	59.93	0.01	A	0.01	47.43	0.01
					B	2.03	2.05	2.31	B	2.45	2.22	2.50
					C	0.00	0.81	0.01	C	1.01	0.66	0.00
					TDS	0.01	0.01	0.69	TDS	0.01	0.01	1.02
	$t_{trip}^{defined}$	0.01 s	$t_{trip}^{defined}$	0.01 s	The tripping time of the relay for the typical fault at Bus 3:							
					t_{trip}	0.01 s	0.01 s	0.01 s	t_{trip}	0.01 s	0.01 s	0.01 s

4.5. Protection Coordination Results in Scenario 4

Scenario 4 is a distribution network topology of the third FISR plan after the 3ph-G fault at Bus 2 is cleared. As referred to the two dashed lines (c) and (d) in Figure 10, the REC Hao-Vo recloser is opened to isolate the fault at Bus 2. The LBS Can-Thanh 97 is then closed to restore the power of interrupted customers while the REC Can-Thanh–Hao-Vo recloser and the CB GEN circuit breaker are closed to allow the interlinking operation of both the feeder supplied by the 110/22 kV An-Nghia power substation and the other feeder supplied by the DG sources. The optimal protection coordination results of the OCRs in Scenario 4 which are calculated by the GA, GSA and hybrid PSO-GSA are indicated in Table 7. In this table, I_f is the tripping-current threshold selected for the 50P and 50G functions of the OCRs including the REC Can-Thanh 36 OCR, the REC Can-Thanh–Hao-Vo OCR, the CB GEN OCR and the REC Can-Thanh 163 OCR. $t_{trip}^{defined}$ is the pre-defined tripping time of the 50P and 50G functions of these OCRs. Moreover, the TDS multiplier and *A*, *B*, and *C* coefficients of the 51P and 51G OC functions of the above OCRs are shown in the table. The CTI is still selected to 0.30~0.35 s. It is worth noting that the parameter settings of the REC Can-Thanh 163 OCR must be simultaneously updated to be adaptable to Scenario 4 instead of Scenario 1, Scenario 2, or Scenario 3. To validate the optimal setting parameters obtained by the ASPC system for the above OCRs, a 3ph-G fault is given at Bus 3. The REC Can-Thanh 163 OCR operates as the primary protection to detect and clear this fault while the CB GEN OCR operates as the backup protection on the feeder supplied by the DG sources and the REC Can-Thanh 36 OCR and the REC Can-Thanh–Hao-Vo OCR are the backup relays on the other feeder supplied by the 110/22 kV An-Nghia power substation. Their tripping time t_{trip} to clear the 3ph-G fault at Bus 3 are also indicated in Table 7. The achieved results have demonstrated that the CTI constraints from Equations (3)–(5) are completely satisfied.

Figure 13 shows the optimal protection characteristic curves of the OC relays in Scenario 4 of the real 22 kV DG-contained DN. Regarding the obtained coordination results of 51P/G and 50P/G functions, Relay 1 represents the OC protection function embedded into the REC Can-Thanh 163 recloser; Relay 2 represents the OC protection function integrated into the REC Can Thanh–Hao Vo recloser; and Relay 3 indicates the OC protection function used to control the REC Can-Thanh 36 OCR recloser. When the 3ph-G fault occurs at Bus 3, the maximum 3ph-G fault current is 2.9 kA as indicated in Table 3, so the primary operation time of the REC Can-Thanh 163 OCR is 0.01 s. The backup operation times of the REC Can-Thanh Hao-Vo OCR and the REC Can-Thanh 36 OCR are about 0.31~0.33 s and 0.63~0.65 s, respectively, according to results of three different algorithms as seen in the table and Figure 13. The backup operation time of the CB GEN OCR is about 0.35 s. Additionally,I_{trip} represents the tripping-current threshold of 50P/G protection functions of the REC Can-Thanh 163 OCR. By further considering the REC Can-Thanh 163 OCR, the tripping current I_f of its 50G function is set to 2640 A as shown in Table 7. Thus, this tripping current is considered as a terminated point of the inverse-time characteristic curve of the 51G function in the same REC Can-Thanh 163 OCR. Similarly, the tripping current I_f of the 50G function of the REC Can-Thanh Hao-Vo OCR which is 2980 A is considered as a finishing point of the inverse-time characteristic curve of its 51G function, as clearly illustrated by the curve of Relay 2, i.e., the red curves in Figure 13.

Table 7. Optimal protection coordination results of the OCRs in Scenario 4 and their tripping time when the fault is at Bus 3.

The OCRs	50P		50G		51P				51G			
						GA	GSA	PSO-GSA		GA	GSA	PSO-GSA
The REC Can-Thanh 36 OCR					A	10.84	42.61	15.58	A	21.36	70.52	149.95
	I_f	18,791 A	I_f	20,376 A	B	1.24	2.39	1.47	B	2.21	2.40	2.50
					C	1.08	0.39	1.12	C	0.94	0.60	0.08
					TDS	0.13	0.14	0.12	TDS	0.19	0.09	0.05
	$t_{trip}^{defined}$	0.01s	$t_{trip}^{defined}$	0.01 s	The tripping time of the relay for the typical fault at Bus 3:							
					t_{trip}	0.64 s	0.63 s	0.65 s	t_{trip}	0.64 s	0.63 s	0.65 s
The REC Can-Thanh–Hao-Vo OCR					A	8.43	72.23	0.01	A	18.43	77.10	1.52
	I_f	1798 A	I_f	2980 A	B	2.42	2.44	0.02	B	2.31	2.37	1.83
					C	0.50	0.62	0.29	C	0.97	0.77	0.01
					TDS	0.26	0.05	0.50	TDS	0.12	0.04	1.16
	$t_{trip}^{defined}$	0.01 s	$t_{trip}^{defined}$	0.01 s	The tripping time of the relay for the typical fault at Bus 3:							
					t_{trip}	0.32 s	0.31 s	0.33 s	t_{trip}	0.32 s	0.31 s	0.33 s
The REC Can-Thanh 163 OCR					A	1.10	53.82	0.01	A	1.93	39.27	0.01
	I_f	1672 A	I_f	2640 A	B	1.73	1.87	0.71	B	2.37	2.28	1.31
					C	0.00	0.65	0.01	C	0.11	0.77	0.00
					TDS	0.04	0.01	0.52	TDS	0.03	0.01	2.98
	$t_{trip}^{defined}$	0.01 s	$t_{trip}^{defined}$	0.01 s	The tripping time of the relay for the typical fault at Bus 3:							
					t_{trip}	0.01 s	0.01 s	0.01 s	t_{trip}	0.01 s	0.01 s	0.01 s
A CB GEN OCR					A	4.39	57.46	0.01	A	0.23	67.95	0.01
	I_f	816 A	I_f	910 A	B	2.41	2.17	2.50	B	2.45	2.24	2.25
					C	1.18	0.59	0.05	C	0.76	0.80	0.80
					TDS	0.09	0.01	7.16	TDS	0.35	0.01	0.45
	$t_{trip}^{defined}$	0.01 s	$t_{trip}^{defined}$	0.01 s	The tripping time of the relay for the typical fault at Bus 3:							
					t_{trip}	0.35 s	0.32 s	0.35 s	t_{trip}	0.35 s	0.32 s	0.35 s

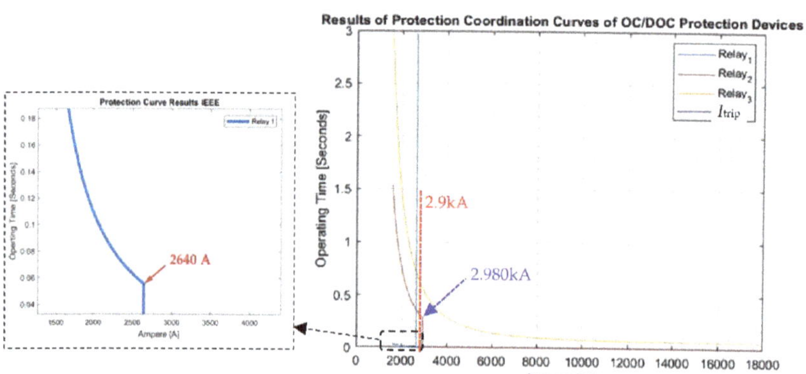

(**a**) The OC protection coordination results of 50G and 51G functions are obtained by the GSA.

Figure 13. *Cont.*

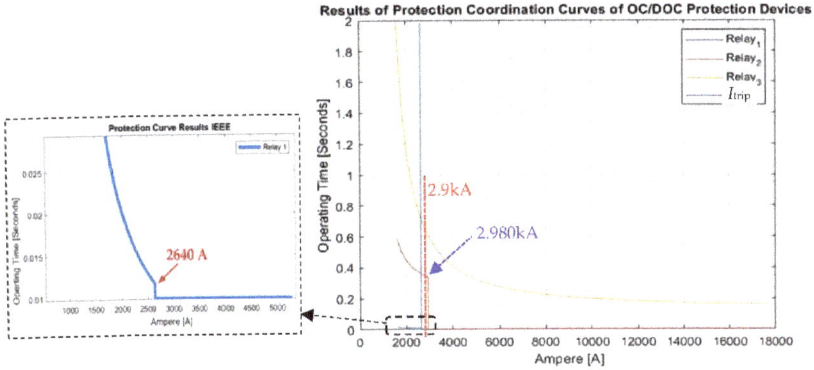

(b) The OC protection coordination results of 50G and 51G functions are obtained by the hybrid PSO-GSA.

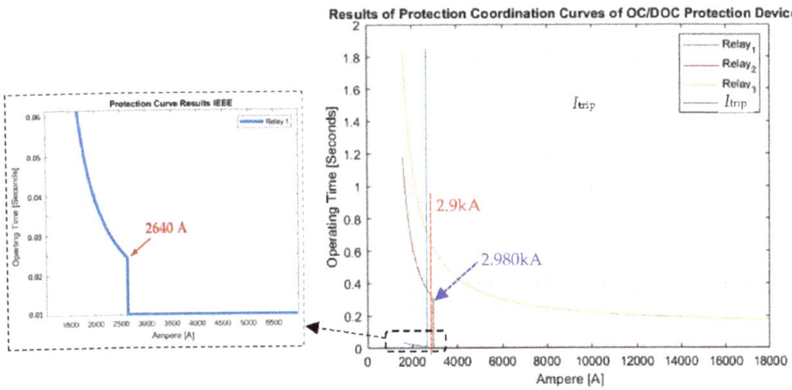

(c) The OC protection coordination results of 50G and 51G functions are obtained by the GA.

Figure 13. Optimal protection characteristic curves of the OC relays in Scenario 4 of the 22 kV distribution system.

In summary, the optimal protection coordination results of the related OCRs regarding four different operation scenarios of the real 22 kV DG-based distribution system have been analyzed and validated in this section. These relay coordination results of Scenarios 1, 2, 3 and 4 can be briefly concluded as follows.

- Three meta-heuristic algorithms, GSA, hybrid PSO-GSA and GA have efficiently solved the adaptable and scalable protection coordination problem of the OCRs in the DG-integrated DN as demonstrated in Figures 11–13. The GSA can be a better optimization solution than others because its processing time is quite fast and the difference in the tripping time between the desired one and the calculated one of the related OCRs for Scenarios 1, 2, 3 and 4 is also small as indicated in Table 8. Moreover, although the computation time of the GA is the fastest, the tripping time difference between the desired one and the calculated one of the GA is higher than that of the GSA.
- By applying the proposed ASPC approach for the OCRs, the optimal setting parameters of 50P, 50G, 51P, and 51G protection functions in the same OCR have been properly calculated and updated by the SCADA system whenever the 22 kV DG-integrated DN topology is changed by the faults.
- For each scenario of the DN, a 3ph-G fault has been assumed to occur at the remote end of the feeder, which is the basis to evaluate the obtained relay coordination results. The CTI constraints have been checked for both the OCR pairs and the OCR triples. More

clearly, the primary and backup operation times of the related OCRs regarding the 3ph-G fault at the remote end of the feeder have been already analyzed in Tables 4–7.

Table 8. Comparison in the calculated tripping time of the related OCRs and the computation time of GA, GSA and hybrid PSO-GSA algorithms for Scenarios 1, 2, 3 and 4.

(a) Scenario 1

Overcurrent relays	Desired tripping time	Calculated tripping time (in seconds) of the related OCRs for the 3ph-G fault at Bus 5 of Scenario 1			Computation time (in seconds) of the optimization algorithms		
		GA	GSA	Hybrid PSO-GSA	GA	GSA	Hybrid PSO-GSA
The REC Can-Thanh 163 OCR	0.01 s	0.01 s	0.01 s	0.01 s	5 s	8 s	14 s
The REC Hao-Vo OCR	0.31 s	0.32 s	0.31 s	0.31 s			
The 471 Hao-Vo OCR	0.61 s	0.62 s	0.61 s	0.62 s			
The highest difference in the tripping time between the desired one and the calculated one among the OCRs		0.01 s	0 s	0.01 s			

(b) Scenario 2

Overcurrent relays	Desired tripping time	Calculated tripping time (in seconds) of the related OCRs for the 3ph-G fault at Bus 3 of Scenario 2			Computation time (in seconds) of the optimization algorithms		
		GA	GSA	Hybrid PSO-GSA	GA	GSA	Hybrid PSO-GSA
The REC Can-Thanh 163 OCR	0.01 s	0.01 s	0.01 s	0.01 s	6 s	7 s	24 s
The REC Can-Thanh Hao-Vo OCR	0.31 s	0.33 s	0.32 s	0.33 s			
The REC Can-Thanh 36 OCR	0.61 s	0.65 s	0.62 s	0.65 s			
The highest difference in the tripping time between the desired one and the calculated one among the OCRs		0.04 s	0.01 s	0.04 s			

(c) Scenario 3

Overcurrent relays	Desired tripping time	Calculated tripping time (in seconds) of the related OCRs for the 3ph-G fault at Bus 3 of Scenario 3			Computation time (in seconds) of the optimization algorithms		
		GA	GSA	Hybrid PSO-GSA	GA	GSA	Hybrid PSO-GSA
The REC Can-Thanh 163 OCR	0.01 s	0.01 s	0.01 s	0.01 s	3 s	6 s	7 s
The CB GEN OCR	0.31 s	0.35 s	0.32 s	0.35 s			
The highest difference in the tripping time between the desired one and the calculated one among the OCRs		0.04 s	0.01 s	0.04 s			

(d) Scenario 4

Overcurrent relays	Desired tripping time	Calculated tripping time (in seconds) of the related OCRs for the 3ph-G fault at Bus 3 of Scenario 4			Computation time (in seconds) of the optimization algorithms		
		GA	GSA	Hybrid PSO-GSA	GA	GSA	Hybrid PSO-GSA
The REC Can-Thanh 163 OCR	0.01 s	0.01 s	0.01 s	0.01 s	5 s	7.8 s	16 s
The REC Can-Thanh Hao-Vo OCR	0.31 s	0.32 s	0.31 s	0.33 s			
The CB GEN OCR	0.31 s	0.35 s	0.32 s	0.35 s			
The REC Can-Thanh 36 OCR	0.61 s	0.64 s	0.63 s	0.65 s			
The highest difference in the tripping time between the desired one and the calculated one among the OCRs		0.04 s	0.02 s	0.04 s			

5. Conclusions

The protection coordination of OCRs becomes more complicated when a distribution network allows high penetration of DGs. Therefore, this paper has proposed an adaptive and scalable protection coordination (ASPC) system for the OCRs in the DG-based DN with two main performance stages. The first stage is to improve the reliability of fault-current calculation results through determining the min-max confidence interval of fault current for each different fault type including 3ph-G, 2ph-G, ph-ph, and 1ph-G faults. The maximum and minimum confidence thresholds of fault current are used to select tripping and pick-up thresholds of definite-time and inverse-time OC functions in the same OCR, respectively. This contribution is to increase the scalability of the relay protection coordination system. In addition, the proposed ASPC approach for the OCRs has high adaptability to different fault current contributions from IBDGs and RBDGs and the available 'on-grid' and 'off-grid' operation modes for these DG units. The second stage is to calculate the *A*, *B*, and *C* certain coefficients and the Time-Dial-Setting (TDS) multiplier for the inverse-time OC functions in the OCR. Three common algorithms, GSA, hybrid PSO-GSA and GA, have been used to find the optimal setting parameters of the OCRs that are satisfied with all coordination constraint conditions. Moreover, the relay coordination results have shown an effective protection combination of the definite-time OC functions (50P and 50G) and the inverse-time OC functions (51P and 51G) in the same OCR to get the adaptable and scalable DN protection system. A real 22 kV DG-integrated distribution network has been simulated by ETAP software to analyze the min-max fault-current values of each fault type. In other words, this real 22 kV DN is used as a test-bed to validate the relay coordination results obtained by the ASPC system. The relay coordination results also show the visibility and superiority of the GSA because it has the fast computation time and the highest stability in comparison with the hybrid PSO-GSA and the GA. The tripping time difference between the desired one and the calculated one of the GSA is also the lowest. Last but not least, it could be interesting that this study has presented the protection coordination results of the related OCRs corresponding to all possible FISR plans of the DG-contained DN after a typical fault is cleared.

Author Contributions: D.M.B., T.P.N., and H.N. contribute conceptualization, methodology, and writing—review and editing of the paper. P.D.L. contributes investigation, data collection and analysis, software, and writing—original draft preparation of the paper. All authors have read and agreed to the published version of the manuscript.

Funding: This research was funded by the Ministry of Education and Training (MOET) and Vietnamese-German University (VGU), a grant number of B2020-VGU-01 and an issued number of 103/QĐ-BGDDT by 21st February 2020.

Institutional Review Board Statement: Not applicable.

Informed Consent Statement: Not applicable.

Acknowledgments: The authors sincerely thank to the Ministry of Education and Training (MOET), Vietnamese-German University (VGU), and Ho Chi Minh City Power Corporation for data supply, administrative and technical support in this research paper.

Conflicts of Interest: The authors declare no conflict of interest.

References

1. Bui, D.M.; Lien, K.Y.; Yu, S.L.; Chen, Y.C.; Lu, C.M.; Chan, C.M.; Chang, Y.R. Investigate dynamic and transient charac-teristics of microgrid operation and develop a fast-scalable-adaptable algorithm for fault protection system. *Electr. Power Syst. Res.* **2015**, *120*, 214–233. [CrossRef]
2. Bui, D.M.; Chen, S.L. Fault protection solutions appropriately proposed for ungrounded low-voltage AC microgrids: Review and proposals. *Renew. Sustain. Energy Rev.* **2017**, *75*, 1156–1171. [CrossRef]
3. Lien, K.-Y.; Bui, D.M.; Chen, S.-L.; Zhao, W.-X.; Chang, Y.-R.; Lee, Y.-D.; Jiang, J.-L. A novel fault protection system using communication-assisted digital relays for AC microgrids having a multiple grounding system. *Int. J. Electr. Power Energy Syst.* **2016**, *78*, 600–625. [CrossRef]

4. Parhizi, S.; Lotfi, H.; Khodaei, A.; Bahramirad, S. State of the art in research on microgrids: A review. *IEEE Access.* **2015**, *3*, 890–925. [CrossRef]
5. Esreraig, M.; Mitra, J. Microgrid protection using system observer and minimum measurement set. *Int. Trans. Electr. Energy Syst.* **2014**, *25*, 607–622. [CrossRef]
6. Bui, D.M. Simplified and automated fault-current calculation for fault protection system of grid-connected low-voltage AC microgrids. *Int. J. Emerg. Electr. Power Syst.* **2017**, *18*, 2. [CrossRef]
7. Laaksonen, H. Protection Principles for Future Microgrids. *IEEE Trans. Power Electron.* **2010**, *25*, 2910–2918. [CrossRef]
8. Esreraig, M.; Lasseter, R.H. Microgrids. In Proceedings of the IEEE Power Engineering Society Winter Meeting, Singapore, 23–27 January 2002; pp. 27–31.
9. Firouz, Y.; Lobry, J.; Vallée, F.; Durieux, O. Numerical comparison of the effects of different types of distributed gener-ation units on overcurrent protection systems in MV distribution grid. *Int. J. Renew. Energy* **2014**, *69*, 271–283.
10. Le, D.P.; Bui, D.M.; Ngo, C.C.; Le, A.M.T. FLISR Approach for Smart Distribution Networks Using E-Terra Soft-ware—A Case Study. *Energies* **2018**, *11*, 3333. [CrossRef]
11. Le, P.D.; Bui, D.M.; Ngo, C.C.; Nguyen, T.P.; Huynh, C.P.; Doan, N.M. A Proposed FLISR Approach for Distribution Networks with Distributed Generators. *IEEE Inter. Con. Power Energy* **2020**, 125–130. [CrossRef]
12. Chabanloo, R.; Abyaneh, H.; Agheli, A.; Rastegar, H. Overcurrent relays coordination considering transient behaviour of fault current limiter and distributed generation in distribution power network. *IET Gener. Transm. Distrib.* **2011**, *5*, 903–911. [CrossRef]
13. Pandi, V.R.; Zeineldin, H.H.; Xiao, W. Determining Optimal Location and Size of Distributed Generation Resources Considering Harmonic and Protection Coordination Limits. *IEEE Trans. Power Syst.* **2012**, *28*, 1245–1254. [CrossRef]
14. Zeineldin, H.; Sharaf, H.; Ibrahim, D.; El-Zahab, E.D.A. Optimal protection coordination for meshed distribution sys-tems with DG using dual setting directional over-current relays. *IEEE Trans. Smart Grid* **2015**, *6*, 115–123. [CrossRef]
15. Mahat, P.; Chen, Z.; Bak-Jensen, B.; Bak, C.L. A Simple Adaptive Overcurrent Protection of Distribution Systems with Distributed Generation. *IEEE Trans. Smart Grid* **2011**, *2*, 428–437. [CrossRef]
16. Walke, S.B.; Nayana, N.J. Methods for relay coordination. In Proceedings of the 2017 International Conference on Computing Meth-Odologies and Communication (ICCMC), Erode, India, 18–19 July 2017; pp. 1027–1032.
17. Huchel, L.; Hatem, H.Z. Planning the coordination of directional overcurrent relays for distribution systems consid-ering DG. *IEEE Trans. Smart Grid* **2015**, *7*, 1642–1649. [CrossRef]
18. Ma, J.; Li, J.; Wang, Z. An adaptive distance protection scheme for distribution system with distributed generation. In Proceedings of the 2010 5th International Conference on Critical Infrastructure (CRIS), Beijing, China, 20–22 September 2010; pp. 1–4. [CrossRef]
19. Javadian, S.; Haghifam, M.-R.; Bathaee, S.; Firoozabad, M.F. Adaptive centralized protection scheme for distribution systems with DG using risk analysis for protective devices placement. *Int. J. Electr. Power Energy Syst.* **2013**, *44*, 337–345. [CrossRef]
20. Zamani, M.A.; Sidhu, T.S.; Yazdani, A. A protection strategy and microprocessor based relay for low-voltage mi-crogrids. *IEEE Trans. Power Deliv.* **2011**, *26*, 1873–1883. [CrossRef]
21. Nikkhajoei, H.; Lasseter, R.H. Microgrid Fault Protection Based on Symmetrical and Differential Current Components. Wisconsin Power Electronics Research Center, Dept. of Elec. and Com., Eng., University of Wisconsin-Madison. 2006. Available online: https://citeseerx.ist.psu.edu/viewdoc/download?doi=10.1.1.375.2266&rep=rep1&type=pdf (accessed on 1 August 2021).
22. Bin, L.; Yongli, L.; Zhiqian, B.; Klimek, A. Design of protection and control scheme for microgrid systems. In Proceedings of the 2009 44th International Universities Power Engineering Conference (UPEC), Glasgow, UK, 1–4 September 2009; pp. 1–5.
23. Best, R.J.; Morrow, D.J.; Crossley, P.A. Communication assisted protection selectivity for reconfigurable and islanded power networks. In Proceedings of the 2009 44th International Universities Power Engineering Conference (UPEC), Glasgow, UK, 1–4 September 2009; pp. 1–5.
24. Polajzer, B.; Pintaric, M.; Roser, M.; Stumberger, G. Protection of MV Closed-Loop Distribution Networks with Bi-Directional Overcurrent Relays and GOOSE Communications. *IEEE Access* **2019**, *7*, 165884–165896. [CrossRef]
25. Asadi, M.R.; Kouhsari, S.M. Optimal overcurrent relays coordination using particle-swarm-optimization methodology. In Proceedings of 2009 IEEE/PES Power Systems Conference and Exposition, Seattle, WA, USA, 15–18 March 2009; pp. 1–7.
26. Albrecht, R.E.; Nisja, M.J.; Feero, W.E.; Rockefeller, G.D.; Wagner, C.L. Digital Computer Protective Device Co-ordination Program I-General Program Description. *IEEE Trans. Power Appar. Syst.* **1964**, *83*, 402–410. [CrossRef]
27. Jenkins, L.; Khincha, H.; Shivakumar, S.; Dash, P. An application of functional dependencies to the topological analysis of protection schemes. *IEEE Trans. Power Deliv.* **1992**, *7*, 77–83. [CrossRef]
28. Urdaneta, A.J.; Nadira, R.; Jimenez, L.G.P. Optimal coordination of directional overcurrent relays in interconnected power systems. *IEEE Trans. Power Deliv.* **1988**, *3*, 903–911. [CrossRef]
29. Mohammadi, R.; Abyaneh, H.A.; Rudsari, H.M.; Fathi, S.H.; Rastegar, H. Overcurrent Relays Coordination Considering the Priority of Constraints. *IEEE Trans. Power Deliv.* **2011**, *26*, 1927–1938. [CrossRef]
30. Noghabi, A.S.; Mashhadi, H.R.; Sadeh, J. Optimal Coordination of Directional Overcurrent Relays Considering Different Network Topologies Using Interval Linear Programming. *IEEE Trans. Power Deliv.* **2010**, *25*, 1348–1354. [CrossRef]
31. Moravej, Z.; Adelnia, F.; Abbasi, F. Optimal coordination of directional overcurrent relays using NSGA-II. *Electr. Power Syst. Res.* **2015**, *119*, 228–236. [CrossRef]

32. Thangaraj, R.; Pant, M.; Abraham, A. New mutation schemes for differential evolution algorithm and their application to the optimization of directional over-current relay settings. *Appl. Math. Comput.* **2010**, *216*, 532–544. [CrossRef]
33. Thangaraj, R.; Pant, M.; Deep, K. Optimal coordination of over-current relays using modified differential evolution algorithms. *Eng. Appl. Artif. Intell.* **2010**, *23*, 820–829. [CrossRef]
34. Sirinivas, S.P.T.; Swarup, K.S.H. A new mixed integer linear programing formulation for protection relay coordination using disjunctive inequalities. *IEEE Power Energy Technol. Syst. J.* **2019**, *6*, 104–112. [CrossRef]
35. Hui, Y.; Fushuan, W.; Zhuhai, G.L. Optimal coordination of overcurrent relays in distribution systems with distributed generators based on differential evolution algorithm. *Eur. Trans. Elect. Power* **2013**, *23*, 1–12.
36. Yazdaninejadi, A.; Nazarpour, D.; Talavat, V. Optimal coordination of dual-setting directional over-current relays in multi-source meshed active distribution networks considering transient stability. *IET Gener. Transm. Distrib.* **2018**, *13*, 157–170. [CrossRef]
37. Saleh, K.; Zeineldin, H.H.; Al-Hinai, A.; El-Saadany, E.F. Optimal Coordination of Directional Overcurrent Relays Using a New Time–Current–Voltage Characteristic. *IEEE Trans. Power Deliv.* **2014**, *30*, 537–544. [CrossRef]
38. Dehghanpour, E.; Karegar, H.K.; Kheirollahi, R.; Soleymani, T. Optimal Coordination of Directional Overcurrent Relays in Microgrids by Using Cuckoo-Linear Optimization Algorithm and Fault Current Limiter. *IEEE Trans. Smart Grid* **2016**, *9*, 1365–1375. [CrossRef]
39. Shrivastava, A.; Tripathi, J.M.; Krishan, R.; Parida, S. Optimal Coordination of Overcurrent Relays using Gravitational Search Algorithm with DG Penetration. *IEEE Trans. Ind. Appl.* **2017**, 1. [CrossRef]
40. Kida, A.A.; Gallego, L.A. A high-performance hybrid algorithm to solve the optimal coordination of overcurrent relays in radial distribution networks considering several curve shapes. *Electr. Power Syst. Res.* **2016**, *140*, 464–472. [CrossRef]
41. Bouchekara, H.; Zellagui, M.; Abido, M. Optimal coordination of directional overcurrent relays using a modified electromagnetic field optimization algorithm. *Appl. Soft Comput.* **2017**, *54*, 267–283. [CrossRef]
42. Moirangthem, J.; Krishnanand, K.R.; Dash, S.S.; Ramaswami, R. Adaptive differential evolution algorithm for solving non-linear coordination problem of directional overcurrent relays. *IET Gener. Transm. Distrib.* **2013**, *7*, 329–336. [CrossRef]
43. Shih, M.Y.; Salazar, C.A.C.; Enriquez, A.C. Adaptive directional overcurrent coordination using ant colony opti-mization. *IET Gener. Transm. Distrib.* **2015**, *9*, 1240–2049.
44. Bedekar, P.P.; Bhide, S.R. Optimum Coordination of Directional Overcurrent Relays Using the Hybrid GA-NLP Approach. *IEEE Trans. Power Deliv.* **2010**, *26*, 109–119. [CrossRef]
45. Sueiro, J.A.; Diaz-Dorado, E.; Mguez, E.; Cidrs, J. Coordination of directional overcurrent relay using evolutionary algorithm and linear programming. *Elect. Power Energy Syst.* **2012**, *42*, 299–305. [CrossRef]
46. Al-basri, F.A.; Al-roomi, A.R.; Talaq, J.H. Optimal coordination of directional overcurrent relays using biogeogra-phy-based optimization algorithms. *IEEE Trans. Power Del.* **2015**, *30*, 1810–1820. [CrossRef]
47. Saberi, H.; Amraee, T. Coordination of directional over-current relays in active distribution networks using generalized benders decomposition. *IET Gener. Transmiss. Distrib.* **2017**, *11*, 4078–4086. [CrossRef]
48. Saha, D.; Datta, A.; Das, P. Optimal coordination of directional overcurrent relays in power systems using Symbiotic Organism Search Optimisation technique. *IET Gener. Transm. Distrib.* **2016**, *10*, 2681–2688. [CrossRef]
49. Amraee, T. Coordination of Directional Overcurrent Relays Using Seeker Algorithm. *IEEE Trans. Power Deliv.* **2012**, *27*, 1415–1422. [CrossRef]
50. Ezzeddine, M.; Kaczmarek, R.; Iftikhar, M.U. Coordination of directional overcurrent relays using GA novel method to select their settings. *IET Gener. Transmiss. Distrib.* **2011**, *5*, 743–750. [CrossRef]
51. Baghaee, H.R.; Mirsalim, M.; Gharehpetian, G.B.; Talebi, H.A. MOPSO/FDMT-based Pareto-optimal solution for co-ordination of overcurrent relays in interconnected networks and multi-DER microgrids. *IET Gener. Transmiss. Distrib.* **2018**, *12*, 2871–2886. [CrossRef]
52. Shih, M.Y.; Conde, A.; Leonowicz, Z.; Martirano, L. An Adaptive Overcurrent Coordination Scheme to Improve Relay Sensitivity and Overcome Drawbacks due to Distributed Generation in Smart Grids. *IEEE Trans. Ind. Appl.* **2017**, *53*, 5217–5228. [CrossRef]
53. Nascimento, J.P.; Brito, N.S.; Souza, B.A. An adaptive overcurrent protection system applied to distribution systems. *Comput. Electr. Eng.* **2020**, *81*, 106545. [CrossRef]
54. Samadi, A.; Chabanloo, R.M. Adaptive coordination of overcurrent relays in active distribution networks based on independent change of relays' setting groups. *Int. J. Electr. Power Energy Syst.* **2020**, *120*, 106026. [CrossRef]
55. Ates, Y.; Uzunoglu, M.; Karakas, A.; Boynuegri, A.R.; Nadar, A.; Dag, B. Implementation of adaptive relay coordination in distribution systems including distributed generation. *J. Clean. Prod.* **2016**, *112*, 2697–2705. [CrossRef]
56. Ahmadi, S.; Karami, H.; Sanjari, M.; Tarimoradi, H.; Gharehpetian, G. Application of hyper-spherical search algorithm for optimal coordination of overcurrent relays considering different relay characteristics. *Int. J. Electr. Power Energy Syst.* **2016**, *83*, 443–449. [CrossRef]
57. IEEE Power and Energy Society. "C37.112-2018—IEEE Standard for Inverse-Time Characteristics Equations for Overcurrent Relays—IEEE Standard; 2018; Volume C37. Available online: https://ieeexplore.ieee.org/document/8635630 (accessed on 1 August 2021)
58. IEC. *Electrical Relays-Part 3: Single Input Energizing Quantity Measuring Relays with Dependent or Independent Time*; International Electrotechnical Commission: Geneva, Switzerland, 1989; pp. 60255–60263.

59. Majumder, R.; Dewadasa, M.; Ghosh, A.; Ledwich, G.; Zare, F. Control and protection of a microgrid connected to utility through back-to-back converters. *Electr. Power Syst. Res.* **2011**, *81*, 1424–1435. [CrossRef]
60. Dewadasa, M.; Ghosh, A.; Ledwich, G.; Wishart, M. Fault isolation in distributed generation connected distribution networks. *IET Gener. Transm. Distrib.* **2011**, *5*, 1053–1061. [CrossRef]
61. Darabi, A.; Bagheri, M.; Gharehpetian, G.B. Highly sensitive microgrid protection using overcurrent relays with a novel relay characteristic. *IET Renew. Power Gener.* **2020**, *14*, 1201–1209. [CrossRef]
62. Jamali, S.; Borhani-Bahabadi, H. Recloser time–current–voltage characteristic for fuse saving in distribution networks with DG. *IET Gener. Transm. Distrib.* **2017**, *11*, 272–279. [CrossRef]
63. Jamali, S.; Borhani-Bahabadi, H. Self-Adaptive Relaying Scheme of Reclosers for Fuse Saving in Distribution Networks with DG. *Int. J. Power Energy Res.* **2017**, *1*. [CrossRef]
64. Kılıçkıran, H.C.; Akdemir, H.; Sengor, I.; Kekezoğlu, B.; Paterakis, N.G. A Non-Standard Characteristic Based Protection Scheme for Distribution Networks. *Energies* **2018**, *11*, 1241. [CrossRef]
65. Alam, M.N.; Gokaraju, R.; Chakrabarti, S. Protection coordination for networked microgrids using single and dual setting overcurrent relays. *IET Gener. Transm. Distrib.* **2020**, *14*, 2818–2828. [CrossRef]
66. Mahindara, V.R.; Rodriguez, D.F.C.; Pujiantara, M.; Priyadi, A.; Purnomo, M.H.; Muljadi, E. Practical Challenges of Inverse and Definite-Time Overcurrent Protection Coordination in Modern Industrial and Commercial Power Distribution System. *IEEE Trans. Ind. Appl.* **2020**, *57*, 187–197. [CrossRef]
67. Bedekar, P.P.; Bhide, S.R. Optimum coordination of overcurrent relay timing using continuous genetic algorithm. *Expert Syst. Appl.* **2011**, *38*, 11286–11292. [CrossRef]
68. Gokhale, S.; Kale, V. An application of a tent map initiated Chaotic Firefly algorithm for optimal overcurrent relay coordination. *Int. J. Electr. Power Energy Syst.* **2016**, *78*, 336–342. [CrossRef]
69. Zeineldin, H.; El-Saadany, E.; Salama, M. Optimal coordination of overcurrent relays using a modified particle swarm optimization. *Electr. Power Syst. Res.* **2006**, *76*, 988–995. [CrossRef]
70. Uthitsunthorn, D.; Pao-La-Or, P.; Kulworawanichpong, T. Optimal overcurrent relay coordination using artificial bees colony algorithm. In Proceedings of the 8th Electrical Engineering/Electronics, Computer, Telecommunications and Information Technology (ECTI) Association of Thailand-Conference, Khon Kaen, Thailand, 17–19 May 2011; pp. 901–904.
71. Alaee, P.; Amraee, T. Optimal Coordination of Directional Overcurrent Relays in Meshed Active Distribution Network Using Imperialistic Competition Algorithm. *J. Mod. Power Syst. Clean Energy* **2021**, *9*, 416–422. [CrossRef]
72. Singh, M.; Panigrahi, B.; Abhyankar, A. Optimal coordination of directional over-current relays using Teaching Learning-Based Optimization (TLBO) algorithm. *Int. J. Electr. Power Energy Syst.* **2013**, *50*, 33–41. [CrossRef]
73. Barzegari, M.; Bathaee, S.; Alizadeh, M. Optimal coordination of directional overcurrent relays using harmony search algorithm. In Proceedings of the IEEE Transactions on Power Delivery, Prague, Czech Republic, 16–19 May 2010; pp. 321–324. [CrossRef]
74. Benabid, R.; Zellagui, M.; Chaghi, A.; Boudour, M. Application of Firefly Algorithm for Optimal Directional Overcurrent Relays Coordination in the Presence of IFCL. *Int. J. Intell. Syst. Appl.* **2014**, *6*, 44–53. [CrossRef]
75. Blackburn, J.L.; Domin, T.J. *Protective Relaying: Principles and Applications*; CRC Press: Boca Raton, FL, USA, 2015.
76. Jalilian, A.; Hagh, M.T.; Hashemi, S.M. An Innovative Directional Relaying Scheme Based on Postfault Current. *IEEE Trans. Power Deliv.* **2014**, *29*, 2640–2647. [CrossRef]
77. Eissa, M.M. Current directional protection technique based on polarizing current. *Int. J. Electr. Power Energy Syst.* **2013**, *44*, 488–494. [CrossRef]
78. Ukil, A.; Deck, B.; Shah, V.H. Current-Only Directional Overcurrent Relay. *IEEE Sensors J.* **2010**, *11*, 1403–1404. [CrossRef]
79. Ukil, A.; Deck, B.; Shah, V.H. Current-Only Directional Overcurrent Protection for Distribution Automation: Challenges and Solutions. *IEEE Trans. Smart Grid* **2012**, *3*, 1687–1694. [CrossRef]
80. Nojavan, M.; Seyedi, H.; Mehdinejad, M. A novel scheme for current-only directional overcurrent relay. *Int. J. Electr. Power Energy Syst.* **2016**, *82*, 252–263. [CrossRef]
81. Le, P.D.; Bui, D.M.; Ngo, C.C.; Huynh, C.D.; Nguyen, T.H. Effectively Determine Confidence Intervals of Bus Voltages Served for Setting Overcurrent Relays in Distribution Networks. In Proceedings of the IEEE International Conference on Power and Energy (PECon), Penang, Malaysia, 7–8 December 2020. [CrossRef]
82. Bui, D.M.; Le, P.D.; Cao, M.T.; Pham, T.T.; Pham, D.A. Accuracy improvement of various short-term load forecasting models by a novel and unified statistical data-filtering method. *Int. J. Green Energy* **2020**, *17*, 382–406. [CrossRef]
83. Distribution Network Analysis Functions (DNAF) Analyst and Configuration Editor User's Guide. ALSTOM Grid Inc.: Redmond, WA, USA, 2014.
84. Teng, J.H.; Lin, W.M. Current-based power flow solutions for distribution systems. In Proceedings of the IEEE Intetnatioal Conference Power System Technology, Beijing, China, 5–9 December 1994; pp. 414–418.
85. Chen, T.-H.; Chen, M.-S.; Hwang, K.-J.; Kotas, P.; Chebli, E. Distribution system power flow analysis-a rigid approach. *IEEE Trans. Power Deliv.* **1991**, *6*, 1146–1152. [CrossRef]
86. Teng, J.-H. A direct approach for distribution system load flow solutions. *IEEE Trans. Power Deliv.* **2003**, *18*, 882–887. [CrossRef]
87. Samoylenko, V.O.; Korkunova, O.L.; Pazderin, A.; Novikov, N.N. Overcurrent protection adjustment when connecting synchronous generation to power supply systems. In Proceedings of the IEEE International Conference on Industrial Technology (ICIT), Seville, Spain, 17–19 March 2015; pp. 2368–2373. [CrossRef]

88. *1547.2-2008-IEEE Application Guide for IEEE Standard 1547™, IEEE Standard for Interconnecting Distributed Resources with Electric Power Systems*; IEEE Standards: Piscataway, NJ, USA, 2009; pp. 1–217. [CrossRef]
89. Sortomme, E.; Mapes, G.J.; Foster, B.A.; Venkata, S.S. Fault analysis and protection of a microgrid. In Proceedings of the 40th North American Power Symposium, Calgary, AB, Canada, 28–30 September 2008; pp. 1–6.
90. Johnston, W.; Katiraei, F. *Impact and Sensitivity Studies of PV Inverters Contribution to Faults based on Generic PV In-Verter Models*; Ontario Grid Connection Study 2012; Canadian Solar Industries Association/L'Association des Industries Solaires du Canada: Ottawa, ON, Canada, 2012.
91. IEEE. *IEEE Recommended Practice for Protection and Coordination of Industrial and Commercial Power Systems (IEEE Buff Book)*; IEEE Standards: Piscataway, NJ, USA, 2008; pp. 1–710. [CrossRef]
92. Geidl, M. *Protection of Power Systems with Distributed Generation: State of the Art*; Swiss Federal Institute of Technology (ETH): Zurich, Switzerland, 2005. [CrossRef]
93. Horowitz, S.; Phadke, A. Power System Relaying [Books and Reports]. *IEEE Power Eng. Rev.* **1993**, *13*, 38. [CrossRef]
94. Rashedi, E.; Nezamabadi-Pour, H.; Saryazdi, S. GSA: A Gravitational Search Algorithm. *Inf. Sci.* **2009**, *179*, 2232–2248. [CrossRef]
95. Kennedy, J.; Eberhart, R. Particle swarm optimization. In Proceedings of the IEEE International Conference Neural Networks IV, Perth, Australia, 27 November–1 December 1995; pp. 1942–1948. [CrossRef]
96. Srivastava, A.; Tripathi, J.M.; Mohanty, S.R.; Panda, B. Optimal Over-Current Relay Coordination with Distributed Generation Using Hybrid Particle Swarm Optimization–Gravitational Search Algorithm. *Electr. Power Compon. Syst.* **2016**, *44*, 1–12. [CrossRef]
97. Ahmad, M.; Ali, W.; Farooq, H.; Jamil, M.; Ali, M.; Rehman, A.U. Solving the Problem of Economic Load Dispatch for a Small Scale Power System Using a Novel Hybrid PSO-GSA Algorithm. In Proceedings of the 2018 International Symposium on Recent Advances in Electrical Engineering (RAEE), Islambad, Pakistan, 17–18 October 2018; pp. 1–6. [CrossRef]
98. Goldberg, D.E. *Genetic Algorithms in Search, Optimization, and Machine Learning*; Dorling Kindersley (India) Pvt. Ltd.: New Delhi, India, 2008.
99. Asadi, M.R.; Askarian Abyaneh, H.; Mahmoodan, M.; Naghizadeh, R.A.; Koochaki, A. Optimal overcurrent relays coordination using genetic algorithm. In Proceedings of the 2008 11th International Conference on Optimization of Electrical and Electronic Equipment, Brasov, Romania, 22–24 May 2008; pp. 197–202.

applied sciences MDPI

Article

Power-Based Concept for Current Injection by Inverter-Interfaced Distributed Generations during Transmission-Network Faults

Boštjan Polajžer [1,*], Bojan Grčar [1], Jernej Černelič [2] and Jožef Ritonja [1]

[1] Faculty of Electrical Engineering and Computer Science, University of Maribor, SI-2000 Maribor, Slovenia; bojan.grcar@um.si (B.G.); jozef.ritonja@um.si (J.R.)
[2] MAHLE Electric Drives Slovenija d.o.o., SI-5290 Šempeter pri Gorici, Slovenia; cernelic@gmail.com
* Correspondence: bostjan.polajzer@um.si

Abstract: This paper analyzes the influence of inverter-interfaced distributed generations' (IIDGs) response during transmission network faults. The simplest and safest solution is to switch IIDGs off during network faults without impacting the network voltages. A more elaborate and efficient concept, required by national grid codes, is based on controlling the IIDGs' currents, involving positive- and negative-sequence voltage measured at the connection point. In this way the magnitude and phase of the injected currents can be adjusted, although the generated power will depend on the actual line voltages at the connection point. Therefore, an improved concept is proposed to adjust IIDGs' fault current injection through the required active and reactive power, employing the same voltage characteristics. The proposed, i.e., power-based concept, is more definite than the current-based one, since the required power will always be generated. The discussed concepts for the fault current injection by IIDGs were tested in different 110-kV networks with loop and radial topologies, and for different short-circuit capabilities of the aggregated network supply. Based on extensive numerical calculations, the power-based concept during transmission networks faults generates more reactive power compared to the current-based concept. However, the voltage support by IIDGs during transmission networks faults, regardless of the concept being used, is influenced mainly by the short-circuit capability of the aggregated network supply. As regards distance protection operation, it is influenced additionally by the network topology, i.e., in radial network topology, the remote relay's operation can be delayed due to a largely seen impedance.

Keywords: distributed generation; fault current injection; voltage support; distance protection

Citation: Polajžer, B.; Grčar, B.; Černelič, J.; Ritonja, J. Power-Based Concept for Current Injection by Inverter-Interfaced Distributed Generations during Transmission-Network Faults. *Appl. Sci.* **2021**, *11*, 10437. https://doi.org/10.3390/app112110437

Academic Editors: Federico Barrero and Mario Bermúdez

Received: 2 October 2021
Accepted: 4 November 2021
Published: 6 November 2021

1. Introduction

1.1. Motivation and Literature Review

The system operator can specify that inverter-interfaced distributed generation (IIDG) should provide a fast fault current (FFC) during network faults. The main idea behind this requirement is to generate a predominantly reactive current to identify a fault by network protection, and to support the network voltages [1]. All the static and dynamic FFC requirements have already been implemented in national grid codes, e.g., the Slovenian requirements are similar to the German ones [2]. However, the inverter currents are limited, typically to 120% of its rated value, which impacts the operation of the network protection. Thus, FFC injection by IIDGs raises questions, not only on inverter control during network faults, but also on the network protection operation. The motivation for this paper is to evaluate the impact of IIDGs with FFC injection on the voltage support and distance protection operation during transmission network faults.

New generations of generic IIDG models incorporate balanced FFC (BFFC) injection based on a voltage sag logic [3–5]. Improved BFFC is proposed in [6] by an inverter's DC-link compensation term, which reduces active power during network faults. However, a large body of literature proposes control approaches for unbalanced FFC (UFFC) injection during asymmetrical network faults [7–20]; a comprehensive review is given in [7]. These control approaches all have several objectives, i.e., the quality and limitation of the injected FFC, mitigation of the inverter's DC-link voltage ripple, damping of instantaneous power oscillations, control of phase voltages within predefined safety limits, and computation of the reactive power reference, depending on the resistive–inductive network impedance model and reference voltage sequences. A commonly used flexible voltage support during network faults is proposed in [8], applying instantaneous symmetrical component-based FFC injection. Ref. [9] investigates an FFC dynamic response, which can reach up to five times the nominal current in cases of sub-optimal control parameters. To prevent the inverter's transient overcurrents, Ref. [10] proposes a model-predictive control, whereas [11,12] propose flexible power control. The approach proposed in [13] can balance positive and negative-sequence active and reactive powers while keeping the injected FFC to a predefined maximum value. The FFC injection proposed in [14] uses a mixed-potential function to control the DC-link voltage compensation term to reduce power oscillations. Ref. [15] proposes an approach for controlling phase voltages within predefined safety limits by setting the positive and negative reactive power references based on an equivalent impedance network model. The IIDG control proposed in [16] boosts short-term voltage stability based on the negative-sequence voltage at the point of common coupling, and controls the phase voltages within predefined safety limits. Ref. [17] adjusts the FFC injection to network conditions, whereas [18] proposes unified reactive power support with an active power curtailment controller, which can distinguish between islanding and voltage transient events, and provides effective voltage unbalance compensation along with zero active power oscillations and FFC with a controlled peak. Another flexible control approach is proposed in [19], where the FFC reference is determined by the fault condition and predefined constants from the grid code. Furthermore, the approach proposed in [20] guarantees safe operation of the inverter during voltage sags by calculating the appropriate reference FFC according to the equivalent impedance and the voltage sag characteristics, avoiding active power oscillations, and limiting FFC to the maximum value allowed by the inverter.

The IIDG models discussed in [3–20] are too complicated for protection studies in large networks. Therefore, IIDGs are, typically, modeled simply as a constant current source, showing the impact of their capacity and location on the coordination of overcurrent relays in distribution networks and micro-grids with different topologies [21–25]. More elaborate is the BFFC-based model proposed in [26], as a current limiting element in parallel with the filter capacitor. Ref. [27] uses this model to develop a protection strategy for microgrids. Furthermore, Ref. [28] uses current-limiting characteristics for grid-following and grid-forming IIDGs as constraints for the protection coordination in microgrids. Other protection studies [29–33] incorporate IIDG models with instantaneous FFCs. A BFFC injection, required by the Spanish grid code, is used in [29], presenting a communication-based directional relay system. Ref. [30] discussed protection coordination in distribution networks while considering UFFC injection. Ref. [31] enhances overcurrent protection coordination using the IIDG model that provides limited power by using orthogonal components of the grid voltages, while a current limitation approach has been proposed in [32] to restrict FFC injection. Ref. [33] uses an IIDG model that complies fully with the network code requirements for BFFC injection, to perform a sensitivity analysis of a test network's transient short circuit current response. Furthermore, Ref. [34] summarizes the FFC contribution by IIDGs and the network protection operation.

1.2. Contribution and Structure of the Paper

This paper analyzes two FFC concepts for IIDGs based on voltage characteristics for positive and negative sequence components, as required in [2]. The standard, i.e., current-based concept adjusts the magnitude and phase of the line currents injected by an IIDG. The proposed concept adjusts FFC through active and reactive power employing the same voltage characteristics, which is the first contribution of this paper. Both FFC concepts, i.e., current- and power-based, are formulated in the phasor-domain, which is the second contribution of this paper. Thus, a simple IIDG model is obtained that complies with grid codes and is suitable for protection studies in large networks; moreover, BFFC and UFFC are both possible. The gap between complicated instantaneous-based FFC calculation approaches and the simplified constant current approach is filled in this way. Furthermore, several protection studies have already reported the impact of IIDGs on protection operation in distribution networks and microgrids, focusing on overcurrent protection coordination. However, a comprehensive analysis of the operation of transmission-network protection impacted by IIDGs is missing. Analysis of IIDG responses and distance protection operation during transmission-network faults is another contribution of this paper.

The rest of this paper is organized as follows. Section 2 proposes the phasor-based IIDG model and describes the discussed FFC concepts based on reference current and power. The test network is presented in Section 3, including relevant data, topologies, and the different SSC of an aggregated network supply. The active and reactive power of all the IIDGs during the network faults are analyzed, together with the voltages at the connection point and distance protection operation. Section 4 presents a case study, i.e., a transmission network related to the 110-kV northeastern part of the Slovenian power system, considering the expected IIDG's capacities. The power generated by IIDGs, voltage profiles, and distance protection operation are analyzed for different topologies during the faults at different locations. Section 5 raises a discussion on IIDG models for FFC injection, and indicates future work on protection studies, while Section 6 concludes the paper.

2. IIDG Model for FFC Injection

A phasor-based model is proposed; Figure 1 shows the basic model structure. Two concepts are used to calculate references for line current phasors, which involve voltage characteristics, whereas the model also enables selection between BFFC and UFFC injection. Ideal three-phase star-connected current sources were used, which can also operate as a sink, enabling the simulation of a battery energy storage (BES) unit in the discharging and charging regime. Thus, the dynamics of current controllers and an inverter were neglected, which is suitable for protection studies in large networks.

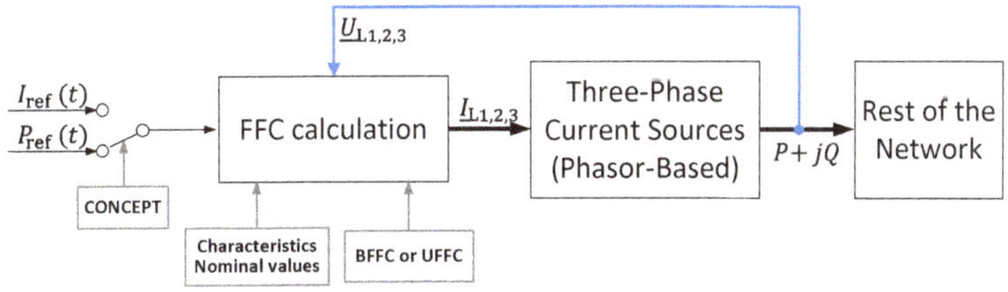

Figure 1. Structure of an IIDG model for FFC injection.

2.1. Voltage Characteristics

The proposed characteristics are shown in Figure 2, where f^{+}_{Re}, f^{+}_{Im}, and f^{-}_{Im} denote functions for the positive and negative sequence components. U^{+}, U^{-}, and U_n are, respectively, positive and negative sequence voltage magnitudes and nominal line voltage. Three independent parameters describe each characteristic, i.e., a voltage dead-band, a slope, and the maximal value. The parameter limits proposed by the grid code [2] are ±20% for the voltage dead-band, a slope $\Delta I/\Delta U \in [2, 6]$ pu/pu, whereas the maximal value should not exceed 80% of the IIDG's overcurrent protection setting. Furthermore, a reactive current is prioritized, while an active current should be readjusted continually.

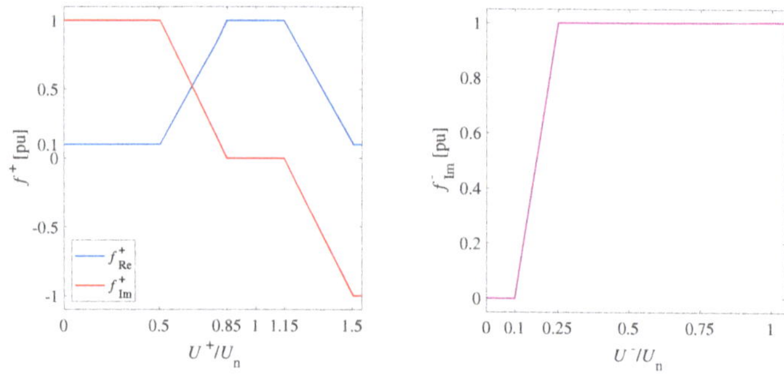

Figure 2. Voltage characteristics for positive and negative sequence components.

2.2. Current- and Power-Based Concept

Applying the standard, i.e., current-based concept, is the reference line current phasor $\underline{I}_{L1}$ calculated as

$$\underline{I}_{L1} = I_{Re} + jI_{Im} = I_n\left[f^+_{Re}(U^+/U_n)I_{ref,pu}(t) + j\,f^+_{Im}(U^+/U_n) + j\,f^-_{Im}(U^-/U_n)\right]^* \quad (1)$$

where $I_{ref,pu}(t)$ refers to the time-dependent normalized reference current, I_n is the IIDG's nominal line current, and $(\cdot)^*$ stands for the complex conjugate. Note that this concept complies with the requirements given in [2]. Figure 3 shows the functional scheme for the current-based concept, where $\underline{U}_{L1,2,3}$ denote line voltage phasors at the IIDG's connection point, $\underline{U}^+$,$\underline{U}^-$ denote positive and negative sequence voltage phasors, while I^+_{Re},I^+_{Im}, and I^-_{Im} are real and imaginary parts of the positive and negative sequence current phasors. Furthermore, the magnitude of the line current phasors is limited as $I_{L1,2,3} \leq 1.2I_n$. Moreover, the voltage at the connection point is a nonlinear function of operating conditions and reference currents, which are also a function of voltages at the same connection point. A delay is introduced in the calculation by a first-order system with a time constant of 1 ms to break the algebraic loop.

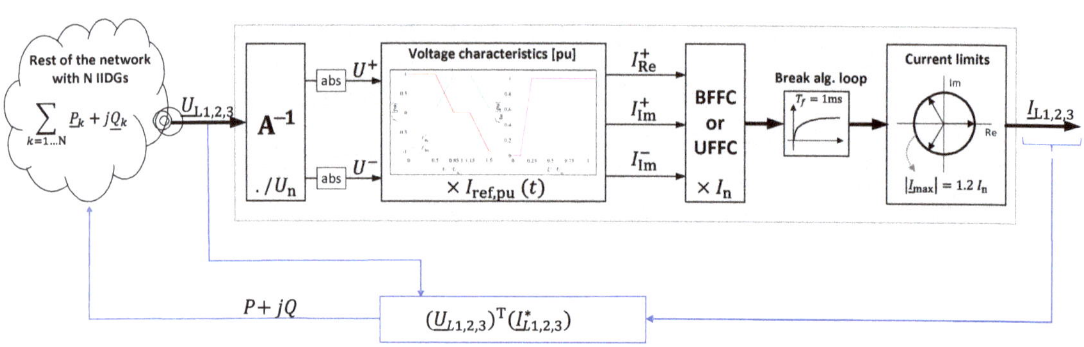

Figure 3. Functional scheme for calculating IIDG's reference line current phasors—current-based concept.

The limitation of the current-based concept is that the generated power will depend on actual line voltages at the connection point. Therefore, active and reactive power are introduced using the same voltage characteristics instead of currents. Thus, applying the proposed, i.e., power-based concept is the reference line current phasor $\underline{I}_{L1}$ calculated as

$$\underline{S} = P + jQ = P_n\left[f^+_{Re}(U^+/U_n)P_{ref,pu}(t) + j\,f^+_{Im}(U^+/U_n) + j\,f^-_{Im}(U^-/U_n)\right] \quad (2)$$

$$\underline{I}_{L1} = \frac{2}{3}\left(\frac{\underline{S}}{\underline{U}^{+}}\right)^{*} \quad (3)$$

where $P_{\mathrm{ref,pu}}(t)$ refers to the time-dependent normalized reference power, P_{n} is the IIDG's nominal capacity, and $(\cdot)^{*}$ stands for the complex conjugate. Furthermore, the functional scheme for the reference phasor currents calculation (Figure 3) should be modified according to the power-based concept (2) and (3). Note that this concept also complies with the requirements given in [2]; however, its behavior is more definite than the current-based.

2.3. BFFC and UFFC

Selection between BFFC and UFFC affects the calculation of line current phasors $\underline{I}_{\mathrm{L2}}$ and $\underline{I}_{\mathrm{L3}}$. When using the BFFC, they are calculated as $\underline{I}_{\mathrm{L2}} = \underline{I}_{\mathrm{L1}} \exp(-2\pi/3)$ and $\underline{I}_{\mathrm{L3}} = -(\underline{I}_{\mathrm{L1}} + \underline{I}_{\mathrm{L2}})$. When using the UFFC are line current phasors $\underline{I}_{\mathrm{L2}}$ and $\underline{I}_{\mathrm{L3}}$ calculated by the transformation matrix for symmetrical components **A**.

3. Test Network

A 4-bus test network with a loop topology was considered, as is shown in Figure 4. The IIDGs were connected at substations B, C, and D. The nominal power of each IIDG was set to 10 MW, which covers local loads at each of the substations. Both concepts were tested, i.e., current- and power-based, while injecting only BFFC.

3.1. Reference Tracking

Arbitrary time-profiles of IIDGs were assumed during normal voltage conditions, where IIDG D was assumed to operate as a BES unit. Note that normalized references $P_{\mathrm{ref,pu}}(t)$ and $I_{\mathrm{ref,pu}}(t)$ were changed in the same way, separately for each of the discussed concepts; Figure 5a shows almost perfect active power reference tracking for the power-based concept, especially for a BES unit during the transition between charging and discharging operating regimes. Furthermore, a small amount of reactive power was generated during a time interval when the voltages deviated from the voltage dead-band. The current-based concept showed exact tracking of the reference current; however, the resulting active power was lower when compared with the power-based concept, as shown in Figure 5b. The power reduction was due to the off-nominal initial voltages that were approximately 0.93 pu. The obtained results show that the power-based concept is more definite, since the required power will be generated.

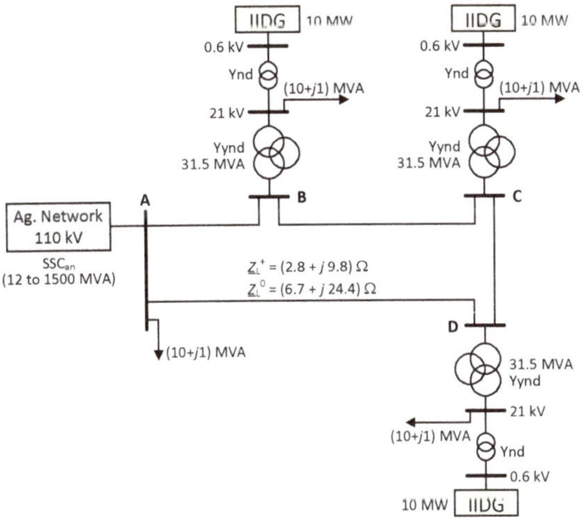

Figure 4. Single line diagram of the test network with relevant data.

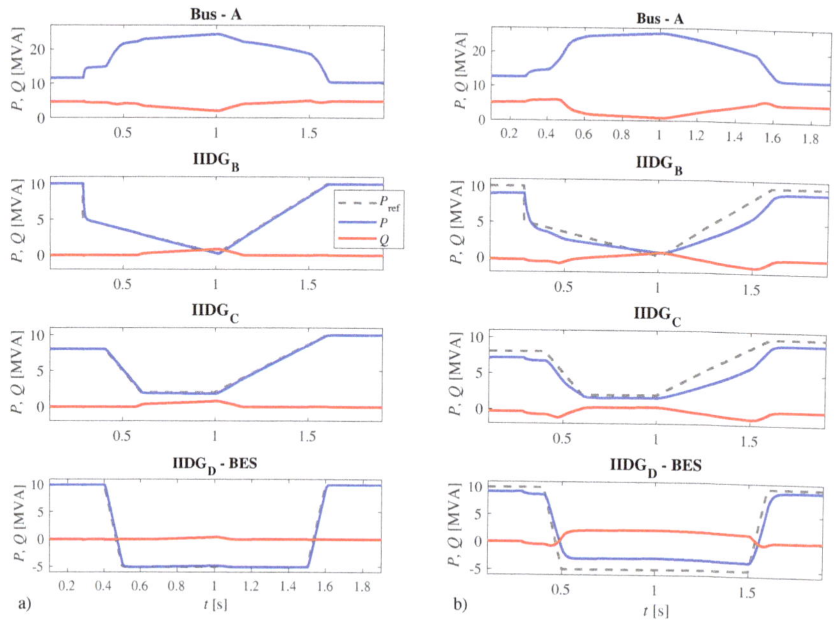

Figure 5. Comparison in power generation assuming reference tracking of the power-based FFC (**a**), and current-based FFC (**b**).

3.2. Responses during Network Faults

The next test assumes a phase-to-phase (Ph-Ph) fault with a fault resistance of 2 Ω in the middle of the C-D line. The fault was incepted at a simulation time of 1 s, while at 1.2 s, the faulted line was disconnected. All IIDGs operated with a nominal power of 10 MW, while the Short-Circuit Capability (SSC) of an aggregated network supply was set as 50 MVA. Figure 6 shows power generation for both the discussed FFC concepts. More reactive power was generated when using the power-based concept, which is a consequence of considering actual voltages (positive and negative components) at the IIDG's connection points to calculate generated line currents. Consequently, a larger reduction in the active power was noticed when using the power-based concept.

Figure 7 shows steady-state line voltages at the IIDGs' connection point during the fault. Along with both FFC concepts, a concept was also considered where the IIDGs were switched off during the fault. The current-based concept resulted in higher line voltages in all three phases, but the differences were minimal compared to the power-based concept. However, when the IIDG's were switched off during a fault, the voltages were considerably lower than when employing the FFC.

3.3. Impact of SSC of an Aggregated Network Supply

Additional tests were performed for a phase-to-phase (Ph-Ph) fault with a fault resistance of 2 Ω in the middle of the C-D line for different SSC values of an aggregated network supply. Tables 1 and 2 give the steady-state values during the fault for the generated power and line voltages. The obtained results show that an increase in the aggregated network SSC decreased the differences in generated power between both FFC concepts. The generated reactive power was not affected significantly by an aggregated network SSC when using the power-based FFC concept. Compared with the current-based concept, the power-based concept also generated more reactive power and less active power. Furthermore, the generated apparent power was approximately equal for both the discussed FFC concepts. Minor differences in voltage support by both FFC concepts were seen only for small values of an aggregated network SSC.

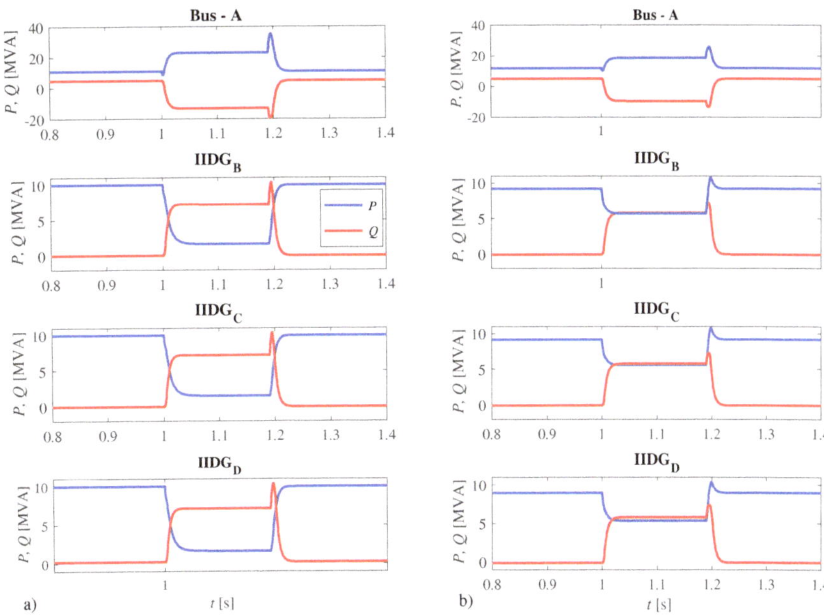

Figure 6. Comparison in power generation during a Ph-Ph fault (L2-L3) in the middle of the C-D line, employing the power-based FFC (**a**), and current-based FFC (**b**).

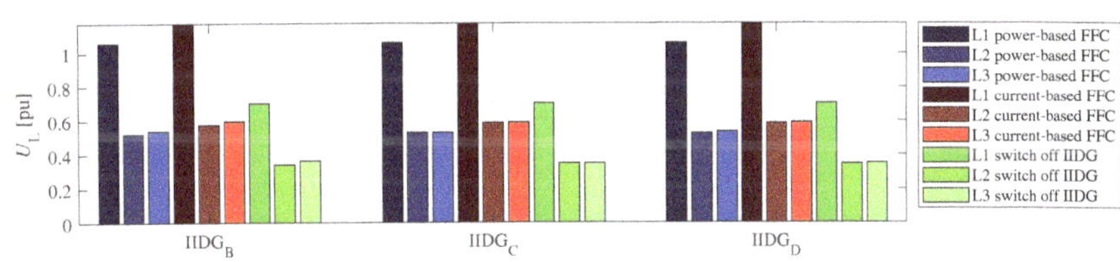

Figure 7. Comparison of line voltages during a Ph-Ph fault (L2-L3) in the middle of the C-D line, employing the power-based FFC, current-based FFC, and when switching-off IIDGs.

Table 1. Steady-state results during a Ph-Ph fault (L2-L3) in the middle of the C-D line, employing the power-based FFC.

SSC_{an} [MVA]	IIDG	P [MW]	Q [MVAr]	S [MVA]	U_{L1} [pu]	U_{L2} [pu]	U_{L3} [pu]
50	B	1.9	7.4	7.6	1.0581	0.5204	0.5394
	C	1.8	7.3	7.5	1.0608	0.5306	0.5306
	D	2.7	7.5	8.0	1.0581	0.5251	0.5336
100	B	2.6	7.6	8.0	1.1092	0.5537	0.5603
	C	2.3	7.5	7.8	1.1118	0.5607	0.5521
	D	2.0	7.2	7.5	1.1092	0.5570	0.5538
1000	B	3.2	7.7	8.3	0.9643	0.5718	0.4984
	C	1.7	7.2	7.4	0.9670	0.5295	0.4531
	D	0.5	6.2	6.2	0.9641	0.5378	0.4622

Table 2. Steady-state results during a Ph-Ph fault (L2-L3) in the middle of the C-D line, employing the current-based FFC.

SSC_{an} [MVA]	IIDG	P [MW]	Q [MVAr]	S [MVA]	U_{L1} [pu]	U_{L2} [pu]	U_{L3} [pu]
	B	5.6	5.8	8.1	1.1712	0.5767	0.5965
50	C	5.5	5.8	8.0	1.1732	0.5853	0.5884
	D	5.4	6.1	8.1	1.1712	0.5828	0.5891
	B	4.6	6.5	8.0	1.1281	0.5638	0.5638
100	C	4.4	6.5	7.8	1.1305	0.5694	0.5621
	D	3.8	6.5	7.5	1.1282	0.5673	0.5625
	B	4.0	7.4	8.4	0.9643	0.5718	0.4983
1000	C	2.6	6.9	7.4	0.9672	0.5292	0.4535
	D	0.6	6.1	6.1	0.9642	0.5382	0.4619

3.4. Distance Protection Operation

Only distance protection was analyzed, since it is the primary protection for 110-kV networks. The criterion for the operation of distance protection is the positive-sequence impedance, which was calculated using line voltage and current phasors measured on both sides of the faulted line [35]. The cases discussed in the previous section show a small error in calculated impedance from both sides of the faulted line, regardless of which FFC concept was employed. The error in calculated impedance was up to 20% in the R-direction and only up to 2.5% in the X-direction. Consequently, the correct operation of distance protection was expected, because the fault was supplied equally from both sides of the line.

To test distance protection operation properly, two more challenging tests were assumed. In the first test, the SSC of the aggregated network supply was reduced to only 12 MVA and the line A-D was out of service, thus leading to the radial topology. Different fault types, i.e., three-phase (3-Ph), phase-to-phase (Ph-Ph), and phase-to-ground (Ph-G), were considered on the line C-D at the distances that correspond to 20% and 80% from the bus C. Furthermore, fault resistance of 2 Ω was considered. Figure 8 shows the impedances seen by the relays C and D, i.e., $\underline{Z}_{C-D}$ and $\underline{Z}_{D-C}$, respectively. Furthermore, line impedance $\underline{Z}_L$ is also shown, together with protection zones. Note that the first zone covers 85% of the line length, while the second zone is extended to the next line by covering 120% of the protected line length. The protection operation for faults in the first zone is instantaneous, while it is delayed for faults in the second zone to achieve selectivity [35]. As shown in Figure 8, the distance relay at side C would operate correctly in all cases, since the impedance $\underline{Z}_{C-D}$ was seen in the first zone. The operation of the relay at side D would be delayed in some situations, due to the impedance $\underline{Z}_{D-C}$ seen in the second zone. However, after the faulted line was disconnected from side C, the measured impedance of the relay at side D moved into the first operating zone in all cases, as indicated by the dotted arrow line. To avoid delays in distance protection operation, the introduction of communication with the appropriate logic scheme (tripping or blocking) is necessary [36].

In the last test, the SSC of an aggregated network was increased significantly to 1500 MVA. Again, the line A-D was out of service, representing the typical situation of a strong network that operates against a weak network with a significant impact on distance protection. Different fault types were assumed on line C-D at a distance of 20% from bus D. Again, fault resistance of 2 Ω was considered. The distance protection relay at side D would fail to operate in cases of 3-Ph and Ph-Ph faults, as is shown in Figure 9, since the seen impedance $\underline{Z}_{D-C}$ was out of all the protection zones. Only after the faulted line was disconnected from side C, the relay at side D calculated short-circuit impedance correctly, as indicated by the dotted arrow line. This would cause some unacceptable delay in the fault clearance; therefore, the build-in function "Echo and Tripping in the Event of Weak Infeed" should be implemented in such cases [36].

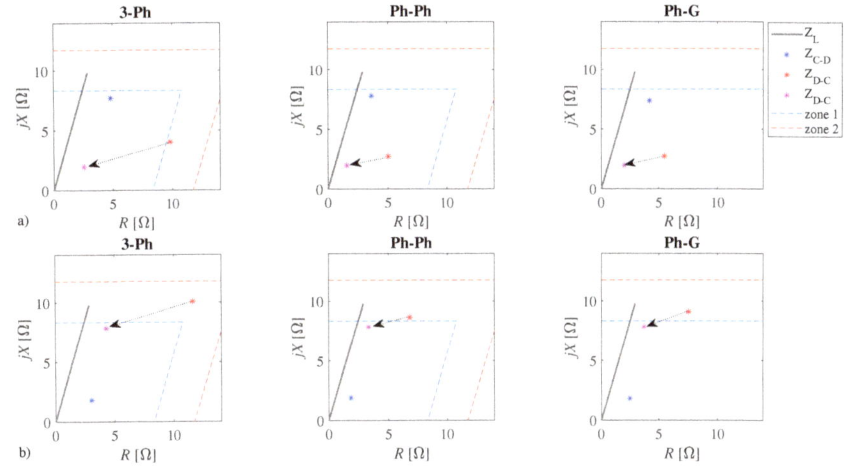

Figure 8. Distance protection operation in radial topology for different fault types on the C-D line at 20% from bus C (**a**), and at 80% from bus C (**b**), where SSC_{an} = 12 MVA (the dotted arrow line indicates disconnection at the side C).

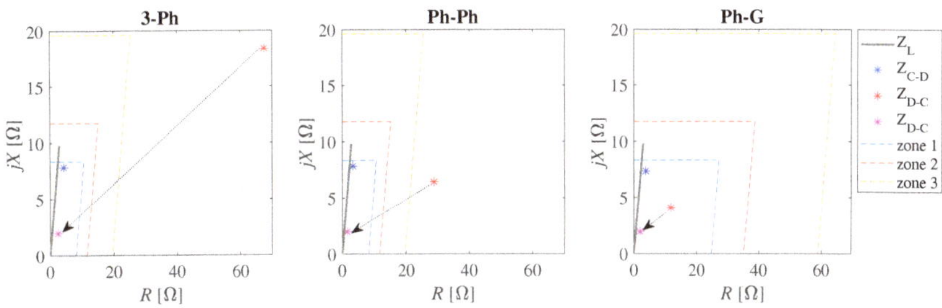

Figure 9. Distance protection operation in radial topology for different fault types on the C-D line at 20% from bus D, where SSC_{an} = 1500 MVA (the dotted arrow line indicates disconnection at the side C).

4. Case Study: The Northeast Part of the Slovenian Transmission System

The northeast part of the Slovenian transmission system is discussed, as shown in Figure 10. An individual substation (SS) was modeled by a TR 110 kV/X, an aggregated constant load, and an aggregated PV unit with the discussed FFC concepts. Furthermore, two BES units with the discussed FFC concepts and two hydropower plants (HPPs) were also considered in the network, along with five different aggregated network supplies. The main data of all discussed SSs are given in Table 3. Maximal load $P_{L,max}$ was calculated according to the measurements for a period of three months during the summer, considering only the day-time. Furthermore, according to realistic plans, the PV and BES units' capacity $P_{IIDG,n}$ will increase approximately up to four times by 2030. All PVs and BESs were assumed to operate with their maximal capacity expected by 2030, except the PVs in SSs Sobota and Mackovci and BES in Kidricevo.

Numerous short-circuit analyses were performed for different fault types and locations on different 110-kV lines. The faults were incepted at a simulation time of 0.3 s, while at 0.42 s, the faulted line was disconnected. Three states were observed, i.e., normal operation (State 1), fault (State 2), and disconnection of the faulted line (State 3). Only two cases are presented, with the most indicative results.

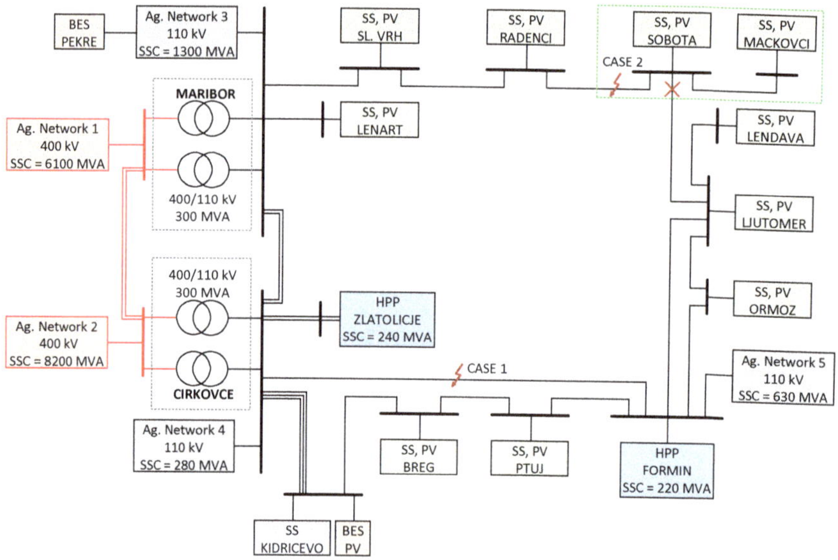

Figure 10. Single line diagram of the Slovenian northeast transmission network.

Table 3. Main substation data with current capacities of the PV and BES units.

Substation	TR 110/X [kV]	$P_{TR,n}$ [MW]	$P_{L,max}$ [MW]	$P_{IIDG,n}$ [MW]
Lenart	21	2 × 20	8.2	5.9
Sl. Vrh	21	2 × 31.5	13.7	2.3
Radenci	21	2 × 31.5	15	7.5
Sobota	21	2 × 40	25.3	12.6
Mackovci	21	1 × 31.5	4.9	1.5
Lendava	21	2 × 31.5	15.8	3.3
Ljutomer	21	2 × 31.5	9.2	4.5
Ormoz	21	2 × 20	7.0	2.1
Ptuj	21	2 × 40	18.7	7.6
Breg	21	2 × 31.5	13.6	16.1
Kidricevo—PV	10.5	1 × 31.5	9.7	4.0
Kidricevo—BES	10.5	2 × 31.5	-	16.0
Kidricevo—load	10.5	2 × 60	60	-
Pekre—BES	21	1 × 15	-	6.0

4.1. Case 1—Faults in Loop Topology with a Strong Network Supply

In the first case, faults were assumed on the Formin–Cirkovce line, as shown in Figure 10. Power generation of IIDGs and line voltage magnitudes at the 110-kV level are shown in Figures 11 and 12 for a Ph-Ph fault. Due to the loop-network topology, all PVs and BESs changed their power during the fault, i.e., active power was reduced, while reactive power was increased. Minor responses were noticed by BES Pekre and by PVs Lenart and Sl. Vrh, since they were the more distant from the fault location, but close to the SS 400/110 kV Maribor. The largest responses were noticed by PVs Lendava, Ljutomer, and Ormoz, which were the more distant from both SSs 400/110 kV. Other PVs show moderate responses in generated power. PVs Ptuj, Breg and BES Kidricevo also showed moderate responses, even though they are the closest to the fault location; however, they are also electrically close to the SS 400/110 kV Maribor. Voltages during the fault were less reduced at PV and BES locations with smaller changes in the generated power, which shows the dominant impact of the aggregated network supplies. Moreover, compared to the concept when

PV and BES units would be switched off during the fault, the voltage profile was slightly improved by the FFC-based IIDGs at the buses near the fault location. Regarding the distance protection operation for different fault types, no issues were observed, regardless of the IIDG concept (FFC or switching off), which is entirely due to the dominant SSC of aggregated network supplies.

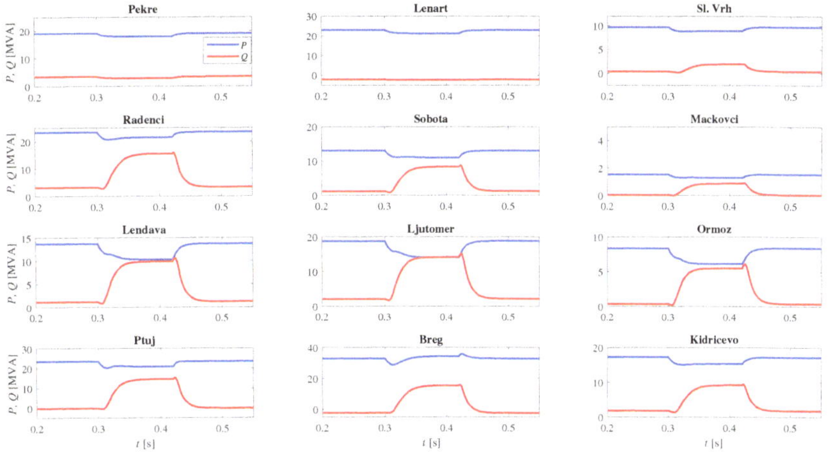

Figure 11. Generated power for case 1 during normal operation, Ph-Ph fault, and after the faulted line was disconnected.

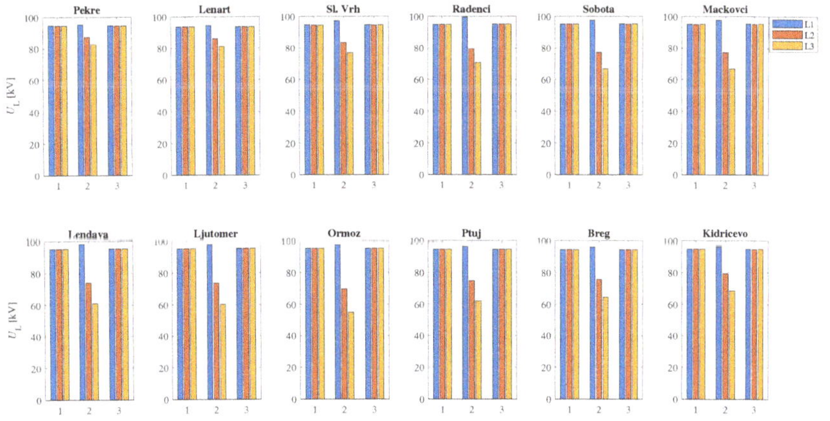

Figure 12. Line voltage magnitudes for case 1 during normal operation (State 1), Ph-Ph fault L2-L3 (State 2), and after the faulted line was disconnected (State 3).

4.2. Case 2—Faults in Radial Topology with Weak IIDGs

The second case assumes faults on the Radenci–Sobota line, while the Sobota–Ljutomer line was out of service, leading to a radial topology between Maribor and Mackovci, as shown in Figure 10. Consequently, a typical situation was obtained, where a strong network (SS 400/110 kV Maribor) operates against a weak network (PVs Sobota and Mackovci) during the faults. Furthermore, when the faulted line was disconnected, the SSs Sobota and Mackovci operated as an island. Figures 13 and 14 show the power generation of the PV and BES units and line voltage magnitudes at the 110-kV level for a Ph-Ph fault. Since the loop was opened between SSs Sobota and Ljutomer, the PVs Lendava, Ljutomer, Ormoz, Ptuj, Breg, and BES Kidricevo showed minor changes in generated power. Consequently, the voltages during the fault at these locations practically did not change. PVs Lenart, Sl. Vrh and BES Pekre showed minor responses in the generated power, since the fault was supplied mainly from the SS 400/110 kV Maribor. The largest repose in generated power was shown by PVs Radenci, Sobota, and Mackovci, which were very close to the fault location. Voltages at those PVs were reduced by approximately 50% during the fault; however, the reduction in voltages at PVs closer to the SS 400/110 kV Maribor (PVs Sl. Vrh and Lenart) was smaller. Furthermore, it can be observed that the voltages at Mackovci and Sobota increased slightly after the disconnection of the faulted Radenci–Sobota line, since these SSs operate as an island. The distance relay at the side of the strong network (SS Radenci) operated correctly for all fault types and admissible locations. On the other hand, the distance relay from the weak side (SS Sobota) failed to operate in all cases, due to the limited PV currents in Sobota and Mackovci.

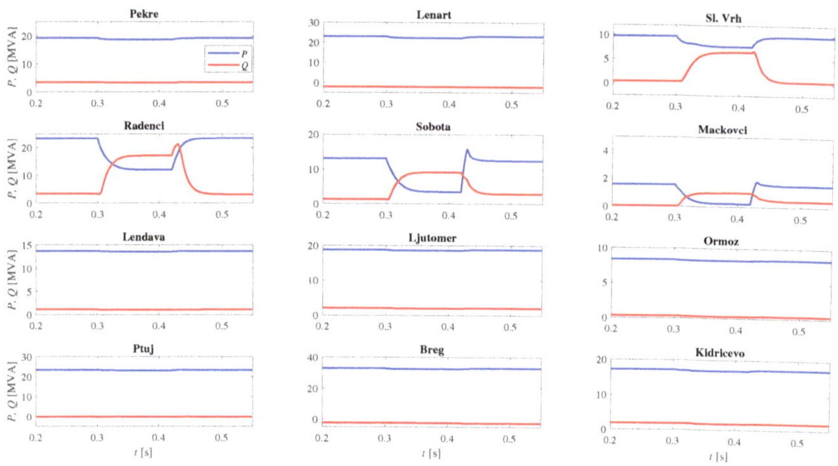

Figure 13. Generated power for case 2 during normal operation, Ph-Ph fault, and after the faulted line was disconnected.

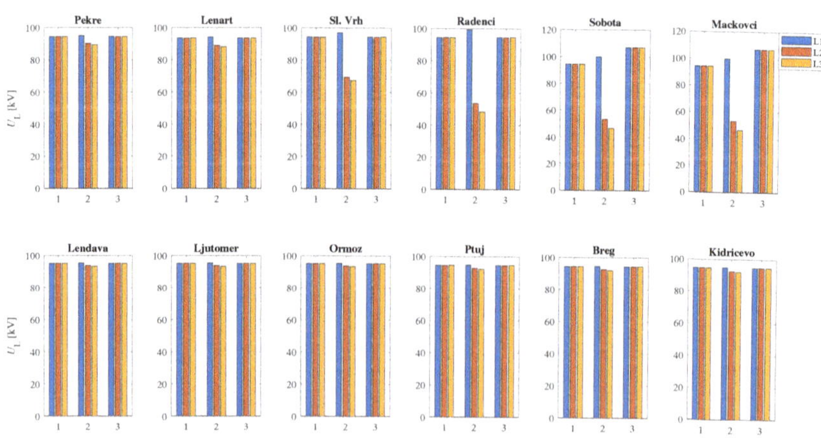

Figure 14. Line voltage magnitudes for case 2 during normal operation (State 1), Ph-Ph fault L2-L3 (State 2), and after the faulted line was disconnected (State 3).

5. Discussion and Future Work

An increased share of the IIDGs will undoubtedly influence the operation of a power system protection. Therefore, a simple phasor-based IIDG model that complies with grid codes is proposed, which can be used to perform protection studies in large networks. Moreover, the proposed model offers flexibility in choosing between the power- and current-based concept, and between the BFFC and UFFC injection. A comparison of relevant IIDG models is given in Table 4. A natural extension of this work would be implementing the proposed IIDG model for FFC injection in overcurrent protection coordination in medium voltage distribution networks and microgrids.

Table 4. Comparison of relevant IIDG models suitable for protection studies.

	Calculation Domain	Unbalanced FFC	FFC Concept	Max Number of Used IIDGs
Proposed	Phasor	Optional	Current/Power	13
[26,27]	Phasor	No	Current	7
[28]	Phasor	No	Current	10
[29]	Instantaneous	No	Current	3
[30]	Instantaneous	Yes	Power	2
[31]	Instantaneous	No	Power	1
[32]	Instantaneous	Yes	Power	5
[33]	Instantaneous	No	Current	3

6. Conclusions

This paper proposes a phasor-based IIDG model for FFC injection, which is suitable for protection studies in large networks. A comparison is made between the power- and current-based FFC concepts for a testing network with three IIDGs and a case study, i.e., the northeast part of the Slovenian transmission system. The obtained results show that both the discussed FFC concepts during network faults increased generated reactive power while decreasing the active power. The responses of generated power were more expressed when using the power-based FFC concept. Furthermore, reactive power generation was not affected considerably by the SSC of an aggregated network supply, and was more significant when using the power-based concept. Moreover, the largest reactive power generation is noticeable at IIDGs electrically close to the fault location and IIDGs distant from the strong network supplies.

Regarding overall voltage support during network faults, a minor impact of FFC injection was noticed only when the IIDGs' overall capacity was comparable to the SSC of the aggregated network supply. IIDGs might also influence the operation of distance protection in combination with special cases of network topology and the SSC of an aggregated network supply. When the capacity of IIDGs is comparable to the SSC of the aggregated network supply, then an impact could be expected on distance protection operation. However, the most critical situation in radial network topologies will occur when a strong network operates against a weak network during the faults.

Author Contributions: Supervision, B.P. and J.R.; Writing—original draft, B.P. and B.G.; Writing—review and editing, B.P. and J.Č. All authors have read and agreed to the published version of the manuscript.

Funding: This work was supported by ARRS under Projects P2-0115 and J2-1742.

Institutional Review Board Statement: Not applicable.

Informed Consent Statement: Not applicable.

Data Availability Statement: This study did not report any data.

Acknowledgments: The authors would like to thank ELES, Ltd., Electricity Transmission System Operator.

Conflicts of Interest: The authors declare no conflict of interest.

Abbreviations

The following abbreviations are used in this manuscript:

IIDG	Inverter-Interfaced Distributed Generation
FFC	Fast Fault Current
BFFC	Balanced FFC
UFFC	Unbalanced FFC
SSC	Short-Circuit Capability
BES	Battery Energy Storage
PV	Photovoltaic
SS	Substation
3-Ph	Three-Phase
Ph-Ph	Phase-to-Phase
Ph-G	Phase-to-Ground

References

1. EUR-Lex. *Establishing a Network Code on Requirements for Grid Connection of Generators;* Commission Regulation (EU) 2016/631; European Union: Brussels, Belgium, 2016.
2. *VDE-AR-N 4120:2017-05, Technical Requirements for the Connection and Operation of Customer Installations to the High-Voltage Network (TCC High-Voltage);* VDE Verlag GmbH: Berlin, Germany, 2017.
3. Modeling and Validation Work Group; Renewable Energy Modeling Task Force. *WECC Wind Plant Dynamic Modeling Guidelines;* WECC: Salt Lake City, UT, USA, 2014.
4. WECC Modeling and Validation Work Group. *Solar Photovoltaic Power Plant Modeling and Validation Guideline;* WECC: Salt Lake City, UT, USA, 2019.
5. Farantatos, E. *Model User Guide for Generic Renewable Energy System Models;* Technical Update 3002014083; EPRI: Palo Alto, CA, USA, 2018.
6. Mirhosseini, M.; Pou, J.; Agelidis, V.G. Single- and Two-Stage Inverter-Based Grid-Connected Photovoltaic Power Plants with Ride-Through Capability under Grid Faults. *IEEE Trans. Sustain. Energy* **2015**, *6*, 1150–1159. [CrossRef]
7. Celik,D.; Meral, M.E. Voltage Support Control Strategy of Grid-Connected Inverter System under Unbalanced Grid Faults to Meet Fault Ride through Tequirements. *IET Gener. Transm. Distrib.* **2020**, *14*, 3198–3210. [CrossRef]
8. Camacho, A.; Castilla, M.; Miret, J.; Vasquez, J.C.; Alarcon-Gallo, E. Flexible Voltage Support Control for Three-Phase Distributed Generation Inverters under Grid Fault. *IEEE Trans. Power Electron.* **2013**, *60*, 1429–1441. [CrossRef]
9. Popadic, B.; Dumnic, B.; Strezoski, L. Modeling of Initial Fault Response of Inverter-Based Distributed Energy Resources for Future Power System Planning. *Int. J. Electr. Power Energy Syst.* **2020**, *117*, 105722. [CrossRef]
10. Meyer, R.; Zlotnik, A.; Mertens, A. Fault Ride-Through Control of Medium-Voltage Converters with LCL Filter in Distributed Generation Systems. *IEEE Trans. Ind. Appl.* **2014**, *50*, 3448–3456. [CrossRef]
11. Guo, X.; Liu, W.; Lu, Z. Flexible Power Regulation and Current-Limited Control of the Grid-Connected Inverter under Unbalanced Grid Voltage Faults. *IEEE Trans. Ind. Electron.* **2017**, 64, 7425–7432. [CrossRef]
12. Afshari, E.; Moradi, G.R.; Rahimi, R.; Farhangi, B.; Yang, Y.; Blaabjerg, F.; Farhangi, S. Control Strategy for Three-Phase Grid-Connected PV Inverters Enabling Current Limitation Under Unbalanced Faults. *IEEE Trans. Ind. Electron.* **2017**, *64*, 8908–8918. [CrossRef]
13. Camacho, A.; Castilla, M.; Miret, J.; Borrell, A.; Garcia de Vicuna, L. Active and Reactive Power Strategies with Peak Current Limitation for Distributed Generation Inverters during Unbalanced Grid Faults. *IEEE Trans. Ind. Electron.* **2015**, *62*, 1515–1525. [CrossRef]
14. Wen, H.; Fazeli, M. A Low-Voltage Ride-Through Strategy Using Mixed Potential Function for Three-Phase Grid-Connected PV Systems. *Electr. Power Syst. Res.* **2019**, *173*, 271–280. [CrossRef]
15. Camacho, A.; Castilla, M.; Miret, J.; Guzman, R.; Borrell, A. Reactive Power Control for Distributed Generation Power Plants to Comply with Voltage Limits During Grid Faults. *IEEE Trans. Power Electron.* **2014**, *29*, 6224–6234. [CrossRef]
16. Islam, M.; Nadarajah, M.; Hossain, M.J. A Grid-Support Strategy with PV Units to Boost Short-Term Voltage Stability under Asymmetrical Faults. *IEEE Trans. Power Syst.* **2020**, *35*, 1120–1131. [CrossRef]
17. Shuvra, M.A.; Chowdhury, B. Distributed Dynamic Grid Support Using Smart PV Inverters during Unbalanced Grid Faults. *IET Renew. Power. Gener.* **2019**, *13*, 598–608. [CrossRef]
18. Khan, H.; Chacko, S.J.; Fernandes, B.G.; Kulkarni, A. Reliable and Effective Ride-Through Controller Operation for Smart PV Systems Connected to LV Distribution Grid under Abnormal Voltages. *IEEE J. Emerg. Sel. Top. Power Electron.* **2020**, *8*, 2371–2384. [CrossRef]
19. Taul, M.G.; Wang, X.; Davari, P.; Blaabjerg, F. Current Reference Generation Based on Next-Generation Grid Code Requirements of Grid-Tied Converters during Asymmetrical Faults. *IEEE J. Emerg. Sel. Top. Power Electron.* **2020**, *8*, 3784–3797. [CrossRef]
20. Garnica, M.; Garcia de Vicuna, L.; Miret, J.; Castilla, M.; Guzman, R. Optimal Voltage-Support Control for Distributed Generation Inverters in RL Grid-Faulty Networks. *IEEE Trans. Power Electron.* **2020**, *67*, 8405–8415. [CrossRef]

21. Zeineldin, H.H.; Sharaf, H.M.; Ibrahim, D.K.; El-Zahab, E.E.-D.A. Optimal Protection Coordination for Meshed Distribution Dystems with DG Using Dual Setting Directional Over-Current Relays. *IEEE Trans. Smart Grid* **2015**, *6*, 115–123. [CrossRef]
22. Huchel, L.; Zeineldin, H.H. Planning the Coordination of Directional Overcurrent Relays for Distribution Systems Considering DG. *IEEE Trans. Smart Grid* **2016**, *7*, 1642–1649. [CrossRef]
23. Pereira, K.; Pereira, B.R.; Contreras, J.; Mantovani, J.R.S. A Multiobjective Optimization Technique to Develop Protection Systems of Distribution Networks with Distributed Generation. *IEEE Trans. Power Syst.* **2018**, *33*, 7064–7075. [CrossRef]
24. Katyara, S.; Staszewski, L.; Leonowicz, Z. Protection Coordination of Properly Sized and Placed Distributed Generations-Methods, Applications and Future Scope. *Energies* **2018**, *11*, 2627. [CrossRef]
25. Bui, D.M.; Le, P.D.; Nguyen, T.P.; Nguyen, H. An adaptive and scalable protection coordination system of overcurrent relays in distributed-generator-integrated distribution networks. *Appl. Sci.* **2021**, *11*, 8454. [CrossRef]
26. Plet, C.A.; Green, T.C. Fault Response of Inverter Interfaced Distributed Generators in Grid-Connected Applications. *Electr. Power Syst. Res.* **2014**, *106*, 21–28. [CrossRef]
27. Pinto, J.O.C.P.; Moreto, M. Protection Strategy for Fault Detection in Inverter-Dominated Low Voltage AC Microgrid. *Electr. Power Syst. Res.* **2021**, *190*, 106572. [CrossRef]
28. Barnes, A.K.; Tabarez, J.E.; Mate, A.; Bent, R.W. Optimization-Based Formulations for Short-Circuit Studies with Inverter-Interfaced Generation in PowerModelsProtection.jl. *Energies* **2021**, *14*, 2160. [CrossRef]
29. Bakkar, M.; Bogarra, S.; Corcoles, F.; Iglesias, J. Overcurrent Protection Based on ANNs for Smart Distribution Networks with Grid-Connected VSIs. *IET Gener. Transm. Distrib.* **2020**, *15*, 1159–1174. [CrossRef]
30. Zarei, S.F.; Khankalantary, S. Protection of Active Distribution Networks with Conventional and Inverter-Based Distributed Generators. *Int. J. Electr. Power Energy Syst.* **2021**, *129*, 106746. [CrossRef]
31. Salem, M.M.; Elkalashy, N.I.; Atia, Y.; Kawady, T.A. Modified Inverter Control of Distributed Generation for Enhanced Relaying Coordination in Distribution Networks. *IEEE Trans. Power Deliv.* **2017**, *32*, 78–87. [CrossRef]
32. Yousaf, M.; Muttaqi, K.M.; Sutanto, D. A Control Strategy to Mitigate the Sensitivity Deterioration of Overcurrent Protection in Distribution Networks with the Higher Concentration of the Synchronous and Inverter-Based DG Units. *IEEE Trans. Ind. Appl.* **2021**, *57*, 2298–2306. [CrossRef]
33. Aljarrah, R.; Marzooghi, H.; Yu, J.; Terzija, V. Sensitivity Analysis of Transient Short Circuit Current Response to the Penetration Level of Non-Synchronous Generation. *Int. J. Electr. Power Energy Syst.* **2021**, *125*, 106556. [CrossRef]
34. ENTSO-E. *Short Circuit Contribution of New Generating Units Connected with Power Electronics and Protection Behaviour*; ENTSO-E: Brussels, Belgium, 2019.
35. Ziegler, G. *Numerical Distance Protection: Principles and Applications*, 4th ed.; Publicis Publishing: Erlangen, Germany, 2011.
36. Werstiuk, C. *The Relay Testing Handbook: Principles and Practice*; Valence Electrical Training Services: Littleton, CO, USA, 2012.

applied sciences

MDPI

Article

Maintenance Management of a Transmission Substation with Optimization

Peter Kitak [1,*], Lovro Belak [2], Jože Pihler [1] and Janez Ribič [1]

[1] Faculty of Electrical Engineering and Computer Science, University of Maribor, 2000 Maribor, Slovenia; joze.pihler@guest.um.si (J.P.); janez.ribic@um.si (J.R.)
[2] Slovenian Transmission System Operator ELES, 1000 Ljubljana, Slovenia; lovro.belak@eles.si
* Correspondence: peter.kitak@um.si

Abstract: The paper deals with the reliability-centered maintenance (RCM) of a transmission substation. The process of the planning and actual performance of maintenance was carried out using an optimization algorithm. This maintenance procedure represents the maintenance management and included reliability of the power system operation, maintenance costs, and associated risks. The originality of the paper lies in the integrated treatment of all maintenance processes that are included in the pre-processing and used in the optimization process for reliability-centered maintenance. The optimization algorithm of transmission substation maintenance was tested in practice on the equipment and components of an existing 400/110–220/110 kV substation in the Slovenian electricity transmission system. A comparison analysis was also carried out of the past time-based maintenance (TBM) and the new reliability-centered maintenance (RCM), on the basis of the optimization algorithm.

Keywords: reliability-centered maintenance; optimization; transmission substation; condition monitoring

Citation: Kitak, P.; Belak, L.; Pihler, J.; Ribič, J. Maintenance Management of a Transmission Substation with Optimization. *Appl. Sci.* **2021**, *11*, 11806. https://doi.org/10.3390/app112411806

Academic Editors: Federico Barrero and Mario Bermúdez

Received: 9 November 2021
Accepted: 9 December 2021
Published: 12 December 2021

1. Introduction

Maintenance is a combination of technical, administrative, and managerial actions during the lifetime of a device, the purpose of which is to keep it in, or bring it back to, the condition that enables performing of its functions. It is, therefore, a usual process needed by every device for normal operation. The maintenance in transmission substations (TS) is crucial for secure and reliable operation of the electric power system (EPS). The maintenance terminology covers two types of maintenance: preventive and corrective. In the field of electricity transmission devices, preventive maintenance still prevails [1]. This type of maintenance can be either time-based maintenance or related to the state of devices, condition-based maintenance (CBM) [2]. From 2009 onwards, the health index [3–5] has been used in the field of condition-based maintenance to evaluate indicators of maintenance.

However, new trends in the field of maintenance, i.e., reliability -centered maintenance (RCM), are being introduced in the wider area of engineering [6,7], as well as in the field of maintenance of devices in the electricity transmission system [1,8–10]. These trends are also accelerated by the standardization in the field of maintenance and asset management [11,12].

The reliability of operation and associated maintenance is, in the majority of cases, based on the reliability calculation using the Markov model [9,13–15]. An adequate efficiency of the determination of system criteria for maintenance can be achieved with the selection of various algorithms, such as the best–worst method (BWM) [16], where numerous possibilities are assessed on the basis of determination with regard to various attributes, and the best maintenance criteria are selected.

In the maintenance process, authors have also included optimization procedures dealing with economy and reliability [10,17–19]. The authors usually deal with individual elements, such as overhead lines [5,20] and transformers [21,22]. References [9,23] deal with maintenance of devices in a TS. The authors describe a maintenance method that is based on the technical condition of the devices. Certain authors have also investigated time scheduling of maintenance tasks [14,24] or determination of the optimal inspection interval of equipment in the TS, taking into account its age [25]. The authors of the presented paper, in [26], discussed the strategic maintenance of switching substations where reliability indices were not optimized.

The contribution of this paper is in the calculations of reliability indicators using the failure effect analysis (FEA) method, and not with the calculations of Markov chains for an individual device. In this method, any change in the state of an individual device affects the entire TS, which can, consequently, change the state of other devices as well. The maintenance processes of the devices in the substation are closely related to the reliability indicators (one of them is importance) for the entire TS, and, at the same time, to the condition of the individual devices, which is one of the article's contributions.

The novelty presented in the present paper lies in the integrated treatment of TS maintenance based on the RCM concept, using an optimization algorithm. The modified differential evolution optimization algorithm with self-adaptation (SA) of the control parameters was used to select the optimal maintenance period of TS devices (revisions are carried out every year, every two years, or every three years), while the maintenance is carried out when the significant operational state of the EPS is the most suitable for it. This state is defined by generation, consumption, and power flows in the EPS. The method of EPS elements' maintenance used until now is TBM, where their maintenance periods are defined by the TSO's internal rules. The main objective of this approach is to reduce maintenance costs while keeping the operational reliability level and improving the maintenance system constantly. This procedure is possible only with the inclusion of optimization algorithms in the maintenance process. Devices that are less important in the system, however, affect maintenance processes with reduced intensity.

The maintenance processes can be influenced by computation of the importance and technical condition of EPS elements, which is based on historical data of operation, events, and monitoring of EPS elements' technical condition. This represents the main hypothesis of this article. The inclusion of optimization processes in the analysis of maintenance procedures enables a reduction of maintenance costs, which represents the second hypothesis.

The methodology of computation of frequency of EPS elements' maintenance influences the maintenance processes and is based on available historical operation data and statistics of events on EPS elements. For the entire EPS, we computed the availability of elements in connection with the transmissioned energy. The availability of elements and transmissioned energy to final customers are reliability indicators. The values of these indicators for individual elements were then compared with the reliability indicators of similar elements operated by other system operators. The importance of elements were computed on the basis of these data. In connection with the technical conditions of elements that are performed in practice through monitoring, it is possible to influence the maintenance processes on the basis of the RCM maintenance concept.

The optimization process that uses all possible data on historical events of operation and maintenance costs yields the results that can be used to manage the maintenance processes in the future. This represents the main originality of this paper. The existing method of assessing the technical condition of EPS elements was upgraded and included actively in the optimization process, where we predicted a dynamic changing of the technical condition of elements due to the interventions to the existing maintenance process.

The significant states in the system are determined on the basis of EPS operation data that include generation, nodal loads, and load flows between the nodes. Their inclusion in the optimization process enables that maintenance activities are performed in the most suitable significant state. Thus, the maintenance activities have the least impact on the EPS operation. Indirectly, this also influences operational reliability and maintenance costs. The optimization process in our article affects the frequency of transitions between maintenance processes.

The past data on the operation of transmission elements that are covered by statistics of events were the basis for calculation of reliability indicators of these elements and their unavailability. The values of these indicators for individual elements were compared with reliability indicators of similar elements operated by other system operators.

The paper consists of six sections. Section 2 presents the RCM concept-based maintenance model. A switching bay was represented with an index of technical condition and an index of importance. The costs of the existing maintenance concept were analyzed in detail. Section 3 provides the basic data for the optimization model. The key data in the RCM maintenance process were the values of technical condition and reliability indicators for every EPS element. The data are also given on the expected costs of outages and maintenance costs for the example presented in this article. Section 4 presents the link between the maintenance model and the optimization algorithm. Three objective functions and one penalty function were defined to reach the optimal RCM maintenance concept. The results of the optimization process are given in Section 5. The optimal method and period of maintenance were defined accurately for each EPS element. A comparison was also given of maintenance costs with the existing time-based maintenance and with the new RCM maintenance concept. Section 6 provides the discussion and outlines the future work.

2. Maintenance Model Based on the RCM Concept

RCM is a maintenance concept that, in addition to the basic maintenance methodologies, is focused on reliability. The objective of the RCM concept is management of maintenance costs and associated risks, having in mind the provision of adequate reliability of operation of the EPS. It is important not to treat reliability at the level of individual elements, but at the level of the entire system. Maintenance tasks carried out during maintenance are as follows: periodical inspection of elements performed on the energized state of the equipment; revision intended to retain elements' functionality, performed on the de-energized state of equipment; and repair or replacement of elements. Elements with lower importance and good technical condition are maintained in the RCM concept using TBM in a longer time period. The main aim of RCM in this paper was to determine the maintenance periods of EPS elements with regard to their importance and condition, considering reliability and costs. This is the main change with regard to the TBM concept, where maintenance periods are determined in advance. The RCM concept in the paper is based on [1], while the basic concept uses the standard [11].

Section 2 is dedicated entirely to the RCM concept and consists of seven subsections. The elements are defined in Section 2.1 EPS. This includes a general set of elements and the term switching bay. The structure of the model of input data for optimization in the maintenance process is presented in Section 2.2. Since we wished to include ecology in the optimization process, Section 2.3 defines the characteristic variable of diagnostics of drops in pressure of the SF_6 insulation gas. One of the key parameters in the RCM process is the index of technical condition. For each type of EPS element, there is a methodology for assessing their technical condition, which is described in Section 2.4. The next key parameter in the RCM concept is the index of importance of an EPS element, which is described in detail in Section 2.5. The cost-related part of our concept is dealt with in Sections 2.6 and 2.7.

2.1. Definition of EPS Elements

The subject of our analyses is the EPS. It comprises generation units, lines, and substations. In substations, there are disconnectors, circuit breakers, power transformers, and other elements. The elements of the EPS represent a set of elements $S_{\mathrm{el}} = \{el_k; 1 \leq k \leq n_{\mathrm{el}}; k \in \mathrm{N}\}$, where n_{el} is the number of elements in the entire EPS, *el* is the element in the EPS, and *k* is the counter of the EPS elements. All these elements form a database with characteristic variables, such as estimation of technical condition, economic indicators, reliability indicators, and indicator of importance. A set of elements $S_{\mathrm{kel}} = \{kel_r; 1 \leq r \leq n_{\mathrm{kel}}; r \in \mathrm{N}\}$ is created, where n_{kel} is the number of kinds of elements used in the EPS, *kel* is the kind of elements used in the EPS, and r is the counter of kinds of elements used in the EPS. Data on the entire EPS are needed for the computation of the reliability indicators of the EPS, technical condition of elements, and maintenance costs. From the set of elements S_{el}, only those elements were observed that are a part of the TS.

The model enables scheduling of maintenance tasks with regard to the state of the system, taking into consideration economic effects and reliability of operation of individual switching bays in the TS.

The analysis includes only switching bays and transformers that are elements of the observed TS. A switching bay is a set of elements in the substation that belongs to the main element (transformer or transmission line). A switching bay is an assembly of switchgear elements (disconnectors, circuit breakers, etc.), and can be either line bay, transformer bay, or bus coupler bay [27]. Bays in the TS form a set of bays $S_{\mathrm{SB}} = \{S_{\mathrm{SB},l};\ S_{\mathrm{SB},l} \subset S_{\mathrm{el}};\ 1 \leq l \leq n_{\mathrm{SB}};\ l \in \mathrm{N}\}$, where n_{SB} is the number of switching bays. The lth bay in the TS comprises a certain number of elements and forms the set of elements of the lth bay $S_{\mathrm{SB},l} = \{el_{\mathrm{SB},lm};\ 1 \leq m \leq n_{\mathrm{SB},l};\ m \in \mathrm{N}\}$, being a subset of the set of elements S_{el}, where $n_{\mathrm{SB},l}$ is the number of elements in the lth bay of the TS and m is the counter of elements in the TS bay. A switching bay can be considered as a new EPS element. The reason for this is that, during revision, the entire bay is always in a de-energized state and is maintained as a whole. The analysis of a TS also includes power transformers. The set of bays in the TS is, thus, case extended to the number of transformers n_{T}. It is defined as $S_{\mathrm{T}} = \{TR_{ll};\ 1 \leq ll \leq n_{\mathrm{T}};\ ll \in \mathrm{N}\}$, where ll is the counter of transformers in the TS. The set of TS elements, therefore, consists of switching bays and transformers, and is defined as $S_{\mathrm{SU}} = \{S_{\mathrm{SU},q};\ S_{\mathrm{SU}} = S_{\mathrm{SB}} \cup S_{\mathrm{T}};\ 1 \leq q \leq n_{\mathrm{SU}};\ q \in \mathrm{N};\ n_{\mathrm{SU}} = n_{\mathrm{SB}} + n_{\mathrm{T}}\}$, where q is the counter of switching bays and transformers and n_{SU} is the number of observed switching bays and transformers of the TS.

2.2. The Structure of the Model

Figure 1 shows a block diagram of the preparation of input data for the optimization of maintenance tasks in transmission substations.

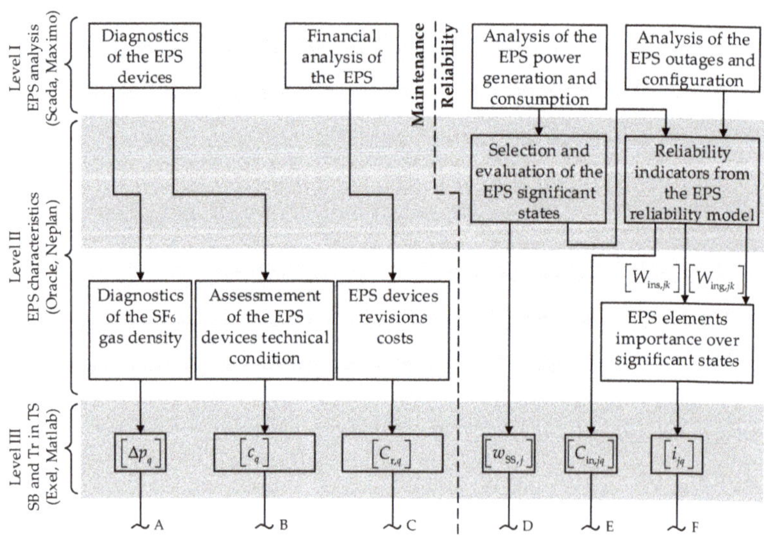

Figure 1. Block diagram of the RCM concept-based maintenance model.

The following data are needed for this optimization: drop in pressure of SF_6 in circuit breakers (CB), estimation of the technical condition of substation elements, financial data on switching bays' maintenance, and reliability indicators for elements of switching bays and transformers.

2.3. Diagnostics of the Density of SF_6 Gas

Diagnostics of the EPS devices is performed as a part of regular maintenance activities of EPS elements, and includes inspections, measurements, etc. One of the tasks of diagnostic activities are measurements of drops in pressure Δp of the SF_6 insulation gas in circuit breakers. Circuit breakers in the switching bays are a part of the set of all SF_6 circuit breakers in the EPS: $S_{SF6} = \{CB_{SF6,kk}; CB_{SF6,kk} \subset S_{el}; 1 \leq kk \leq n_{SF6}; kk \in \mathrm{N}\}$. The characteristic variable of the set of circuit breakers S_{SF6} is the vector of drop in pressure in the SF_6 circuit breakers $\Delta \boldsymbol{p}_{SF6} = [\Delta p_{kk}]$, where kk is the counter of SF_6 circuit breakers and n_{SF6} is the number of SF_6 circuit breakers in the TS.

If there is an SF_6 circuit breaker in the lth bay (e.g., the kkth element), the drop in pressure Δp_{kk} can be considered as the characteristic variable of this element. In this case, we can write $\Delta p_l = \Delta p_{kk}$, otherwise $\Delta p_l = 0$. All the bays in the TS form a vector of drops in pressure in circuit breakers $\Delta \boldsymbol{p}_{SB} = [\Delta p_{SB,l}]$. The vector of drops in pressure of SF_6 for all TS elements is, thus, defined as $\Delta \boldsymbol{p} = [\Delta p_q]$, where $\Delta p_q = \Delta p_{SB,l}$ for $1 \leq q \leq n_{SB}$, $l = q$ and $\Delta p_q = \Delta p_{T,ll} = 0$ for $n_{SB} + 1 \leq q \leq n_{SB} + n_T$ and $ll = q - n_{SB}$. These are the input data for the optimization procedure.

2.4. Index of Technical Condition c

Diagnostics of the EPS devices are also the basis for estimation of the technical condition of the EPS devices. For each kind of EPS element, there is an adopted methodology for estimation of the technical condition of devices c_k [26]. The index of the technical condition is determined on the basis of the Slovenian transmission system operator internal application for determination of technical condition c_k, which contains a set of 18 criteria with an adequate weighting factor (from 1 to 10), and each rating factor (from 1—good to 10—bad). The rating factor is determined by the engineer responsible for monitoring the switching substation for each corresponding criterion of the element.

The values of the technical condition of an element can lie in the range $0 \leq c_k \leq 100$. The higher the value is, the worse the technical condition of the EPS element. The technical condition of all EPS elements is evaluated in this way and is included in $\boldsymbol{c}_s = [c_k]$. The lth switching bay of the TS contains a certain number of elements $n_{SB,l}$ and forms the set of elements of this bay. The characteristic variable of the set $S_{SB,l}$ is the vector of technical condition of the elements of this set, which is defined as $\boldsymbol{c}_{SB,l} = [c_{SB,lm}]$. The technical condition c_l of the lth bay can be defined as the maximum estimated value of technical condition of this bay's elements, i.e., as $c_l = \max(\boldsymbol{c}_{SB,l})$. The vector $\boldsymbol{c}_{SB} = [c_l]$ can be defined for all the bays in the substation. The values of technical condition for power transformers are defined as $\boldsymbol{c}_T = [c_{T,ll}]$. The vector of technical condition for the entire TS, including both bays and transformers, $\boldsymbol{c} = [c_q]$, is obtained using the identical procedure, as described in Section 2.3.

2.5. Index of Importance i

To calculate the importance of the EPS elements, the reliability indicators need to be calculated first. The first step in this calculation is a detailed analysis of the EPS as a whole. For the calculation of reliability indicators, we used a powerful program tool for steady-state calculations in the EPS, Neplan [26]. The Neplan program tool for calculation of reliability indicators requires data on outages of all EPS elements, as well as hourly data on generation and consumption for all EPS nodes. The analysis begins with a statistical survey of hourly data on generation and consumption in all EPS nodes, as well as an analysis of power flows in certain parts of the EPS in a certain period (one year or more). A clustering approach is used on the basis of historical data. Average values of power (μ_P) and deviations from the average value (σ_P) are calculated for generation, consumption, and power flows. Combinations of strings of hourly data for the total power P_{SYS} are selected, which is defined as the total power of generation, consumption, and power flows for the entire EPS. The n_{SS} sets of significant states (SS) are selected for generation, consumption, and power flows for the combinations that appear most often. SS are states in the EPS with a certain quantity of generation, load flows, and consumption. Significant states are each a static snapshot that combines generation, consumption, and transmission. SS can be a variable (SS).

$S_{SS} = \{h_{SS,j}; 1 \leq j \leq n_{SS}; j \in \mathrm{N}\}$, where $h_{SS,j}$ is the characteristic hour from the selected jth set of states. Table 1 presents 9 selected SS for generation, consumption, and power flows that, to a certain extent, reflect the state of the EPS in the observed period. This is important in deciding when to implement maintenance of the TS elements.

Table 1. Selected significant states for EPS.

P_{SYS} (MW)	SS_1	SS_2	SS_3	SS_4	SS_5	SS_6	SS_7	SS_8	SS_9
Power generation	+	−	+	++	+	++	++	+	++
Consumption	−	−	+	+	+	−	−	−	−
Power flow	−	+	−	−	−	+	−	++	−
$w_{SS,j}$	0.315	0.086	0.029	0.020	0.167	0.221	0.079	0.001	0.003

Legend: + mean $P_{SYS} > P_{avg}$; ++ mean P_{peak}; − mean $P_{SYS} < P_{avg}$.

It is also possible to calculate the shares of frequency of occurrence of SS that are characterized by generation and consumption in the selected hours of the observed period. Thus, the vector of shares of selected states in the EPS $\boldsymbol{w}_{SS} = [w_{SS,j}]$; $0 < w_{SS,j} < 1$ is calculated (Table 1). The weight $w_{SS,j}$ depends on the share of hours in an individual set $S_{SS,j}$ of the whole set of hours.

For a certain past period, the data on outages of EPS elements are collected and modified in the reliability model of the Neplan program tool. In the model, it is also necessary to make possible changes of the EPS configuration. For each chosen jth SS reliability indicators are calculated in Neplan using the failure effect analysis (FEA) [28]. For the subsequent calculations, the costs are used of expected long duration outages of the EPS elements. The data on energy not supplied to the consumers $W_{ins,jk}$ in the case of an outage of a certain EPS element are included in the matrix $\boldsymbol{w}_{ins} = [w_{ins,jk}]$, while the matrix $\boldsymbol{w}_{ing} = [w_{ing,jk}]$ contains data on energy not produced in the generation units $W_{ing,jk}$. $W_{ins,jk}$ and $W_{ing,jk}$ for every significant state and every certain element of EPS are calculated with FEA in the program tool Neplan. They are reliability indicators.

For each element r from the set of elements S_{el}, the subset $S_{el,jrv}$ is created for each significant state j, kind of elements r, and a certain voltage level v $S_{el,jrv} = \{W_{ins,jkrv}, W_{ing,jkrv}; 1 \leq k \leq n_{kel}; 1 \leq v \leq 3; k, v \in \mathrm{N}\}$. For each significant state j and kind of elements r, $W_{inmax,rv}$ is determined, which is the maximum value from the set $S_{el,jrv}$, and for a certain voltage level v from the set S_{VL}; S_{VL} = {110 kV, 220 kV, 400 kV}. On the basis of these data, we are able to calculate the importance of the kth EPS element for the jth SS using kind of element r, and certain voltage level v is calculated as

$$i_{jk,rv} = \frac{0.8 \cdot W_{ins,\, jkrv} + 0.2 \cdot W_{ing,\, jkrv}}{W_{inmax,rv}} \quad (1)$$

which is based on experiences of the transmission system operator (TSO) ELES, and foresees that energy not supplied contributes 80% to the importance, while energy not produced contributes the remaining 20%, where v is a counter of voltage levels between $1 \leq v \leq 3$. Thus, the matrix of importance $\boldsymbol{I}_{SSEPS} = [i_{jk}]$ is determined for all SS and the entire EPS.

The matrix of importance for the lth switching bay of the TS can be created for all significant states as $\boldsymbol{I}_{SB,l} = [i_{SB,ljm}]$. From the matrix $\boldsymbol{I}_{SB,l}$ for the jth SS, the vector $\boldsymbol{i}_{SB,jl} = [i_{SB,jlm}]$; $1 \leq m \leq n_{SB,l}$ is created, and, for the jth SS, the total importance of the lth bay is determined using the equation $i_{jl} = \max(\boldsymbol{i}_{SB,jl})$. From these vectors, the matrix of importance for all significant states $\boldsymbol{I}_{SB} = [i_{jl}]$ is created for all TS bays. In a similar way, the matrix of importance is created for power transformers in the TS $\boldsymbol{I}_{\mathrm{T}} = [i_{\mathrm{T},jll}]$. The total importance of all TS elements is obtained by joining the matrices $\boldsymbol{I}_{\mathrm{SB}}$ and $\boldsymbol{I}_{\mathrm{T}}$ to $\boldsymbol{I} = [i_{jq}]$, where $i_{jq} = i_{\mathrm{SB},jl}$ for $1 \le j \le n_{\mathrm{SS}}$, $1 \le q \le n_{\mathrm{SB}}$ and $l = q$, and $i_{jq} = i_{\mathrm{T},jll}$ for $1 \le j \le n_{\mathrm{SS}}$ and $n_{\mathrm{SB}} + 1 \le q \le n_{\mathrm{SB}} + n_{\mathrm{T}}$, and $ll = q - n_{\mathrm{SB}}$.

2.6. Past Maintenance Cost

It is necessary to perform a detailed financial analysis of the entire EPS for the determination of maintenance costs, which includes analysis of incomes, maintenance costs, and replacements of TS elements. This analysis is carried out by a unified information system (IBM Maximo). From the database, it is possible to obtain revision costs for all EPS elements $\boldsymbol{C}_{\mathrm{rEPS}} = [C_{\mathrm{r},k}]$ for the past period in EUR/year. For the lth bay of the TS, it is possible to create the vector of revision costs, defined as $\boldsymbol{C}_{\mathrm{rSB},l} = [C_{\mathrm{rSB},lm}]$. The total annual revision costs for the lth bay are defined as the sum of costs for all its elements:

$$C_{\mathrm{r},l} = \sum_{m=1}^{n_{\mathrm{SB},l}} C_{\mathrm{rSB},lm} \tag{2}$$

Using (2), the vector is obtained of all revision costs by switching bays $\boldsymbol{C}_{\mathrm{rSB}} = [C_{\mathrm{r},l}]$. The vector of revision costs for power transformers $\boldsymbol{C}_{\mathrm{rT}} = [C_{\mathrm{rT},ll}]$ is obtained in a similar way. The total revision costs for all TS elements $\boldsymbol{C}_{\mathrm{r}}$ are obtained using the above-described procedure for determination of the vector $\Delta\boldsymbol{p}$.

2.7. Expected Costs of Outages

The computation of the reliability model in the Neplan program tool yields as the final result FEA for each element k and each SS j the total anticipated costs of outage of all consumers by power $C_{\mathrm{inP},jk}$ in EUR/year, caused by an outage of the element k in SS j (which are affected by outage of the element k in SS j). The expected costs depend on individual contracts between the TSO and each consumer, the power of consumption, and the duration of each customer's outage that is affected by an outage of element k in SS j. Following the FEA procedure, the maximum value of reliability indicators is obtained as a result. This method captures all possible changes in the state of devices in the TS.

The scaled costs of total energy not supplied due to an outage of the element k in SS j as $C_{\mathrm{inWs},jk} = c_{\mathrm{inWs}} \cdot W_{\mathrm{ins},jk}$ (EUR/year) are also computed, where c_{inWs} are specific costs of energy not supplied, which are defined by the TSO (c_{inWs} = 5000 EUR/MWh). The last scaled costs are the costs of energy not generated due to an outage of the element k in SS j as $C_{\mathrm{inWg},jk} = c_{\mathrm{inWg}} \cdot W_{\mathrm{ing},jk}$ (EUR/year), where c_{inWg} are specific costs of generated energy, defined by the TSO (c_{inWg} = 60 EUR/MWh). The total costs of expected outages due to an outage of the element k in SS j are calculated as the sum of costs $C_{\mathrm{in},jk} = C_{\mathrm{inP},jk} + C_{\mathrm{inWs},jk} + C_{\mathrm{inWg},jk}$ (EUR/year). They are, for all significant states, collected in the matrix $\boldsymbol{C}_{\mathrm{ins}} = [C_{\mathrm{in},jk}]$.

For the lth bay of the TS, the matrix of expected outage costs could be created for all significant states $\boldsymbol{C}_{\mathrm{inSB},l} = [C_{\mathrm{inSB},ljm}]$. The total expected costs of outages are calculated using (3).

$$C_{\mathrm{in},\,jl} = \sum_{m=1}^{n_{\mathrm{SB},l}} C_{\mathrm{inSB},ljm} \tag{3}$$

The matrix of costs of expected outages $\boldsymbol{C}_{\mathrm{inSB}} = [C_{\mathrm{in},jl}]$ could be created for all significant states and TS bays. The matrix of costs of expected outages for power transformers is $\boldsymbol{C}_{\mathrm{inT}} = [C_{\mathrm{inT},jll}]$. The matrix of costs of expected outages for all TS elements $\boldsymbol{C}_{\mathrm{in}} = [C_{\mathrm{in},jq}]$ is obtained using the same procedure as described above for the calculation of importance $\boldsymbol{I}$.

3. Data for the Maintenance Model of an Existing 400/110–220/110 kV Transmission Substation

The optimization of maintenance tasks is shown on a practical example of a Slovenian transmission substation. The analysis included all primary devices on 400, 220, and 110 kV levels in the substation. There were altogether n_{SB} = 26 bays, 7 of them on 400 kV, 5 on 220 kV, and 14 on a 110 kV level. In addition to the bays, there were also n_{T} = 5 power transformers, two of them 220/110 kV, one 400/110 kV, and two components of a 400/400 kV phase-shifting transformer. The TS model therefore comprised n_{SU} = 31 elements.

3.1. Data on Technical Condition c and Importance i

Table 2 shows the calculated data on the importance of individual significant states of the RCM model of the TS i_{jq}, average importance of bays by various significant states i_q, technical condition (technical indicator) c_q, and average deviation d_q for all bays, calculated using (5).

Table 2. TS elements technical and reliability indicators.

q	Code	Voltage Level (kV)	Importance i_{qj} SS_1	SS_2	SS_3	SS_4	SS_5	SS_6	SS_7	SS_8	SS_9	Average Importance $i_{avg,q}$	Condition c_q	Distance d_q
							Switching bay arrangements							
1	T401	400	0.96	0.68	0.49	0.94	1.55	0.75	0.16	0.14	10.29	0.86	24.78	18.13
2	L401	400	4.44	0.60	0.53	0.91	2.06	3.65	2.60	100	100	3.24	60.22	44.87
3	L402	400	4.61	0.53	1.13	0.73	2.15	4.07	2.22	0.10	1.36	2.98	24.78	19.63
4	L403	400	0.96	0.68	0.49	0.95	1.55	0.75	0.22	0.14	11.14	0.87	24.78	18.14
5	C401	400	5.99	0.71	1.00	0.94	5.24	2.12	4.73	100	1.50	3.82	25.65	20.84
6	T402	400	4.60	0.60	0.52	0.91	2.12	4.05	0.72	100	5.46	2.95	17.39	14.39
7	T403	400	4.60	0.60	0.52	0.91	2.12	4.05	0.72	100	5.46	2.95	17.39	14.39
8	T201	220	42.31	100	0.34	0.76	2.36	0.24	0.38	4.83	0.43	22.43	30.65	37.54
9	T202	220	42.31	100	0.34	0.76	2.36	0.24	0.38	4.83	0.43	22.43	41.74	45.38
10	L201	220	42.31	100	0.34	0.76	12.16	0.24	0.38	4.83	0.43	24.07	20.87	31.78
11	L202	220	69.13	100	0.34	0.76	9.09	0.24	0.38	4.83	0.43	32.01	20.87	37.39
12	L203	220	42.31	100	0.34	0.76	17.91	0.24	0.38	4.83	0.43	25.03	20.87	32.46
13	T101	110	0.02	1.87	4.89	2.64	0.24	0.02	2.35	100	0.00	0.69	42.39	30.46
14	T102	110	0.02	1.76	4.75	2.51	0.24	0.02	2.37	100	0.00	0.68	50.43	36.14
15	L101	110	0.02	1.91	5.46	2.96	0.70	0.03	1.81	100	0.00	0.75	52.17	37.43
16	T103	110	0.01	0.35	0.97	0.58	0.01	0.02	0.06	100	0.00	0.18	42.83	30.41
17	T104	110	0.02	1.64	4.71	2.46	0.04	0.02	2.29	100	0.00	0.63	49.78	35.64
18	L102	110	0.02	2.24	5.84	3.10	3.37	0.02	0.83	99.99	0.04	1.16	53.04	38.33
19	L103	110	0.02	2.36	6.00	3.24	3.37	0.02	0.85	99.99	0.04	1.18	53.04	38.34
20	L104	110	0.02	1.91	4.94	2.68	0.29	0.02	2.29	100	0.00	0.70	43.7	31.39
21	L105	110	0.02	1.81	4.79	2.55	0.29	0.02	2.31	100	0.00	0.69	48.26	34.61
22	T105	110	0.02	7.31	9.71	8.47	3.35	0.05	2.82	100	0.01	1.98	22.61	17.39
23	L106	110	0.02	2.24	4.95	3.04	0.65	0.04	1.85	100	0.00	0.77	38.7	27.91
24	L107	110	0.02	2.50	5.19	3.45	0.82	0.06	1.61	100	0.00	0.82	40.65	29.32
25	L108	110	0.02	2.42	5.12	3.25	0.72	0.05	1.76	100	0.00	0.80	48.26	34.69
26	C101	110	0.01	0.45	1.07	0.70	0.02	0.09	0.06	100	0.00	0.22	24.13	17.22
							Transformers							
27	Tr211	220/110	88.01	69.22	16.54	13.60	63.83	8.73	29.52	13.38	100	49.66	49.29	69.97
28	Tr212	220/110	100	83.41	26.44	19.53	90.99	11.90	21.84	19.61	86.64	59.66	38.69	69.54
29	Tr411	400/110	63.43	100	100	100	100	73.31	17.00	1.57	87.78	67.99	29.29	68.78
30	Tr441	400/400	100	38.31	75.34	37.09	1.82	100	100	100	100	68.42	6.67	53.10
31	Tr442	400/400	100	38.31	75.34	37.09	1.82	100	100	100	100	68.42	6.67	53.10

The average importance of TS bays by each SS is determined in a similar way with (4).

$$i_{avg} = I \cdot w_{SS}^{T} \tag{4}$$

The common index d is defined on the basis of the index of technical condition c and index of importance i_{avg}. It represents the uniform participation of technical condition and importance of c and i_{avg} of the device. The index d encompasses the bays and power transformers of the TS and is defined by (5).

$$d = \frac{c + i_{avg}}{\sqrt{2}} \tag{5}$$

The average value of costs of expected outages by each SS for all TS bays is calculated using (6) as the vector of costs of expected outages $\mathbf{C}_{\text{in,avg}}$.

$$\mathbf{C}_{\text{in,avg}} = \mathbf{C}_{\text{in}} \cdot \boldsymbol{w}_{SS}^{T} \tag{6}$$

All these values from Table 2 are input data of the optimization process, and the basis for the $c - i_{\text{avg}}$ diagram for all 31 bays (Figure 2b). In the $c - i_{\text{avg}}$ diagram, the x-axis represents the index of importance i, and the y-axis the index of technical condition c. For both indexes c and i to be considered equally, the line x needs to be rotated by the angle $\beta = 45°$. The point on the $c - i_{\text{avg}}$ diagram represents a couple c and i_{avg} for the switching bay q_s.

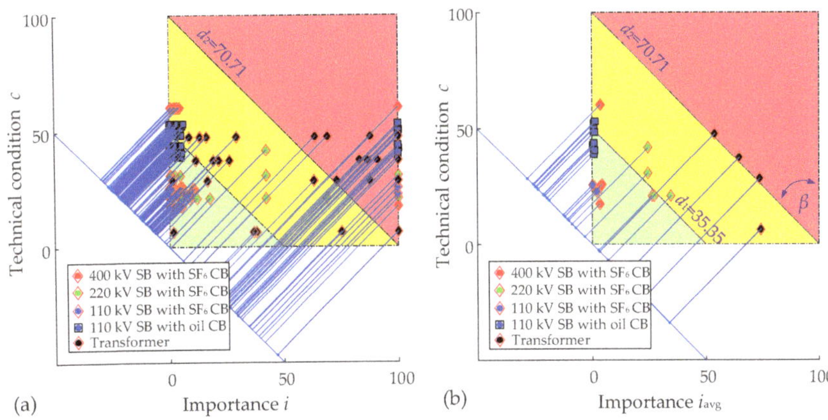

Figure 2. (**a**) $c - i$ diagram for all bays and transformers in all significant states; (**b**) $c - i_{\text{avg}}$ diagram for all bays and transformers with average value of importance.

The RCM diagram in Figure 2 represents a graphical representation of the main data, which are the importance and technical conditions of elements from Table 2. The importance of elements was averaged, due to the need for a better visualization and request for a uniform solution for the determination of maintenance frequency of elements.

The bay code in Table 2 consists of one letter and three digits. The letter represents the type of bay: line bay (L), transformer bay (T), or bus coupler bay (C).

The first digit indicates the voltage level of the bay (1–110 kV, 2–220 kV, and 4–400 kV). The last two digits represent the sequence number of the bay. The code for power transformers consists of two letters (Tr) and three digits. The first digit indicates the voltage level of the primary winding, the second is the secondary winding, and the third is the sequence number of the transformer.

Figure 2a shows the $c - i$ diagram with the condition of all elements in all significant states of the index i_{jq}, while Figure 2b shows the $c - i_{\text{avg}}$ diagram with the average index of the element i_q. In comparison to the diagram in Figure 2a, the diagram in Figure 2b shows only one point on the $c - i$ diagram for the TS elements. d_1 and d_2 are bounds between the areas of action and are defined by the TSO. These areas represent risk of failures on devices.

The colors in the $c - i$ diagram describe the kind of maintenance tasks. In the green area, there are prescribed regular inspections of HV devices; in the yellow area, there are maintenance interventions that require equipment revision; and in the red area, there are the cases where the equipment needs to be replaced. The deviation d_q defines as to which area an individual TS element belongs [1,26].

From the diagram (Figure 2b), it is evident that 110 kV bays were in the green-yellow border area (110 kV circuit breakers). The 220 kV bays were in the same area, only their importance was higher. The 400 kV bays were, due to their importance in the transmission system and recent replacement, in the green area. The transformers were in the yellow-red area, due mostly to their age and importance for the EPS. Their replacement was planned for the near future. Δp was negligible for all SF_6 SB in TS. It was set to $\Delta \boldsymbol{p} = [\Delta p_q = 0]$.

3.2. Costs

Table 3 contains the data on expected costs of outages and maintenance costs for the example presented in this paper. The nine columns for SSs present the costs $C_{\mathrm{in},jq}$ of long duration outages for all significant states, defined in Section 2.7. The last two columns contain total costs $C_{\mathrm{in},q}$ of outages for each element, calculated using (3), and revision costs $C_{\mathrm{r},q}$ for each element defined in Section 2.6.

Table 3. TS elements' costs.

q	Code	Voltage Level (kV)	Costs of Expected Outages of the EPS Element $C_{\mathrm{in},qj}$ (EUR/Year)									Averaged Outages Costs $C_{\mathrm{in},q}$	Revision Costs $C_{\mathrm{r},q}$
			SS_1	SS_2	SS_3	SS_4	SS_5	SS_6	SS_7	SS_8	SS_9		
			Switching bay arrangements										
1	T401	400	6.02	8.52	8.68	15.17	21.66	10.88	1.57	57.53	6.03	9.40	693.45
2	L401	400	30.45	7.76	9.45	14.94	277,083	56.29	37.42	1,848,919	262,999	48,937	2779
3	L402	400	31.49	2.40	4.32	1.63	9.34	62.31	33.67	69.09	19.39	28.40	6000
4	L403	400	6.63	8.22	8.39	14.76	22.08	10.84	1.65	59.73	6.27	9.64	4005
5	C401	400	41.41	8.80	16.76	9.90	9566	81.07	3663	17,688	9079	1964.30	1108
6	T402	400	60.30	9.65	12.78	15.60	19,132	115.5	3609	35,472	18,162	3616.28	445.05
7	T403	400	60.30	9.65	12.78	15.60	19,132	115.5	3609	35,472	18,162	3616.28	469.2
8	T201	220	0.68	5.61	1.82	5.67	1.87	1.66	2.55	24.05	2.77	1.78	1669
9	T202	220	0.68	5.61	1.82	5.67	1.87	1.66	2.55	24.05	2.77	1.78	2656
10	L201	220	0.69	5.62	1.83	5.68	1.98	1.71	2.58	24.38	2.80	1.81	4052
11	L202	220	0.73	5.67	1.89	5.77	1.87	1.68	2.57	24.18	2.78	1.81	4390
12	L203	220	0.69	5.87	1.92	5.92	1.91	1.70	2.63	24.97	2.85	1.83	3065
13	T101	110	12.20	0.09	0.03	0.09	0.04	0.09	0.07	0.03	0.15	3.88	2089
14	T102	110	6.39	0.03	0.03	0.03	0.03	0.03	0.03	0.04	0.04	2.03	1277
15	L101	110	6.26	0.07	0.24	0.40	0.10	0.12	0.28	0.09	0.12	2.06	2835
16	T103	110	0.23	0.00	0.00	0.00	0.00	0.00	0.00	0.00	0.00	0.07	895.85
17	T104	110	6.28	0.08	0.03	0.09	0.02	0.08	0.03	0.02	0.11	2.01	1392
18	L102	110	5.76	0.30	0.78	0.50	1.95	0.02	1.54	0.98	3.65	2.34	2458
19	L103	110	5.55	0.31	0.79	0.52	1.96	0.02	1.54	0.99	3.67	2.28	2459
20	L104	110	6.17	0.01	0.01	0.02	0.01	0.01	0.02	0.02	0.01	1.95	1336
21	L105	110	6.38	0.01	0.01	0.01	0.01	0.01	0.01	0.01	0.00	2.01	4303
22	T105	110	12.14	2.52	2.51	3.45	2.31	1.02	0.11	0.36	0.24	4.80	991.3
23	L106	110	6.31	0.31	0.34	0.47	0.16	0.28	0.26	0.16	0.29	2.14	2063
24	L107	110	6.07	0.42	0.49	0.72	0.25	0.44	0.39	0.27	0.42	2.15	2899
25	L108	110	6.09	0.37	0.40	0.55	0.19	0.32	0.31	0.19	0.34	2.10	2930
26	C101	110	0.45	0.00	0.01	0.01	0.01	0.00	0.01	0.02	0.01	0.14	1779
			Transformers										
27	Tr211	220/110	2.65	1.71	0.08	1.57	0.25	1.61	1.28	0.09	3.24	1.52	1099
28	Tr212	220/110	3.03	2.05	0.84	2.16	0.46	2.11	0.72	0.51	2.74	1.81	1368
29	Tr411	400/110	1.03	3.27	3.26	4.51	2.91	1.15	0.37	0.67	0.48	1.56	1496
30	Tr441	400/400	1.67	1.22	1.96	1.58	0.47	1.50	2.03	21.21	0.92	1.31	46,969
31	Tr442	400/400	1.67	1.22	1.96	1.58	0.47	1.50	2.03	21.21	0.92	1.31	47,590

4. Optimization Process of Maintenance Activity

The concept of optimization is based upon seeking for the minimum maintenance costs in the observed period, which is usually one year with regard to the TSO, and taking into consideration the reliability of operation.

A block diagram of the optimization process and analysis of the optimal data is shown in Figure 3. The diagram is linked to the diagrams in Figure 1 through the connections A–F.

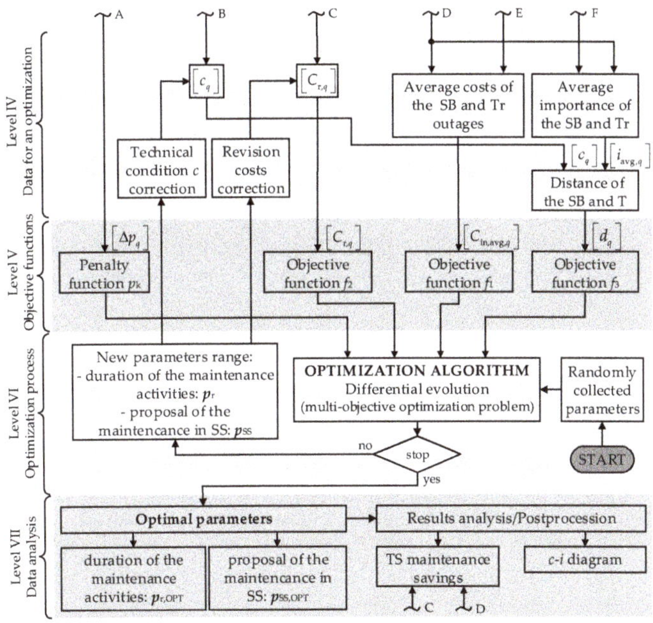

Figure 3. Algorithm of the optimization process of maintenance.

4.1. Objective and Penalty Function

The maintenance model is linked to the optimization algorithm with objective and penalty functions. Three objective functions and one penalty function were defined to achieve the reliability-based maintenance concept. Since the objective functions are related to different quantities, they are normalized in the optimization process to enable merging of criteria.

The first objective function f_1 is related to the expected costs of outage during the maintenance for all significant states. Figure 4a shows the fuzzy function (bell-shaped membership function) used to normalize the expected costs of non-transmitted energy $C_{in,q}$. The concept of normalizing is such that the costs of non transmitted energy, which in "usual" significant states lie between x_l = 5 and x_u = 300 EUR/year, represent 90% variability of the normalized value. The costs of expected outages $C_{in,q}$ that occur several times per year amount to approximately EUR 1,000,000 per year (Table 3) and have no more impact on the value of the objective function.

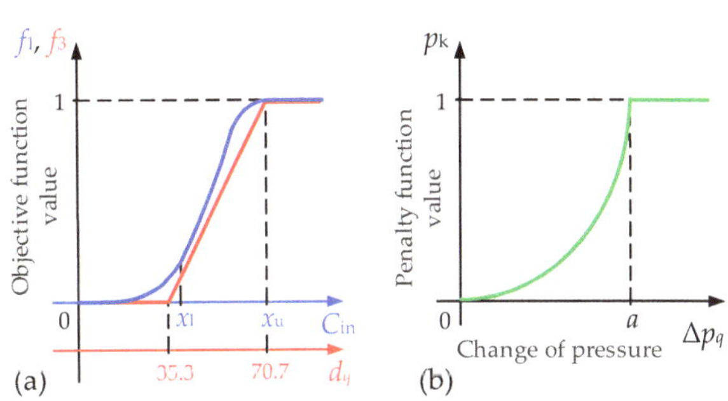

Figure 4. (**a**) Presentation of normalization of the objective functions f_1 (expected costs of outages) and f_3 (*c-i* indices); (**b**) presentation of normalizing of the penalty function p_k (leaking of SF_6).

The second objective function f_2 in the optimization process that represents the revision costs for a certain TS element $C_{r,q}$ for all tasks that require de-energizing of an individual bay. Since the range of these costs is smaller than for the first objective function, there is no need for the use of fuzzy function-based normalization. The normalization is performed using (7)

$$f_2 = \sum_{q=1}^{n_{SB}} \frac{C_{r,q}}{y_q} / \sum_{q=1}^{n_{SB}} C_{r,q};\ n_{SB} = 31;\ y \in \{1,\ 2,\ 3\} \quad (7)$$

where the objective function encompasses the total costs of all the 31 transmission substation bays and transformers, where the revision time period (1, 2, or 3 years) is taken into consideration for each bay separately.

The third objective function f_3 is related to the indices of the RCM maintenance concept. In the block diagram of input data (Figure 3), the RCM indices c and i are included in the common index d. In the denotation of the vector, the weighting factor w_{SSj} needs to be taken into consideration. For the determination of the objective function f_3, the common part d_q is used, which includes all the above-mentioned factors. An easier insight into the situation is enabled by the $c - i_{avg}$ diagram, shown in Figure 2b. If the value of d_q is lower than 35.3 (green area), this objective function yields $f_3 = 0$. If the value of d_q lies between 35.3 and 70.7 (yellow area), the value of the objective function f_3 is linear between 0 and 1. If the value exceeds 70.7 the value of the objective function is $f_3 = 1$. The third objective function is illustrated graphically in Figure 4a.

Penalty function p_k is related to the variations of pressure of the greenhouse SF_6 gas in the circuit breakers. The penalty function describes that the leaking of SF_6 in the period Δt is the reason for more frequent maintenance interventions. The higher the change of pressure in the period Δt, the steeper is the penalty function p_k, i.e., the sooner it approaches the value of 1 (Figure 4b). If the pressure drops below the factory-set value a, the protection disables operation of the HV device, which requires a maintenance intervention—replacement of the element. The equation of the penalty function (8) can be derived from Figure 4b.

$$p_k = \begin{cases} 1;\ (\Delta p_q > a) \Rightarrow \text{replacement} \\ 2^{\frac{\Delta p_q}{a}} - 1;\ \Delta p_q < a \end{cases} \quad (8)$$

The optimization algorithm for maintenance of transmission substations, designed in the above-described way, is referred to as a multiobjective optimization algorithm. The most commonly used approach for solving the multiobjective optimization problem is the use of the weighted sum method $f = \gamma_1 \cdot f_1 + \gamma_2 \cdot f_2 + \gamma_3 \cdot f_3 + p_k$, where the optimization algorithm finds an unambiguous solution with regard to the selected weights γ_1, γ_2, and γ_3 of the objective function. Weights are chosen empirically: $\gamma_1 = 0.25$, $\gamma_2 = 0.25$, and $\gamma_3 = 0.5$. The highest weight is assigned to the third objective function, due to the higher importance of the RCM concept. Equal weights are assigned to the first and the second objective functions that deal with maintenance and operation costs. The reason for this decision is in the final reduction of overall costs, since we believe that performing of maintenance activities based on the RCM concept contributes significantly to the reduction of associated risks and rational use of financial funds.

4.2. Optimization Parameters

In the optimization procedure, two optimization parameters are selected for each bay n_{SU}. The first optimization parameter represents the periods of revision that are performed on every TS bay. Three periods were selected: every year, every two years, and every three years. The vector of revision $\boldsymbol{p}_r$, which is defined by $\boldsymbol{p}_r = [p_{r,q}]$; $p_{r,q} \in \{1,2,3\}$, can be created for all bays and transformers. The second parameter of the optimization procedure represents the SS-based maintenance (Table 1). This parameter defines for each TS bay and transformer the most suitable SS for performing of revision. The vector of maintenance by SS p_{SS} is created for all TS bays and transformers. The codomain of this vector is the significant states themselves. The parameter is defined by $p_{SS} = [p_{SS,q}]$; $p_{SS,q} \in \{j; 1 \leq j \leq n_{SS}\}; 1 \leq q \leq n_{SU}; j,q \in \mathrm{N}$.

4.3. Optimization Process with the Modified Differential Evolution Algorithm

The modified differential evolution algorithm, which belongs to the group of evolutionary computation methods, is used in the optimization process. In the process of seeking the optimal solution of the objective function, which represents the maintenance model of a substation on the basis of the RCM concept, we introduced a modification of the algorithm, representing self-adaptation (SA) of the control parameters CR in F [29,30]. The following three possibilities were analyzed: optimization without SA (basic DE algorithm), optimization with CR and F self-adapted for the entire population, and optimization with CR and F self-adapted for each individual in the population.

These improvements ensured adequate robustness of the algorithm since the optimization problem is extremely complex and comprehensive (62 optimization parameters).

To ensure successful operation of the DE optimization algorithm, we selected the control parameters of differential evolution that are presented in Table 4. The control parameters have a significant impact on performance of the search process in DE (convergence of the optimization process and accuracy of computation of the optimal solution).

Table 4. Control parameters of DE.

Control Parameter	Value	Description
VTR	1×10^{-3}	value-to-reach
D	62	number of parameters
NP	600	population size
Strategy	7	strategy DE/rand/1/bin
CR	0.7/self-adaptive	crossover probability
F	0.6/self-adaptive	differential weighting factor
itermax	1000	maximum number of iterations

5. Results of Optimal RCM Maintenance

The use of an adequate optimization tool can ensure preservation of the existing level of maintenance quality, despite the cost reduction achieved by the use of the modified differential evolution algorithm.

5.1. Optimization Process

The progress of optimization process versus iterations is shown in Figure 5. It can be seen that the optimization process converged to the optimal solution after approximately 700 iterations.

Seeking an optimal solution in the optimization process is conditioned by the values of individual objective functions and their weights. Individual objective functions, of course, cannot converge to their own optimums, since the optimization process is oriented to the common criterion that represents the correlation of individual criteria.

Ten independent calculations of optimization algorithm runs were performed for each SA variant. The average value and standard deviation of the objective function were computed on the basis of these runs (Table 5). The use of an optimization algorithm without SA indicates a dispersion of the results of 10 independent runs. The results of both other variants with the use of SA, on the other hand, indicate the reliability and robustness of the optimization algorithm, since there was no dispersion of the results of 10 independent optimization runs (standard deviation equals zero). All results are related to the optimization with SA.

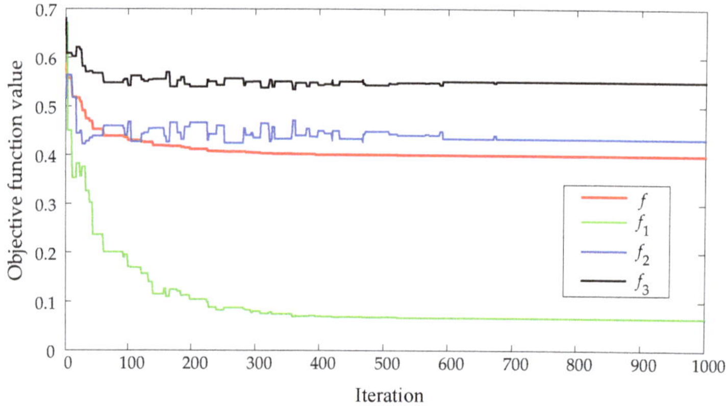

Figure 5. Progress of the optimization process of common objective function and partial objective functions.

Table 5. Mean values and standard deviation of objective function for variants of SA.

	No.	Without SA	*F* and *CR* for the Whole Population	*F* and *CR* for Each Individual in the Population
Value of objective function	1	0.410812	0.400430	0.400430
	2	0.409546	0.400430	0.400430
	3	0.407953	0.400430	0.400430
	4	0.409546	0.400430	0.400430
	5	0.407953	0.400430	0.400430
	6	0.409546	0.400430	0.400430
	7	0.407953	0.400430	0.400430
	8	0.409546	0.400430	0.400430
	9	0.407953	0.400430	0.400430
	10	0.409179	0.400430	0.400430
Mean value of objective function		0.408999	0.400430	0.400430
StD of objective function		0.000573	0	0

5.2. Optimal Maintenance for the Observed TS

The optimization was performed to obtain optimal parameters representing the frequency of maintenance activities $\boldsymbol{p}_{r,OPT}$ and the optimal SS $\boldsymbol{p}_{SS,OPT}$, in which the maintenance should be carried out. The maintenance of each TS bay is, therefore, conditioned by two optimization parameters.

Figure 6 shows the results of optimal maintenance for the observed TS, where Figure 6a shows the optimization parameters from the vector $\boldsymbol{p}_{r,OPT}$ (in years) for the revision performed in each individual switching bay or power transformers (the existing TBM maintenance methodology—red line, proposed optimal methodology—blue histogram). Figure 6b shows the optimal data of the second set of optimization parameters $\boldsymbol{p}_{SS,OPT}$, which represents the proposal for maintenance. It provides the information about the selected SS of the EPS (possible future combination of generation, consumption, and power flows in the TS) when it would be most suitable to de-energize the TS elements and perform revision.

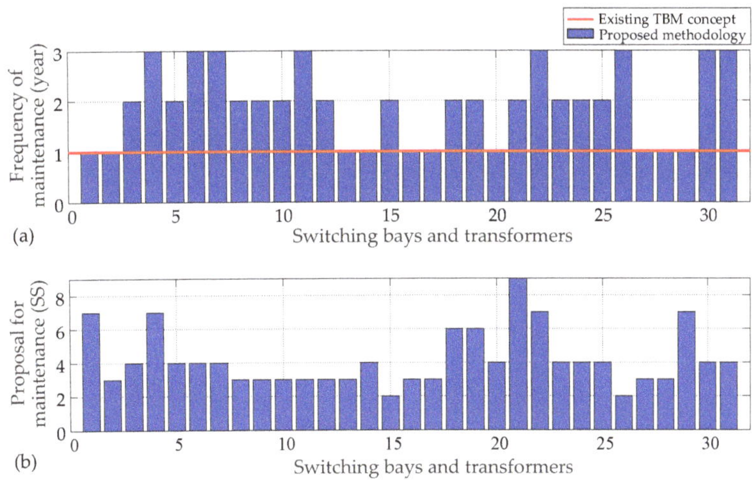

Figure 6. Results of optimization. (**a**) Period of maintenance works (the existing TBM concept—red line, proposed optimal methodology—blue histogram); (**b**) proposal of maintenance in possible significant states.

The summaries of optimal parameters are in the form of histograms, shown in Figure 7. They show that it would be most suitable to perform revision on 10 elements every year, on 13 every two years, and on 8 every three years. With the proposed maintenance methodology, only approximately one-third of switchyard elements kept yearly frequency of maintenance activities. For the existing TBM concept, all 31 elements were found to have the frequency of maintenance activities every year. Despite the optimally selected maintenance periods, it would be necessary to perform revision on 10 elements every year. A comparison of the selected significant states in Table 1 and the histogram of the most suitable significant states led to the conclusion that the most suitable significant states for de-energizing state are the TS elements.

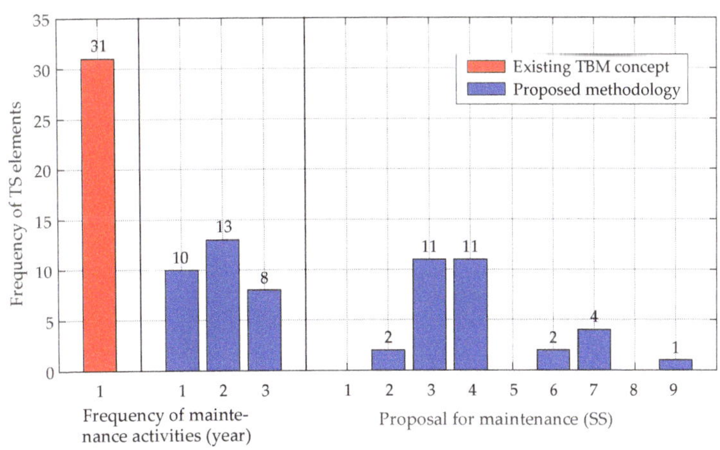

Figure 7. Histogram of existing TBM concept and proposed optimal maintenance methodology.

The optimization results of RCM maintenance methodology are presented in the $c - i_{avg}$ diagrams in Figure 8. The $c - i_{avg}$ diagram after the optimization (Figure 8a) was compared with the $c - i_{avg}$ diagram before optimization (Figure 2b). The comparison of both $c - i_{avg}$ diagrams showed that the points only moved vertically. Differences between the values of technical condition index c before and after the optimization for individual switching bays ranged between 0 and 18.

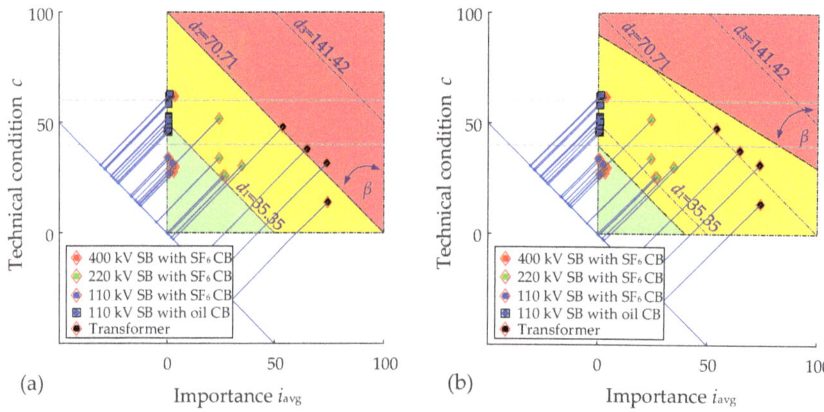

Figure 8. (**a**) $c - i_{avg}$ diagram after optimization; (**b**) $c - i_{avg}$ diagram with changed maintenance areas.

In the optimization procedure, we also took into consideration the fact that the boundaries between maintenance areas of the bays on the $c - i_{avg}$ diagram could also be changed by varying the distances d_1 and d_2 or the angle β (Figure 8b). The determination of boundaries of individual areas in the $c - i_{avg}$ diagram depends on risks (outages of TS elements, financial risks) that the transmission system operator is willing to accept. Changing the boundaries and angle β in Figure 8b influences the maintenance concept. Any change of boundaries in the $c - i_{avg}$ diagram also influences boundary conditions of the optimization and requires a new run of the optimization tool.

The optimization process is performed with a feedback loop, which enables correction of new parameters of technical condition (index c) with regard to the period of the maintenance works, characteristic maintenance states, and performance of tasks. This ensures feedback to the input parameters through the index of technical condition c and a feedback loop to the optimization model.

The time window in the optimization model is three years, although the optimization is performed every year. This means that the time window is, every year, shifted by one year to the future. In the case of major investments to the TS the model is updated, and a new optimization run is performed. If there are no extraordinary events, the state of elements in principle does not change; otherwise, a correction is made that is considered in the optimization model for the time window of the next three years.

The purpose of introducing the optimization procedure in maintenance is not only to optimize costs by providing the highest reliability possible, but also in planning of maintenance tasks for the whole year.

5.3. Postprocessing of the Results

A comparison analysis of maintenance costs with the existing time-based maintenance and with the new RCM maintenance concept (Table 6) showed that the costs of the existing maintenance method amount to EUR 568,423, to which inspections contributed the main share. The optimization also influenced the revision activities in the three-year period.

Table 6. Revision and total costs with and without optimization.

Maintenance Tasks	Year	Without Optimization (EUR/Year)	With Optimization (EUR/Year)	Saving (EUR/Year)	Saving (%)
Revisions	x	159,574	14,429	145,145	90.96
	$x+1$	159,574	52,933	106,641	66.83
	$x+2$	159,574	121,069	38,505	24.13
Inspections		405,935	405,935	colspan="2"	–
Replacements		2914	2914	–	–
Total	x	**568,423**	**423,278**	**145,145**	**25.53**
	$x+1$	**568,423**	**461,782**	**106,641**	**18.76**
	$x+2$	**568,423**	**529,918**	**38,505**	**6.77**

The savings were the highest in the first year, and amounted to 90.96% of the anticipated revision costs, or 25.53% of the total maintenance budget for the observed TS. In the second year, the savings were lower, and amounted to 66.83% of the anticipated revision costs, or 18.76% of the total maintenance budget for the observed TS. In the last year of the three-year period, the savings were the lowest, amounting to 24.13% of the anticipated revision costs, or 6.77% of the total maintenance budget for the observed TS.

The costs and savings shown above depend on the selection of weights γ_1, γ_2, and γ_3 in the objective function. The selection of weights can provide more importance, either to the introduction of RCM maintenance or costs, i.e., reliability, or savings. The TSO needs to decide which aspect is more important. Table 6 shows the costs and savings for the selection of weights, described in Section 4.1 ($\gamma_1 = 0.25$, $\gamma_2 = 0.25$, and $\gamma_3 = 0.5$), where higher importance is given to RCM maintenance, i.e., reliability. For comparison, Table 7 shows costs and savings for the following selection of weights: $\gamma_1 = 0.5$, $\gamma_2 = 0.25$, and $\gamma_3 = 0.25$.

Table 7. Revision and total costs with and without optimization for changed objective function weights in the optimization process.

Maintenance Tasks	Year	Without Optimization (EUR/year)	With Optimization (EUR/Year)	Saving (EUR/Year)	Saving (%)
Revisions	x	159,574	7347	152,227	95.40
	$x+1$	159,574	28,520	131,054	82.13
	$x+2$	159,574	131,054	28,520	17.87
Inspections		405,935	405,935	–	–
Replacements		2914	2914	–	–
Total	x	**568,423**	**416,196**	**152,227**	**26.78**
	$x+1$	**568,423**	**437,369**	**131,054**	**23.06**
	$x+2$	**568,423**	**539,903**	**28,520**	**5.02**

The selection of weights in Table 7 brings in the first year a 95.50% saving of predicted revision costs and 26.78% of the total maintenance budget of the substation. In the second year, the savings were lower, amounting to 82.13% of the predicted revision costs and 23.06% of the total maintenance budget. In the third year, the savings were the lowest, amounting to 17.87% of the predicted revision costs and 5.02% of the total maintenance budget. The maximum difference between savings in the revision costs from Tables 6 and 7 was 15.3%.

6. Discussion

In the field of maintenance of electric power system devices and networks, time-based maintenance (TBM) is still used widely. Nevertheless, a large number of companies are gradually introducing condition-based maintenance (CBM). An upgrade of such maintenance system is RCM, wherein the frequency of maintenance interventions is defined with regard to the state and importance of the EPS element. RCM maintenance methodology considers the reliability of the entire system and not only individual elements, which is the key advantage of this maintenance methodology.

An RCM model was developed and tested for transmission substations. For this purpose, in the first step, input data were defined for the indicators of operational reliability, technical condition of devices, and maintenance costs. These data were obtained on the basis of results of diagnostics and monitoring of elements, as well as from the calculations of parameters for the entire Slovenian EPS. The RCM model was upgraded with an optimization algorithm that enables it to find the optimal maintenance costs at the guaranteed reliability of operation and the most suitable combination of generation, consumption, and power flows (SS). Improper maintenance of the EPS elements causes various risks. These risks can be managed by the introduction of maintenance optimization. For the maintenance costs to be reduced, it is necessary to know the limit of acceptable risks. The maintenance optimization mode presented in this paper was tested practically in one of the Slovenian transmission substations. The developed algorithm is universal and can be used for any TS and included in the entire process of EPS maintenance, i.e., in the asset management of any transmission system operator.

A limitation of use of the proposed algorithm can, in the case of very large maintenance systems (e.g., a few thousands of elements), be problematic. In such a case, the use of conventional optimization algorithms, such as the one used in this paper, could jeopardize the stability of the optimization process and the accuracy of the obtained solutions.

With the proposed new algorithm for calculating maintenance periods using the optimization process, we confirmed the hypothesis that analyzing the past data of operation, events, and maintenance costs enables a direct impact on maintenance process for the future periods. With this maintenance concept, we maintain the reliability of operation of EPS elements and of their technical condition. At the same time, we reduce maintenance costs. A comparison of the existing (TBM) maintenance concept and the proposed one confirms savings. The proposed computations showed 25.53% saving in the first year, 18.76% in the second year, and 6.77% in the third year. This was the confirmation of the second hypothesis.

Further research work in the field of EPS elements' maintenance enables extensions of the optimization model on all substations and all transmission lines in the transmission system. Testing with different maintenance periods could be performed for more extensive comparisons. It would also be possible to perform research of the use and testing of different optimization algorithms, as well as introduction of new approaches, such as the deep learning maintenance process.

Author Contributions: Conceptualization, P.K. and J.R.; methodology, P.K. and L.B.; software, P.K. and J.R.; validation, L.B., J.P., and J.R.; formal analysis, J.P.; investigation, L.B. and J.P.; resources, L.B. and J.P.; data curation, L.B.; writing—original draft preparation, P.K.; writing—review and editing, J.P. and J.R.; visualization, P.K.; supervision, J.P. and J.R.; project administration, J.P. All authors have read and agreed to the published version of the manuscript.

Funding: This research received no external funding.

Institutional Review Board Statement: Not applicable.

Informed Consent Statement: Not applicable.

Data Availability Statement: Not applicable.

Conflicts of Interest: The authors declare no conflict of interest.

References

1. Balzer, G. Condition assessment and reliability centered maintenance of high voltage equipment. In Proceedings of the 2005 International Symposium on Electrical Insulating Materials (ISEIM 2005), Kitakyushu, Japan, 5–9 June 2005; pp. 259–264.
2. Birkner, P. Field experience with a condition-based maintenance program of 20-kV XLPE distribution system using IRC-analysis. *IEEE Trans. Power Deliv.* **2004**, *19*, 3–8. [CrossRef]
3. Bohatyrewicz, P.; Płowucha, J.; Subocz, J. Condition Assessment of Power Transformers Based on Health Index Value. *Appl. Sci.* **2019**, *9*, 4877. [CrossRef]
4. Zhang, F.; Wen, Z.; Liu, D.; Jiao, J.; Wan, H.; Zeng, B. Calculation and Analysis of Wind Turbine Health Monitoring Indicators Based on the Relationships with SCADA Data. *Appl. Sci.* **2020**, *10*, 410. [CrossRef]
5. Liu, Y.; Xv, J.; Yuan, H.; Lv, J.; Ma, Z. Health Assessment and Prediction of Overhead Line Based on Health Index. *IEEE Trans. Ind. Electron.* **2019**, *66*, 5546–5557. [CrossRef]
6. Fang, F.; Zhao, Z.; Huang, C.; Zhang, X.; Wang, H.; Yang, Y. Application of Reliability-Centered Maintenance in Metro Door System. *IEEE Access* **2019**, *7*, 186167–186174. [CrossRef]
7. Vera-García, F.; Pagán Rubio, J.A.; Hernández Grau, J.; Albaladejo Hernández, D. Improvements of a Failure Database for Marine Diesel Engines Using the RCM and Simulations. *Energies* **2020**, *13*, 104. [CrossRef]
8. Zhang, X.; Zhang, J.; Gockenbach, E. Reliability Centered Asset Management for Medium-Voltage Deteriorating Electrical Equipment Based on Germany Failure Statistics. *IEEE Trans. Power Syst.* **2009**, *24*, 721–728. [CrossRef]
9. Ge, H.; Asgarpoor, S. Reliability and Maintainability Improvement of Substations With Aging Infrastructure. *IEEE Trans. Power Deliv.* **2012**, *27*, 1868–1876. [CrossRef]
10. de França Moraes, C.C.; Pinheiro, P.R.; Rolim, I.G.; da Silva Costa, J.L.; Junior, M.D.S.E.; Andrade, S.J.M.D. Using the Multi-Criteria Model for Optimization of Operational Routes of Thermal Power Plants. *Energies* **2021**, *14*, 3682. [CrossRef]
11. IEC Standard 60300-3-11. In *IEC Dependability Management-Part 3-11: Application Guide-Reliability Centred Maintenance*; IEC: Geneva, Switzerland, 2009.
12. IEC Standard TS 62775. In *IEC Application Guidelines-Technical And Financial Processes for Implementing Asset Management Systems*; IEC: Geneva, Switzerland, 2016.
13. Mohd Selva, A.; Azis, N.; Yahaya, M.S.; Ab Kadir, M.Z.A.; Jasni, J.; Yang Ghazali, Y.Z.; Talib, M.A. Application of Markov Model to Estimate Individual Condition Parameters for Transformers. *Energies* **2018**, *11*, 2114. [CrossRef]
14. Yang, F.; Chang, C.S. Multiobjective Evolutionary Optimization of Maintenance Schedules and Extents for Composite Power Systems. *IEEE Trans. Power Syst.* **2009**, *24*, 1694–1702. [CrossRef]
15. Rafiei, M.; Khooban, M.-H.; Igder, M.A.; Boudjadar, J. A novel approach to overcome the limitations of reliability centered maintenance implementation on the smart grid distance protection system. *IEEE Trans. Circuits Syst. II Express Briefs* **2020**, *67*, 320–324. [CrossRef]
16. Bahrami, S.; Rastegar, M.; Dehghanian, P. An FBWM-TOPSIS Approach to Identify Critical Feeders for Reliability Centered Maintenance in Power Distribution Systems. *IEEE Syst. J.* **2020**, *15*, 3893–3901. [CrossRef]
17. Hilber, P.; Miranda, V.; Matos, M.A.; Bertling, L. Multiobjective Optimization Applied to Maintenance Policy for Electrical Networks. *IEEE Trans. Power Syst.* **2007**, *22*, 1675–1682. [CrossRef]
18. Arno, R.; Dowling, N.; Schuerger, R. Equipment Failure Characteristics and RCM for Optimizing Maintenance Cost. *IEEE Trans. Ind. Appl.* **2016**, *52*, 1257–1264.
19. Piasson, D.; Bíscaro, A.A.P.; Leao, F.B.; Mantovani, J.R.S. A new approach for reliability-centered maintenance programs in electric power distribution systems based on multiobjective genetic algorithm. *Electr. Power Syst. Res.* **2016**, *137*, 41–50. [CrossRef]
20. Zhang, D.; Li, W.; Xiong, X. Overhead line preventive maintenance strategy based on condition monitoring and system reliability assessment. *IEEE Trans. Power Syst.* **2014**, *29*, 1839–1846. [CrossRef]
21. Suwnansri, T. Asset management of power transformer: Optimization of operation and maintenance costs. In Proceedings of the 2014 International Electrical Engineering Congress (iEECON), Chonburi, Thailand, 19–21 March 2014; pp. 1–4.
22. Dong, M.; Zheng, H.; Zhang, Y.; Shi, K.; Yao, S.; Kou, X.; Ding, G.; Guo, L. A Novel Maintenance Decision Making Model of Power Transformers Based on Reliability and Economy Assessment. *IEEE Access* **2019**, *7*, 28778–28790. [CrossRef]
23. Hinow, M.; Mevissen, M. Substation Maintenance Strategy Adaptation for Life-Cycle Cost Reduction Using Genetic Algorithm. *IEEE Trans. Power Deliv.* **2011**, *26*, 197–204. [CrossRef]
24. Pandzic, H.; Conejo, A.J.; Kuzle, I.; Caro, E. Yearly Maintenance Scheduling of Transmission Lines Within a Market Environment. *IEEE Trans. Power Syst.* **2012**, *27*, 407–415. [CrossRef]
25. Zhong, J.; Li, W.; Wang, C.; Yu, J.; Xu, R. Determining Optimal Inspection Intervals in Maintenance Considering Equipment Aging Failures. *IEEE Trans. Power Syst.* **2017**, *32*, 1474–1482. [CrossRef]
26. Belak, L.; Marusa, R.; Ferlic, R.; Pihler, J. Strategic Maintenance of 400-kV Switching Substations. *IEEE Trans. Power Deliv.* **2013**, *28*, 394–401. [CrossRef]
27. McDonald, J.D. *Electric Power Substations Engineering*, 2nd ed.; CRC Press: Boca Raton, FL, USA, 2006.
28. Wang, F.; Tuinema, B.W.; Gibescu, M.; Meijden, M.A.M.M. Reliability evaluation of substations subject to protection system failures. In Proceedings of the 2013 IEEE Grenoble Conference, Grenoble, France, 16–20 June 2013.

29. Brest, J.; Greiner, S.; Boskovic, B.; Mernik, M.; Zumer, V. Self-adapting control parameters in differential evolution: A comparative study on numerical benchmark problems. *IEEE Trans. Evol. Computat.* **2006**, *10*, 646–657. [CrossRef]
30. Glotić, A.; Glotić, A.; Kitak, P.; Pihler, J.; Tičar, I. Parallel Self-Adaptive Differential Evolution Algorithm for Solving Short-Term Hydro Scheduling Problem. *IEEE Trans. Power Syst.* **2014**, *29*, 2347–2358. [CrossRef]

applied sciences

MDPI

Article

Performance Analysis of Direct Torque Controllers in Five-Phase Electrical Drives

Mario Bermúdez [1], Federico Barrero [2,*], Cristina Martín [1] and Manuel Perales [2]

[1] Department of Electrical Engineering, University of Seville, 41092 Seville, Spain; mbermudez4@us.es (M.B.); cmartin15@us.es (C.M.)
[2] Department of Electronic Engineering, University of Seville, 41092 Seville, Spain; mperales@us.es
* Correspondence: fbarrero@us.es

Abstract: The industrial application of electric machines has grown in the last decades, thanks to the development of microprocessors and power converters, which have permitted their use as variable-speed drives. Although three-phase machines are the common trend, the interest of the research community has recently focused on machines with more than three phases, known as multiphase machines. The principal reason lies in the exploitation of their advantages in terms of reliability, i.e., post-fault operating capability. Additionally, multiphase machines provide a better current distribution among phases, and lower current harmonic production in the power converter, than conventional three-phase machines. However, multiphase drive applications require the development of complex controllers to regulate the torque (or speed) and flux of the machine. In this regard, direct torque controllers have appeared as a viable alternative due to their easy formulation and high flexibility to incorporate control objectives. However, these controllers face some peculiarities and limitations in their use that require attention. This work aims to tackle direct torque control as a viable alternative for the regulation of multiphase drives. Special attention will be paid to the development of the control technique and the expected benefits and limitations in the obtained results. Case examples based on symmetrical five-phase induction machines with distributed windings in the motoring mode of operation will be used to this end.

Keywords: direct torque control; multiphase; electrical drives

Citation: Bermúdez, M.; Barrero, F.; Martín, C.; Perales, M. Performance Analysis of Direct Torque Controllers in Five-Phase Electrical Drives. *Appl. Sci.* **2021**, *11*, 11964. https://doi.org/10.3390/app112411964

Academic Editor: Gaetano Zizzo

Received: 18 November 2021
Accepted: 7 December 2021
Published: 16 December 2021

1. Introduction

It is expected that about 80% of all the produced energy will be used by electric drives by 2030, playing a major role in the automotive field, where they will dominate almost 50% of the market by that year. Furthermore, electric drives are the basis of locomotive traction, electric ship propulsion, electric aircraft with various auxiliary functions (e.g., fuel pumps, starter/generator solutions, etc.), and renewable energy production. Although conventional three-phase drives represent the principal choice for industrial applications, multiphase ones have recently aroused the interest of practitioner engineers and researchers in the field. Any energy conversion system formed by a multiphase electric machine and converter and regulated by a certain control technique is called a multiphase drive. The first application of such a system, particularly for a five-phase drive, was used in the late 1960s [1], showing the advantages of multiphase systems over conventional three-phase ones. The main interest in the proposal was that the higher number of phases yields a torque ripple three times lower with respect to the equivalent three-phase case due to a better power distribution per phase, this being one of the most reported problems in conventional drives at that time. However, it was not until the end of the 20th and the beginning of the 21st centuries that the interest of researchers in multiphase machines was renewed due to two main reasons. First, the development of high-power and high-frequency semiconductors and, consequently, the appearance of pulse width modulation (PWM) methods to control the ON and OFF states of these electronic devices, as well as the energy conversion process. Second, there is the development of microelectronic technology and the appearance of powerful electronic devices with the ability to implement control algorithms in real time, such as digital signal processors (DSPs) and field-programmable gate arrays (FPGAs).

Notwithstanding the above, the crucial reason for the renewed interest in multiphase drives can be found in the intrinsic benefits that they provide versus the conventional three-phase ones. These benefits are based on the extra degrees of freedom introduced by the higher number of phases and are principally the following:

- The fault-tolerant capability against a fault situation in the machine and/or the power converter, first presented in [2]. An n-phase machine can operate after one or several fault occurrences without any external equipment, as long as the number of healthy phases remains greater than or equal to three (assuming a single isolated neutral connection). Consequently, the system reliability is enhanced at the expense of a reduction in the post-fault electrical torque production.
- The capability of increasing the power density in healthy operation by injecting specific current harmonics, exposed in [3]. This is possible in certain multiphase machine configurations based on concentrated windings, where the lower current harmonic components can be used to increase the torque production.

These advances and advantages underlie the adoption of multiphase drives in specific industry applications such as variable-speed drives [4,5]. Electric propulsion of ships, electric vehicle traction (hybrid/electric vehicles and locomotives), wind energy generation, and low-power electric systems for more electric aircrafts are fields where research has been focused in the last 20 years [6,7]. The interest of multiphase machines in the cited applications, instead of the conventional three-phase counterparts, arises from the high-torque/current production and/or more robust and cheaper fault-tolerant capability that are usually required. Benchmark solutions adopted by important companies are: the Hyundai ultra-high-speed elevator, based on a 1.1 MW nine-phase electric drive; the 5 MW 12-phase electric drives in the Gamesa wind turbines for onshore and offshore plants; and the 20 MW 15-phase electric drive for ship propulsion introduced by the GE Power Conversion company in the Royal Navy.

To create a body of knowledge, recent research works review advances in the field of multiphase drives including their industrial applications, machine design and modeling, types of converters, modulation techniques, and control strategies; and explore innovative uses of their degrees of freedom (i.e., multimotor drives, battery chargers, post-fault control, or dynamic breaking) [8–11]. Then, these state-of-the-art analyses in the multiphase drives' topic set the advances in the area in the last decades. In general, they show that symmetrical five-phase and asymmetrical six-phase machines with isolated neutrals are the most popular multiphase machine types in the research community, while an evolution in the control techniques has been necessary in order to optimally exploit their inherent advantages. In this regard, asymmetric six-phase machines with isolated neutrals can be considered as two conventional three-phase machines coupled in a common case, while the five-phase drive can be considered the ideal case example to illustrate any study in the multiphase drive field.

Although field-oriented control (FOC) methods, based on decoupled control of the flux and electromagnetic torque and assisted by modulation stages, can be considered as the most popular control technique for conventional and multiphase drives [8,9], direct control techniques have recently been presented as interesting competitors [12–14]. The essence of direct controllers is to eliminate any form of modulation, forcing the states of the power switches to rapidly track a reference value. Then, the meaning of 'direct control' techniques is related to control strategies without the intervention of a pulse width modulation or any other form of modulation, providing control commands that are applied directly to the power converter. As a main consequence, direct controllers, being direct torque controllers (DTC) are the most extended industrial alternative, can favor fast torque responses and control robustness with respect to the variation of the electrical parameters of the machine. In this regard, DTC appears to be a viable (from a commercial perspective) control alternative in conventional three-phase drives due to an easy formulation and high flexibility to incorporate different control objectives. However, the use of DTC in multiphase drives is restricted in normal operation due to the impossibility of regulating more than two degrees of freedom (electrical torque and stator flux).

The objective of our work is to review the main concepts and interest of DTC controllers and to perform a performance analysis in multiphase drives, particularizing to the five-phase case. Different operating conditions, including normal operation and limited electrical/magnetic or faulty situations, are analyzed, where the DTC technique is effectively extended to face the operation of the five-phase induction machine. Experimental tests are provided to show that the speed, torque, and flux references are successfully tracked in all cases.

2. The Case Study: Five-Phase Distributed Windings Induction Motor Drive Using a Conventional Two-Level VSI

The system under study will be analytically examined in this section. It is based on a five-phase Induction Machine (IM) with a squirrel-cage rotor and symmetrically distributed stator windings (spatial equal displacement between windings) fed by a DC power supply through a five-phase two-level voltage source inverter (VSI). A graphical representation of the analyzed system is shown in Figure 1.

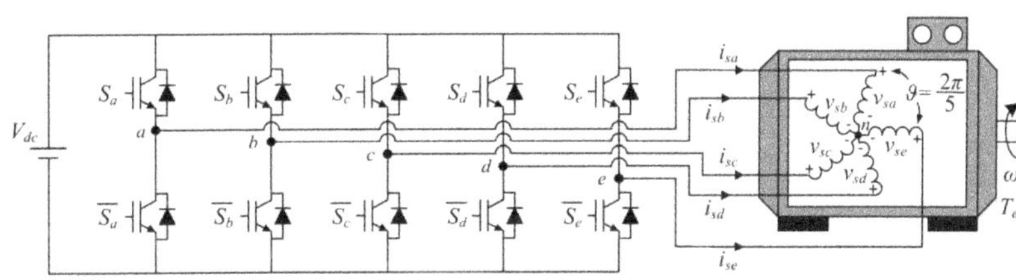

Figure 1. Schematic diagram of the case study.

The five-phase two-level VSI has $2^5 = 32$ different switching states characterized by the switching vector $[S_a\ S_b\ S_c\ S_d\ S_e]^T$, with $S_k = \{0,1\}$. Therefore, the phase voltages generated in the stator (subindex s), $[v_{sa}\ v_{sb}\ v_{sc}\ v_{sd}\ v_{se}]^T$, can be defined as a function of the switching states as follows:

$$\begin{bmatrix} v_{sa} \\ v_{sb} \\ v_{sc} \\ v_{sd} \\ v_{se} \end{bmatrix} = \frac{V_{dc}}{5} \cdot \begin{bmatrix} 4 & -1 & -1 & -1 & -1 \\ -1 & 4 & -1 & -1 & -1 \\ -1 & -1 & 4 & -1 & -1 \\ -1 & -1 & -1 & 4 & -1 \\ -1 & -1 & -1 & -1 & 4 \end{bmatrix} \cdot \begin{bmatrix} S_a \\ S_b \\ S_c \\ S_d \\ S_e \end{bmatrix}$$

Note that the midpoint of the external DC power supply (V_{dc}) is assumed to be the ground of the electrical system and a balanced load is also considered, so the sum of all phase voltages must be equal to zero ($v_{sa} + v_{sb} + v_{sc} + v_{sd} + v_{se} = 0$).

The machine is modeled in the stationary reference frame (a, b, c, d, e) with a set of voltage equilibrium equations obtained from the electromagnetic circuits of the stator and rotor (subindexes s and r, respectively), but then represented in two orthogonal planes, namely, α–β and x–y, plus the homopolar component z as follows:

$$\begin{bmatrix} v_{s\alpha} \\ v_{s\beta} \\ v_{sx} \\ v_{sy} \\ v_{sz} \end{bmatrix} = R_s \cdot \begin{bmatrix} i_{s\alpha} \\ i_{s\beta} \\ i_{sx} \\ i_{sy} \\ i_{sz} \end{bmatrix} + \frac{d}{dt} \begin{bmatrix} \lambda_{s\alpha} \\ \lambda_{s\beta} \\ \lambda_{sx} \\ \lambda_{sy} \\ \lambda_{sz} \end{bmatrix}$$

$$\begin{bmatrix} \lambda_{s\alpha} \\ \lambda_{s\beta} \\ \lambda_{sx} \\ \lambda_{sy} \\ \lambda_{sz} \end{bmatrix} = \begin{bmatrix} L_s & 0 & 0 & 0 & 0 \\ 0 & L_s & 0 & 0 & 0 \\ 0 & 0 & L_{ls} & 0 & 0 \\ 0 & 0 & 0 & L_{ls} & 0 \\ 0 & 0 & 0 & 0 & L_{ls} \end{bmatrix} \cdot \begin{bmatrix} i_{s\alpha} \\ i_{s\beta} \\ i_{sx} \\ i_{sy} \\ i_{sz} \end{bmatrix} + \begin{bmatrix} L_m & 0 & 0 & 0 & 0 \\ 0 & L_m & 0 & 0 & 0 \\ 0 & 0 & 0 & 0 & 0 \\ 0 & 0 & 0 & 0 & 0 \\ 0 & 0 & 0 & 0 & 0 \end{bmatrix} \cdot \begin{bmatrix} i_{r\alpha} \\ i_{r\beta} \\ i_{rx} \\ i_{ry} \\ i_{rz} \end{bmatrix}$$

$$\begin{bmatrix} 0 \\ 0 \\ 0 \\ 0 \\ 0 \end{bmatrix} = R_r \cdot \begin{bmatrix} i_{r\alpha} \\ i_{r\beta} \\ i_{rx} \\ i_{ry} \\ i_{rz} \end{bmatrix} + \frac{d}{dt} \begin{bmatrix} \lambda_{r\alpha} \\ \lambda_{r\beta} \\ \lambda_{rx} \\ \lambda_{ry} \\ \lambda_{rz} \end{bmatrix} + \begin{bmatrix} 0 & \omega_r & 0 & 0 & 0 \\ -\omega_r & 0 & 0 & 0 & 0 \\ 0 & 0 & 0 & 0 & 0 \\ 0 & 0 & 0 & 0 & 0 \\ 0 & 0 & 0 & 0 & 0 \end{bmatrix} \cdot \begin{bmatrix} \lambda_{r\alpha} \\ \lambda_{r\beta} \\ \lambda_{rx} \\ \lambda_{ry} \\ \lambda_{rz} \end{bmatrix}$$

$$\begin{bmatrix} \lambda_{r\alpha} \\ \lambda_{r\beta} \\ \lambda_{rx} \\ \lambda_{ry} \\ \lambda_{rz} \end{bmatrix} = \begin{bmatrix} L_r & 0 & 0 & 0 & 0 \\ 0 & L_r & 0 & 0 & 0 \\ 0 & 0 & L_{lr} & 0 & 0 \\ 0 & 0 & 0 & L_{lr} & 0 \\ 0 & 0 & 0 & 0 & L_{lr} \end{bmatrix} \cdot \begin{bmatrix} i_{r\alpha} \\ i_{r\beta} \\ i_{rx} \\ i_{ry} \\ i_{rz} \end{bmatrix} + \begin{bmatrix} L_m & 0 & 0 & 0 & 0 \\ 0 & L_m & 0 & 0 & 0 \\ 0 & 0 & 0 & 0 & 0 \\ 0 & 0 & 0 & 0 & 0 \\ 0 & 0 & 0 & 0 & 0 \end{bmatrix} \cdot \begin{bmatrix} i_{s\alpha} \\ i_{s\beta} \\ i_{sx} \\ i_{sy} \\ i_{sz} \end{bmatrix}$$

where λ denotes flux variables, ω_r is the electrical equivalent speed of the rotor, the resistances of the stator, and the rotor are R_s and R_r, respectively, the mutual inductance is represented by L_m, while L_{ls} and L_{lr} designate the leakage inductances of the stator and the rotor, respectively. Finally, $L_s = L_{ls} + L_m$ and $L_r = L_{lr} + L_m$ are called stator and rotor inductances. This is called the Clarke decoupled model of the electrical machine because a Clarke transformation (C_5 shown below, with $\vartheta = 2\pi/5$) is used to refer the rotor variables to the stator reference frame, leading to an invariant transformation of voltage and current magnitudes and allowing a considerable simplification of the machine model.

$$[C_5] = \frac{2}{5} \cdot \begin{bmatrix} 1 & \cos(\vartheta) & \cos(2\vartheta) & \cos(3\vartheta) & \cos(4\vartheta) \\ 0 & \sin(\vartheta) & \sin(2\vartheta) & \sin(3\vartheta) & \sin(4\vartheta) \\ 1 & \cos(2\vartheta) & \cos(4\vartheta) & \cos(6\vartheta) & \cos(8\vartheta) \\ 0 & \sin(2\vartheta) & \sin(4\vartheta) & \sin(6\vartheta) & \sin(8\vartheta) \\ \frac{1}{2} & \frac{1}{2} & \frac{1}{2} & \frac{1}{2} & \frac{1}{2} \end{bmatrix}$$

Note that the stator voltages applied to the electrical machine using the power converter must be referred to the same coordinates, which is easily done by multiplying C_5 by voltages in the (a, b, c, d, e) reference frame. The same happens with the stator current and flux, as well as with the rotor magnitudes (voltage, current, and flux).

$$\begin{bmatrix} v_{s\alpha} \\ v_{s\beta} \\ v_{sx} \\ v_{sy} \\ v_{sz} \end{bmatrix} = C_5 \cdot \begin{bmatrix} v_{sa} \\ v_{sb} \\ v_{sc} \\ v_{sd} \\ v_{se} \end{bmatrix}$$

Applying this transformation, 30 active voltage vectors and 2 null vectors can be applied at the connection terminals of the electrical machine. Figure 2 shows the two-dimensional projections obtained for every vector, identified with the decimal number equivalent of their respective switching state $[S_a\ S_b\ S_c\ S_d\ S_e]^T$ expressed in binary logic (1 or 0), being S_a and S_e the most and the least significant bits, respectively. These vectors uniformly divide the space that they occupy in 10 sectors with a separation of $\pi/5$ between them. Likewise, active voltage vectors can be classified according to their magnitude in long (0.647 V_{dc}), medium (0.4 V_{dc}), and short (0.247 V_{dc}) vectors. The switching states that generate long vectors in the α–β plane correspond to those that generate short vectors in the plane x–y and vice versa. The switching states corresponding to vectors of medium magnitude in the α–β plane, also generate medium vectors in the plane x–y. Null vectors are generated by the same switching states in both planes. This transformation allows for a detailed study of the harmonic components, since they are projected in certain planes. In particular, the fundamental frequency together with the harmonics of order 10 $k \pm 1$ (k = 0, 1, 2, etc.) are mapped in the α–β plane, while the harmonics of order 10 $k \pm 3$ are related to the plane x–y. The homopolar component and harmonics of order 5 k are projected on the z-axis.

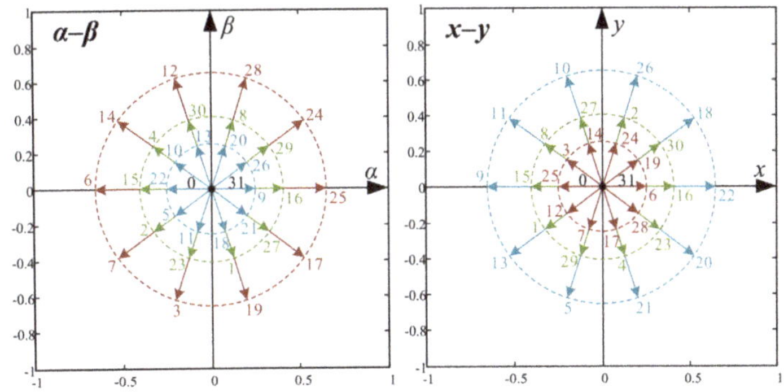

Figure 2. Mapping of the phase stator voltages of the two-level five-phase VSI in the α–β (**left** graph) and x–y (**right** graph) planes.

In distributed winding induction machines, as the one used in our study, only the components of the α–β plane are involved in torque production and a new transformation of variables is desirable (see Figure 3), moving the α–β plane to a new rotating reference frame d–q (at synchronous speed ω_a), where the components are not oscillating, are constant in steady state and vary only in transient state. The basis of this transformation is the P_{s5} and P_{r5} operators,

$$[P_{s5}] = \begin{bmatrix} \cos(\theta_a) & \sin(\theta_a) & 0 & 0 & 0 \\ -\sin(\theta_a) & \cos(\theta_a) & 0 & 0 & 0 \\ 0 & 0 & 1 & 0 & 0 \\ 0 & 0 & 0 & 1 & 0 \\ 0 & 0 & 0 & 0 & 1 \end{bmatrix} \quad [P_{r5}] = \begin{bmatrix} \cos(\delta) & \sin(\delta) & 0 & 0 & 0 \\ -\sin(\delta) & \cos(\delta) & 0 & 0 & 0 \\ 0 & 0 & 1 & 0 & 0 \\ 0 & 0 & 0 & 1 & 0 \\ 0 & 0 & 0 & 0 & 1 \end{bmatrix}$$

$$\theta_a = \int_0^t \omega_a \cdot dt$$

$$\delta = \theta_a - \theta = \int_0^t (\omega_a - \omega_r) \cdot dt = \int_0^t \omega_{sl} \cdot dt$$

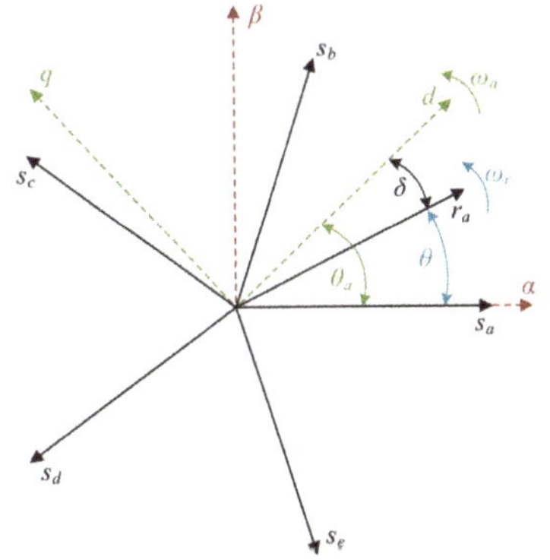

Figure 3. Reference frames for modeling the electrical machine.

Giving the following mathematical representation of the system:

$$\begin{bmatrix} v_{sd} \\ v_{sq} \\ v_{sx} \\ v_{sy} \\ v_{sz} \end{bmatrix} = R_s \cdot \begin{bmatrix} i_{sd} \\ i_{sq} \\ i_{sx} \\ i_{sy} \\ i_{sz} \end{bmatrix} + \frac{d}{dt} \begin{bmatrix} \lambda_{sd} \\ \lambda_{sq} \\ \lambda_{sx} \\ \lambda_{sy} \\ \lambda_{sz} \end{bmatrix} + \begin{bmatrix} 0 & -\omega_a & 0 & 0 & 0 \\ \omega_a & 0 & 0 & 0 & 0 \\ 0 & 0 & 0 & 0 & 0 \\ 0 & 0 & 0 & 0 & 0 \\ 0 & 0 & 0 & 0 & 0 \end{bmatrix} \cdot \begin{bmatrix} \lambda_{rd} \\ \lambda_{rq} \\ \lambda_{rx} \\ \lambda_{ry} \\ \lambda_{rz} \end{bmatrix}$$

$$\begin{bmatrix} 0 \\ 0 \\ 0 \\ 0 \\ 0 \end{bmatrix} = R_r \cdot \begin{bmatrix} i_{rd} \\ i_{rq} \\ i_{rx} \\ i_{ry} \\ i_{rz} \end{bmatrix} + \frac{d}{dt} \begin{bmatrix} \lambda_{rd} \\ \lambda_{rq} \\ \lambda_{rx} \\ \lambda_{ry} \\ \lambda_{rz} \end{bmatrix} + \begin{bmatrix} 0 & -\omega_{sl} & 0 & 0 & 0 \\ \omega_{sl} & 0 & 0 & 0 & 0 \\ 0 & 0 & 0 & 0 & 0 \\ 0 & 0 & 0 & 0 & 0 \\ 0 & 0 & 0 & 0 & 0 \end{bmatrix} \cdot \begin{bmatrix} \lambda_{rd} \\ \lambda_{rq} \\ \lambda_{rx} \\ \lambda_{ry} \\ \lambda_{rz} \end{bmatrix}$$

$$\begin{bmatrix} \lambda_{sd} \\ \lambda_{sq} \\ \lambda_{sx} \\ \lambda_{sy} \\ \lambda_{sz} \end{bmatrix} = \begin{bmatrix} L_s & 0 & 0 & 0 & 0 \\ 0 & L_s & 0 & 0 & 0 \\ 0 & 0 & L_{ls} & 0 & 0 \\ 0 & 0 & 0 & L_{ls} & 0 \\ 0 & 0 & 0 & 0 & L_{ls} \end{bmatrix} \cdot \begin{bmatrix} i_{sd} \\ i_{sq} \\ i_{sx} \\ i_{sy} \\ i_{sz} \end{bmatrix} + \begin{bmatrix} L_m & 0 & 0 & 0 & 0 \\ 0 & L_m & 0 & 0 & 0 \\ 0 & 0 & 0 & 0 & 0 \\ 0 & 0 & 0 & 0 & 0 \\ 0 & 0 & 0 & 0 & 0 \end{bmatrix} \cdot \begin{bmatrix} i_{rd} \\ i_{rq} \\ i_{rx} \\ i_{ry} \\ i_{rz} \end{bmatrix}$$

$$\begin{bmatrix} \lambda_{rd} \\ \lambda_{rq} \\ \lambda_{rx} \\ \lambda_{ry} \\ \lambda_{rz} \end{bmatrix} = \begin{bmatrix} L_r & 0 & 0 & 0 & 0 \\ 0 & L_r & 0 & 0 & 0 \\ 0 & 0 & L_{lr} & 0 & 0 \\ 0 & 0 & 0 & L_{lr} & 0 \\ 0 & 0 & 0 & 0 & L_{lr} \end{bmatrix} \cdot \begin{bmatrix} i_{rd} \\ i_{rq} \\ i_{rx} \\ i_{ry} \\ i_{rz} \end{bmatrix} + \begin{bmatrix} L_m & 0 & 0 & 0 & 0 \\ 0 & L_m & 0 & 0 & 0 \\ 0 & 0 & 0 & 0 & 0 \\ 0 & 0 & 0 & 0 & 0 \\ 0 & 0 & 0 & 0 & 0 \end{bmatrix} \cdot \begin{bmatrix} i_{sd} \\ i_{sq} \\ i_{sx} \\ i_{sy} \\ i_{sz} \end{bmatrix}$$

On the other hand, the generated electromagnetic torque (T_{em}) can be obtained in different reference frames using the following equations, where $n = 5$ and p is the pair of poles of the electrical machine,

$$T_{em} = p \cdot \frac{n}{2} \cdot L_m \cdot \left(i_{r\alpha} \cdot i_{s\beta} - i_{r\beta} \cdot i_{s\alpha}\right)$$

$$T_{em} = p \cdot \frac{n}{2} \cdot \left(\lambda_{s\alpha} \cdot i_{s\beta} - \lambda_{s\beta} \cdot i_{s\alpha}\right)$$

$$T_{em} = p \cdot \frac{n}{2} \cdot \left(i_{r\alpha} \cdot \lambda_{r\beta} - i_{r\beta} \cdot \lambda_{r\alpha}\right)$$

$$T_{em} = p \cdot \frac{n}{2} \cdot \frac{L_m}{L_r} \cdot \left(\lambda_{r\alpha} \cdot i_{s\beta} - \lambda_{r\beta} \cdot i_{s\alpha}\right)$$

$$T_{em} = p \cdot \frac{n}{2} \cdot L_m \cdot \left(i_{rd} \cdot i_{sq} - i_{rq} \cdot i_{sd}\right)$$

$$T_{em} = p \cdot \frac{n}{2} \cdot \left(\lambda_{sd} \cdot i_{sq} - \lambda_{sq} \cdot i_{sd}\right)$$

$$T_{em} = p \cdot \frac{n}{2} \cdot \left(i_{rd} \cdot \lambda_{rq} - i_{rq} \cdot \lambda_{rd}\right)$$

$$T_{em} = p \cdot \frac{n}{2} \cdot \frac{L_m}{L_r} \cdot \left(\lambda_{rd} \cdot i_{sq} - \lambda_{rq} \cdot i_{sd}\right)$$

It is in turn mechanically coupled to the load applied on the machine shaft, verifying the following differential equation:

$$J_m \cdot \frac{d\omega_m}{dt} = T_{em} - T_L - B_m \cdot \omega_m$$

With ω_m, the mechanical speed of the rotor shaft ($\omega_r = p \cdot \omega_m$); T_L, the load torque applied to the machine; J_m, the rotational inertial constant; and B_m, the friction coefficient of the rotor load bearings.

3. DTC in Five-Phase Drives

Direct Torque Control is a well-known strategy for three-phase electrical drives. It was presented in the mid-1980s by Takahashi [15] and Depenbrock [16], showing fast flux and torque responses, as well as more robustness with respect to the variation of the electrical parameters of the machine and generating a high-torque/flux ripple and harmonic current content, compared to the more standard field-oriented control technique. The operating principle is based on an off-line look-up table, which is used to select the stator voltage to be applied to the machine. The selection is made taking into account the position of the flux vector and the stator flux and electromagnetic torque error signals, obtained from the difference between reference and estimated values and processed using hysteresis comparators. Another disadvantage of DTC that should be considered from the analysis of its operating principle is that it does not generate a constant switching frequency. In fact, this switching frequency is variable and depends on the operating point and the bandwidth of the hysteresis controllers. Note, however, that DTC schemes have been proposed to also be used with PI regulators and space vector PWM methods (see [17]), to compensate for the variable switching frequency and reduce the torque and flux ripple.

DTC has been commercialized [18] and extended to the case of multiphase drives in recent times, considering different types of machines [19,20], machine neutral connections [21,22], and drives without speed sensors [23]. In the case of multiphase drives, since the controller has only two freedom degrees (stator flux and electromagnetic torque), there is no chance of regulating the current and voltage components in the orthogonal α–β and x–y planes. In this sense, some DTC strategies have been developed that satisfy this additional requirement, controlling the current and voltage components in the α–β plane while reducing at the same time the current and voltage components in the x–y plane. For example, in [24,25], a modification of the traditional control scheme is proposed, performing a two-step search to minimize the effect of low-order harmonics. Alternatively, the use of virtual vectors has been suggested to reduce current distortion [23]. Some criteria have also been included in the selection process within the look-up table to improve its performance in the low-speed region and an optimization between the two zero vectors to minimize the average switching frequency obtained [26]. On the other hand, and based on the virtual vectors defined in [23], different DTC schemes are presented defining new virtual vectors and avoiding the use of the zero vector to reduce the common-mode voltage generated by the VSI in [27,28], to improve open-phase fault operations in [29], or to avoid any reconfiguration of the controller when open-phase faults appear [30].

In our case example, the five phase IM with distributed windings, the control goal is reduced to the α–β plane (torque and flux regulation), while the x–y components are nullifying. The analysis is based on the machine model in the stationary reference frame, which defines the variation of the stator flux in the α–β subspace using a vector notation as:

$$\Delta\vec{\lambda}_{s\alpha\beta} = \int_0^t \left(\vec{v}_{s\alpha\beta} - R_s \cdot \vec{i}_{s\alpha\beta}\right) \cdot dt$$

If the voltage drop in the stator resistance is ignored for simplicity, it can be assumed that the variation in the modulus of the flux vector in a sampling period (T_s) depends on the voltage vector as follows:

$$\Delta\vec{\lambda}_{s\alpha\beta} \approx \vec{v}_{s\alpha\beta} \cdot T_s$$

The electromagnetic torque behavior can be determined by the following expression:

$$\begin{aligned} T_{em} &= \tfrac{5}{2} \cdot \tfrac{p \cdot k_r}{\sigma \cdot L_s} \cdot \left(\vec{\lambda}_{r\alpha\beta} \times \vec{\lambda}_{s\alpha\beta}\right) = \tfrac{5}{2} \cdot \tfrac{p \cdot k_r}{\sigma \cdot L_s} \cdot \left\|\vec{\lambda}_{r\alpha\beta}\right\| \cdot \left\|\vec{\lambda}_{s\alpha\beta}\right\| \cdot \sin\gamma \\ \sigma &= 1 - \left(\tfrac{L_m^2}{L_s \cdot L_r}\right) \\ k_r &= \tfrac{L_m}{L_r} \end{aligned}$$

where γ is the angle of load (angle between the flux vectors of the stator and the rotor). Then, a change in γ, obtained by applying a voltage vector, produces an increase or decrease in T_{em}. Note that the rotor time constant is greater than the stator time constant, so it can be assumed that a slower variation of the rotor flux compared to the stator flux, and therefore the rotor flux, can be considered constant in a sampling time. Note also that there is an impact of a spatial voltage vector on the magnitude of the stator flux. Different results are obtained depending on the applied stator voltage (32 alternatives using the five-phases two-level VSI, being 30 active vectors and 2 null vectors; see Figure 4). In short, the applied voltage vector can be divided into a tangential and a radial component with respect to the flux. The tangential component produces a change in machine torque, increasing or decreasing the sine of the angle γ, while the radial component modifies the magnitude of the stator flux, increasing or decreasing its modulus. Hence, the flux and the electromagnetic torque are controlled simultaneously using DTC.

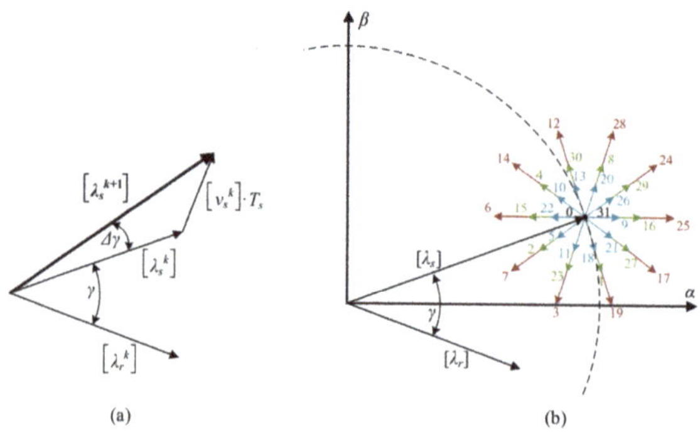

Figure 4. Impact of voltage vectors on the stator flux and the load angle for the application of DTC. (**a**) Estimated phasor diagram. (**b**) Alternatives available using the 2-level 5-phase VSI.

The application of the voltage vector controls the flux and electromagnetic torque and affects not only the α–β plane, but also the plane x–y, where losses and unwanted voltage and current harmonics are generated in the case study machine. Figures 2 and 4 show that 32 voltage vectors with 4 different magnitudes can be applied: 2 null vectors (0 volts applied), 10 short vectors (0.247 V_{dc} volts applied), 10 medium vectors (0.4 V_{dc} volts applied), and 10 long vectors (0.647 V_{dc} volts applied). Note also (see Figure 2) that switching states that represent long vectors in the α–β plane, symbolize short vectors in the x–y plane (and vice versa), while switching states that generate medium vectors in the α–β plane also cause medium vectors in the x–y plane. In addition, long and medium voltage vectors with the same direction in the α–β subspace, are equivalent to medium and short vectors with opposite directions in the x–y subspace. The same happens with medium and short vectors in the α–β plane, since they are equivalent to medium and long vectors with opposite directions in the plane x–y. These geometrical characteristics make possible the definition of a kind of voltage vector, called virtual voltage vector or VV_i [23], which minimizes currents in the x–y plane. Each virtual vector is based on the application of two available voltage vectors ($v1$ and $v2$) during adequate dwell time ratios (K_{v1} and K_{v2}) to generate zero average volts per second in the x–y subspace. It is then possible to define in each sector a long virtual vector (formed by a long and a medium vector in the α–β plane) and a short virtual vector (formed by a medium and a short vector in the α–β plane), as shown in the following equations and in Figures 5 and 6:

$$VV_{Li} = v_{Long} \cdot K_{v1} + v_{Medium} \cdot K_{v2}$$

$$VV_{Si} = v_{Medium} \cdot K_{v1} + v_{Short} \cdot K_{v2}$$

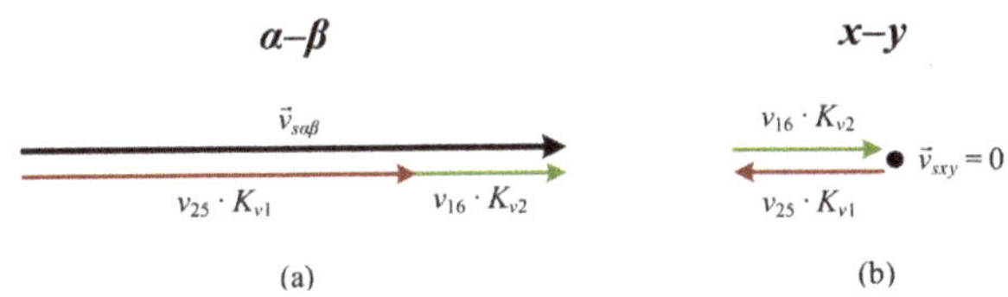

Figure 5. VV_{L1} projections in the (**a**) α–β plane and (**b**) x–y plane.

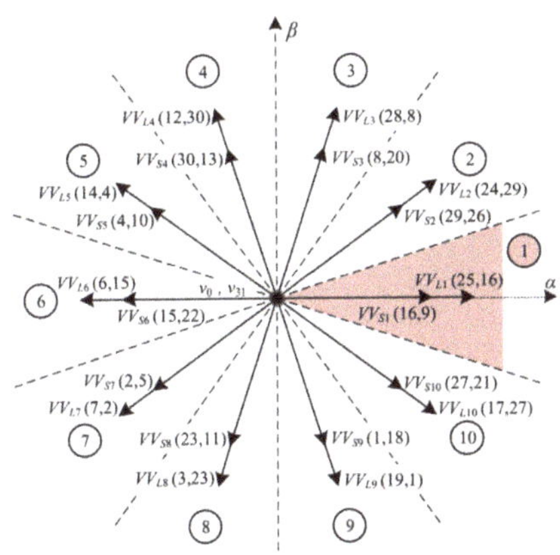

Figure 6. Virtual voltage vectors in the α–β plane.

For example, consider the long virtual vector in sector number one, formed by the voltage vectors 25 and 16 (v_{25} and v_{16}, respectively), which are in the same direction in the α–β subspace and are opposite in the subspace x–y. By selecting adequate values of K_{v1} and K_{v2}, it is obtained zero average volts-per-second in the x–y subspace (see Figure 5). Figure 6 shows all the virtual vectors in the α–β plane, VV_{Li} being the long virtual vectors and VV_{Si} the short ones. Furthermore, the dwell times of each vector in each sampling time T_s to achieve the minimization of the x–y currents are K_{v1} = 0.618 T_s and K_{v2} = 0.382 T_s. However, it should be noted that even in the case where virtual voltage vectors are used, x–y currents are controlled with an open-loop strategy. Consequently, the machine must have low asymmetries and spatial harmonic content and/or high impedance in the x–y plane to effectively limit the circulation of x–y currents. It is important to note that the x–y components do not contribute to the torque in our case study machine, but they increase power losses in the electromechanical system. A good control practice is therefore to minimize the x–y components.

The general implementation of the DTC technique, proposed in [31,32], can be done using the scheme shown in Figure 7, where flux and torque estimators are required (normal drives do not include flux and torque sensors, and their values must be estimated). The stator flux is regulated close to a reference value, using a two-level comparator with hysteresis band. Electromagnetic torque is controlled using a five-level comparator with a hysteresis band. Finally, a two-level comparator with a hysteresis band is applied to differentiate between low and normal speed operations of the drive (the effect of neglecting the voltage drop in the stator resistance cannot be neglected because it produces an appreciable drop in the stator flux vector when placed at the limit of some sectors [26]). To sum up, the estimated torque and flux values are used to compare with the reference values to select the applied stator voltage vector, using a correction term that depends on the speed operation and the following mathematical expressions:

$$\lambda_s^* > \left\| \overrightarrow{\lambda}_{s\alpha\beta} \right\| \quad d\lambda = +1$$

$$\lambda_s^* \leq \left\| \overrightarrow{\lambda}_{s\alpha\beta} \right\| \quad d\lambda = -1$$

$$\begin{array}{cc} T_{em}^* \geq T_{em} + \frac{\Delta T_{em}}{2} & dT = +2 \\ T_{em} + \frac{\Delta T_{em}}{2} > T_{em}^* > T_{em} + \frac{\Delta T_{em}}{4} & dT = +1 \\ T_{em} + \frac{\Delta T_{em}}{4} \geq T_{em}^* \geq T_{em} - \frac{\Delta T_{em}}{4} & dT = 0 \\ T_{em} - \frac{\Delta T_{em}}{4} < T_{em}^* < T_{em} - \frac{\Delta T_{em}}{2} & dT = -1 \\ T_{em}^* \leq T_{em} - \frac{\Delta T_{em}}{2} & dT = -2 \end{array}$$

$$\begin{array}{cc} \omega_m > \omega_{mth} & d\omega = +1 \\ \omega_m \leq \omega_{mth} & d\omega = -1 \end{array}$$

With $d\lambda$, dT, and $d\omega$ the outputs of the flux, torque, and speed comparators, respectively, and ω_{mth} the considered low-speed threshold. Selection of the applied VV is made using the lookup table shown in Table 1, as stated in [26,31,32].

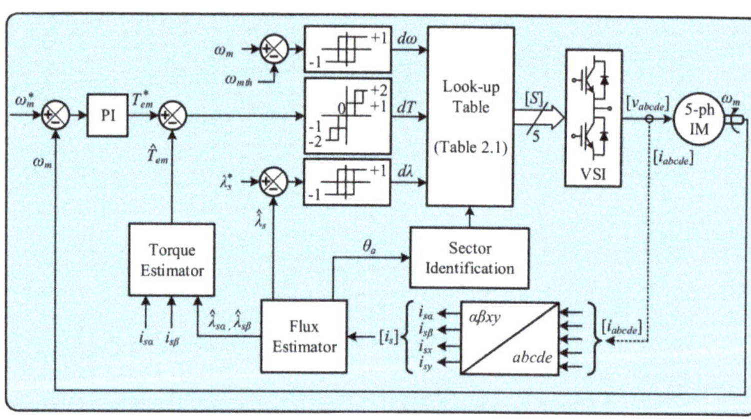

Figure 7. DTC scheme.

Note that the flux and torque estimator considers the discrete-time state-space model of the machine to calculate the variables in the α–β plane through the following equation:

$$\begin{pmatrix} \hat{\lambda}_{s\alpha} \\ \hat{\lambda}_{s\beta} \end{pmatrix} = \begin{pmatrix} \sigma \cdot L_s & 0 & \frac{L_m}{L_r} & 0 \\ 0 & \sigma \cdot L_s & 0 & \frac{L_m}{L_r} \end{pmatrix} \cdot \begin{pmatrix} i_{s\alpha} \\ i_{s\beta} \\ \hat{\lambda}_{r\alpha} \\ \hat{\lambda}_{r\beta} \end{pmatrix}$$

$$\hat{\lambda}_s = \left\| \hat{\lambda}_{s\alpha\beta} \right\| = \sqrt{\hat{\lambda}_{s\alpha}^2 + \hat{\lambda}_{s\beta}^2}$$

$$\hat{T}_{em} = p \cdot \frac{5}{2} \cdot \left(\hat{\lambda}_{s\alpha} \cdot i_{s\beta} - \hat{\lambda}_{s\beta} \cdot i_{s\alpha} \right)$$

Table 1. Lookup control table.

$d\lambda$	dT	$d\omega$	Position of the Stator Flux (Sector) 1	2	3	4	5	6	7	8	9	10
+1	+2	+1	VV_{L3}	VV_{L4}	VV_{L5}	VV_{L6}	VV_{L7}	VV_{L8}	VV_{L9}	VV_{L10}	VV_{L1}	VV_{L2}
		−1	VV_{L2}	VV_{L3}	VV_{L4}	VV_{L5}	VV_{L6}	VV_{L7}	VV_{L8}	VV_{L9}	VV_{L10}	VV_{L1}
	+1	+1	VV_{S3}	VV_{S4}	VV_{S5}	VV_{S6}	VV_{S7}	VV_{S8}	VV_{S9}	VV_{S10}	VV_{S1}	VV_{S2}
		−1	VV_{S2}	VV_{S3}	VV_{S4}	VV_{S5}	VV_{S6}	VV_{S7}	VV_{S8}	VV_{S9}	VV_{S10}	VV_{S1}
	0	+1	v_0	v_{31}	v_0	v_{31}	v_0	v_{31}	v_0	v_{31}	v_0	v_{31}
		−1	v_0	v_{31}	v_0	v_{31}	v_0	v_{31}	v_0	v_{31}	v_0	v_{31}
	−1	+1	VV_{S9}	VV_{S10}	VV_{S1}	VV_{S2}	VV_{S3}	VV_{S4}	VV_{S5}	VV_{S6}	VV_{S7}	VV_{S8}
		−1	VV_{S10}	VV_{S1}	VV_{S2}	VV_{S3}	VV_{S4}	VV_{S5}	VV_{S6}	VV_{S7}	VV_{S8}	VV_{S9}
	−2	+1	VV_{L9}	VV_{L10}	VV_{L1}	VV_{L2}	VV_{L3}	VV_{L4}	VV_{L5}	VV_{L6}	VV_{L7}	VV_{L8}
		−1	VV_{L10}	VV_{L1}	VV_{L2}	VV_{L3}	VV_{L4}	VV_{L5}	VV_{L6}	VV_{L7}	VV_{L8}	VV_{L9}
−1	+2	+1	VV_{L4}	VV_{L5}	VV_{L6}	VV_{L7}	VV_{L8}	VV_{L9}	VV_{L10}	VV_{L1}	VV_{L2}	VV_{L3}
		−1	VV_{L5}	VV_{L6}	VV_{L7}	VV_{L8}	VV_{L9}	VV_{L10}	VV_{L1}	VV_{L2}	VV_{L3}	VV_{L4}
	+1	+1	VV_{S4}	VV_{S5}	VV_{S6}	VV_{S7}	VV_{S8}	VV_{S9}	VV_{S10}	VV_{S1}	VV_{S2}	VV_{S3}
		−1	VV_{S5}	VV_{S6}	VV_{S7}	VV_{S8}	VV_{S9}	VV_{S10}	VV_{S1}	VV_{S2}	VV_{S3}	VV_{S4}
	0	+1	v_{31}	v_0	v_{31}	v_0	v_{31}	v_0	v_{31}	v_0	v_{31}	v_0
		−1	v_{31}	v_0	v_{31}	v_0	v_{31}	v_0	v_{31}	v_0	v_{31}	v_0
	−1	+1	VV_{S8}	VV_{S9}	VV_{S10}	VV_{S1}	VV_{S2}	VV_{S3}	VV_{S4}	VV_{S5}	VV_{S6}	VV_{S7}
		−1	VV_{S7}	VV_{S8}	VV_{S9}	VV_{S10}	VV_{S1}	VV_{S2}	VV_{S3}	VV_{S4}	VV_{S5}	VV_{S6}
	−2	+1	VV_{L8}	VV_{L9}	VV_{L10}	VV_{L1}	VV_{L2}	VV_{L3}	VV_{L4}	VV_{L5}	VV_{L6}	VV_{L7}
		−1	VV_{L7}	VV_{L8}	VV_{L9}	VV_{L10}	VV_{L1}	VV_{L2}	VV_{L3}	VV_{L4}	VV_{L5}	VV_{L6}

The look-up table of the DTC controller selects the appropriate virtual voltage vector in each sampling time according to the outputs of the hysteresis controllers and the sector where the stator flux is currently located, which is estimated as follows:

$$\theta_a = \arctan\left(\frac{\hat{\lambda}_{s\beta}}{\hat{\lambda}_{s\alpha}}\right)$$

$$Sector = \begin{cases} 1 & if\ -\frac{\pi}{10} \leq \theta_a \leq \frac{\pi}{10} \\ 2 & if\ \frac{\pi}{10} \leq \theta_a \leq \frac{3\pi}{10} \\ 3 & if\ \frac{3\pi}{10} \leq \theta_a \leq \frac{5\pi}{10} \\ 4 & if\ \frac{5\pi}{10} \leq \theta_a \leq \frac{7\pi}{10} \\ 5 & if\ \frac{7\pi}{10} \leq \theta_a \leq \frac{9\pi}{10} \\ 6 & if\ \frac{9\pi}{10} \leq \theta_a \leq \frac{11\pi}{10} \\ 7 & if\ \frac{11\pi}{10} \leq \theta_a \leq \frac{13\pi}{10} \\ 8 & if\ \frac{13\pi}{10} \leq \theta_a \leq \frac{15\pi}{10} \\ 9 & if\ \frac{15\pi}{10} \leq \theta_a \leq \frac{17\pi}{10} \\ 10 & if\ \frac{17\pi}{10} \leq \theta_a \leq \frac{19\pi}{10} \end{cases}$$

4. Results and Discussion

To assess the performance of the DTC scheme, different experimental tests were performed on a conventional three-phase induction machine with 30 slots that was redesigned to have five phases and three pole pairs. The electrical and mechanical parameters of the multiphase electrical machine are depicted in Table 2. The reference stator flux (λ_s*) is set to its rating value (0.4 Wb) during the experiments, the applied sampling frequency is fixed to 10 kHz and the hysteresis bands of the torque and flux regulators are programmed to be 1% of the rating values (this value was experimentally obtained for our test rig using a trial and error method). The maximum reference torque is set to 3.25 N·m, according to machine limits. The multiphase power converter is based on two conventional three-phase VSIs from SEMIKRON® (two SKS-22F modules in which five power legs are used). The DC link voltage is set to 300 V using an external DC power supply. The electronic control unit is based on a MSK28335 board and a Texas Instruments® TMS320F28335 digital signal processor. A digital encoder (GHM510296R/2500) and the enhanced quadrature encoder pulse peripheral of the DSP are used to measure the rotor mechanical speed ω_m. The load torque (T_L), which is demanded in the tests, is set by an independently controlled DC machine that is mechanically coupled to the five-phase machine. The experimental test rig is shown in Figure 8, where some photographs of the real system are included.

Table 2. Electrical and mechanical parameters of the five-phase IM.

Parameter	Value	Unit
Stator resistance, R_s	12.85	Ω
Rotor resistance, R_r	4.80	Ω
Stator leakage inductance, L_{ls}	79.93	mH
Rotor leakage inductance, L_{lr}	79.93	mH
Mutual inductance, M	681.7	mH
Rotational inertia, J_m	0.02	kg-m^2
Number of pairs of poles, p	3	-

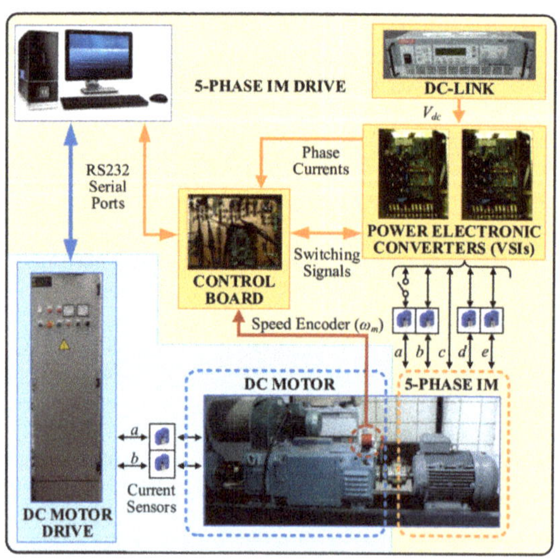

Figure 8. Experimental test rig.

4.1. Steady-State Operation

First, the performance of the system in steady-state operation at 500 rpm is analyzed in Figures 9 and 10, where different load torques are applied (1 N·m in Figure 9 and 2.75 N·m in Figure 10). The reference and measured values are colored red and blue, respectively. The speed and electrical torque responses are shown in the upper rows, where it is appreciated that the controller works well and the mechanical speed is successfully maintained in the reference value. Note that the electrical torque is mathematically estimated using the machine model, which produces some estimation errors. The reference and estimated stator flux in the regulated α–β plane are then shown in the second row, where it can be observed that the estimated stator flux values coincide with their references. Lastly, the measured stator currents are depicted in the last two rows, where it is appreciated that the α–β stator current vector describes a circular trajectory, with nearly null x–y stator current components. Therefore, the control goals are met using the DTC controller in steady-state operation because the results obtained can be extended to different reference speed and load torques.

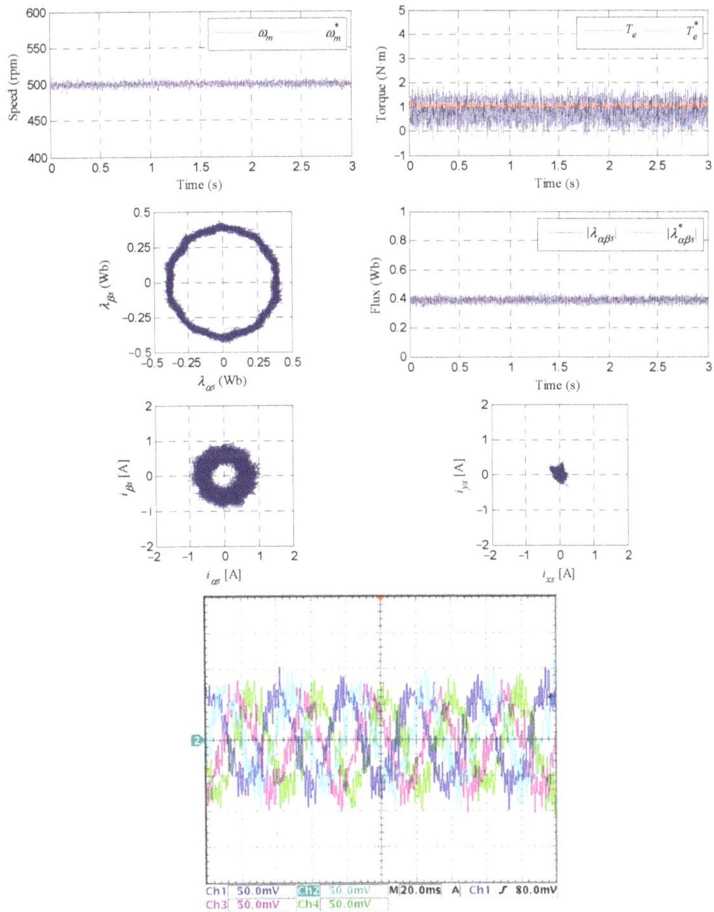

Figure 9. Experimental steady-state operation test where the reference speed is settled at 500 rpm and a load torque of 1 N·m is applied. Upper row: speed and torque responses. Second row: stator flux waveforms. Third row: current trajectories of the stator in the α–β and x–y planes. Last row: stator phase currents.

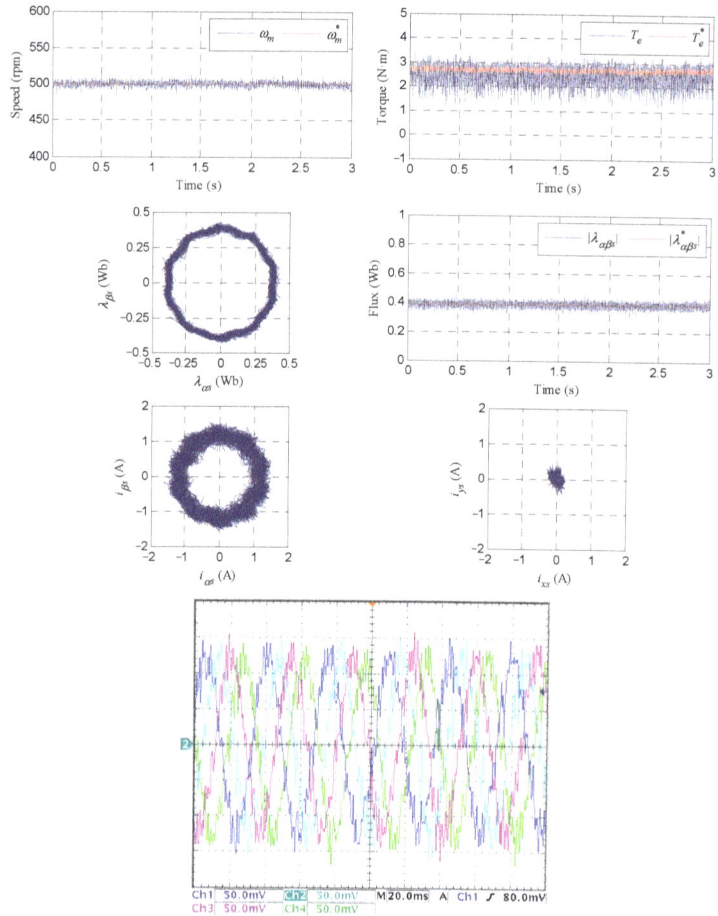

Figure 10. Experimental steady-state operation test where the reference speed is set at 500 rpm and a load torque of 2.75 N·m is applied. Upper row: speed and torque responses. Second row: stator flux waveforms. Third row: current trajectories of the stator in the α–β and x–y planes. Last row: stator phase currents.

4.2. Load Torque Rejection

Then load torque rejection tests were performed. The results obtained are summarized in Figure 11, where the reference speed is 500 rpm and the coupled DC machine imposes a heavy load torque within the system limits at t = 0.5 s. A drop in the speed is observed when the load is suddenly applied; although, the controller successfully manages this disturbance, upper-left plot. The estimated electrical torque is also regulated to be the referred one in steady and transient states, as shown in the upper right figure, while stator phase currents increase to manage the increment in the load (see bottom-left timing diagram). The estimated stator flux value is regulated in steady and transient states to coincide with the references, as can be appreciated in the bottom right figure. Then, these results, which summarize the ones obtained under different operating conditions, prove a controlled electrical torque in the multiphase drive. Note that flat lines are also observed in the x–y plane polar diagrams of the stator current, similar to the ones shown in Figures 9 and 10.

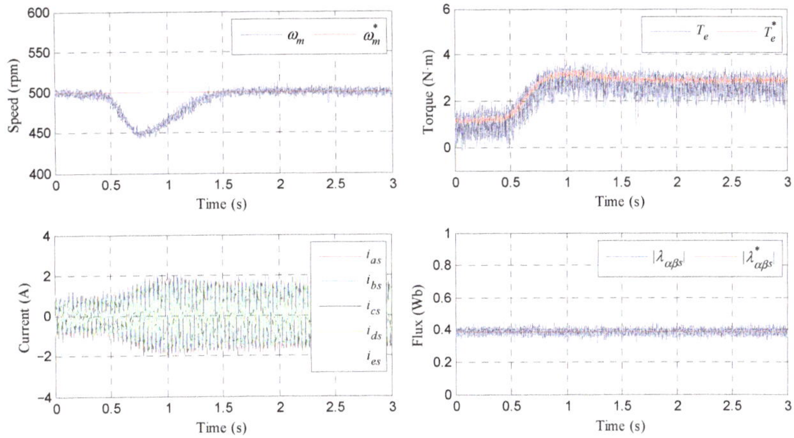

Figure 11. Experimental response of the controlled system in a load torque rejection test. The reference speed is 500 rpm and a load torque is applied at 0.5 s. The upper row shows the speed and torque responses. The lower row shows the stator current waveform and modulus of the stator flux during the test.

4.3. Step-Speed and Reverse Speed Tests

Then, experimental tests were conducted to evaluate the step-speed response of the system. Figure 12 reviews the results obtained. In the experiment, the machine is initially magnetized and the reference speed is changed from 0 to 500 rpm at t = 0.2 s. The performance of the mechanical speed (upper row, left plot), the electrical torque (upper row, right figure), stator phase currents (bottom row, left illustration), and the stator flux (bottom row, right drawing) is shown. Again, it can be observed that the DTC scheme provides accurate tracking of the reference speed and electrical torque, while the stator flux is maintained constant and equal to the reference value.

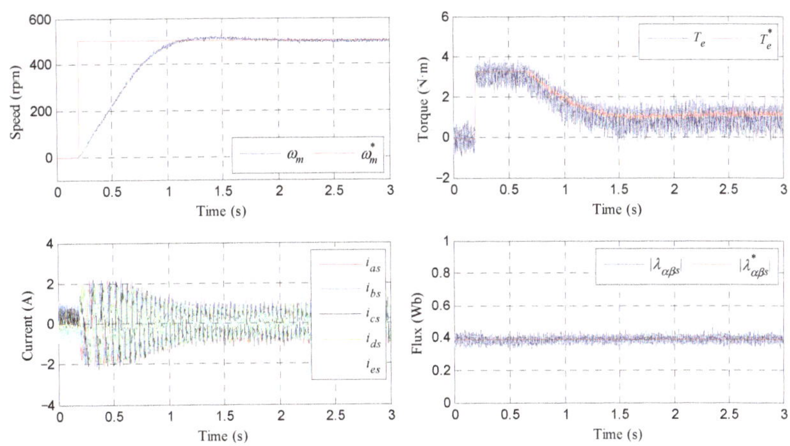

Figure 12. Experimental response of the controlled system in a step speed test. The reference speed is changed from 0 to 500 rpm at 0.2 s. The upper row shows the speed and torque responses. The lower row shows the stator current waveform and modulus of the stator flux during the test.

Finally, a reverse speed test is generated, where a change in the reference speed from 500 to −500 rpm is forced at $t = 0.2$ s. The obtained results are shown in Figure 13. Appropriate tracking is observed in the speed and the electrical torque, as can be seen in the upper row plots, left and right ones, respectively, while the expected stator phase current performance is observed close to the zero-speed crossing operating point (see lower row, left plot). Note that a saturation value of the reference torque (T_{MAX}) is set to 3.25 N·m, being the stator flux constant and equal to the reference value.

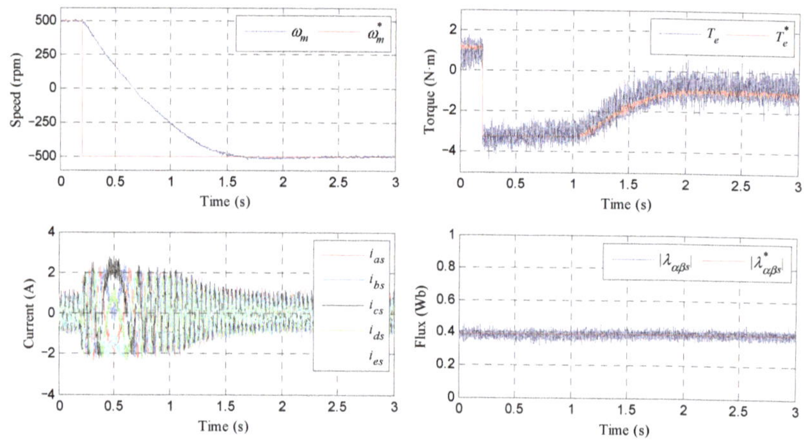

Figure 13. Experimental response of the controlled system in a speed reversal test. The reference speed is changed from 500 to −500 rpm at 0.2 s. The upper row shows the speed and torque responses. The lower row shows the stator current waveform and modulus of the stator flux during the test.

4.4. Fault-Tolerant Capability

In our work, the fault-tolerant capability of the system using the DTC controller is also analyzed. First, we consider an open-phase fault operation and later take into account the operation with two consecutive and nonconsecutive faults. Figures 14–16 show the performance of the mechanical speed and healthy stator currents when faults appear at t = 0.2 s. The system is operated as explained above with a load torque of 2.75 N·m. Figure 14 summarizes the behavior when phase a is open, while Figures 15 and 16 show the performance when consecutive phases a and b and non-consecutive a and c, respectively, are open. It is interesting to note that there is no degradation of the speed tracking in all cases; although, the stator phase currents show unequal peak values in the faulty system that becomes worse if the number of faulty phases increases.

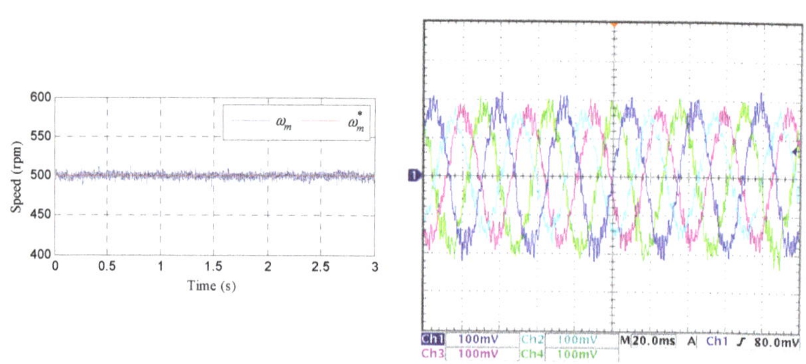

Figure 14. Experimental response of the controlled system when stator phase a is open at t = 0.2 s and the system is controlled at 500 rpm. The upper row shows the speed response and the lower one shows the stator current waveforms corresponding to phases b, c, d, and e during the test.

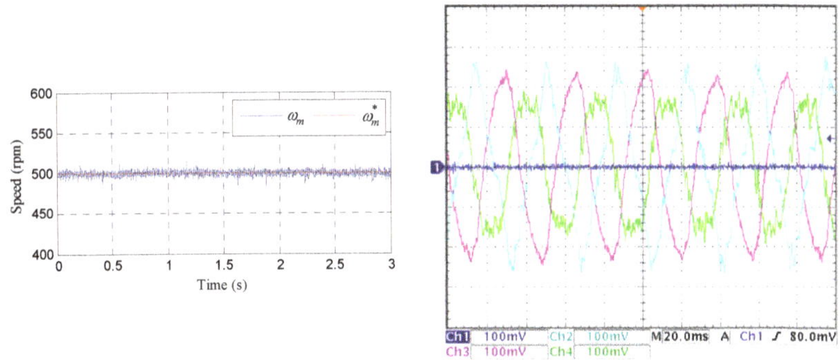

Figure 15. Experimental response of the controlled system when stator phases *a* and *b* are open at t = 0.2 s and the system is controlled at 500 rpm. The upper row shows the speed response, and the lower one depicts the stator current waveforms corresponding to phases *c*, *d*, and *e* during the test.

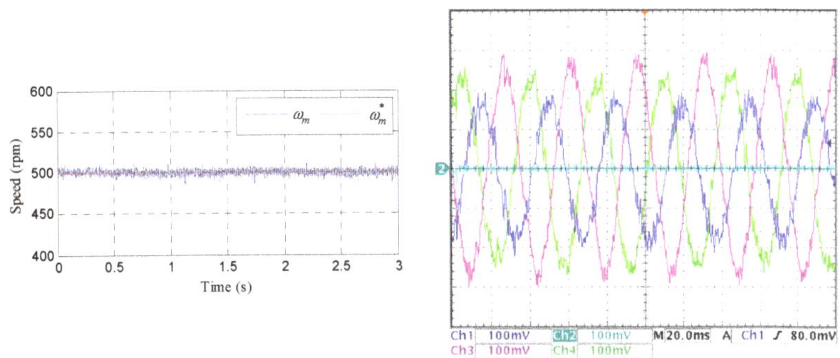

Figure 16. Experimental response of the controlled system when stator phase *a* and *c* are open at t = 0.2 s and the system is controlled at 500 rpm. The upper row shows the speed response, and the lower one depicts the stator current waveforms corresponding to phases *b*, *d*, and *e* during the test.

Although the use of DTC is not habitual in the field of multiphase drives due to the intrinsic limitations of a method that only manages two degrees of freedom, it has been experimentally validated that it is possible to apply it to five-phase IM as long as a series of virtual voltage vectors is created to impose zero voltage in the non-controllable plane of x–y. This allows DTC to be successfully applied when the machine does not have notable asymmetries in the design, where DTC can be considered as a quite competitive alternative to conventional and more complex field-oriented controllers. The results obtained, which can be summarized in Table 3, can be easily extended to multiphase drives with a higher number of phases, but the required virtual voltage vectors must be redefined to minimize stator voltage vectors in the x–y planes, which justifies considering the DTC technique as a stimulating strategy for the multiphase case.

Table 3. Qualitative analysis of the performance of the 5-phase IM drive using DTC.

Closed-Loop System Performance	DTC
Speed tracking error when the fault appears	Negligible
Torque tracking loss in control during the delay	No
Robustness against fault detection delay	High
Computational cost	Low
Harmonic content in stator currents	High

5. Conclusions

Direct torque control emerged in the late 20th century to become an alternative to vector controllers in conventional three-phase drives, where robustness, simplicity, and computational cost were desirable regulation characteristics. However, it does not seem to be an adequate control technique if the number of phases of the drive increases, due to its ability to command two electrical magnitudes in the drive: the electrical torque and the stator flux. We note that interest in DTC has been barely explored or experimentally analyzed in multiphase drives, this paper therefore tries to reduce this gap, proposing DTC in managing normal and faulty operations of five-phase induction motor drives, where the application of DTC goes from the introduction of the virtual voltage vector concept, to resemble the number of freedom degrees of the system, to the conventional three-phase drive case. Virtual voltage vectors can easily extend the use of DTC to multiphase drives with a higher number of phases, as long as they are chosen to nullify the stator voltage and current components in all x–y planes. Experimental evaluations were carried out using a test bed, and operating the drive in normal and faulty conditions (one and two open phases), with speed, torque, and flux references successfully tracked in all cases. The results obtained reveal that DTC is viable and can be extended to the case of multiphase drives, particularly where the required control goals are robustness, computational cost, and natural fault-tolerant capability of the drive in post-fault mode, without changing the control structure and/or current references.

Author Contributions: Conceptualization, M.B., F.B., C.M. and M.P.; methodology, M.B., F.B., C.M. and M.P.; software, M.B.; validation, M.B., F.B., C.M. and M.P.; formal analysis, M.B., F.B., C.M. and M.P.; investigation, M.B., F.B., C.M. and M.P.; resources, M.B., F.B., C.M. and M.P.; data curation, M.B. and C.M.; writing—original draft preparation, F.B.; writing—review and editing, M.B., F.B., C.M. and M.P.; visualization, M.B., F.B., C.M. and M.P.; supervision, F.B.; project administration, F.B.; funding acquisition, F.B. All authors have read and agreed to the published version of the manuscript.

Funding: This research received no external funding.

Conflicts of Interest: The authors declare that they have no conflict of interest.

References

1. Warg, E.E.; Härer, H. Preliminary investigation of an inventor-fed 5-phase induction motor. *Proc. Inst. Electr. Eng.* **1969**, *116*, 980–984.
2. Jahns, T.M. Improved reliability in solid-state AC drives by means of multiple independent phase drive units. *IEEE Trans. Ind. Appl.* **1980**, *IA-16*, 321–331. [CrossRef]
3. Toliyat, H.A.; Lipo, T.A.; White, J.C. Analysis of a concentrated winding induction machine for adjustable speed drive applications. I. Motor analysis. *IEEE Trans. Energy Convers.* **1991**, *6*, 679–683. [CrossRef]
4. Levi, E.; Bojoi, R.; Profumo, F.; Toliyat, H.A.; Williamson, S. Multiphase induction motor drives—A technology status review. *IET Electr. Power Appl.* **2007**, *1*, 489–516. [CrossRef]
5. Levi, E. Multiphase electric machines for variable-speed applications. *IEEE Trans. Ind. Electron.* **2008**, *55*, 1893–1909. [CrossRef]
6. Bojoi, R.; Rubino, S.; Tenconi, A.; Vaschetto, S. Multiphase electrical machines and drives: A viable solution for energy generation and transportation electrification. In Proceedings of the 2016 International Conference and Exposition on Electrical and Power Engineering (EPE), Iasi, Romania, 20–22 October 2016; pp. 632–639.
7. Cao, W.; Mecrow, B.C.; Atkinson, G.J.; Bennett, J.W.; Atkinson, D.J. Overview of electric motor technologies used for more electric aircraft (MEA). *IEEE Trans. Ind. Electron.* **2012**, *59*, 3523–3531.
8. Barrero, F.; Duran, M.J. Recent advances in the design, modeling, and control of multiphase machines—Part I. *IEEE Trans. Ind. Electron.* **2016**, *63*, 449–458. [CrossRef]
9. Duran, M.J.; Barrero, F. Recent advances in the design, modeling, and control of multiphase machines—Part II. *IEEE Trans. Ind. Electron.* **2016**, *63*, 459–468. [CrossRef]
10. Duran, M.J.; Levi, E.; Barrero, F. Multiphase electric drives: Introduction. In *Wiley Encyclopedia of Electrical and Electronics Engineering*; Wiley: Hoboken, NJ, USA, 2017.
11. Barrero, F.; González, I. *Control of Multiphase Machines and Drives*; MDPI: Basel, Switzerland, 2020; ISBN 978-3-03928-136-7.
12. Bermúdez, M.; Martín, C.; González, I.; Duran, M.J.; Arahal, M.R.; Barrero, F. Predictive current control in electrical drives: An illustrated review with case examples using a five-phase induction motor drive with distributed windings. *IET Electr. Power Appl.* **2020**, *14*, 1291–1310. [CrossRef]

13. Arahal, M.R.; Martin, C.; Barrero, F.; Duran, M.J. Assessing variable sampling time controllers for five-phase induction motor drives. *IEEE Trans. Ind. Electron.* **2020**, *67*, 2523–2531. [CrossRef]
14. Arahal, M.R.; Satué, M.G.; Barrero, F.; Ortega, M.G. Adaptive Cost Function FCSMPC for 6-Phase IMs. *Energies* **2021**, *14*, 5222. [CrossRef]
15. Takahashi, I.; Noguchi, T. A New Quick-Response and High-Efficiency Control Strategy of an Induction Motor. *IEEE Trans. Ind. Appl.* **1986**, *IA-22*, 820–827. [CrossRef]
16. Depenbrock, M. Direct self-control (DSC) of inverter-fed induction machine. *IEEE Trans. Power Electron.* **1988**, *3*, 420–429. [CrossRef]
17. Fei, Y.; Xiaofeng, Z.; Minzhong, Q.; Chengdong, D. The direct torque control of multiphase permanent magnet synchronous motor based on low harmonic space vector PWM. In Proceedings of the IEEE International Conference on Industrial Technology (ICIT-2008), Chengdu, China, 21–24 April 2008; pp. 1–5.
18. ABB Group. DTC: A Motor Control Technique for All Seasons. ABB White Paper. 2015. Available online: http://library.e.abb.com/public/0e07ab6a2de30809c1257e2d0042db5e/ABB_WhitePaper_DTC_A4_20150414.pdf (accessed on 27 October 2021).
19. Hadiouche, D.; Razik, H.; Rezzoug, A. On the modeling and design of dual-stator windings to minimize circulating harmonic currents for VSI fed AC machines. *IEEE Trans. Ind. Appl.* **2004**, *40*, 506–515. [CrossRef]
20. Buja, G.S.; Kazmierkowski, M.P. Direct torque control of PWM inverter-fed AC motors—A survey. *IEEE Trans. Ind. Electron.* **2004**, *51*, 744–757. [CrossRef]
21. Kianinezhad, R.; Nahid, B.; Betin, F.; Capolino, G.A. A novel Direct Torque Control (DTC) method for dual three phase induction motors. In Proceedings of the IEEE International Conference on Industrial Technology (ICIT 2006), Mumbai, India, 15–17 December 2006; pp. 939–943.
22. Kianinezhad, R.; Alcharea, R.; Nahid, B.; Betin, F.; Capolino, G.A. A novel direct torque control (DTC) for six-phase induction motors with common neutrals. In Proceedings of the International Symposium on Power Electronics, Electrical Drives, Automation and Motion (SPEEDAM 2008), Ischia, Italy, 11–13 June 2008; pp. 107–112.
23. Zheng, L.; Fletcher, J.E.; Williams, B.W.; He, X. A Novel Direct Torque Control Scheme for a Sensorless Five-Phase Induction Motor Drive. *IEEE Trans. Ind. Electron.* **2011**, *58*, 503–513. [CrossRef]
24. Parsa, L.; Toliyat, H.A. Sensorless Direct Torque Control of Five-Phase Interior Permanent-Magnet Motor Drives. *IEEE Trans. Ind. Appl.* **2007**, *43*, 952–959. [CrossRef]
25. Gao, Y.; Parsa, L. Modified Direct Torque Control of Five-Phase Permanent Magnet Synchronous Motor Drives. In Proceedings of the 22nd Annual IEEE Applied Power Electronics Conference and Exposition (APEC 2007), Anaheim, CA, USA, 25 February–1 March 2007; pp. 1428–1433.
26. Gao, L.; Fletcher, J.E.; Zheng, L. Low-Speed Control Improvements for a Two-Level Five-Phase Inverter-Fed Induction Machine Using Classic Direct Torque Control. *IEEE Trans. Ind. Electron.* **2011**, *58*, 2744–2754. [CrossRef]
27. Riveros, J.A.; Duran, M.J.; Barrero, F.; Toral, S. Direct torque control for five-phase induction motor drives with reduced common-mode voltage. In Proceedings of the 38th Annual Conference on IEEE Industrial Electronics Society (IECON 2012), Montreal, QC, Canada, 25–28 October 2012; pp. 3616–3621.
28. Tatte, Y.N.; Aware, M.V. Direct Torque Control of Five-Phase Induction Motor with Common-Mode Voltage and Current Harmonics Reduction. *IEEE Trans. Power Electron.* **2017**, *32*, 8644–8654. [CrossRef]
29. Chikondra, B.; Muduli, U.R.; Behera, R.K. An Improved Open-Phase Fault-Tolerant DTC Technique for Five-Phase Induction Motor Drive Based on Virtual Vectors Assessment. *IEEE Trans. Ind. Electron.* **2021**, *68*, 4598–4609. [CrossRef]
30. Barrero, F.; Bermúdez, M.; Durán, M.J.; Salas, P.; González-Prieto, I. Assessment of a Universal Reconfiguration-less Control Approach in Open-Phase Fault Operation for Multiphase Drives. *Energies* **2019**, *12*, 4698. [CrossRef]
31. Bermúdez, M.; González-Prieto, I.; Barrero, F.; Guzmán, H.; Duran, M.J.; Kestelyn, X. Open-Phase Fault-Tolerant Direct Torque Control Technique for Five-Phase Induction Motor Drives. *IEEE Trans. Ind. Electron.* **2017**, *64*, 902–911. [CrossRef]
32. Bermúdez, M.; González-Prieto, I.; Barrero, F.; Guzmán, H.; Kestelyn, X.; Duran, M.J. An Experimental Assessment of Open-Phase Fault-Tolerant Virtual-Vector-Based Direct Torque Control in Five-Phase Induction Motor Drives. *IEEE Trans. Power Electron.* **2018**, *33*, 2774–2784. [CrossRef]

applied sciences

MDPI

Article

Optimal Management of Reactive Power Considering Voltage and Location of Control Devices Using Artificial Bee Algorithm

Hassan Shokouhandeh [1], Sohaib Latif [2], Sadaf Irshad [2], Mehrdad Ahmadi Kamarposhti [3,*], Ilhami Colak [4] and Kei Eguchi [5]

1 Department of Electrical Engineering, Semnan University, Semnan, Iran; shokouhandeh@semnan.ac.ir
2 School of Mathematics and Big Data, Department of Computer Science, Anhui University of Science and Technology, Huainan 232001, China; sohaib.latif@aust.edu.cn (S.L.); sadafirshad09@gmail.com (S.I.)
3 Department of Electrical Engineering, Jouybar Branch, Islamic Azad University, Jouybar, Iran
4 Department of Electrical and Electronics Engineering, Faculty of Engineering and Architectures, Nisantasi University, Istanbul, Turkey; ilhcol@gmail.com
5 Department of Information Electronics, Fukuoka Institute of Technology, Fukuoka 811-0295, Japan; eguti@fit.ac.jp
* Correspondence: mehrdad.ahmadi.k@gmail.com or m.ahmadi@jouybariau.ac.ir

Abstract: Reactive power compensation is one of the practical tools that can be used to improve power systems and reduce costs. These benefits are achieved when the compensators are installed in a suitable place with optimal capacity. This study solves the issues of optimal supply and the purchase of reactive power in the IEEE 30-bus power system, especially when considering voltage stability and reducing total generation and operational costs, including generation costs, reserves, and the installation of reactive power control devices. The modified version of the artificial bee colony (MABC) algorithm is proposed to solve optimization problems and its results are compared with the artificial bee colony (ABC) algorithm, the particle swarm optimization (PSO) algorithm and the genetic algorithm (GA). The simulation results showed that the minimum losses in the power system requires further costs for reactive power compensation. Also, optimization results proved that the proposed MABC algorithm has a lower active power loss, reactive power costs, a better voltage profile and greater stability than the other three algorithms.

Keywords: control devices; artificial bee colony algorithm; voltage stability; reactive power

Citation: Shokouhandeh, H.; Latif, S.; Irshad, S.; Ahmadi Kamarposhti, M.; Colak, I.; Eguchi, K. Optimal Management of Reactive Power Considering Voltage and Location of Control Devices Using Artificial Bee Algorithm. *Appl. Sci.* **2022**, *12*, 27. https://doi.org/10.3390/app12010027

Academic Editors: Federico Barrero and Mario Bermúdez

Received: 18 November 2021
Accepted: 16 December 2021
Published: 21 December 2021

1. Introduction

Reactive power compensation located near the place of consumption is one of the most effective ways to modify power system performances [1–3]. The use of reactive power compensators such as capacitor banks, flexible AC transmission system (FACTS) devices, as well as on-load tap changers (OLTC) adjustments are common methods used to control the reactive current flowing in the power system. Reactive power compensations generate required reactive power near the load instead of being supplied from a distant power plant [4]. As a result, the power factor is corrected, loss is reduced and line capacity is increased. But these benefits can only be achieved when reactive power compensators are optimally sized and properly controlled. Therefore, the supply and purchase of reactive power is an essential issue in novel grids. In recent studies, voltage stability and optimal reactive power flow are integrated.

Many algorithms have been proposed to solve the problem of optimal reactive power compensation. In [5], a new method based on the multi-objective genetic algorithm (NS-GAII) was proposed for capacitor placement in a harmonic polluted system. The optimization results showed that the proposed method can converge to a set of optimal results in the search space. In [6], an innovative hybrid artificial bee colony and particle swarm optimization (ABC-PSO) algorithm was proposed for capacitor bank sizing in a radial distribution feeder. The optimization results indicated that the proposed ABC-PSO algorithm performed better than the PSO and ABC algorithms. In the proposed method in [7], an integrated approach of loss sensitivity factor (LSF) and voltage stability index (VSI) were used to determine the optimal placement of reactive power compensators. Also, the genetic algorithm (GA) was used to determine the optimal size of the compensators. In [8], optimal Static Var Compensator (SVC) placement was done by the ABC algorithm. The objective function was the reduction of ohmic losses and the improvement of the power quality. In [9], optimal thyristor-controlled series compensator (TCSC) placement in IEEE 57 and IEEE 118 Bus Systems was achieved. The ABC algorithm was used for optimization to reduce transmission loss. The results of optimization by the ABC algorithm and comparison with some other metaheuristic algorithms proved the superiority of ABC algorithm. In [10], the fuzzy based ABC algorithm was used to solve the optimization problem of voltage regulator and capacitor placement. The use of the fuzzy method allowed simultaneous minimization of losses in the radial distribution system and voltage deviation. In [11], a new modified version of the ABC algorithm for optimal location of shunt compensators was proposed. In order to validate the proposed ABC algorithm, a number of power systems of different sizes and complexities were used and also compared with meta-heuristic optimization algorithms. The results showed that the proposed method was able to provide better solutions.

Usually, a reactive power supply aims to minimize the active power loss or increase voltage stability. In general, it can be said that it aims to reduce the costs of reactive power. Reactive power resources have to supply and purchase reactive power in the grid [12]. The separate costs of grid-connected generators, line charging capacitors, capacitors installed on the buses and reactive power compensators are all worth mentioning, as each one presents an independent problem. It should be noted that the costs of reactive power generation resources depend on their operation conditions, while the cost of compensators depends on their capacity and installation location, which should be modeled accurately.

In this paper, a modified version of the artificial bee colony algorithm is proposed to solve the optimization problem. The percentage of the adjustable on-load tap changers (OLTC), capacity of the capacitor banks, and optimal location and capacity of the reactive power compensators are considered as control parameters. Changing these parameters affects the reactive power distribution, voltage profile and stability. Also, the effect of using different objective functions to determine the location and size of various reactive power compensators in the power system is evaluated.

The rest of this paper is organized as follows. Section 2 presents the reactive power supply problem. Section 3 discusses reactive power supply costs. Section 4 formulates the optimization problem. Section 5 introduces the MABC algorithm. Finally, Section 6 presents the numerical simulation and compares the results.

2. Reactive Power Supply

This paper aims to achieve an optimized voltage profile of the grid, reducing power system loss and minimizing reactive compensation costs by optimizing the reactive power supply. Due to the inherent relationship between reactive power and the voltage profile, optimizing the reactive power flow and reactive power supply improves the voltage profile. Thus, the reactive power supply should be determined using the best method. Maximizing the output reserve reactive power of the generators, maximizing the line charge and minimizing costs are the three main goals of reactive power compensation. Three different methods have been presented in which the costs corresponding to the best operation mode of the network are determined. The costs include the total cost of supplying reactive power by various resources, including generators, the line charging capacitance, installed capacitors and reactive power compensators, and the costs of active power loss. The three methods of the reactive power supply include:

(a) Maximizing the output reserve reactive power of the generators: increasing the reactive reserve power or reducing the reactive output power that increases the dynamic voltage stability.
(b) Maximizing the line charge: increasing the reactive output power of the inherent reactive power generation resources or the lines' capacitance.
(c) Minimizing the costs: the total cost of reactive power supply might be minimized by observing the system constraints.

Reducing voltage deviation, increasing voltage stability and reducing active power loss are three methods for improving the voltage profile [13].

3. Mathematical Formulation of the Reactive Power Supply

The total cost corresponding to active losses, which is supplied by the generators, can be determined only when the share of losses corresponding to each generator is known. Similarly, the costs of investing in reactive power demand and reactive power loss should be determined. The total cost is described as follows [14]:

$$Cost = Cost_{QG} + Cost_{QCap} + Cost_{Qcharg} + Cost_{QFact} \tag{1}$$

where $Cost_{QG}$, $Cost_{QCap}$, $Cost_{Qcharg}$ and $Cost_{QFact}$ are the reactive power generation costs of the generators, capacitors, line charge and compensators respectively. The total cost will be minimized when the cost of each reactive power compensation method is at its lowest value.

3.1. The Reactive Power Generation Cost of the Generators

In Equation (2), the coefficients a_i, b_i, c_i are the cost coefficients of the active power generation of the generators, N_g is the number of the generators and $\sin\theta$ are proportional to the power factor of each generator.

$$Cost_{QG} = \sum_{i=1}^{N_g} \left(C_i + b_i \sin\theta QG_i + a_i \sin\theta^2 QG_i^2 \right) \tag{2}$$

3.2. The Cost of Reactive Power Generation of the Capacitors Installed on Buses

The reactive power generation cost of the installed capacitors that is spent on the buses depends on the initial investment used to install the capacitance of interest on the bus. Also, to convert the costs to \$/h, the lifetime of the capacitance per hour should be considered. Equation (3) can be used to calculate the capacitor bank cost.

$$Cost_{QCap} = \sum_{i=1}^{N_C} \frac{QCap_I.c_C}{L.U} \tag{3}$$

where N_c is the number of the capacitor, c_c is the price per kVAr of the installed capacitor and Q_{CapI} is the capacitance of the ith capacitor. In addition, the parameter U is the performance factor and describes a ratio of the capacitor lifetime that is spent on generating reactive power. Also, L is the life time factor.

3.3. The Cost of Generating Reactive Power from the Line Charge

The total cost of generating reactive power to the total capacitors of the line charge is equal to the total cost of each line capacitance, which is the product of reactive power generation and cost of each unit's reactive power generation, as given in Equations (4) and (5).

The parameter C is equal to the average cost of generating each unit's reactive power of all grid resources. Figure 1 shows the π-model of the transmission lines with tap changer.

$$Q_{charg} = \sum_{charg=1}^{N_{charg}} Q_{chargh_{i-j}} = \sum_{charg=1}^{N_{charg}} \left(V_i^2 \frac{yc_{ij}}{2} + V_j^2 \frac{yc_{ij}}{2} \right) \tag{4}$$

$$Cost_{Qcharg} = Q_{charg} * C_{ave} \tag{5}$$

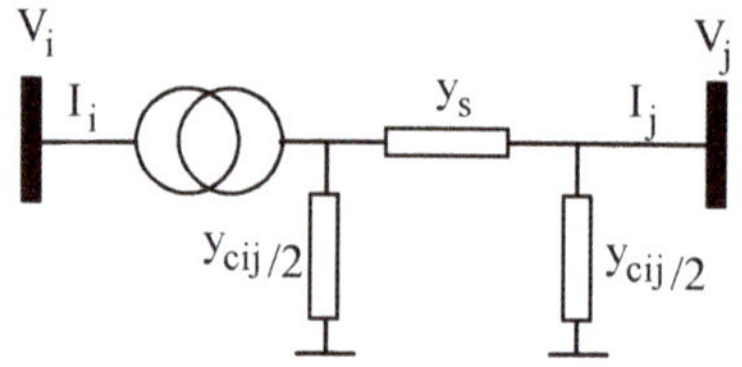

Figure 1. The π-model of the transmission lines with tap changer.

In Equation (4), Q_{charg} is the reactive power generated by line capacitor charge, V_i and V_j are the voltage amplitude of buses i and j and the line admittance. In general, if a transformer exists at the beginning of a bus, the line charge admittance is calculated as follows:

$$I_i = V_i \left(y_s(t-1)t + 0.5y_{c.ij}t^2 \right) + y_s t (V_i - V_j) \tag{6}$$

$$I_j = V_j \left(y_s(1-t) + 0.5y_{c.ij} \right) + y_s t (V_i - V_j) \tag{7}$$

where, V_i and V_j are ith and jth terminal voltage, y_s and $y_{c,ij}$ are series admittance and susceptance, I_i and I_j are line current.

3.4. Cost of Reactive Power Generation of the Compensators

Reactive power compensators of interest are SVC and TCSC, which are connected in series and parallel. In this section, these two devices are modeled and then, similar to previous sections, the cost of generating each reactive power unit is calculated. It should be pointed out that the optimal location and capacity of these two devices in the network is not known, and their generation costs and the total cost of reactive power supply should be minimized by selecting the proper location and capacitance.

3.4.1. SVC

This device is connected in parallel with buses that might operate in capacitive or inductive modes. It can be modeled as a capacitor with variable capacitance ranging from positive to negative. In power flow equations, the bus with SVC is considered as a generator bus with zero active power generation [12]. Figure 2 shows the SVC model.

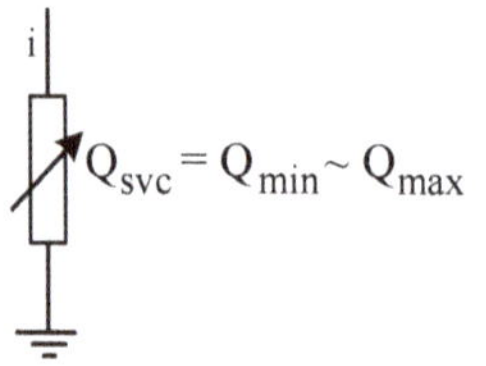

Figure 2. Single-line diagram of a Static Var Compensator model.

The total cost of SVC is a second-order function of its nominal capacity. The SVC cost can be calculated by Equation (8) [13].

$$C_{svc} = 0.0003s^2 - 0.3051S + 127.38\left(\frac{US\$}{K}var\right) \tag{8}$$

In the above equation, S is the rated capacity of the SVC in MVAr. Considering the 5-year lifetime, the cost is calculated in $/h.

3.4.2. TCSC

This device is modeled as a reactance in series with the transmission line and can operate in both capacitive and inductive ranges. The capacity of the TCSC depends on the reactance of the transmission line on which it is installed [12]. As mentioned in Equation (9), the equivalent impedance is the sum of the line impedance and the TCSC impedance. In most cases, the TCSC is in capacitive mode and its reactance is negative. Therefore, the equivalent impedance is less than the transmission line impedance. Figure 3 shows the TCSC model.

$$X_{ij} = X_{line} + X_{TCSC} = (1 + r_{TCSC})X_{line} \tag{9}$$

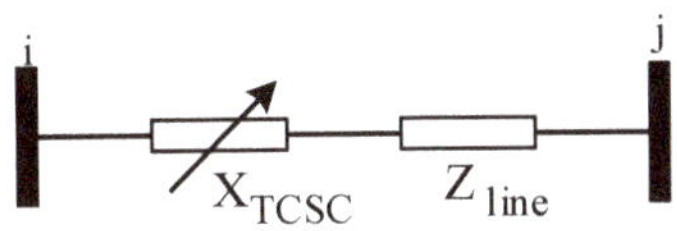

Figure 3. Single-line diagram of a Thyristor Controlled Series Compensator model.

In the above equation, X_{line} is the reactance of the transmission line and r_{TCSC} is the compensation factor. The compensation factor usually varies between 0.2 and −0.8. The cost of the TCSC, similar to the SVC, is a second-order function of its reactive power injection.

The cost function [13] is as follows:

$$C_{TCSC} = 0.0015s^2 - 0.7130S + 153.75\left(\frac{US\$}{K}var\right) \tag{10}$$

In this equation, S is the reactive power flowing through the transmission line before and after installing the TCSC.

4. Defining the Optimization Objectives

The mathematical definition of the reactive power supply is presented in this section. This section defines the problem presented in Section 2 mathematically. Four new objective functions with two classic objective functions are studied here. The constraints are considered as a penalty factor (P) in the objective functions:

$$P_{QG} = \sum_{i=1}^{Ng}[-min\left(1 - \frac{Q_G}{Q_G}, 0\right) + min\left(\frac{|Q_G|}{|Q_G^{min}|} - 1, 0\right)] \tag{11}$$

$$P_V = \sum_{i=1}^{Nbus}[-min\left(1 - \frac{V}{V^{max}}, 0\right) + min\left(\frac{V}{V^{min}} - 1\right)] \tag{12}$$

$$P_{tap} = \sum_{i=1}^{Ntap}[-min\left(1 - \frac{t}{t^{max}}, 0\right) + min\left(\frac{t}{t^{min}} - 1, 0\right)] \tag{13}$$

$$P_{cap} = \sum_{i=1}^{Ncap}[-min\left(1 - \frac{Q_{cap}}{Q_{cap}}, 0\right) + min\left(\frac{|Q_{cap}|}{|Q_{cap}^{min}|} - 1, 0\right)] \tag{14}$$

4.1. The Proposed Objective Functions

The first objective function is reducing the output power of the synchronous generators of the network (increasing the reserve of the reactive power generated by the generators) that generate reactive power.

$$Q_G = \sum_{i=1}^{NG} Q_{G_i}\,,\ \forall Q_{G_i} > 0 \tag{15}$$

$$min f_1 = Q_G \times (1 + \alpha P_{QG} + \beta P_V + \lambda P_{tap} + \delta P_{Cap}) \tag{16}$$

The second objective function is increasing the output power of the reactive power generation resources or the line charging capacitors. To achieve these conditions, the voltage of the buses should be increased, which improves the voltage profile.

$$max f_2 = Q_{charg} \times (1 + \alpha P_{QG} + \beta P_V + \lambda P_{tap} + \delta P_{Cap}) \tag{17}$$

The third objective function is minimizing the costs of supply and the purchase of the total reactive power

$$min f_3 = Cost_{QG} \times (1 + \alpha P_{QG} + \beta P_V + \lambda P_{tap} + \delta P_{Cap}) \tag{18}$$

The fourth objective function is minimizing the supply and purchase costs of reactive power considering the reactive power generation costs of the compensators installed in the network. The total costs of this function should be lower than the cost calculated in the previous section so that installing new reactive power compensators is cost-effective. Thus, this objective function can be represented as follows:

$$P_{loss} = \sum_{i=1}^{NG} P_{G_i}\,,\ \sum_{j=1}^{Nbus} PD_j \tag{19}$$

$$min f_4 = P_{loss} \times (1 + \alpha P_{QG} + \beta P_V + \lambda P_{tap} + \delta P_{Cap}) \tag{20}$$

In which P_{G_i} is the total power generation of the network generators, and PD_j is the real load connected to each bus.

4.2. The Classic Objective Functions

4.2.1. Minimizing the L-Index

In the following equation, *i* is the index related to G generator buses and j is the index of the load buses. The matrix is calculated using the following equations [15].

$$L_{max} = max(L_j),\ L_j = \left| 1 - \sum_{i=1}^{NG} \left(\frac{V_i}{V_j} \right) \right| \tag{21}$$

$$min f_5 = L_{max} \times (1 + \alpha P_{QG} + \beta P_V + \lambda P_{tap} + \delta P_{Cap}) \tag{22}$$

$$\begin{bmatrix} I_G \\ I_L \end{bmatrix} = \begin{bmatrix} Y_{GG} & Y_{GL} \\ Y_{LG} & Y_{LL} \end{bmatrix} \begin{bmatrix} V_G \\ V_L \end{bmatrix} \tag{23}$$

In the above equations $V_G, I_G, V_L,\ I_L$ are the current and voltage vectors of the generator buses and the load. By sorting the above equations, Equation (22) is obtained:

$$\begin{bmatrix} V_L \\ I_G \end{bmatrix} = \begin{bmatrix} Z_{LL} & F_{LG} \\ K_{GL} & Y_{GG} \end{bmatrix} \begin{bmatrix} I_L \\ V_G \end{bmatrix} \tag{24}$$

Finally:

$$[F_{LG}] = -[Y_{LL}]^{-1}[Y_{LG}] \tag{25}$$

To improve voltage stability, the L-index for each load bus should not exceed its upper limit, which is one. In other words, the L-index describes the system stability and the maximum value among L_j load buses. This index is a numerical quantity that estimates the real distance of the system from its stability boundary.

4.2.2. Improving the Voltage Profile

Another aim of reactive power compensation is to minimize buses voltage deviation. The voltage deviation function is calculated as Equation (26).

$$\Delta V = \frac{1}{N_{bus}} \sum_{i=1}^{N_{bus}} |V_{Nom} - V_i| \tag{26}$$

where N_{bus} is the number of buses, V_i is the ith bus voltage amplitude and V_{nom} is the nominal voltage equal to 1^{pu}. Voltage deviation is minimized when the voltage amplitude for all buses is close to the nominal value. The sixth objective function can be represented as Equation (27).

$$minf_6 = \Delta V \times \left(1 + \alpha P_{QG} + \beta P_V + \lambda P_{tap} + \delta P_{Cap}\right) \tag{27}$$

5. Modified Artificial Bee Colony (MABC) Algorithm

The algorithm used in this paper for solving the problem is the modified version of the artificial bee colony (MABC) algorithm. The ABC algorithm is an algorithm based on the random movement of bees, which provides higher quality and accuracy.

5.1. ABC

In this algorithm, the food sources solve the problem, and the extractable nectar represents the corresponding fitness. In this algorithm, the bees are divided into three groups: employed, onlooker and scout bees. The performance of this algorithm is as follows [16]:

(a) Primary preparation: similar to any optimization algorithm, this algorithm starts with an initial population, and the problem evaluation requires determining the population based on the random selection of the upper and lower limit of the problem.

(b) Repeat: the scout bees select each food source based on the probability obtained from Equation (28) and corresponding to the source.

$$p_i = \frac{f_i}{\sum_{n=1}^{N} f_n} \tag{28}$$

In the above equation, f_i is the fitness of each source, and N is the number of employed bees. In this step, to extract more nectar from each source, new sources are created by combining old food sources so that a source's performance can be represented.

$$v_{ij} = x_{ij} + \varphi_{ij}\left(x_{ij} - x_{kj}\right) \tag{29}$$

In which φ_{ij} is a random number between -1 and 1, and j is an integer random number between one and a variable decision-making dimension.

(c) Replacing the bees: in this algorithm, if a food source cannot be improved (exceed the determined value), the food source is left and replaced by a new random population selected by the onlooker bees. After generating and evaluating each food source, its performance is compared with the previous food sources. If the value of the objective function for a new source is less than an old one, it will replace an old source. Otherwise, the old food source is used.

$$x_i^j = x_{min}^j + rand(0,1) * \left(x_{max}^j - x_{min}^j\right) \tag{30}$$

5.2. Improved ABC

Although the basic case of this algorithm is efficient for solving large problems with various modalities, if the problem is constrained, its convergence rate is decreased. Three modifications are therefore applied to improve this algorithm. Changing the repeat pattern: in this case, instead of changing only one cell of the population, all cells of a group might change considering the following rule:

$$v_{ij} = \begin{cases} x_{ij} + \varphi_{ij}\left(x_{ij} - x_{kj}\right) & R_{ij} < MR \\ x_{ij} & otherwise \end{cases} \tag{31}$$

MR is an arbitrary number between [0, 1] in the above equation and R_{ij} is a random number in the same interval. According to Equation (31), φ_{ij} is a random number in the range of $[-1, 1]$ indicating that the movement step from one source to another is constant, while larger or smaller steps might be required in some problems; thus, φ_{ij} is selected in the range of $[-a, a]$ considering the problem conditions before selection. Adaptive modification of the repeat steps, as mentioned above, requires more extractions for various food sources or exploring new sources, which can be implemented if the movement steps of each source change adaptively. To this end, the Rosenberg law is used. This law, known as the $0/2$ law, operates in each m iteration of the main loop [9].

$$a(t+1) = \begin{cases} a(t) \times 0.85, \; if \; \Upsilon(m) < 0.2 \\ a(t) \times 0.85, \; if \; \Upsilon(m) > 0.2 \\ a(t) \qquad\quad , \; if \; \Upsilon(m) = 0.2 \end{cases} \tag{32}$$

In the above equation, t is the loop's counter, $\Upsilon(m)$ is the number of successful repetitions to the total repetitions in iteration m. To better understand the presented definitions of the MABC algorithm in solving the optimization problems, its flowchart is shown in Figure 4.

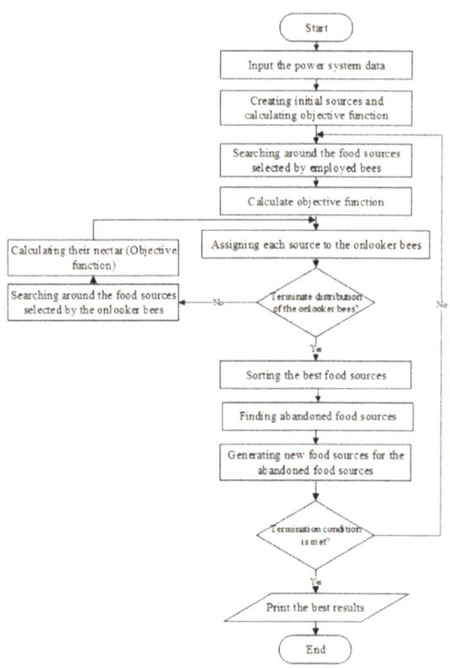

Figure 4. Flowchart of solving the problem.

6. Simulation Results

The decision-making variables in the above equations include the generator bus voltage, tap changers, reactive power generated by the capacitors, optimal location and capacity of the SVC and TCSC. The reactive power flow also changes; thus, ABC can be used to achieve the most optimal case. To evaluate the performance of this method, it is applied to the standard 30-bus network represented in the attachment. Information on the network is found in [17]. There are two capacitor banks that are also installed on buses 5 and 24 with a maximum value of 4.3 and 19 MVAr, respectively. The voltage magnitude in generator buses and load buses varies between 0.96 to 1.05 and 0.94 to 1.06, respectively. To determine the optimal location and capacity, one SVC and TCSC are installed, which are considered as a sequence with three cells for the optimization problem, where the first two cells represent the location (SVC bus number and line number for the TCSC) and the third cell represents capacity. According to the generator bus model for the SVC, its optimal reactive capacity is calculated using the power flow [18–34]. The capacitance range of the SCV is between −50 to 50 MVAr. The simulation results of the 30-bus network that include the system's decision-making variables and performance parameters for the proposed algorithm (MABC), the artificial bee colony (ABC) algorithm, the particle swarm optimization (PSO) algorithm and also the genetic algorithm (GA) are given in Table 1.

The optimal solutions of the six objective functions mentioned for the proposed modified artificial bee colony (MABC) algorithm are compared with the optimal solutions of the GA, PSO and ABC algorithms. In all cases, the MABC algorithm has a lower loss and costs and also a better voltage profile than three other algorithms. The power system total active power loss is 12.36 MW for the MABC algorithm when the f1 function is considered as an objective function. The total losses are 12.83 MW, 12.94 MW and 12.85 MW for the PSO, GA and ABC algorithms respectively. Also, the reactive power supply costs are 394.5 \$/h for PSO, 397.2 \$/h for GA, 387.2 \$/h for ABC and 381.6 \$/h for MABC. The average voltage of the buses for the second objective function, f_2, is maximum. However, the value of this parameter for the other functions is acceptable. The third objective function (f_3), which is the cost of supplying reactive power, is the minimum of 350.1 \$/h, while it is 481 and 553 \$/h for f_5 and f_6 for the proposed MABC algorithm. Also, in this case, the reactive power supply cost by using MABC is less than the PSO, GA and ABC algorithms. By using loss as the objective function (f_4), the loss is reduced to 11.52 MW for the MABC. The losses are 12.77 MW for PSO, 12.32 MW for GA and 11.89 MW for the ABC algorithm. In the objective function (f_4), the reactive output power of the generators is reduced, and the reactive output power of the capacitor banks and the SVC is increased to reduce the costs. As the power flowing through the lines is reduced, the loss is also reduced.

Table 1. Optimization res.

		f1				f2				f3			
		PSO	**GA**	**ABC**	**MABC**	**PSO**	**GA**	**ABC**	**MABC**	**PSO**	**GA**	**ABC**	**MABC**
Loss (MW)		12.83	12.94	12.85	12.36	12.92	12.98	12.88	12.39	13.07	13.13	12.94	12.72
Line charging reactive power (MVAr)		27.45	26.23	27.14	28.32	32.21	31.47	32.27	33.14	30.24	30.19	30.54	31.03
Generator reactive reserve (MVAr)		161.8	159.3	160.9	162.7	147.6	145.1	149.2	149.9	157.3	151.4	157.2	158.7
Reactive power supply cost ($/h)		394.5	397.2	387.2	381.6	437.7	430.8.7	435	426.7	357.6	359.6	353.2	350.1
Average voltage of the bus		0.964	0.953	0.972	0.991	1.014	1.016	1.012	1.011	0.984	1.003	0.992	1
Capacitors (MVar)	Bus.5	18.4	19.7	18.2	19.1	19.3	19.8	19.4	18.9	19.1	19.3	19.3	19.8
	Bus.24	4.3	4.23	4.11	3.91	4.3	3.86	4.11	4.05	4.3	4.3	4.09	4.3
SVC	Bus number	7	7	24	28	27	27	27	27	9	8	28	28
	(MVar) capacity	44.6	48.2	49.3	49.7	40.4	42.7	40.6	45.1	41.2	38.5	40.6	40.3
TCSC	Line number	35	28	36	41	34	27	32	16	11	32	12	8
	compensation percentage	−0.63	−0.52	−0.82	−0.51	−0.68	−0.83	−0.82	−0.86	−0.84	−0.23	−0.41	−0.32
		f4				f5				f6			
		PSO	**GA**	**ABC**	**MABC**	**PSO**	**GA**	**ABC**	**MABC**	**PSO**	**GA**	**ABC**	**MABC**
Loss (MW)		12.77	12.32	11.89	11.52	13.21	13.18	13.09	13.02	12.88	12.65	12.23	12.1
Line charging reactive power (MVAr)		28.4	28.9	29.2	29.5	28.3	29.3	28.2	30.4	34.2	33.1	34.9	34.9
Generator reactive reserve (MVAr)		133.7	135.2	136.4	137.2	137.4	138.1	137.9	138.6	143.2	146.9	147.8	149
Reactive power supply cost ($/h)		463.2	465.1	460.1	457.8	493.1	490.4	487.3	481.6	568.2	569.4	561.5	553.2
Average voltage of the bus		0.996	0.996	0.997	0.998	0.993	0.992	0.996	0.997	1.04	1.02	1.01	1
Capacitors (MVAr)	Bus.5	19.2	18.4	19.6	17.2	3.45	3.67	3.88	2.79	19.2	18.4	18.7	15.7
	Bus.24	4.2	4.1	4.3	4.3	0	0	0	0	4.2	3.9	4.1	3.6
SVC	Bus number	23	23	25	24	23	23	25	21	13	12	27	28
	(MVar) capacity	16.5	14.23	15.61	13.58	50.9	51.1	50.8	48.4	41.4	36.8	45.2	37.7
TCSC	Line number	7	9	12	3	32	35	31	34	12	32	14	40
	compensation percentage	−0.64	−0.43	−0.23	−0.18	−0.83	−0.78	−0.65	−0.81	−0.81	−0.41	−0.78	−0.47

Using two 19.8 and 4.3 MVAr capacitor banks, one 40.3 MVAr SVC on bus 28 and a TCSC with 0.32% compensation on line 8, the cost of supplying the reactive power of the network is reduced and voltage deviation is at a minimum.

The voltage profile is shown in Figure 5 to evaluate the voltage stability and the voltage profile changes in the different objective functions.

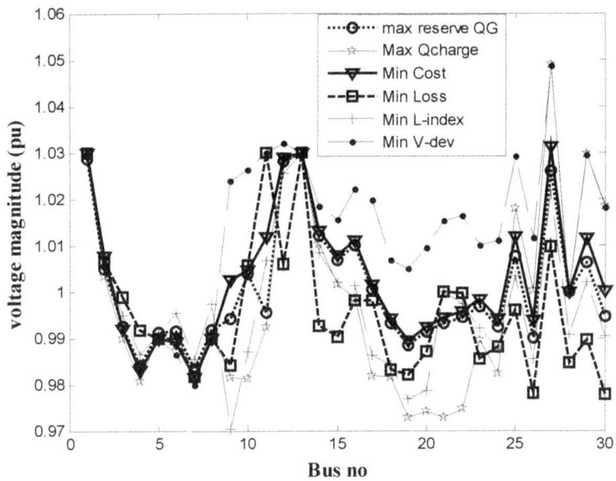

Figure 5. Voltage profile for different objective functions.

As can be seen, the best voltage profile is obtained for the objective function that minimizes the voltage violation. The voltage profile that minimizes the total purchase cost of reactive power is also favorable.

7. Conclusions

This paper studies reactive power supply to supply the reactive power demand, aiming to improve the voltage profile and voltage stability. All parameters and variables that affect the reactive power distribution of the network, such as the voltage of the generators, tap changer of the transformers and capacity of the capacitor banks, are considered in the equations. Also, to increase the system's flexibility and improve its efficiency in terms of optimal capacity, location, voltage, and SVC and TCSC devices are examined. By employing the modified artificial bee colony algorithm and the third objective function, the supply and purchase of reactive power is minimized. The simulation results indicate that the minimum active power losses in the power system require more payment to reactive power compensation. Minimum loss in the power system is about 11.52 MW, so 457.8$ should be spend for reactive power compensation. Also, optimization results prove that, in all the scenarios, the proposed MABC algorithm has a lower active power loss, reactive power cost and better voltage profile than the PSO, GA and ABC algorithms.

Author Contributions: Conceptualization, H.S. and M.A.K.; methodology, H.S. and M.A.K.; software, H.S. and M.A.K.; validation, H.S., S.L., S.I. and M.A.K.; formal analysis, H.S., S.L., S.I. and M.A.K.; investigation, H.S., S.L., S.I., M.A.K., I.C. and K.E.; resources, H.S., M.A.K., I.C. and K.E.; data curation, H.S. and M.A.K.; writing—original draft preparation, H.S. and M.A.K.; writing—review and editing, H.S. and M.A.K.; visualization, H.S., S.L., S.I., M.A.K., I.C. and K.E.; supervision, H.S., M.A.K., I.C. and K.E.; project administration, H.S., S.L., S.I., M.A.K., I.C. and K.E.; funding acquisition, H.S., S.L., S.I., M.A.K., I.C. and K.E. All authors have read and agreed to the published version of the manuscript.

Funding: This research received no external funding.

Institutional Review Board Statement: Not applicable.

Informed Consent Statement: Not applicable.

Data Availability Statement: Not applicable.

Conflicts of Interest: The authors declare no conflict of interest.

References

1. Mehmood, K.; Li, Z.; Tahir, M.F.; Cheema, K.M. Fast excitation control strategy for typical magnetically controllable reactor for reactive power compensation. *Int. J. Electr. Power Energy Syst.* **2021**, *129*, 106757. [CrossRef]
2. Montoya, O.; Chamorro, H.; Alvarado-Barrios, L.; Gil-González, W.; Orozco-Henao, C. Genetic-Convex Model for Dynamic Reactive Power Compensation in Distribution Networks Using D-STATCOMs. *Appl. Sci.* **2021**, *11*, 3353. [CrossRef]
3. Shokouhandeh, H.; Jazaeri, M. Robust design of fuzzy-based power system stabiliser considering uncertainties of loading conditions and transmission line parameters. *IET Gener. Transm. Distrib.* **2019**, *13*, 4287–4300. [CrossRef]
4. Karami, M.; Mariun, N.B.; Ab-Kadir, M.Z.A.; Misron, N.; Mohd Radzi, M.A. Motor Current Signature Analy-sis-based Non-invasive Recognition of Mixed Eccentricity Fault in Line Start Permanent Magnet Synchronous Motor. *Electr. Power Compon. Syst.* **2021**, *49*, 133–145. [CrossRef]
5. Moghadam, M.E.; Falaghi, H.; Farhadi, M. A Novel Method of Optimal Capacitor Placement in the Presence of Harmonics for Power Distribution Network Using NSGA-II Multi-Objective Genetic Optimization Algorithm. *Math. Comput. Appl.* **2020**, *25*, 17. [CrossRef]
6. Sharma, S.; Ghosh, S. FIS and hybrid ABC-PSO based optimal capacitor placement and sizing for radial distribution networks. *J. Ambient. Intell. Humaniz. Comput.* **2020**, *11*, 901–916. [CrossRef]
7. Öner, A.; Abur, A. Computationally efficient method for placing reactive power sources against contingencies. *Int. J. Electr. Power Energy Syst.* **2021**, *133*, 107253. [CrossRef]
8. Alimuddin, A.; Nurhalim, G.; Arafiyah, R. Optimization Placement Static Var Compensator (SVC) Using Artificial Bee Colony (ABC) Method on PT PLN (Persero) Jawa-Bali, Indonesia. In Proceedings of the IEEE 2018 1st International Conference on Computer Applications & Information Security (ICCAIS), Riyadh, Saudi Arabia, 4–6 April 2018; pp. 1–4.
9. Rao, R.; Reddy, V. Optimal Placement of Thyristor Controlled Series Compensator in IEEE 57 and IEEE 118 Bus Systems to Reduce Transmission Loss Using ABC Algorithm. *Int. J. Simul. Syst. Sci. Technol.* **2019**, *20*, 1–8. [CrossRef]
10. Addisu, M.; Salau, A.O.; Takele, H. Fuzzy logic based optimal placement of voltage regulators and capacitors for distribution systems efficiency improvement. *Heliyon* **2021**, *7*, 07848. [CrossRef]
11. Yamaçlı, V.; Abacı, K. A New Method Based on Artificial Bee Colony Algorithm for the Optimal Location of Shunt Connected Facts Devices. *J. Eng. Technol. Appl. Sci.* **2018**, *3*, 17–34. [CrossRef]
12. Shokouhandeh, H.; Jazaeri, M.; Sedighizadeh, M. On-time stabilization of single-machine power system connected to infinite bus by using optimized fuzzy-PID controller. In Proceedings of the IEEE 2014 22nd Iranian Conference on Electrical Engineering (ICEE), Tehran, Iran, 20–22 May 2014; pp. 768–773.
13. Naderi, E.; Narimani, H.; Pourakbari-Kasmaei, M.; Cerna, F.V.; Marzband, M.; Lehtonen, M. State-of-the-Art of Optimal Active and Reactive Power Flow: A Comprehensive Review from Various Standpoints. *Processes* **2021**, *9*, 1319. [CrossRef]
14. Acosta, M.; Gonzalez-Longatt, F.; Andrade, M.; Torres, J.; Chamorro, H. Assessment of Daily Cost of Reactive Power Procurement by Smart Inverters. *Energies* **2021**, *14*, 4834. [CrossRef]
15. Karbalaei, F.; Abbasi, S. L-index based contingency filtering for voltage stability constrained reactive power planning. *Turk. J. Electr. Eng. Comput. Sci.* **2018**, *26*, 3156–3167. [CrossRef]
16. Kamarposhti, M.A.; Shokouhandeh, H.; Colak, I.; Band, S.S.; Eguchi, K. Optimal Location of FACTS Devices in Order to Simultaneously Improving Transmission Losses and Stability Margin Using Artificial Bee Colony Algorithm. *IEEE Access* **2021**, *9*, 125920–125929. [CrossRef]
17. Available online: http://labs.ece.uw.edu/pstca/pf30/pg_tca30bus.htm (accessed on 1 December 2020).
18. Guo, C.; Ye, C.; Ding, Y.; Wang, P. A Multi-State Model for Transmission System Resilience Enhancement Against Short-Circuit Faults Caused by Extreme Weather Events. *IEEE Trans. Power Deliv.* **2021**, *36*, 2374–2385. [CrossRef]
19. Zhou, H.; Xu, C.; Lu, C.; Jiang, X.; Zhang, Z.; Wang, J.; Xiao, X.; Xin, M.; Wang, L. Investigation of transient magnetoelectric response of magnetostrictive/piezoelectric composite applicable for lightning current sensing. *Sens. Actuators A Phys.* **2021**, *329*, 112789. [CrossRef]
20. Xu, X.; Niu, D.; Xiao, B.; Guo, X.; Zhang, L.; Wang, K. Policy analysis for grid parity of wind power generation in China. *Energy Policy* **2020**, *138*, 111225. [CrossRef]
21. Yang, Y.; Peng, J.C.-H.; Ye, C.; Ye, Z.; Ding, Y. A Criterion and Stochastic Unit Commitment towards Frequency Resilience of Power Systems. *IEEE Trans. Power Syst.* **2021**, 1. [CrossRef]
22. Hu, J.; Ye, C.; Ding, Y.; Tang, J.; Liu, S. A Distributed MPC to Exploit Reactive Power V2G for Real-Time Voltage Regulation in Distribution Networks. *IEEE Trans. Smart Grid* **2021**, 1. [CrossRef]
23. Ding, L.; Huang, L.; Li, S.; Gao, H.; Deng, H.; Li, Y.; Liu, G. Definition and Application of Variable Resistance Coefficient for Wheeled Mobile Robots on Deformable Terrain. *IEEE Trans. Robot.* **2020**, *36*, 894–909. [CrossRef]
24. Chen, H.; Miao, Y.; Chen, Y.; Fang, L.; Zeng, L.; Shi, J. Intelligent Model-based Integrity Assessment of Nonstationary Mechanical System. *J. Web Eng.* **2021**, *20*, 253–280. [CrossRef]
25. Wu, Z.; Cao, J.; Wang, Y.; Wang, Y.; Zhang, L.; Wu, J. hPSD: A Hybrid PU-Learning-Based Spammer Detection Model for Product Reviews. *IEEE Trans. Cybern.* **2018**, *50*, 1595–1606. [CrossRef] [PubMed]
26. Wu, Z.; Li, C.; Cao, J.; Ge, Y. On Scalability of Association-rule-based Recommendation: A Unified Distributed-computing Framework. *ACM Trans. Web* **2020**, *14*, 1–21. [CrossRef]

27. Lv, Z.; Chen, D.; Lou, R.; Song, H. Industrial Security Solution for Virtual Reality. *IEEE Internet Things J.* **2021**, *8*, 6273–6281. [CrossRef]
28. Chen, X.; Wang, T.; Ying, R.; Cao, Z. A Fault Diagnosis Method Considering Meteorological Factors for Transmission Networks Based on P Systems. *Entropy* **2021**, *23*, 1008. [CrossRef]
29. Wang, T.; Liu, W.; Zhao, J.; Guo, X.; Terzija, V. A rough set-based bio-inspired fault diagnosis method for electrical substations. *Int. J. Electr. Power Energy Syst.* **2020**, *119*, 105961. [CrossRef]
30. Wang, T.; Wei, X.; Wang, J.; Huang, T.; Peng, H.; Song, X.; Cabrera, L.V.; Pérez-Jiménez, M.J. A weighted corrective fuzzy reasoning spiking neural P system for fault diagnosis in power systems with variable topologies. *Eng. Appl. Artif. Intell.* **2020**, *92*, 103680. [CrossRef]
31. Zhang, L.; Wang, X.; Zhang, Z.; Cui, Y.; Ling, L.; Cai, G. An adaptative control strategy for interfacing converter of hybrid microgrid based on improved virtual synchronous generator. *IET Renew. Power Gener.* **2021**. [CrossRef]
32. Zhong, Q.; Yang, J.; Shi, K.; Zhong, S.; Li, Z.; Sotelo, M.A. Event-Triggered H_∞ Load Frequency Control for Multi-Area Nonlinear Power Systems Based on Non-Fragile Proportional Integral Control Strategy. *IEEE Trans. Intell. Transp. Syst.* **2021**, 1–11. [CrossRef]
33. Wu, Y.; Zhu, W. The Role of CSR Engagement in Customer-Company Identification and Behavioral Intention During the COVID-19 Pandemic. *Front. Psychol.* **2021**, *12*, 721410. [CrossRef] [PubMed]
34. Feng, Y.; Zhang, B.; Liu, Y.; Niu, Z.; Dai, B.; Fan, Y.; Chen, X. A 200–225-GHz Manifold-Coupled Multiplexer Utilizing Metal Waveguides. *IEEE Trans. Microw. Theory Tech.* **2021**, *69*, 5327–5333. [CrossRef]

applied sciences

MDPI

Article

Energy Efficiency Improvement of Diesel–Electric Trains Using Solar Energy: A Feasibility Study

Ahmad Fayad [1,*], Hussein Ibrahim [2], Adrian Ilinca [3], Sasan Sattarpanah Karganroudi [1] and Mohamad Issa [4]

1 Technological Institute for Industrial Maintenance, Cegep of Sept-Îles, Sept-Îles, QC G4R 5B7, Canada; sasan.karganroudi@itmi.ca
2 Energy Intelligence Research and Innovation Center, Cegep of Sept-Îles, Sept-Îles, QC G4R 5B7, Canada; hussein.ibrahim@itmi.ca
3 Wind Energy Laboratory, University of Quebec in Rimouski, Rimouski, QC G5L 3A1, Canada; adrian_ilinca@uqar.ca
4 Quebec Maritime Institute, Rimouski, QC G5L 4B4, Canada; missa@imq.qc.ca
* Correspondence: ahmad.fayad@cegepsi.ca

Abstract: Nowadays, productivity challenges in modern manufacturing systems have been the driving force in generating energy-efficient technologies in every industry, including diesel–electric locomotives. The diesel–electric locomotive is one of the most widely used methods in rail transportation, especially in North America. More precisely, the evolution of the electric transmission has allowed the locomotive's effective tractive effort to increase its diesel engine horsepower. In this paper, we study a new way to improve the energy efficiency of diesel–electric trains using photovoltaic solar panels. This solution is suitable for reducing greenhouse gas emissions of the diesel–electric locomotive system, particularly in cold climates. We explore the amount of energy produced by the PV solar panels and compare it with that produced by the auxiliary diesel-generator during a train's journey. This comparison clarifies the actual percentage of energy that solar panels can cover. Thus, this paper presents a validation of feasibility and profitability as a function of the train's specific operating conditions and the meteorological data associated with their routes. Based on the results, the minimum annual fuel reduction of auxiliary generators allowed using PV solar panels is above 50% in all cases and wagon classes, proving this solution's feasibility. Regarding the comparison, case 3 (Sept-Îles to Schefferville) and case 4 (Luxor to Aswan) are the best, with over 100% of the energy provided by PV solar panels in all the wagons' classes. The payback period ranges from 2.5 years to 9.1 years, while the CO_2 emission reduction's revenues range from $460 to $998 per year/wagon.

Keywords: energy efficiency; diesel–electric train; photovoltaic solar panel; emission reduction

Citation: Fayad, A.; Ibrahim, H.; Ilinca, A.; Sattarpanah Karganroudi, S.; Issa, M. Energy Efficiency Improvement of Diesel–Electric Trains Using Solar Energy: A Feasibility Study. *Appl. Sci.* **2022**, *12*, 5869. https://doi.org/10.3390/app12125869

Academic Editor: Fabrice Goubard

Received: 24 May 2022
Accepted: 6 June 2022
Published: 9 June 2022

1. Introduction

Worldwide, energy consumption is a central issue of economic activity. Over the past few months, the COVID-19 pandemic has caused an unprecedented global economic and social crisis. The pandemic has significantly affected all aspects of life, including the energy sector [1]. Due to the pandemic, many governmental decisions were made, such as the imposition of lockdowns worldwide which decimated transport-related demand. Associated events led to an unprecedented collapse in oil demand and, therefore, a drop in the global energy demand, with a rate of 4.5–6% in 2020 relative to 2019 [2,3]. Indeed, recently, the statistical review of world energy in 2021 affirms that the carbon emissions from energy use were falling by over 6% in 2020. However, there are worrying signs that the COVID-induced dip in carbon emissions and the energy demand will be short-lived as the world economy recovers and lockdowns are lifted [3].

Regardless of the lower energy demand in 2020, renewable energy, led by wind and solar, continued to grow. In 2020, renewable energy posted a record increase in production, by 358 TWh compared to 2019 [3]. Moreover, this increase is comparable to those seen in 2017, 2018, and 2019. Thus, we noticed a relative immunity of renewable energy growth to the world political and health issue events in the last years. Remarkably, PV solar capacity in both 'OnGrid' and 'OffGrid' increased over time and reached around 710 GW and 4 GW, respectively, in 2020 as we can see in Table 1. Furthermore, there was a 125.8 GW and 0.234 GW increase in installed capacity from 2019 to 2020. Therefore, renewable energy has formed a crucial part of any viable solution to reduce primary energy consumption over the last years [4–6].

Table 1. Global installed PV power (GW) [5].

Global Installed PV Power (GW)	2016	2017	2018	2019	2020
OnGrid	294.8	389.579	482.912	583.872	709.674
OffGrid	2.278	2.908	3.61	4.059	4.293

In general, renewable energy share, the energy intensity of gross domestic product, and the electrification of final uses of energy have all shown improvements in recent years. Yet, the pace is insufficient to put the world on track to meet the Paris agreement (COP21) goals. This international climate conference marks a turning point in the fight against global warming. It commits all countries to reducing greenhouse gas emissions and keep warming below 2 °C and, preferably, below 1.5 °C by 2100 [7].

Considering ample resource availability, significant market potential, and cost competitiveness, PV solar panels are expected to continue driving overall renewables' growth in several regions over the next decade. From today's levels, IRENA's analysis shows that solar PV power installations could grow almost four-fold over the next ten years, reaching a cumulative capacity of 2840 GW globally by 2030 and rising to 8519 GW by 2050. This growth implies that the total installed capacity in 2050 will be almost twelve times higher than 2020. Around 60% of total solar PV capacity in 2050 would be utility-scale globally, with the remaining 40% distributed (rooftop) [8]. Thus, to achieve the goal by 2050, the use of PV solar panels must increase enormously in all end-use sectors.

One of these sectors is the transport sector, the second-largest energy consumer after the industrial sector, with a rating factor of 14% of global energy consumption [9,10]. The transport sector is divided into six principal modes: road, maritime, air, rail, intermodal, and pipeline [11]. Our study in this paper focuses on using PV solar panels in the rail mode, mainly in diesel–electric trains.

Rail is responsible for 9% of global motorized passenger movement and 7% of freight [12]. Since 2017, rail-based transport on an electric network has been a relatively low carbon dioxide emission and so-called 'green transport' mode, compared with other means of transport (such as planes and cars/trucks) [13]. On the other hand, rail-based primary fuel constitutes an essential reason for the growth of CO_2 emissions globally [14]. Therefore, typically, installing PV solar panels in the rail sector improves energy-savings and emissions reduction in the rail sector, especially those based on primary fuel, and is a step forward to achieving the goal in 2050, mentioned previously.

It is essential to note that railway transport has improved in energy efficiency from 1990 to 2014: 13% less energy is required to move a passenger 1 km and 19% less energy to move a ton load 1 km. As a result, from 1990 to 2009, total CO_2 emissions from the European railways were reduced by 32%. The passenger-specific emissions (per passenger-km) were reduced by 20% and freight-specific emissions (per ton-km) by 38% [15]. However, this is not yet enough, because the International Union of Railways (UIC) and Canadian Electrical Raceways (CER) have already established a new target for 2030. The UIC studied and documented the feasibility of the 2030 targets in the technical report "Moving towards Sustainable Mobility: European Rail Sector Strategy 2030 and beyond". The targets were set in December 2010 by the UIC and CER. Since 2011, the progress toward the 2030 targets has been monitored and reported on yearly by the UIC (technical document: "Monitoring report to 2020–2030 UIC/CER strategy targets") [16].

The new recommendations were as follows:

- Reduce the specific final energy consumption from train operations by 30% compared to 1990.
- Reduce the specific average CO_2 emissions from train operations by 50% compared to 1990.

The railway sector includes diesel (shown in Figure 1) and electric trains. Although electrified vehicles have been in use in railway applications for over 100 years, trains with energy-independent, diesel-based drives are still primarily used in the world's railway networks. For example, until 2020, only 38% of the heavy mainline railways in Bretagne were electrified [17]. Therefore, the work to improve the energy efficiency of diesel–electric presents a fundamental challenge in the world, especially in reducing fuel consumption and GHG emissions.

Figure 1. An example of a diesel–electric train (F7A) [18].

Installation of photovoltaic solar panels on the roof of diesel trains had already begun in 2011. Previously, the profitability calculation was based on the estimation of weather conditions and the theoretical efficiency of solar panels, which led to unsatisfactory results and difficulty in accessing data. The originality of this work is to propose a simple energy approach to calculate the feasibility and profitability of using PV solar panels in diesel–electric trains based on the use of RetScreen software. RetScreen provides easy access to weather conditions associated with the geographical position of the train. In addition, it provides the possibility to the user to select any photovoltaic solar panels and find the actual solar energy produced by the panels.

Moreover, this software can complete feasibility studies by providing technical, financial, and risk analyses related to electricity produced using renewable sources. Nevertheless, in our case studies, the energy produced by PV solar panels will be used to replace the energy produced by the auxiliary generator used to operate the auxiliary equipment. Therefore, the profitability was measured as the percentage of solar energy from the total energy produced by the auxiliary generator of the train, either 0% (the energy produced does not cover anything) or 100% (fully covered). As a result, the RetScreen results are adjusted to the type of application in this study.

This study is divided into six sections. Section 2 presents a brief literature review on using PV solar panels in diesel–electric trains. Section 3 presents the six studied cases of the train's journey worldwide in different climates. Section 4 describes the methodology to evaluate the feasibility of using a PV solar system on the roof of the trains. Furthermore, it determines the power required to operate the auxiliary electric equipment for each of the wagons' classes. Section 5 presents the results extracted from RetScreen and the percentage of each wagon's consumption provided by solar energy, then compares and analyzes the studied cases' results. Finally, the conclusions are in Section 6.

2. Literature Review on the Use of PV Solar Panels in Diesel-Electric Trains

The Indian train UNESCO became, in 2011, the first diesel–electric train that installed solar photovoltaic (PV) panels on the roofs of its coaches. A 3 kW auxiliary generator and its accessories were removed from the train with this PV system. The generated energy from the solar PV panels is stored in batteries and is used to power seven 6 W LED lights in each of the coaches as well as charging mobile phones during a five-hour journey.

In Australia, solar farms were installed along or near the railway in addition to rooftop solar panels. All generated energy is stored in batteries mounted on the diesel–electric train. The required 0.015 GW power needed to operate the train is generated by 8% from the train's rooftops, 58% from the solar farm, and 34% from the solar railway panels [19]. Figure 2 shows the UNESCO train [20]. While Figure 3 shows Australia's train with PV solar panels [21].

Figure 2. The UNESCO train with PV solar panels [20].

Figure 3. Australia's train with PV solar panels [21].

After the success of the experimental tests in India and Australia, the train companies started to build solar panels on the tunnels above trains, such as the Belgian tunnel. This solar tunnel connecting Schoten and Brasschaat was the first European tunnel. It comprises 16,000 PV panels, produces about 3.6 GWh, and reduces the annual CO_2 emissions by approximately 2500 tons [22]. Figure 4 shows the Belgian solar tunnel [23].

Figure 4. The Belgian solar tunnel [23].

Recently, Kapetanović et al. [24] developed and optimized Lithium-ion battery sizing for a hybrid diesel–electric train to decrease fuel consumption and related emissions. Furthermore, Cipek et al. [25] studied a comparative assessment of traditional diesel–electric and hypothetical battery–electric heavy haul locomotive operations regarding emissions reduction and fuel savings potential. In addition, various tandem locomotive configurations have been proposed and validated. Based on the results, fuel cost savings between 22% and 30% may be achieved. Moreover, according to Alfonso et al., Normanyo et al., and de Almeida et al., PV solar panels, as a well-known asset in renewable structure, have been developed. Solar-powered trains are usually put in motion by placing photovoltaic panels close to or on rail lines. This method could provide several financial advantages by improving the energy efficiency of diesel–electric trains [26–28].

With a growing lack of funds and the fact that it is harder to generate sufficient electricity from renewable sources to power the traction system of the trains during the journey, especially in a cold climate, this method remains limited in use. In this paper, and to be more realistic for practical applications, we focus on generating energy for the equipment on board, such as heating and lighting. As such, we are displacing energy otherwise produced using auxiliary diesel generators. The feasibility will, therefore, be assessed as a function of the percentage of displaced energy by the PV solar panels installed on the roof of the trains with different climate conditions. The following sections describe the methodology to evaluate the profitability of using PV solar systems on the roof of the trains and associated opportunities.

3. Case Studies

This section aims to apply the PV solar panels to the train's roof for the six cases shown in Table 2. The first three cases are in cold climate zones, while the last three are in hot climate regions. The feasibility and electricity production by PV solar panels strongly depends on the amount of absorbed light and less on the ambient temperature. The main difference between the cases is, thus, the local solar resource, and its influence appears under the term "Capacity factor" in RetScreen. The other difference between the studied cases is the annual energy consumed by the auxiliary generator. That depends on the travel duration, the number of trips per year, and the type of train wagons. Table 2 shows the characteristics of each case.

Table 2. The case studies and their characteristics [29,30].

Case Study		Location	Duration of Each Travel (h)	Number of Travels per Week
Case 1	London to Edinburgh	Southeast of Great Britain	9.2 h	7
Case 2	Toronto to Montreal	Southwest of the province of Quebec—Ontario, Canada	5.38 h	14
Case 3	Sept-Îles to Schefferville	Northern of the province of Quebec, Canada	10 h	4
Case 4	Luxor to Aswan	South of the capital Cairo, Egypt	3.33 h	14
Case 5	Sydney to Henty	The southeastern coast of Australia	5.33 h	14
Case 6	Champaign to Chicago	Southwestern tip of Lake Michigan-USA	8.33 h	7

As seen in Table 2, we consider three factors: the local solar and meteorological conditions, each travel duration, and the number of trips per week. Unfortunately, detailed data on the wagons' types for each case are unavailable. Therefore, we consider three wagon categories depending on the service quality offered to passengers, which affects the energy consumed by the auxiliary equipment in the train.

The type of wagon depends on each car's electrical equipment, and is different for every class. The electrical equipment's electricity consumption for each class is shown in Table 3 [31].

Table 3. Electrical appliance's power consumption according to the ticket class.

Train's Car Consumption (W)	3rd Class	2nd Class	1st Class
Lights	400	600	800
Heating System	500	1000	1500
Laptops	200	450	600
Internet Router	10	20	20
Phones	125	375	500
Sound System	95	95	95
Coffee machine	800	800	800
Refrigerator	265	265	500
Total	2395	3605	4815

Table 3 shows the power consumption consumed for each class of wagons. The total values are equal to 4815 W, 3605 W, and 2395 W for the first class, second class, and third class, respectively. Next, we calculate each case's annual energy consumed per wagon based on each travel duration and the number of trips per week (Table 4).

As seen in Table 4, the annual energy consumed increases significantly when we pass from the third class to the first class. On the other hand, case 2 has the highest energy consumed per year, while case 3 has the lowest energy consumed per year. This result is related to the multiplication of each travel's duration by the number of trips per week.

Table 4. The annual energy consumed per each class of wagons for the case studies.

Case Study		3rd Class (MWh)	2nd Class (MWh)	1st Class (MWh)
Case 1	**London to Edinburgh**	8.06	12.14	16.21
Case 2	**Toronto to Montreal**	9.59	14.43	19.28
Case 3	**Sept-Îles to Schefferville**	4.98	7.50	10.02
Case 4	**Luxor to Aswan**	5.81	8.74	11.67
Case 5	**Sydney to Henty**	9.29	13.99	18.68
Case 6	**Champaign to Chicago**	7.26	10.93	14.60

4. Methodology

This study evaluates six train journeys worldwide and under various climate conditions. The goal is to analyze the energy and financial profitability of installing PV solar panels on the roof of each the train's wagons for each of its trips. The PV solar panels are a supplementary source for the auxiliary diesel engine. These provide the train's auxiliary energy demand, such as the train lighting and operating electric tools such as TV, AC, refrigerators, and charging phones. Consequently, the study will concentrate on how much of the percentage of each wagon's consumption is provided by solar energy, which depends strongly on the train's location. The results of this study used RetScreen Expert software. This software is a comprehensive platform that intelligently enables professionals and decision-makers to rapidly identify and assess the viability of potential energy efficiency, renewable energy, and cogeneration projects. Therefore, upon using this program, users will be able to:

- Quickly gauge a facility's energy performance using benchmarking and evaluate energy costs and savings, GHG reductions, and financing viability.
- Determine the energy production and savings potential for any location in the world employing Archetypes.
- Verify the performance of implemented projects and find opportunities for further energy improvements.

Several parameters were considered for this study in RetScreen, as follows:

1. The type of solar panels (PV) used: Most solar panels on the market today fit into three categories: monocrystalline solar panels, polycrystalline solar panels, and thin-film solar panels. Each of these types has different characteristics. However, according to Sandy [32], the polycrystalline technology panel presents the balance between costs and performance. As a result, their efficiency increased to 18–20% in the last several years.
2. The number of PV solar panels used: The solar panels are collections of solar cells. Multiple small solar cells spread over a large area can work together to provide enough power to be helpful. In the case of trains, the amount of energy produced depends on each PV solar panel's dimension and the available surface atop each wagon. For this study, one type of wagon was used: the 50-feet standard, which means that 54.49 m^2 are available on the roof of each train's wagon [33].
3. The considered losses: Solar energy is subject to multiple losses in the conversion system. The solar panels' energy reaches the load and the batteries through the charge controller and the inverter. It suffers attenuation in each process whenever it passes through each component. First, solar panels (PV) can convert a small percentage, approximately 18%, of incoming solar energy into electricity. Second, the energy is stored in the solar batteries in the form of chemical energy, which can be used later to run the appliances, when there is no sunlight or during the night. Third, the battery provides energy by converting the stored chemical energy into DC electrical energy,

with additional losses in the conversion process. Finally, the DC electrical energy passes through the inverter. The inverter's essential function converts DC electrical energy into AC electrical energy. Today's inverters have approximately 99% efficiency in converting DC into AC [34].

4. The investment and O&M costs: The initial cost should include all capitalized preoperative and setup costs, as well as the cost of equipment, land, installation expenses, and EPC charges. At the same time, the O&M costs encompass routine maintenance of the equipment and minor part replacement to ensure maximum electricity generation [35].
5. The financial parameters: Investing in a project requires understanding different economic variables to make an informed decision. Here are a few such variables:
    - Inflation rate (%): The projected annual average inflation rate over the project's life.
    - Project life (year): The duration over which the project's financial viability is evaluated.
    - Debt ratio (%): The debt ratio reflects the financial leverage created for a project; the higher the debt ratio, the more significant the financial leverage.
    - Debt interest rate (%): The annual rate of interest paid to the debt holder at the end of each year of the debt term.
    - Debt term (years): The number of years over which the debt is repaid. The debt term is equal to or shorter than the project life [35].
6. The solar electricity price: In general applications, RetScreen uses market electricity price (approximately 0.1 $/kWh) to determine the revenues associated with solar energy production. In this study, the energy produced by the PV solar panels replaces the energy produced by the auxiliary generator. The electricity price is thus considered to be the one produced by a diesel generator at 0.3 $/kWh [36].
7. The GHG emission parameters: The RetScreen software requires the GHG emission factor (excluding transmission and distribution (T&D) losses) for the alternative electricity source. The units are tons of CO_2 per megawatt-hour of end-use electricity delivered (t CO_2/MWh). If any, the user enters an optional GHG reduction credit per equivalent ton of CO_2 (t CO_2). It is used in conjunction with the net GHG reduction to calculate the annual GHG reduction revenue. However, GHG reduction credits' prices per equivalent ton of CO_2 (t CO_2) vary widely depending on how the credit is generated and how it will be delivered [37].

The main results from RetScreen used for analysis are: (1) the ROI of the project (equity payback), which represents the length of time that it takes for the owner to recoup its initial investment (equity) out of the project cash flows generated, and (2) GHG emission reduction revenues per year ($). Then, by using a different software tool (Excel), we determine the profitability of using this method for each wagon in terms of annual energy consumption and the percentage of each wagon's consumption provided by PV solar panels for each studied case. The method and the steps of profitability calculation will be presented in Section 5.1.

5. Results and Discussion

The results of installing PV solar panels on a train's roof, applied on six train trips, will be presented in this section. This section is divided into two parts:

- Base case: We use this case for data calibration and validation. This section identifies the user inputs in RetScreen and presents the steps to extract profitability results for each studied case.
- Comparison and analysis: we compare and analyze the profitability for the six cases in terms of ROI (year), capacity factor (%), i.e., the actual percentage of the solar irradiation transformed into electricity, annual solar production (MWh), annual CO_2 reduction revenue ($), and the percentage of each wagon's consumption provided by solar energy. This value equals the ratio of yearly energy produced by PV solar panels over the wagon's annual energy.

5.1. Base Case

We chose London as the base case of study. Figures 5 and 6 show London's location and some meteorological data from the RET Screen database [32].

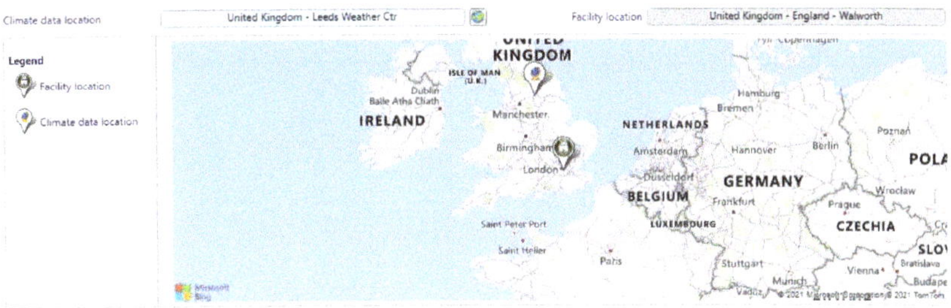

	Unit	Climate data location	Facility location	Source
Latitude		53.8	51.5	
Longitude		-1.5	-0.1	
Climate zone		4A - Mixed - Humid		Ground+NASA
Elevation	m	47	7	Ground - Map
Heating design temperature	°C	-1.1		Ground
Cooling design temperature	°C	24.1		Ground
Earth temperature amplitude	°C	14.2		NASA

Figure 5. London's location, United Kingdom.

Month	Air temperature	Relative humidity	Precipitation	Daily solar radiation - horizontal	Atmospheric pressure	Wind speed	Earth temperature	Heating degree-days 18 °C	Cooling degree-days 10 °C
	°C	%	mm	kWh/m²/d	kPa	m/s	°C	°C-d	°C-d
January	5.1	81.1%	78.74	0.65	99.1	4.5	2.0	400	0
February	5.6	77.2%	58.52	1.32	99.3	5.2	2.3	347	0
March	7.2	74.3%	61.07	2.22	99.2	4.8	4.4	335	0
April	8.5	71.6%	66.00	3.39	99.2	4.1	6.9	285	0
May	12.0	70.9%	62.31	4.42	99.4	3.8	10.4	186	62
June	14.6	71.0%	79.20	4.50	99.4	3.9	13.3	102	138
July	16.8	70.7%	79.05	4.48	99.3	4.1	15.2	37	211
August	16.7	72.1%	86.18	3.85	99.3	3.8	14.7	40	208
September	13.7	76.5%	73.80	2.64	99.3	3.7	12.0	129	111
October	10.6	79.9%	89.90	1.57	99.0	3.9	8.7	229	19
November	7.2	83.8%	83.70	0.82	99.0	3.8	4.9	324	0
December	5.1	83.1%	83.08	0.51	99.1	4.2	2.7	400	0
Annual	**10.3**	**76.0%**	**901.55**	**2.54**	**99.2**	**4.1**	**8.2**	**2,815**	**748**
Source	Ground	Ground	NASA	NASA	NASA	Ground	NASA	Ground	Ground
Measured at					m	10	0		

Figure 6. Meteorological data for London, United Kingdom.

Figure 5 shows that the climate data's location is Leeds Weather Center. Figure 6 shows that the average annual air temperature is 10.3 °C, and the daily solar radiation (horizontal) is 2.54 kWh/m^2 per day. Moving to the energy section, and as shown in Figure 7, we consider the following inputs:

- Fixed panels for the "solar tracking mode", the most straightforward model installed on the train's roof.
- Canadian Solar Poly-Si—CS34-365P-FG-KuDymond for the type of panels, with a total rated power of 365 W and total efficiency of 18.4%. At the same time, its high relative efficiency makes it a good choice.
- 27 PV solar panels are installed on the available area of each wagon's roof. The panels' total area is 53.6 m^2, which is slightly less than the available area of the roof (54.49 m^2).
- 5% and 1% are the losses for the solar panels and inverters. The choice of these values is recommended in the RET Screen guide.
- 2500 $/kW and 33 $/kW-year for the initial costs and O&M costs, respectively. An experienced worker has validated the choice of these values in the field of PV solar panels.
- 0.3 $/kWh for the annual electricity export rate, as explained in the methodology section. This is the cost of diesel generated electricity production, which is avoided.

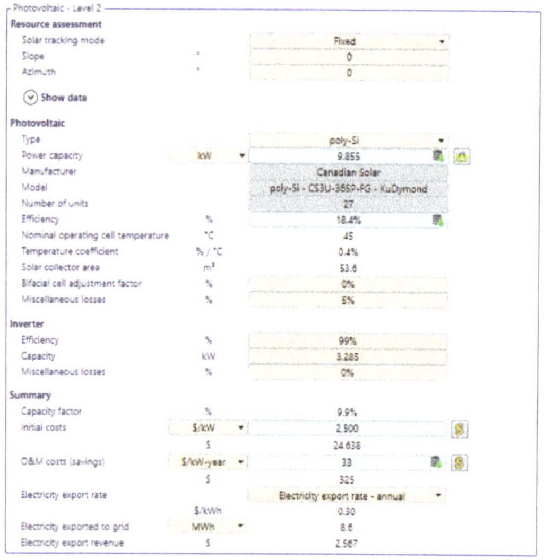

Figure 7. Energy data input for the case of London, United Kingdom.

After the energy section, the GHG emissions reduction and main finance parameters should be inserted. We considered them as follows:

- 1.075 t CO_2/MWh and 50 \$/t CO_2 for the GHG emission factor (excluding T&D) and GHG reduction credit card, respectively, as shown in Figure 8. The emissions correspond to diesel generator electricity production, while the carbon credit corresponds to the market's actual values.
- 2%, 20 years, 50%, 5%, and 10 years are the values used for the inflation rate, project life, debt ratio, debt interest rate, and debt term, respectively, as shown in Figure 9. These values' choice corresponds to actual market conditions and similar PV solar panel projects.

The main results from RetScreen are: (1) the annual solar energy production (MWh), which is the same as the "electricity exported to grid", shown in Figure 6. For London, it is 8.6 MWh, with a 9.9% capacity factor, which is directly affected by the local meteorological conditions; (2) the ROI (payback), which appears at the bottom of Figure 8 and is equal to 9.1 years; (3) and, finally, the GHG (CO_2) reduction value and associated revenues (Figure 8).

On the other hand, considering the duration of travel per day (9.33 h) and the number of trips per year (seven trips/week) for the case of London (shown in Table 2), the required energy is 16.21 MWh, 12.14 MWh, and 8.06 MWh for the first class, second class, and third class wagons, respectively. Therefore, we determined the percentage of each wagon's supply with solar energy as 53.05%, 70.85%, and 106.65% for the first class, second class, and third class wagons, respectively.

Thus, the PV solar panels provide around 50% of the total energy consumed per year for the first class wagon. Therefore, the annual fuel bill of the auxiliary generator is reduced by 50%. In the same context, the PV solar panels provide 100% of the total energy consumed for the third class wagon. An additional 6% can be stored in batteries. This stored energy can be a supplementary source for other train charges.

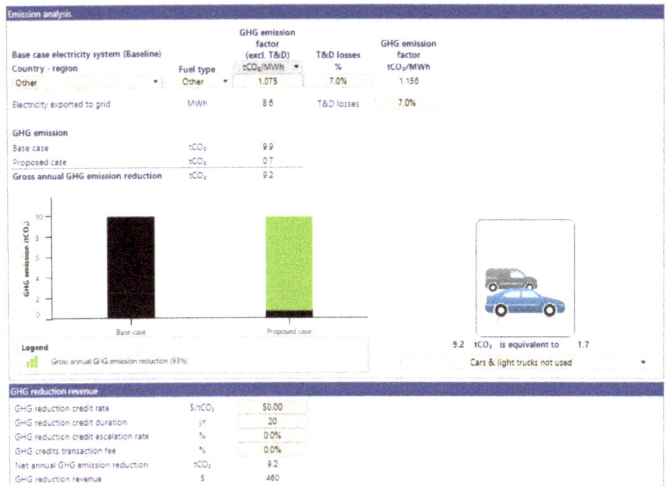

Figure 8. GHG emission reduction's inputs in the case of the London, United Kingdom.

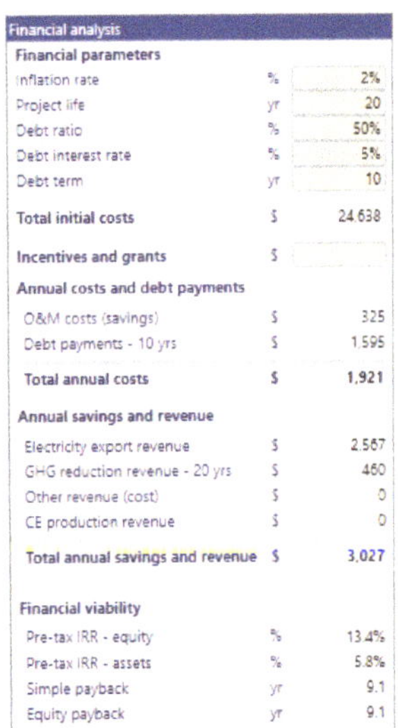

Figure 9. Financial inputs in the case of London, United Kingdom.

In the following sections, we use the same inputs as the base case for all other trains' cases. The results will be affected by the local solar resource and the train operation schedule, i.e., journey duration and frequency.

5.2. Comparison and Analysis

The profitability results for the six cases shown in Table 5 are in terms of ROI (years), capacity factor (%), annual solar production (MWh), and CO_2 emission reduction revenues ($).

Table 5. The main results from RETScreen for installing PV solar panels on the wagon's roof (per wagon).

Case Number	Journey	Capacity Factor (%)	Annual Solar Production (MWh)	ROI (Years)	CO_2 Yearly Emission Reduction Revenues ($)
1	From London to Edinburgh	9.9	8.6	9.1	460
2	From Toronto to Montreal	13.9	11.6	4.9	644
3	From Sept-Îles to Schefferville	12.4	10.7	6.0	574
4	From Luxor to Aswan	21.5	18.6	2.5	998
5	From Sydney to Henty	17.0	14.7	3.5	790
6	From Champaign to Chicago	15.3	13.2	4.2	710

Case 4 has the highest capacity factor and annual solar production (MWh). The location of case 4, in North Africa (Egypt), benefits from a very high solar resource. On the other hand, case 1 has the lowest annual solar production (MWh) in the UK, resulting from the lowest solar energy availability. Additionally, the cases located in cold climates produce slightly less energy than the others, such as cases 2 and 3.

The ROI (years) is inversely proportional to the annual solar production (MWh) as well as CO_2 emission reduction revenues ($). In detail, the ROI values range from 2.5 years (case 4) to 9.1 years (case 1), while the CO_2 emission reduction revenues range from $460 to $998 per year. Therefore, the ROI values are appropriate for such a project, except for the first case, where the ROI is considered very high. Table 5 shows the annual solar energy production for each case study.

As mentioned in the introduction, installing PV solar panels on the train roof reduces fuel consumption associated with auxiliary diesel generators. Figure 10 shows the profitability of using this technique regarding the percentage of each wagon's consumption, as provided by solar energy/journey (ratio). The minimum fuel reduction of auxiliary generators per year is above 0.5, proving this solution's feasibility and reliability.

According to Figure 10, cases 3 and 4 are the best, with over 100% of the energy provided by PV solar panels in all the wagon's classes. This performance is due to the lower energy consumed in these two cases, which depends on the number of trips per week. However, the energy production in case 4 is considerably higher than in case 3 because the available solar resource in Egypt (case 4) is significantly higher than in Quebec (case 3). Inversely, cases 1 and 2 are the worse. Only the third class wagon's energy can be covered at 100 % from PV production due to the higher annual energy consumed in these two cases. Energy production for case 2 is better than for case 1 as solar irradiation is higher in southern Canada (case 2) than in the United Kingdom (case 1). Moreover, case 5 is characterized by high annual energy consumption and significant PV energy production.

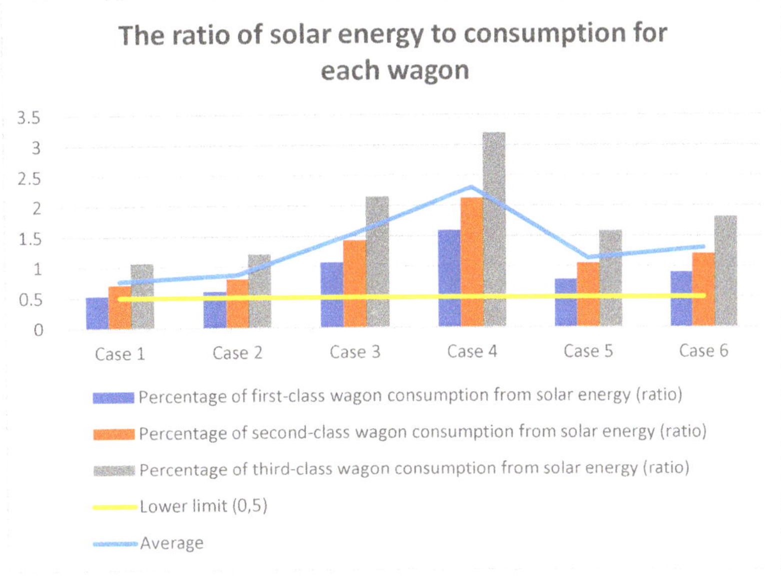

Figure 10. The ratio of solar energy to each wagon's consumption.

6. Conclusions

This paper presents a straightforward approach to evaluating the profitability and the impact of installing PV panels on the wagon's roof on the train's energy efficiency. The solution has been considered for six different journeys worldwide, with various solar resource potentials and meteorological weather conditions. The solar resource and the wagon's usage determine the level of solar energy percentage and economic feasibility. The minimum annual fuel reduction of auxiliary generators allowed using PV solar panels is above 50% in all cases and wagon classes, proving this solution's feasibility.

Moreover, the payback period (ROI, years) is inversely proportional to the annual solar production (MWh) as well as to the CO_2 emission reduction revenues ($). The payback period ranges from 2.5 years to 9.1 years, while the CO_2 emission reduction's revenue values range from $460 to $998 per year/wagon. These results are acceptable and appropriate for such projects.

On the other hand, the effect of the low temperature on the profitability of this technique is insignificant. The profitability is strongly related to the solar resource (light) and travel duration, and the results for the London and Sydney cases prove that. Future research can perform tests on a batch of data using artificial intelligence or statistical analysis to pave the way toward intelligent management of energy systems in the Industry 4.0 era.

Author Contributions: Conceptualization, A.I. and H.I.; methodology, A.F., H.I. and A.I.; validation, H.I., A.I., S.S.K. and M.I.; formal analysis, A.F., H.I., A.I., S.S.K. and M.I.; writing—original draft preparation, A.F., S.S.K. and M.I.; writing—review and editing, A.F., H.I., A.I., S.S.K. and M.I.; visualization, A.F., S.S.K. and M.I.; supervision, H.I. and A.I.; project administration, H.I., A.I. and S.S.K. All authors have read and agreed to the published version of the manuscript.

Funding: This research was funded by the Government of Canada through Transport Canada "Clean Transportation Program—Research and Development".

Institutional Review Board Statement: Not applicable.

Informed Consent Statement: Not applicable.

Data Availability Statement: Not applicable.

Conflicts of Interest: The authors declare no conflict of interest.

Nomenclature and Symbols

PV	Photovoltaic
COP21	21st Conference of Paris
IRENA	International Renewable Energy Agency
CER	Canadian Electrical Railways
O&M	Operation and Maintenance
T&D	Transmission and Distribution
GHG	Greenhouse gas
ROI	Payback period
UIC	International Union of Railways

References

1. Mylenka, T. Impact of COVID-19 on the Global Energy Sector. PV Magazine. Available online: https://www.pv-magazine.com/2020/04/24/impact-of-covid-19-on-the-global-energy-sector/ (accessed on 24 April 2020).
2. Iea. COVID-19 and Energy: Setting the Scene. Sustainable Recovery. 2020. Available online: https://www.iea.org/reports/sustainable-recovery/covid-19-and-energy-setting-the-scene (accessed on 1 June 2020).
3. Bp. Statistical Review of World Energy. 70th Edition. 2021. Available online: https://www.bp.com/content/dam/bp/business-sites/en/global/corporate/pdfs/energy-economics/statistical-review/bp-stats-review-2021-full-report.pdf (accessed on 1 July 2021).
4. IRENA (International Renewable Energy Agency). Global Renewables Outlook: Energy Transformation 2050. 2020. Available online: https://irena.org/-/media/Files/IRENA/Agency/Publication/2020/Apr/IRENA_GRO_Summary_2020.pdf?la=en&hash=1F18E445B56228AF8C4893CAEF147ED0163A0E47 (accessed on 1 June 2020).
5. IRENA (International Renewable Energy Agency). Renewable Energy Statistics 2021. Available online: https://IRENA_Renewable_Energy_Statistics_2021.pdf (accessed on 1 June 2020).
6. IRENA (International Renewable Energy Agency). Renewable Energy Employment by Country. 2020. Available online: https://www.irena.org/Statistics/View-Data-by-Topic/Benefits/Renewable-Energy-Employment-by-Country (accessed on 1 June 2020).
7. Agence Parisienne du Climat. Qu'est-ce que le COP 21? [French Edition]. Available online: https://www.apc-paris.com/COP-21 (accessed on 4 December 2018).
8. IRENA (International Renewable Energy Agency). FUTURE OF SOLAR PHOTOVOLTAIC: Deployment, Investment, Technology, Grid Integration and Socio-Economic Aspects. A Global Energy Transformation Paper. November 2019. Available online: https://irena.org/-/media/Files/IRENA/Agency/Publication/2019/Nov/IRENA_Future_of_Solar_PV_2019.pdf (accessed on 1 November 2019).
9. Eia. Energy Consumption by Sector. 2020. Available online: https://www.eia.gov/totalenergy/data/monthly/pdf/sec2.pdf (accessed on 1 June 2020).
10. UNEP, UNEP DTU Partnership. UN Environment Program: Emissions Gap Report 2020. Available online: https://www.unep.org/emissions-gap-report-2020 (accessed on 9 December 2020).
11. Mihlfeld & Associates. The 6 Modes of Transportation. Available online: https://blog.mihlfeld.com/the-6-modes-of-transportation (accessed on 6 October 2018).
12. Iea. Rail. Tracking Transport 2020. 2020. Available online: https://www.iea.org/reports/tracking-transport-2020/rail (accessed on 1 June 2020).
13. ORR Office of Rail and Road, Rail Emission 2019–2020. Available online: https://dataportal.orr.gov.uk/media/1843/rail-emissions-2019-20.pdf (accessed on 5 November 2020).
14. Adele, B. Solar-Powered Trains: The Future of Rail? Railway Technology. Available online: https://www.railway-technology.com/features/solar-powered-trains/ (accessed on 12 September 2019).
15. Moriarty, P.; Honnery, D. Global transport energy consumption. In *Alternative Energy and Shale Gas Encyclopedia*; Wiley Series on Energy; Jay, H.L., Jack, K., Thomas, B.K., Eds.; John Wiley & Sons: Hoboken, NJ, USA, 2016; pp. 651–656.
16. De Blas, I.; Mediavilla, M.; Capellan-Perez, I.; Duce, C. The Limits of Transport Decarbonization under the Current Growth Paradigm, Energy Strategy Reviews 32, Elsevier. 2020. Available online: https://www.researchgate.net/publication/344255972_The_limits_of_transport_decarbonization_under_the_current_growth_paradigm (accessed on 1 September 2020).
17. ORR Office of Rail and Road, Rail Infrastructure and Assets 2019–2020. Available online: https://dataportal.orr.gov.uk/media/1842/rail-infrastructure-assets-2019-20.pdf (accessed on 5 November 2020).
18. Available online: https://www.reddit.com/r/trains/comments/agayt3/i_found_this_photo_on_americanrailscom_also_i (accessed on 12 September 2019).
19. Vajihi, M. Solar-Powered Light Rail Vehicle and Tram Systems. Master's Thesis, University of Roma, Roma, Italy, 2018.
20. Available online: https://www.engadget.com/2017-07-18-india-first-solar-powered-train.html (accessed on 18 July 2017).
21. Available online: https://createdigital.org.au/worlds-first-solar-train-byron-bay (accessed on 15 February 2018).

22. High-Speed Euro Train Gets Green Boost from Two Miles of Solar Panels. Available online: https://www.theguardian.com/environment/2011/jun/06/tunnel-solar-rail (accessed on 6 June 2011).
23. Available online: https://siteselection.com/issues/2012/mar/belgium.cfm (accessed on 1 March 2012).
24. Kapetanović, M.; Núñez, A.; van Oort, N.; Goverde, R.M.P. Reducing Fuel Consumption and Related Emissions through Optimal Sizing of Energy Storage Systems for Diesel-Electric Trains. *Appl. Energy* **2021**, *294*, 117018. [CrossRef]
25. Cipek, M.; Pavković, D.; Krznar, M.; Kljaić, Z.; Mlinarić, T.J. Comparative analysis of conventional diesel-electric and hypothetical battery-electric heavy haul locomotive operation in terms of fuel savings and emissions reduction potentials. *Energy* **2021**, *232*, 121097. [CrossRef]
26. Afonso, J.L.; Lisboa Cardoso, L.A.; Pedrosa, D.; Sousa, T.J.C.; Machado, L.; Tanta, M.; Monteiro, V. A review on power electronics technologies for electric mobility. *Energies* **2020**, *13*, 6343. [CrossRef]
27. Normanyo, E.; Sulemana, A.R. Using solar-powered 3-phase squirrel cage induction motors as prime movers for synchronous electric generators. *Ghana J. Technol.* **2021**, *5*, 11–20.
28. De Almeida Carneiro, M.; da Fonseca Soares, D. Solar photovoltaic assistance system study for a Brazilian light rail vehicle. *U. Porto J. Eng.* **2020**, *6*, 35–45. [CrossRef]
29. Discover How to Get Anywhere. Available online: https://www.rome2rio.com/fr/ (accessed on 5 June 2022).
30. Tshiuetin Enterprise. Available online: https://www.tshiuetin.net/calendrier-septembre-2020.pdf (accessed on 1 June 2020).
31. Available online: https://generatorist.com/list-of-electric-appliances-their-wattage-usage (accessed on 3 May 2021).
32. Sendy, A. Types of Solar Panels: Which One Is the Best Choice? Available online: https://www.solarreviews.com/blog/pros-and-cons-of-monocrystalline-vs-polycrystalline-so (accessed on 5 May 2021).
33. Railroad Equipment. February 2020. Available online: https://www.csx.com/index.cfm/customers/resources/equipment/railroad-equipment (accessed on 1 June 2020).
34. What Are the Significant Losses in the Solar Power System? 2020. Available online: http://www.adityagreens.com/%20Losses%20in%20the%20solar%20system (accessed on 1 June 2020).
35. RET Screen-NRCan. Available online: http://www.RETScreenExpert.rtx (accessed on 1 June 2020).
36. Chauhan, A.; Saini, R.P. A review on Integrated Renewable Energy System based power generation for stand-alone applications: Configurations, storage options, sizing methodologies and control. *Renew. Sustain. Energy Rev.* **2014**, *38*, 99–120.
37. RET Screen User Manual. Available online: http://www.unfccc.int/%20RET%20Screen%20Software%20Online%20User%20Manual (accessed on 1 June 2020).

applied sciences

MDPI

Article

Power Factor Correction Application Based on Independent Double-Boost Interleaved Converter (IDBIC)

Norbert Csaba Szekely [1], Sorin Ionut Salcu [1], Vasile Mihai Suciu [1], Lucian Nicolae Pintilie [1], Gheorghe Ioan Fasola [2] and Petre Dorel Teodosescu [1,*]

1 Department of Electrical Machines and Drives, Technical University of Cluj-Napoca, 400489 Cluj-Napoca, Romania; norbert.szekely@emd.utcluj.ro (N.C.S.); sorin.salcu@emd.utcluj.ro (S.I.S.); mihai.suciu@emd.utcluj.ro (V.M.S.); lucian.pintilie@emd.utcluj.ro (L.N.P.)
2 BKD Electronic SA, 332005 Petrosani, Romania; fasola.ionut@bkdelectronic.ro
* Correspondence: petre.teodosescu@emd.utcluj.ro

Abstract: In this paper, a Power Factor Correction (PFC) application, based on the novel power stage topology named Independent Double-Boost Interleaved Converter (IDBIC), has been analyzed. The novelty of the proposed PFC rectifier is based on the sum of capabilities, such as supplying three independent output voltage levels with interleaved operation at the input and high voltage gain. The hardware used within this application consists of an AC input L-C-L filter, a single-phase bridge rectifier, the IDBIC power stage, output capacitors group and a group of variable high-power rheostats (resistors) group as DC load. The main purpose of the carried study was to highlight the advantages and disadvantages of the novel power stage topology in the context of a green and modern AC to DC conversion solution. Nowadays, a high level of the efficiency and power factor have become a mandatory feature for the AC to DC conversion solutions to satisfy the international electrical standards. Thus, considering the modern electrical standards and recommendations, the current study tries to better depict the working steps and principles of the modern power stage topology within an AC to DC conversion application. The behavior of the considered power stage described in different detailed working steps (such as the Discontinuous Conduction Mode and Continuous Conduction Mode) may help understand how the energy conversions process of AC to DC becomes more efficient. The high output voltage gain of the considered power stage is the key feature in the Power Factor Correction process. With such a feature, the AC to DC conversion solution/application can also operate at lower input AC voltages (such as 90 [V] and 110 [V]). The proposed solution can be successfully used in the electric vehicle (automotive field) and high-power electrical traction (e.g., trains, high power electrical machines and drives). The same solution can also be used successfully in fast battery charging applications and chemical electrolysis processes.

Keywords: Power Factor Correction (PFC); Independent Double-Boost Interleaved Converter (IDBIC); high voltage gain; low input voltage; electrical traction; electric vehicle; fast battery charging

Citation: Szekely, N.C.; Salcu, S.I.; Suciu, V.M.; Pintilie, L.N.; Fasola, G.I.; Teodosescu, P.D. Power Factor Correction Application Based on Independent Double-Boost Interleaved Converter (IDBIC). *Appl. Sci.* **2022**, *12*, 7209. https://doi.org/10.3390/app12147209

Academic Editors: Federico Barrero and Mario Bermúdez

Received: 8 June 2022
Accepted: 14 July 2022
Published: 18 July 2022

1. Introduction

Modern applications such as smart houses, hybrid microgrids, renewable energy, electric vehicles and energy storage systems demand an increase in quality for their infrastructure, as different policies regarding energy efficiency have been internationally introduced [1–3]. Concerning energy conversion systems and external power supplies, the European Union (EU) Commission has also established efficiency criteria that aims to improve power quality [4]. Since AC/DC converters have become a component of almost all electronic devices used daily, the AC distribution grid may be subjected to poor performance due to the behavior and low power quality of such equipment. Often single-phase rectifiers are required to operate over a wide supply voltage range, with low input current ripple and near-unity power factor in order to meet the present-day standards and market needs [5,6]. Usually, if bidirectional power flow is not required, a common solution is the boost PFC rectifier. In order to improve the performance of this type of converter, reducing the input current ripple and minimizing the volume of passive devices in the interleaved topology are considered to be among the best practice. Studies and analyses have been made in developing and improving the performance and efficiency of such converter topology. In [7], a method for analyzing the input and output currents in the converter is proposed and derives the specific time functions. A coupled inductor approach is presented in [8–11] to improve the inductor current ripple and power density of the interleaved PFC bridge rectifier. In [12], a hybrid topology comprised of a conventional boost PFC rectifier and a semi-bridgeless PFC interleaved rectifier is presented, claiming better efficiency and performance than that of the traditional interleaved boost PFC. Bridgeless configurations, as mentioned in [13], can present an interleaved topology that offers better performance and reduces the size of the magnetic devices needed. Improved efficiency through the soft switching operation of all switching devices is analyzed in [14]. Here, a snubber circuit is integrated into an interleaved PFC converter. Lighting applications [15,16], connected to the mains grid or industrial, can be provided with a voltage driver which in term has PFC features realized with an interleaved boost rectifier topology. Other applications such as electric vehicle chargers [17–19] are also being studied.

The novelty of the proposed PFC rectifier is the capability of supplying three output voltage levels with interleaved operation at the input. The so-called "three-level PWM rectifiers" [20,21] have been studied for their high efficiency power conversion (>98%) [22], large voltage gain and low stress on the semiconductor devices [23], but they lack the interleaved functionality. More, the concept of bipolar DC microgrids used in distributed generation systems [24,25] presents an increasing interest because of its benefits in energy saving, power quality and power electronics control [26]. Hence, the proposed PFC rectifier can easily interconnect a single-phase AC grid with a multilevel DC microgrid, also assuring the interleaved behavior, and it can be successfully used as a solution in the development of distributed generation.

2. Converter Topology Analysis

2.1. General Representation of the Switching States

The proposed PFC application consists of a novel DC–DC power-stage topology named Independent Double Boost Interleaved Converter (IDBIC). The current topology is based on a patent application in reference [27] that describes its operation in DC–DC systems through paper [28]. The basic PFC converter topology is presented in Figure 1, while the characteristic waveforms during operation are introduced in Figure 2, where the continuous conduction mode (CCM) is exemplified for a duty cycle D larger than 0.5 by means of switching states S1, S4 and S5, respectively, for a duty cycle smaller than 0.5 with the help of switching states S1, S2, S3 and S5.

The main operating stages of the converter for positive alternation are shown in Figure 3, where the diodes D_{r1} and D_{r4} are in conduction and five independent switching states (S1–S5) are highlighted. For the negative alternation, the operating stages are the same on the DC stage; the only difference is that the input current flows through the D_{r2} and D_{r3} diodes.

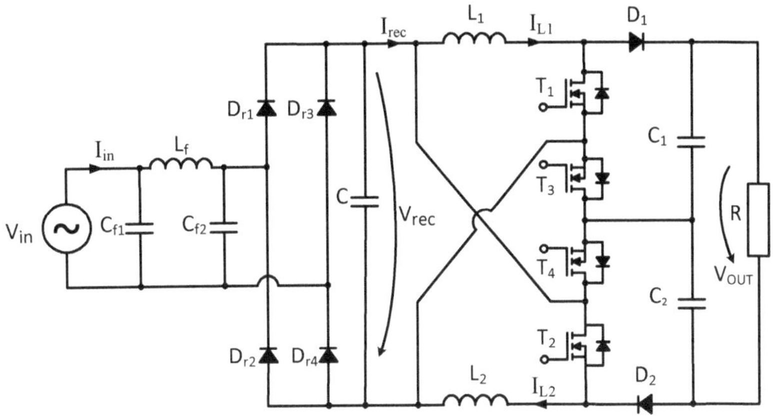

Figure 1. The electronic schematic of the proposed converter. V_{in}—main input voltage; V_{rec}—rectified voltage; V_{out}—DC output voltage; I_{rec}—rectified current; I_{in}—input current; and I_{L1} and I_{L2}–L_1 and L_2 inductor currents.

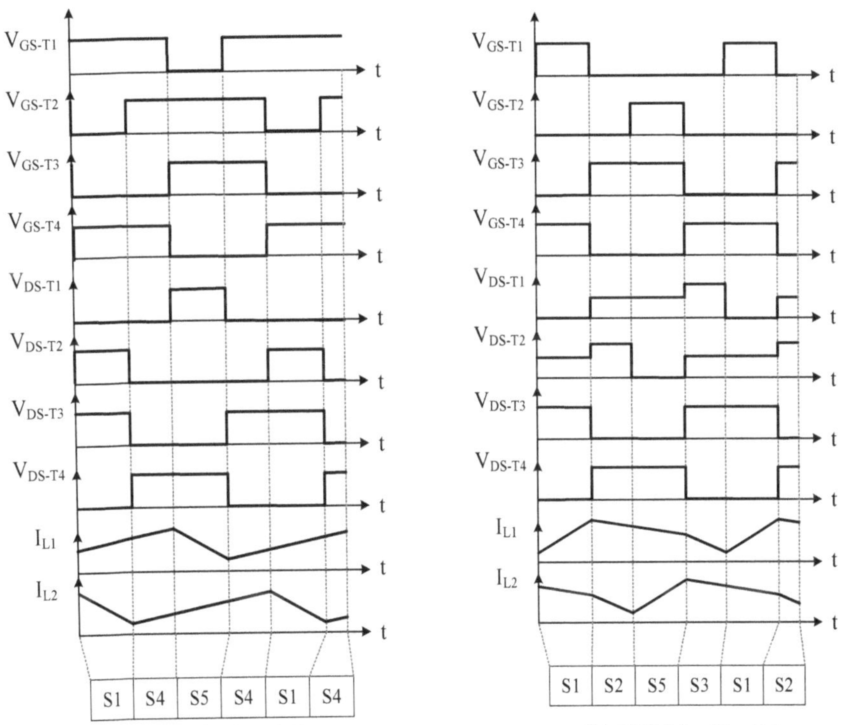

(a) CCM for D > 0.5 (b) CCM for D < 0.5

Figure 2. Presumptive functioning waveform and switching stages correlations. $V_{GS\text{-}T1}$, $V_{GS\text{-}T2}$, $V_{GS\text{-}T3}$ and $V_{GS\text{-}T4}$—Gate-Source voltage for T_1–T_4 transistors; $V_{DS\text{-}T1}$, $V_{DS\text{-}T2}$, $V_{DS\text{-}T3}$ and $V_{DS\text{-}T4}$—Drain-Source voltage for T_1–T_4 transistors; I_{L1} and I_{L2}–L_1 and L_2 inductor currents; D—Duty Cycle; CCM—continuous conduction mode; S1–S5—switching stages from Figure 2.

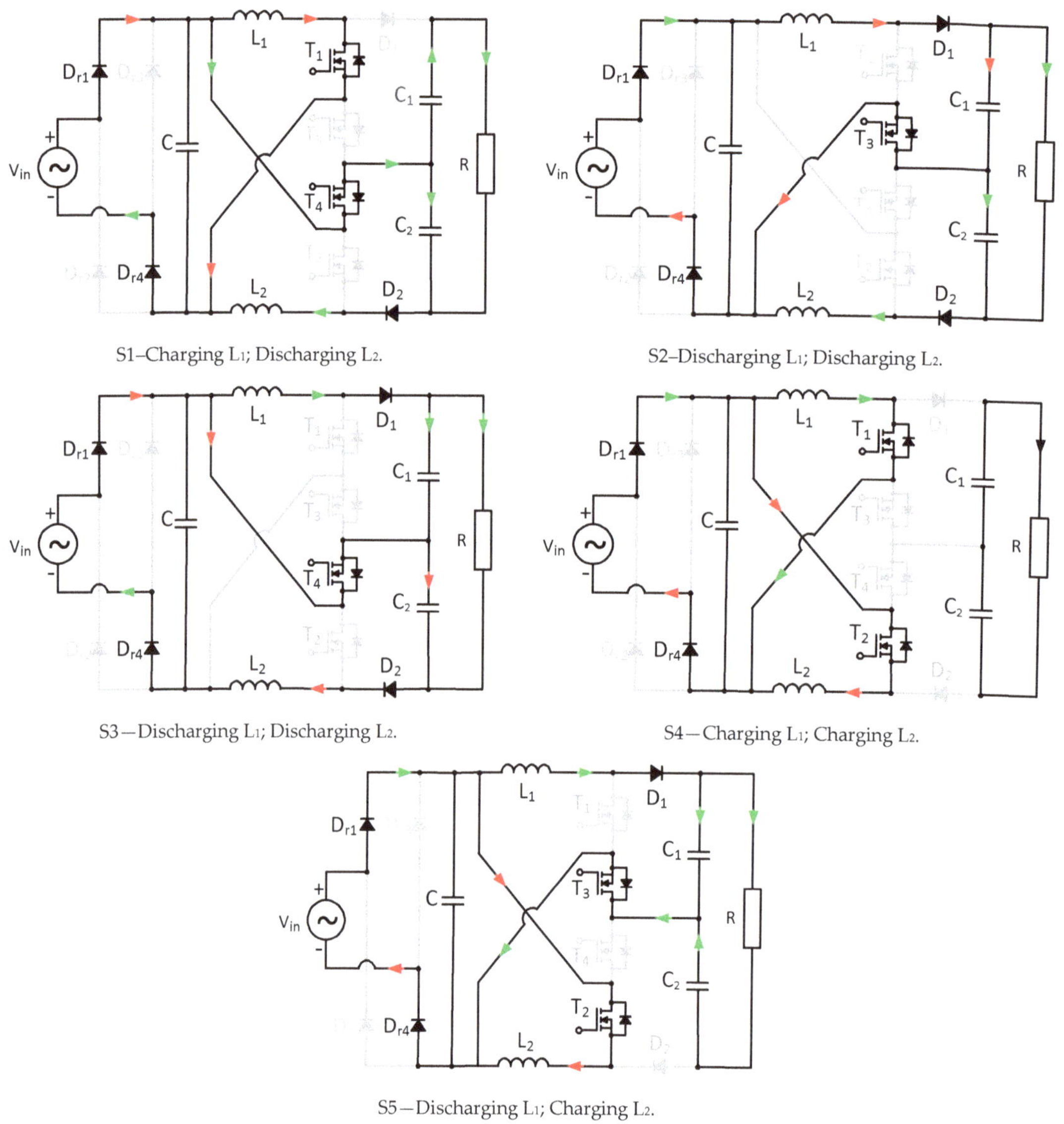

Figure 3. Switching stages for the proposed PFC converter.

2.2. State-Space Modeling of the Converter

Observing the switching states presented in Figure 3, the state-space analysis of the proposed converter has been made [29]. Thus, the expressions that describe the behavior of the converter at CCM were first derived for the operation at a duty cycle smaller than 0.5.

This operation is characterized by the switching stated S1–S2–S5–S3, as can be observed in Figure 2b. Therefore, for the switching state S1, will result:

$$\begin{cases} \frac{di_{L1}}{dt} = \frac{1}{L_1} v_{\text{rec}} \\ \frac{di_{L2}}{dt} = \frac{1}{L_2} v_{\text{rec}} - \frac{1}{L_2} v_{C2} \\ \frac{dv_{C1}}{dt} = \frac{-1}{RC_1} v_{C2} - \frac{1}{RC_1} v_{C1} \\ \frac{dv_{C2}}{dt} = \frac{1}{C_2} i_{L2} - \frac{1}{RC_2} v_{C1} - \frac{1}{RC_2} v_{C2} \end{cases} \tag{1}$$

Considering the next switching state, S2, when operating at the specified conditions, the equations will be:

$$\begin{cases} \frac{di_{L1}}{dt} = \frac{-1}{L_1} v_{C1} + \frac{1}{L_1} v_{rec} \\ \frac{di_{L2}}{dt} = \frac{-1}{L_2} v_{C2} \\ \frac{dv_{C1}}{dt} = \frac{-1}{C_1} i_{L1} + \frac{1}{RC_1} v_{C1} + \frac{1}{RC_1} v_{C2} \\ \frac{dv_{C2}}{dt} = \frac{1}{C_2} i_{L2} - \frac{1}{RC_2} v_{C1} - \frac{1}{RC_2} v_{C2} \end{cases} \tag{2}$$

For the switching states S5 and S3, the resulted expressions can be written as:

$$\begin{cases} \frac{di_{L1}}{dt} = \frac{-1}{L_1} v_{C1} + \frac{1}{L_1} v_0 \\ \frac{di_{L2}}{dt} = \frac{1}{L_2} v_0 \\ \frac{dv_{C1}}{dt} = \frac{1}{C_1} i_{L1} - \frac{1}{RC_1} v_{C1} - \frac{1}{RC_1} v_{C2} \\ \frac{dv_{C2}}{dt} = \frac{-1}{RC_2} v_{C1} - \frac{1}{RC_2} v_{C2} \end{cases} \tag{3}$$

Respectively,

$$\begin{cases} \frac{di_{L1}}{dt} = \frac{-1}{L_1} v_{C1} \\ \frac{di_{L2}}{dt} = \frac{-1}{L_2} v_{C2} + \frac{1}{L_2} v_0 \\ \frac{dv_{C1}}{dt} = \frac{1}{C_1} i_{L1} - \frac{1}{RC_1} v_{C1} - \frac{1}{RC_1} v_{C2} \\ \frac{dv_{C2}}{dt} = \frac{-1}{C_2} i_{L2} + \frac{1}{RC_2} v_{C1} + \frac{1}{RC_2} v_{C2} \end{cases} \tag{4}$$

The general state space representation of the modeled system will be:

$$\begin{cases} \dot{x} = A_i x + B_i u \\ y = C_i x \end{cases} \tag{5}$$

where $\dot{x} = \left[\dot{i_{L1}}\ \dot{i_{L2}}\ \dot{v_{C1}}\ \dot{v_{C2}} \right]$; $x = [i_{L1}\ i_{L2}\ v_{C1}\ v_{C2}]$; $u = v_{rec}$.

Thus, from relation (1), the state-space system for the S1 switching configuration is characterized by the matrices:

$$A_{S1} = \begin{bmatrix} 0 & 0 & 0 & 0 \\ 0 & 0 & 0 & \frac{-1}{L_2} \\ 0 & 0 & \frac{-1}{RC_1} & \frac{-1}{RC_1} \\ 0 & \frac{1}{C_2} & \frac{-1}{RC_2} & \frac{-1}{RC_2} \end{bmatrix};\ B_{S1} = \begin{bmatrix} \frac{1}{L_1} \\ \frac{1}{L_2} \\ 0 \\ 0 \end{bmatrix} \quad (6)$$

The representation of relation (2) under state matrices will yield:

$$A_{S2} = \begin{bmatrix} 0 & 0 & \frac{-1}{L_1} & 0 \\ 0 & 0 & 0 & \frac{-1}{L_2} \\ \frac{-1}{C_1} & 0 & \frac{1}{RC_1} & \frac{1}{RC_1} \\ 0 & \frac{1}{C_2} & \frac{-1}{RC_2} & \frac{-1}{RC_2} \end{bmatrix};\ B_{S2} = \begin{bmatrix} \frac{1}{L_1} \\ 0 \\ 0 \\ 0 \end{bmatrix} \quad (7)$$

For the expressions in (3) and (4), corelated with the switching states S5 and S3, the state-space matrices will be:

$$A_{S5} = \begin{bmatrix} 0 & 0 & \frac{-1}{L_1} & 0 \\ 0 & 0 & 0 & 0 \\ \frac{1}{C_1} & 0 & \frac{-1}{RC_1} & \frac{-1}{RC_1} \\ 0 & 0 & \frac{-1}{RC_2} & \frac{-1}{RC_2} \end{bmatrix};\ B_{S5} = \begin{bmatrix} \frac{1}{L_1} \\ \frac{1}{L_2} \\ 0 \\ 0 \end{bmatrix} \quad (8)$$

and

$$A_{S3} = \begin{bmatrix} 0 & 0 & \frac{-1}{L_1} & 0 \\ 0 & 0 & 0 & \frac{-1}{L_2} \\ \frac{1}{C_1} & 0 & \frac{-1}{RC_1} & \frac{-1}{RC_1} \\ 0 & \frac{-1}{C_2} & \frac{1}{RC_2} & \frac{1}{RC_2} \end{bmatrix};\ B_{S3} = \begin{bmatrix} 0 \\ \frac{1}{L_2} \\ 0 \\ 0 \end{bmatrix} \quad (9)$$

while

$$C_{S1} = C_{S2} = C_{S5} = C_{S3} = \begin{bmatrix} 1 & 0 & 0 & 0 \\ 0 & 1 & 0 & 0 \\ 0 & 0 & 1 & 0 \\ 0 & 0 & 0 & 1 \end{bmatrix} \quad (10)$$

Considering the representation from Figure 2b and the duty cycle as d, the state matrices averaged over one switching period will result the following:

$$\begin{aligned} A_{av} &= A_{S1}\,d + A_{S2}(0.5-d) + A_{S5}\,d + A_{S3}(0.5-d) \\ B_{av} &= B_{S1}\,d + B_{S2}(0.5-d) + B_{S5}\,d + B_{S3}(0.5-d) \\ C_{av} &= C_{S1}\,d + C_{S2}(0.5-d) + C_{S5}\,d + C_{S3}(0.5-d) \end{aligned} \quad (11)$$

To obtain the expressions depicting the continuous, linear behavior of the converter, the small signal analysis is implied to the linearized model. Small variations of the input variables, $\tilde{v}_{rec}$ and duty cycle $\tilde{d}$, around the quiescent operating point of the converter will result in small variations of the output variables $\tilde{i}_{L1}$, $\tilde{i}_{L2}$, $\tilde{v}_{C1}$ and $\tilde{v}_{C2}$.

$$\begin{cases} \dot{\tilde{x}} = A_{av}\,\tilde{x} + B_{av}\,\tilde{v}_{rec} + E_{av}\,\tilde{d} \\ \tilde{y} = C_{av}\,\tilde{x} \end{cases} \tag{12}$$

where $E_{av} = [(A_{S1} + A_{S5}) - (A_{S2} + A_{S3})]X + [(B_{S1} + B_{S5}) - (B_{S2} + B_{S3})]V_{rec}$.

After specific algebraic operation, the Laplace domain solutions of the state vectors will determine the output to input transfer functions. Thus, the solution for the output to input voltage transfer function will be:

$$\frac{\tilde{Y}}{\tilde{V}_{rec}} = C_{av}\,(sI - A_{av})^{-1}\cdot B_{av} \tag{13}$$

and for the output to duty cycle variation:

$$\frac{\tilde{Y}}{\tilde{D}} = C_{av}\,(sI - A_{av})^{-1}\cdot E_{av} \tag{14}$$

The steady-state values of the state variables can be determined as:

$$Y = -C_{av}\,(A_{av})^{-1}\cdot B_{av}\,V_{rec} \tag{15}$$

Concerning the operation at a duty cycle greater than 0.5, the switching states from Figure 2a are partially similar with the ones described by (1) and (3). The only particular switching state for this mode of operation is S4, which is characterized by the equations:

$$\begin{cases} \frac{di_{L1}}{dt} = \frac{1}{L_1}v_0 \\ \frac{di_{L2}}{dt} = \frac{1}{L_2}v_0 \\ \frac{dv_{C1}}{dt} = \frac{-1}{RC_1}v_{C2} - \frac{1}{RC_1}v_{C1} \\ \frac{dv_{C2}}{dt} = \frac{-1}{RC_2}v_{C1} - \frac{1}{RC_2}v_{C2} \end{cases} \tag{16}$$

from which will result the following state matrices:

$$A_{S4} = \begin{bmatrix} 0 & 0 & 0 & 0 \\ 0 & 0 & 0 & 0 \\ 0 & 0 & \frac{-1}{RC_1} & \frac{-1}{RC_1} \\ 0 & 0 & \frac{-1}{RC_2} & \frac{-1}{RC_2} \end{bmatrix};\; B_{S4} = \begin{bmatrix} \frac{1}{L_1} \\ \frac{1}{L_2} \\ 0 \\ 0 \end{bmatrix} \tag{17}$$

Having obtained the state-space model of the converter, a suitable control strategy can be developed.

3. Simulation Results

The simulation was performed based on a MATLAB Simulink model of the described PFC solution within this paper. The whole structure is depicted in Figure 4.

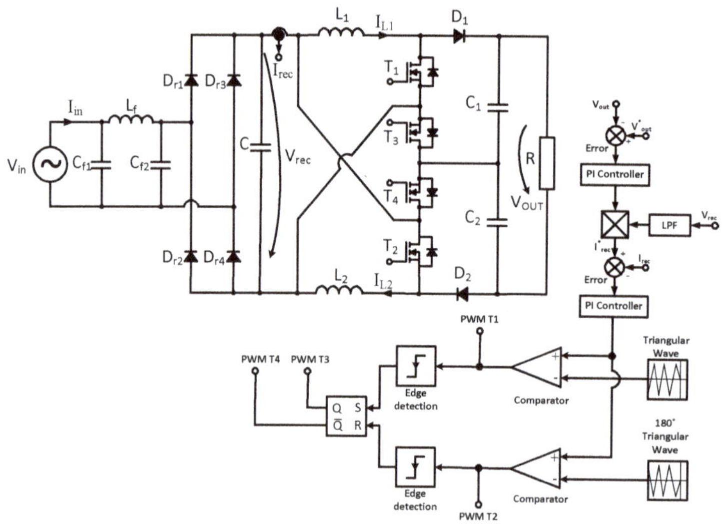

Figure 4. The generic output voltage control loop and PWM generator. V_{in}—AC input voltage; V_{rec}—rectified voltage; V_{out}—DC output voltage; V^*_{out}—reference DC output voltage; I_{rec}—rectified current; I^*_{rec}—reference rectified current; I_{in}—input current; PWM T1, PWM T2, PWM T3 and PWM T4 — Gate-Source voltage for T_1–T_4 transistors.

The "PFC" sub-system represented in Figure 5 contains the control law of the power stage [30]. The reference rectified voltage was obtained from the input voltage waveform by reversing the negative half-cycle. Based on two proportional–integrator (PI) regulators, the control law was implemented, and the output "Duty_Cycle" drive signal was computed as result. The "Voltage_PI" regulator maintains the output DC voltage constant, and the "Current_PI" regulator limits and shapes the input current.

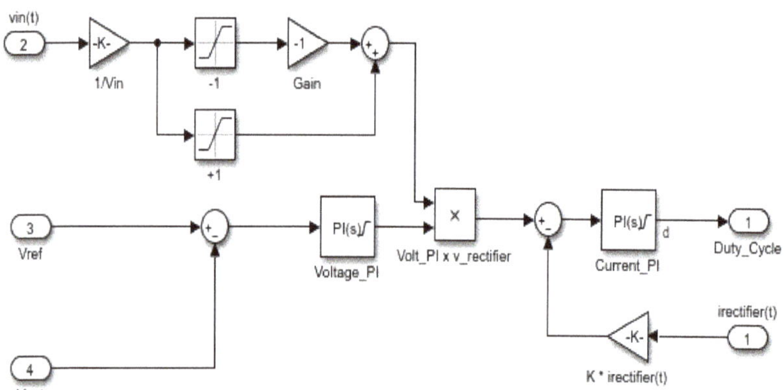

Figure 5. MATLAB Simulink inside "PFC" sub-system. K—Current calibration factor.

Figure 6 represents the sub-system of the "PWM_generator" which, depending on the signals received by the "PFC" subsystem, realizes the control logic of the four transistors.

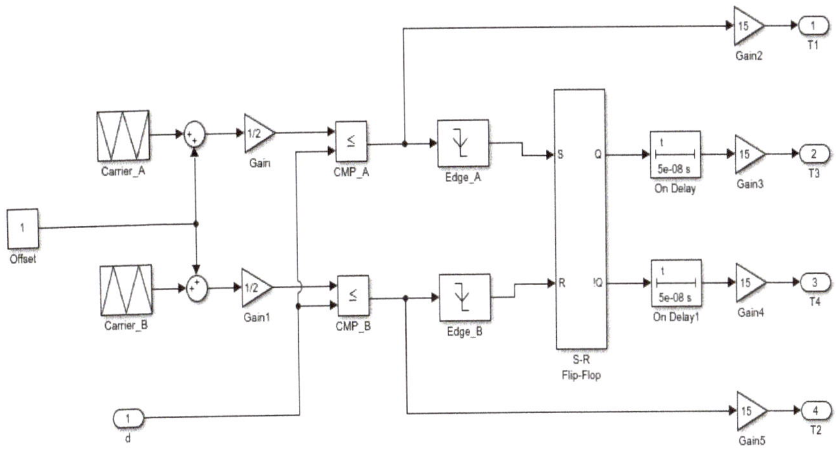

Figure 6. MATLAB Simulink inside "PWM_generator" sub-system.

Following Figure 7, the input current (I_{in}) and voltage (V_{in}) can be analyzed in conjunction with the output voltage (V_{out}). Additionally, the inductor L_1 and L_2 currents (i_{L1} and i_{L2}) synchronized at low frequency are represented, followed by the rectified input voltage (V_{rec}), all these signals being introduced in Figure 4.

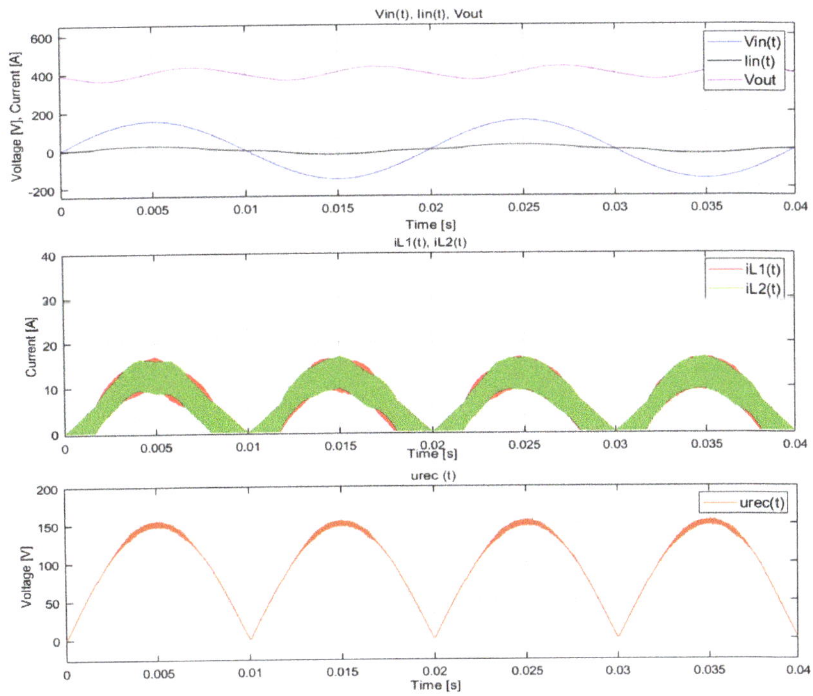

Figure 7. Low—frequency waveforms representation of the input voltage (Vin), input current (Iin), output voltage (V_{out}), inductor L_1 and L_2 currents (iL_1, iL_2) and the input rectified voltage (Vrec).

Figure 8 shows the L_1 and L_2 coils' current evolution at both low- and high-frequency representations. One can see that during a low-frequency cycle, the boost converters are working from DCM (Figure 8a) and BCM (boundary conduction mode, Figure 8b) to CCM. In CCM, two current waveforms can be noticed at duty cycles bigger then 0.5 (Figure 8c) or smaller then 0.5 (Figure 8d).

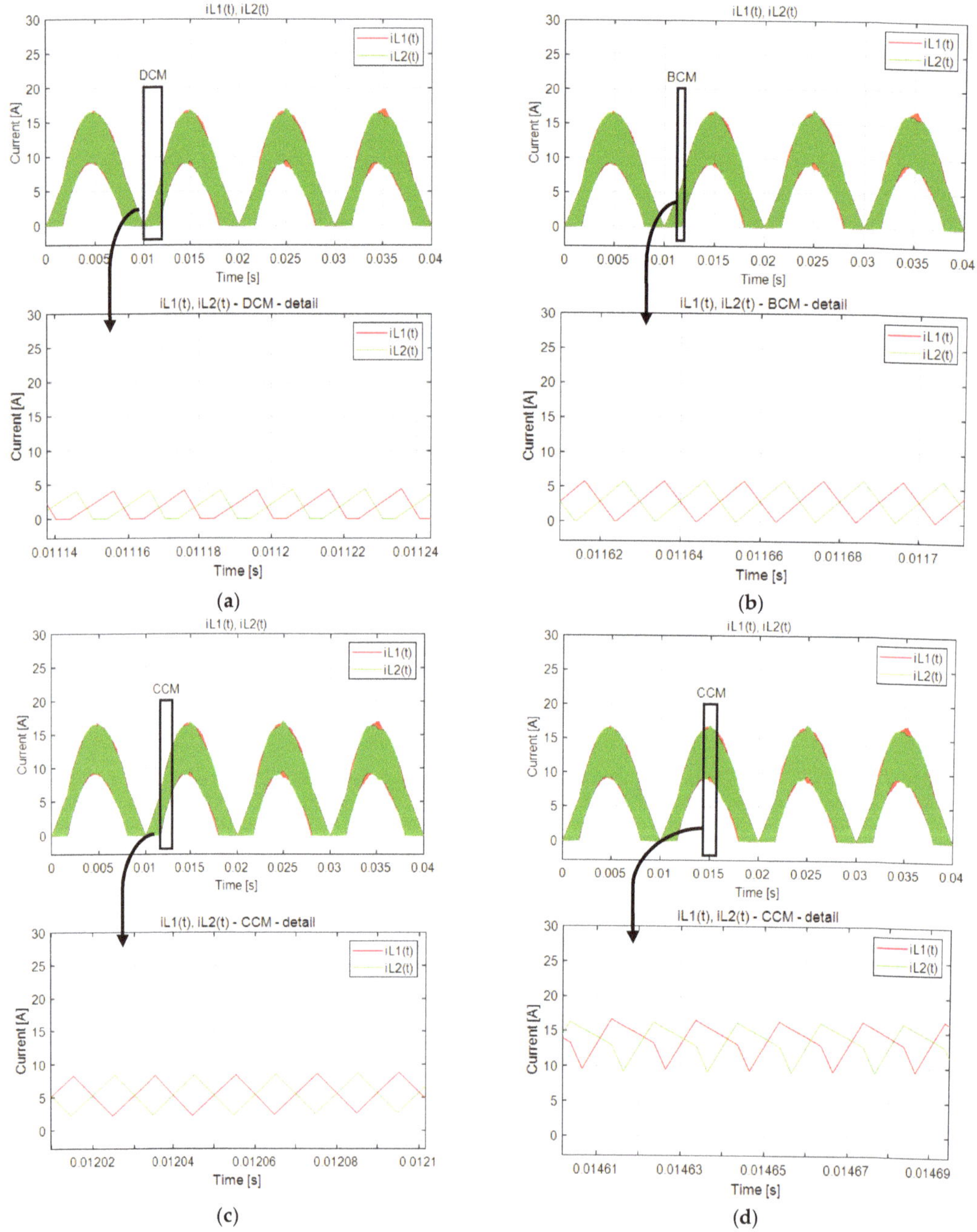

Figure 8. L_1 and L_2 inductor currents' (I_{L1}, I_{L2}) representation at low and high frequency representations; (**a**) DCM; (**b**) BCM; (**c**) CCM for D > 0.5; (**d**) CCM for D < 0.5.

4. Experimental Implementation

The experimental implementation was based on the schematic represented in Figure 4. In this figure, both the control loop and the gate signal generator block structure was depicted. The resultant PWM gate signals are used to drive the four IDBIC power stage transistors (MOSFET IPW60R099CPFKSA1, 600 [V], 31 [A]). This scheme is suitable for a symmetrical control of the two integrated boost converters in which a single reference voltage is used for both, so that each will work independently but identically. In view of this, the voltage on the output capacitors C_1 and C_2 will be regulated at the same value, and there is no need for a further balancing mechanism. For the considered power factor application, in which the total output voltage V_{out} must be regulated, the symmetrical approach of the control loop is sufficient, which means that only a PI controller and a voltage reference V_{ref} are required for the output voltage control loop.

If a three-voltage-levels approach is desired at the output, the asymmetric control of the converter can be applied, in which two reference voltages must be entered and the two integrated boost converters will operate independently and potentially with different duty cycles.

The laboratory test settings are shown in Figure 9. Based on an application with AC input voltages of 90 [V], 110 [V] and 130 [V] and a 350 [V] DC output voltage, the total harmonic distortion (THD) and power factor (PF), performed with the Tektronix PA3000 power analyzer, are illustrated in Figures 10 and 11. The main waveforms taken with the Tektronix MDO3024 oscilloscope are shown in Figure 12, while in Figure 13 one can notice the efficiency measurements.

In Table 1, a comparison is made of the proposed solution with similar converter topologies. Limited three-level boost converter topologies are available for power factor correction applications; thus, the comparison is made also with similar topologies used in the DC–DC converter application.

Table 1. Comparison of the proposed topology with similar approaches.

Ref.	Converter Application	Voltage Stress On		Maximum Efficiency	Components S*/D*/L*/C*/C.I*/T*
		Switches V_S/V_O	Diodes V_D/V_O		
[27]	DC–DC	(M + 1)/2 M	(M + 1)/2 M	91.7	2/2/2/4/2/-
[28]	DC–DC	0.5	0.5	95.9	4/2/2/3/-/-
[29]	DC–DC	0.5	0.5	95	2/3/2/3/-/-
[30]	DC–DC	(M + 1)/4 M (M − 1)/2 M	(M + 1)2 M	95.85	3/4/2/3/-/-
[31]	DC–DC	(1 + 5 M)/6 M	(M + 1)/M	95.9	6/9/6/1/-/-
[32]	DC–DC	0.5	-	94.3	4/0/1/4/-/-
[33]	DC–DC	0.33	0.33	93.9	1/5/1/5/0/-
[34]	DC–DC	(M + 1)/4 M	(M + 1)/2 M	96	2/3/-/3/1/-
[35]	Boost PFC	0.5	0.5	95.8	2/6/1/2/-/-
[36]	Boost PFC	0.5	0.5	94.8	2/4/1/2/-/-
[37]	Boost PFC	$V_{dc}/2$; V_{dc}; $V_O/2$	-	95.1	6/6/2/5/-/1
[38]	Boost PFC	$V_{dc}/2$; $V_O/2$	-	94.2	6/8/2/6/-/1
Proposed	Boost PFC	1/M + 0.5 1/M	0.5	95.8	4/6/2/2/-/-

S*: switch, D*: diode, L*: inductor, C*: capacitor, C.I*: coupled inductor, T*: transformer, V_O: output voltage, V_S: switch voltage, V_D: diode voltage, V_{in}: input voltage, M: voltage gain (V_o/V_{in}), V_{dc}: primary stage DC voltage.

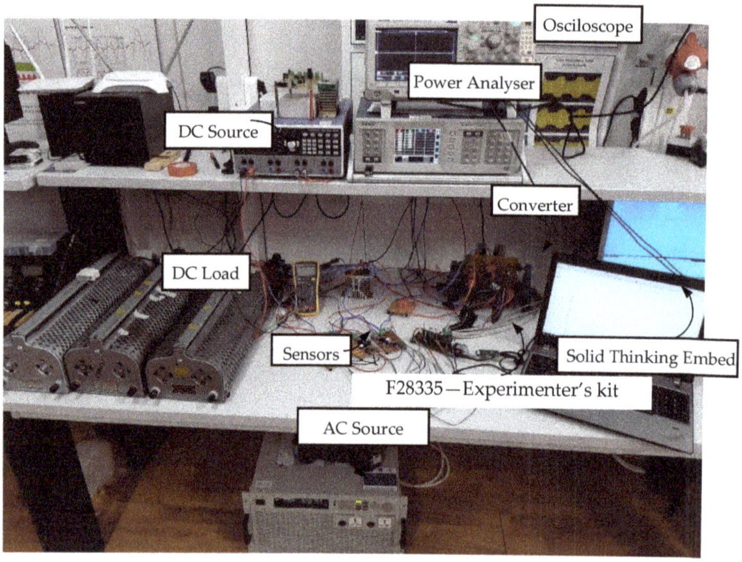

Figure 9. Laboratory test setup.

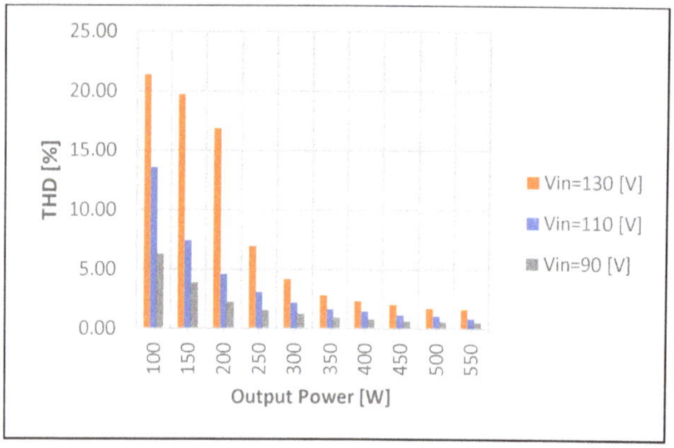

Figure 10. Laboratory practical total harmonic distortion measurements.

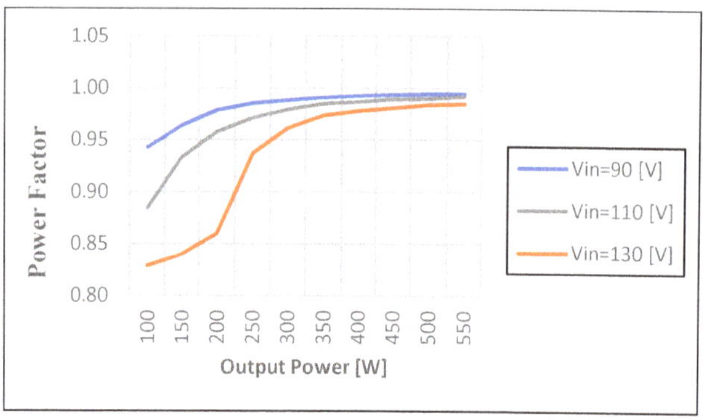

Figure 11. Laboratory practical power factor measurements.

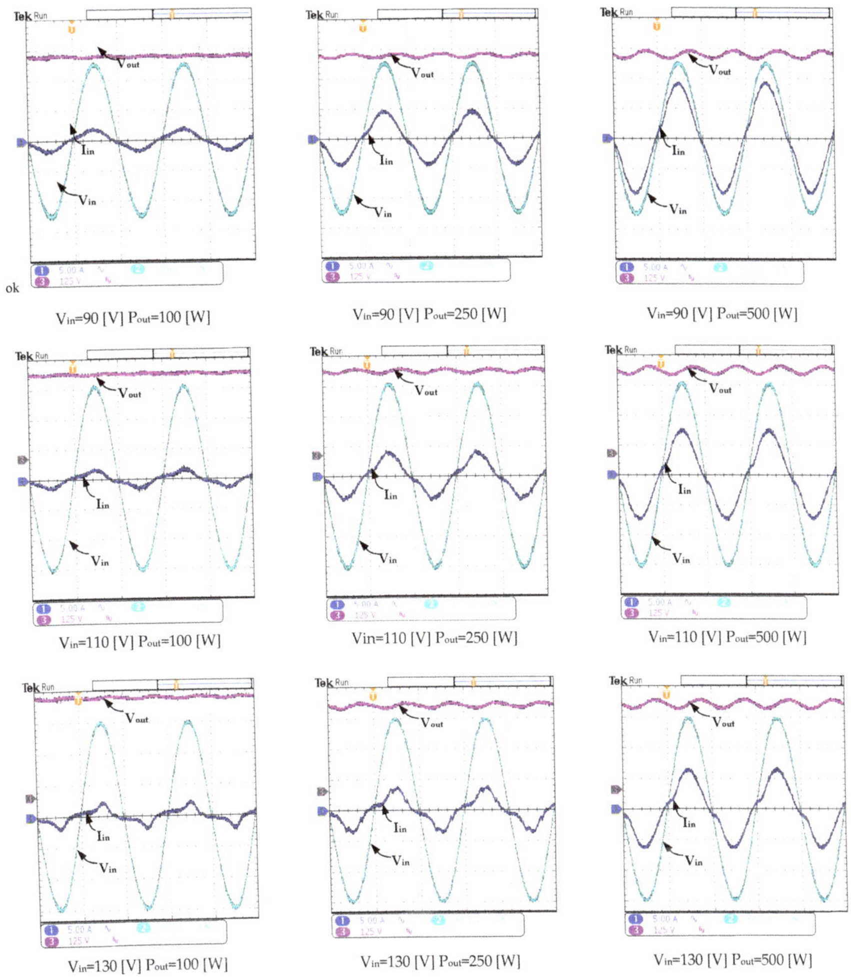

Figure 12. Practical measurements of the proposed converter at different input voltages and different output powers. V_{in}—main input voltage; V_{out}—DC output voltage and I_{in} input current.

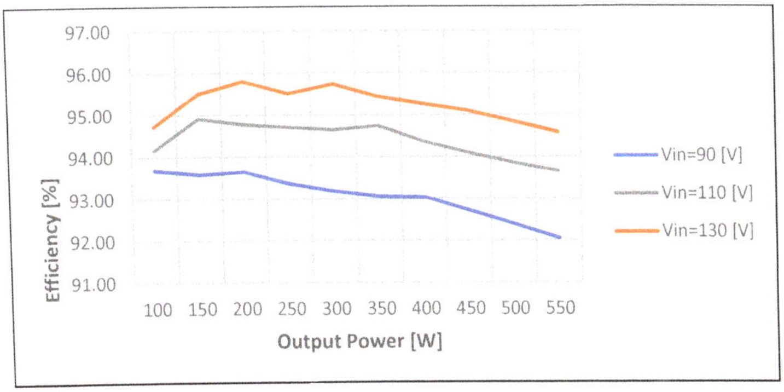

Figure 13. Laboratory practical efficiency measurements.

5. Conclusions

The paper proposes a new type of converter for active Power Factor Correction applications, combining some key functions in one solution. The three voltage levels at the output together with the high gain capability could be favorable assets for future integration in bipolar symmetric/asymmetric DC microgrids. Likewise, the input interleaved operation and low voltage stress on the power semiconductors can be beneficial for efficiency improvements and high-power/high-voltage applications.

The proposed converter has been analyzed through theoretical, simulated and practical approaches to highlight the overall impact of the solution. From the simulation results, the operation of the new converter meets the analytical description presented, where the mathematical model of the converter was developed to help tuning the control loop. According to the experimental part, near the converter rated power, the obtained results in terms of THD and power factor are improving. Another phenomenon that can be observed is that the converter has better performance, in terms of power quality, at higher voltage gains. In terms of efficiency, on the other hand, it has been observed that at low supply voltage, the efficiency is lower. Therefore, the performance is decreasing with the increasing of the converter's gain. In this matter, for future optimization iterations, this tradeoff must be considered, but with the remark that these are common facts in boost PFC applications.

Regarding the converter overall performance attained, the results are promising, considering that this is the first prototype developed, which is not optimized in terms of switching and passive elements criteria. In view of this, future developments are mandatory for power and control optimizations, and more close analyses are needed on the implementation of the converter topology in market applications.

Author Contributions: Conceptualization, N.C.S. and P.D.T.; methodology, S.I.S.; software, L.N.P.; validation, N.C.S. and V.M.S.; formal analysis, V.M.S. and G.I.F.; investigation, S.I.S.; resources, N.C.S.; data curation, L.N.P.; writing—original draft preparation, N.C.S. and S.I.S.; writing—review and editing, P.D.T.; visualization, G.I.F.; supervision, N.C.S.; project administration, P.D.T. All authors have read and agreed to the published version of the manuscript.

Funding: This research was funded by the EUROPEAN REGIONAL DEVELOPMENT FUND through the COMPETITIVENESS OPERATIONAL PROGRAM 2014–2020 Romania, grant number 16/01.09.2016, project title "High power density and high efficiency micro-inverters for renewable energy sources-MICROINV".

Institutional Review Board Statement: Not applicable.

Informed Consent Statement: Not applicable.

Conflicts of Interest: The authors declare no conflict of interest.

References

1. dos Santos, A.H.C.; Fagá, M.T.W.; dos Santos, E.M. The Risks of an Energy Efficiency Policy for Buildings Based Solely on the Consumption Evaluation of Final Energy. *Int. J. Electr. Power Energy Syst.* **2013**, *44*, 70–77. [CrossRef]
2. Bye, B.; Fæhn, T.; Rosnes, O. Residential Energy Efficiency Policies: Costs, Emissions and Rebound Effects. *Energy* **2018**, *143*, 191–201. [CrossRef]
3. Brown, M. Innovative Energy-Efficiency Policies: An International Review. *WIREs Energy Environ.* **2015**, *4*, 1–25. [CrossRef]
4. The European Commission. COMMISION REGULATION (EU) 2019/1782 of 1 October 2019. *Off. J. Eur. Union* **2019**, 92–106.
5. Teixeira, C.A.; Holmes, D.G.; McGrath, B.P. Single-Phase Semi-Bridge Five-Level Flying-Capacitor Rectifier. *IEEE Trans. Ind. Appl.* **2013**, *49*, 2158–2166. [CrossRef]
6. Figueiredo, J.P.M.; Tofoli, F.L.; Silva, B.L.A. A Review of Single-Phase PFC Topologies Based on the Boost Converter. In Proceedings of the 2010 9th IEEE/IAS International Conference on Industry Applications-INDUSCON 2010, São Paulo, Brazil, 8–10 November 2010; pp. 1–6.
7. Zhang, S.; Garner, R.; Zhang, Y.; Bakre, S. Quantification Analysis of Input/Output Current of Interleaved Power Factor Correction (PFC) Boost Converter. In Proceedings of the 2014 IEEE Applied Power Electronics Conference and Exposition-APEC 2014, Fort Worth, TX, USA, 16–20 March 2014; pp. 1902–1908.

8. Hua, C.-C.; Chou, L.-K.; Chuang, C.-W.; Chuang, C.-C. Interleaved Voltage-Doubler Boost PFC with Coupled Inductor. In Proceedings of the 2019 2nd International Conference on High Voltage Engineering and Power Systems (ICHVEPS), Denpasar, Bali, Indonesia, 1–4 October 2019; pp. 1–6.
9. Yang, F.; Ruan, X.; Yang, Y.; Ye, Z. Interleaved Critical Current Mode Boost PFC Converter with Coupled Inductor. *IEEE Trans. Power Electron.* **2011**, *26*, 2404–2413. [CrossRef]
10. Yang, F.; Li, C.; Cao, Y.; Yao, K. Two-Phase Interleaved Boost PFC Converter with Coupled Inductor under Single-Phase Operation. *IEEE Trans. Power Electron.* **2020**, *35*, 169–184. [CrossRef]
11. Yang, F.; Ruan, X.; Yang, Y.; Ye, Z. Design Issues of Interleaved Critical Conduction Mode Boost PFC Converter with Coupled Inductor. In Proceedings of the 2010 IEEE Energy Conversion Congress and Exposition, Atlanta, GA, USA, 12–16 September 2010; pp. 2245–2252.
12. Hu, J.; Xiao, W.; Zhang, B.; Qiu, D.; Ho, C.N.M. A Single Phase Hybrid Interleaved Parallel Boost PFC Converter. In Proceedings of the 2018 IEEE Energy Conversion Congress and Exposition (ECCE), Portland, OR, USA, 23–27 September 2018; pp. 2855–2859.
13. Wang, Q.; Wen, B.; Burgos, R.; Boroyevich, D.; White, A. Efficiency Evaluation of Two-Level and Three-Level Bridgeless PFC Boost Rectifiers. In Proceedings of the 2014 IEEE Applied Power Electronics Conference and Exposition-APEC 2014, Fort Worth, TX, USA, 16–20 March 2014; pp. 1909–1915.
14. Ando, I.; Abe, K.; Ochiai, M.; Ohishi, K. Soft-Switching-Interleaved Power Factor Correction Converter with Lossless Snubber. In Proceedings of the IECON 2013—39th Annual Conference of the IEEE Industrial Electronics Society, Vienna, Austria, 10–13 November 2013; pp. 7216–7221.
15. Cheng, C.-A.; Chang, C.-H.; Cheng, H.-L.; Chang, E.-C.; Chung, T.-Y.; Chang, M.-T. A Single-Stage LED Streetlight Driver with Soft-Switching and Interleaved PFC Features. *Electronics* **2019**, *8*, 911. [CrossRef]
16. Musumeci, S.; Bojoi, R.; Armando, E.; Borlo, S.; Mandrile, F. Three-Legs Interleaved Boost Power Factor Corrector for High-Power LED Lighting Application. *Energies* **2020**, *13*, 1728. [CrossRef]
17. Wang, H.; Dusmez, S.; Khaligh, A. Design Considerations for a Level-2 on-Board PEV Charger Based on Interleaved Boost PFC and LLC Resonant Converters. In Proceedings of the 2013 IEEE Transportation Electrification Conference and Expo (ITEC), Metro Detroit, MI, USA, 16–19 June 2013; pp. 1–8.
18. Indalkar, S.S.; Sabnis, A. An OFF Board Electric Vehicle Charger Based On ZVS Interleaved AC-DC Boost PFC Converter. In Proceedings of the 2019 8th International Conference on Power Systems (ICPS), Jaipur, India, 20–22 December 2019; pp. 1–6.
19. Musavi, F.; Eberle, W.; Dunford, W.G. A High-Performance Single-Phase Bridgeless Interleaved PFC Converter for Plug-in Hybrid Electric Vehicle Battery Chargers. *IEEE Trans. Ind. Appl.* **2011**, *47*, 1833–1843. [CrossRef]
20. Lin, B.-R.; Lu, H.-H. Single-Phase Three-Level PWM Rectifier. In Proceedings of the IEEE 1999 International Conference on Power Electronics and Drive Systems, PEDS'99 (Cat. No.99TH8475), Hong Kong, 27–29 July 1999; Volume 1, pp. 63–68.
21. Ortmann, M.S.; Mussa, S.A.; Heldwein, M.L. Concepts for High Efficiency Single-Phase Three-Level PWM Rectifiers. In Proceedings of the 2009 IEEE Energy Conversion Congress and Exposition, San Jose, CA, USA, 20–24 September 2009; pp. 3768–3775.
22. Kim, J.-S.; Lee, S.-H.; Cha, W.-J.; Kwon, B.-H. High-Efficiency Bridgeless Three-Level Power Factor Correction Rectifier. *IEEE Trans. Ind. Electron.* **2017**, *64*, 1130–1136. [CrossRef]
23. Su, B.; Zhang, J.; Wen, H.; Lu, Z. Low Conduction Loss and Low Device Stress Three-level Power Factor Correction Rectifier. *IET Power Electron.* **2013**, *6*, 478–487. [CrossRef]
24. Kakigano, H.; Miura, Y.; Ise, T. Low-Voltage Bipolar-Type DC Microgrid for Super High Quality Distribution. *IEEE Trans. Power Electron.* **2010**, *25*, 3066–3075. [CrossRef]
25. Rivera, S.; Lizana, F.R.; Kouro, S.; Dragičević, T.; Wu, B. Bipolar DC Power Conversion: State-of-the-Art and Emerging Technologies. *IEEE J. Emerg. Sel. Top. Power Electron.* **2021**, *9*, 1192–1204. [CrossRef]
26. Wang, R.; Feng, W.; Xue, H.; Gerber, D.; Li, Y.; Hao, B.; Wang, Y. Simulation and Power Quality Analysis of a Loose-Coupled Bipolar DC Microgrid in an Office Building. *Appl. Energy* **2021**, *303*, 117606. [CrossRef]
27. Li, Q.; Huangfu, Y.; Xu, L.; Wei, J.; Ma, R.; Zhao, D.; Gao, F. An Improved Floating Interleaved Boost Converter With the Zero-Ripple Input Current for Fuel Cell Applications. *IEEE Trans. Energy Conv.* **2019**, *34*, 2168–2179. [CrossRef]
28. Ganjavi, A.; Ghoreishy, H.; Ahmad, A.A. A Novel Single-Input Dual-Output Three-Level DC–DC Converter. *IEEE Trans. Ind. Electron.* **2018**, *65*, 8101–8811. [CrossRef]
29. Chen, J.; Hou, S.; Sun, T.; Deng, F.; Chen, Z. A New Interleaved Double-Input Three-level Boost Converter. *J. Power Electron.* **2016**, *16*, 925–935. [CrossRef]
30. Marzang, V.; Hosseini, S.H.; Rostami, N.; Alavi, P.; Mohseni, P.; Hashemzadeh, S.M. A High Step-Up Nonisolated DC–DC Converter with Flexible Voltage Gain. *IEEE Trans. Power Electron.* **2020**, *35*, 10489–10500. [CrossRef]
31. Maheri, H.M.; Babaei, E.; Sabahi, M.; Hosseini, S.H. High Step-Up DC-DC Converter with Minimum Output Voltage Ripple. *IEEE Trans. Ind. Electron.* **2017**, *64*, 3568–3575. [CrossRef]
32. Zhang, Y.; Gao, Y.; Zhou, L.; Sumner, M. A Switched-Capacitor Bidirectional DC-DC Converter with Wide Voltage Gain Range for Electric Vehicles with Hybrid Energy Sources. *IEEE Trans. Power Electron.* **2018**, *33*, 9459–9469. [CrossRef]
33. Haji-Esmaeili, M.M.; Babei, E.; Sabahi, M. High Step-up Quasi-Z Source DC-DC Converter. *IEEE Trans. Power Electron.* **2018**, *33*, 10563–10571. [CrossRef]

34. Maroti, P.K.; Ranjana, M.S.B.; Prabhakar, D.K. A Novel High Gain Switched Inductor Multilevel Buck-Boost DC-DC Converter for Solar Applications. In Proceedings of the IEEE 2nd International Conference on Electrical Energy Systems (ICEES), Chennai, India, 7–9 January 2014; pp. 152–156.
35. Lee, M.; Lai, J.-S. Unified Voltage Balancing Feedforward for Three-Level Boost PFC Converter in Discontinuous and Critical Conduction Modes. *IEEE Trans. Circuits Syst. II Express Briefs* **2021**, *68*, 441–445. [CrossRef]
36. Su, B.; Zhang, J.; Lu, Z. Single inductor three-level boost bridgeless PFC rectifier with nature voltage clamp. In Proceedings of the 2010 International Power Electronics Conference-ECCE ASIA, Sapporo, Japan, 8–10 November 2010; pp. 2092–2097.
37. Aldosari, O.; Rodriguez, L.A.G.; Oggier, G.G.; Balda, J.C. A High-Efficiency Isolated PFC AC-DC Topology with Reduced Number of Semiconductor Devices. *IEEE J. Emerg. Sel. Top. Power Electron.* **2021**. [CrossRef]
38. Aldosari, O.; Rodriguez, L.A.G.; Oggier, G.G.; Balda, J.C. A Three-Level Isolated AC–DC PFC Power Converter Topology with a Reduced Number of Switches. *IEEE J. Emerg. Sel. Top. Power Electron.* **2021**, *9*, 1052–1063. [CrossRef]

applied sciences

MDPI

Article

Research on the Distribution Characteristics of Transformer Axial Vibration under Short-Circuit Conditions Considering Damping Parameters

Shuguo Gao *, Lu Sun, Yuan Tian and Hongliang Liu

Electric Power Research Institute of State Grid Hebei Electric Power Co., Ltd., Shijiazhuang 050021, China
* Correspondence: dyy_gaosg@he.sgcc.com.cn; Tel.: +86-158-3010-5382

Featured Application: This research provides theoretical support for the design and improvement of transformer winding resistance to a short circuit and has certain guiding significance for real-time monitoring of transformer winding acceleration.

Abstract: The existing research on the distribution characteristics of displacement and acceleration of the transformer axial vibration under short-circuit conditions is based on ignoring the damping parameters. An accurate description of the axial distribution characteristics of the windings, especially for the axial vibration of the winding under short-circuit conditions, has a poor effect. In this paper, the damping, stiffness, and mass parameters between windings are comprehensively considered, and the classical "mass-spring-damping" axial vibration mathematical model of transformer windings is established. After solving, the natural frequency, main mode shape, displacement, and acceleration of each wire cake of the multi-degree-of-freedom (multi-DOF) vibration system were quickly obtained. The relationship between the axial displacement and acceleration of the wire cake and the axial deformation of the transformer winding was discussed, and the transformer winding axis was summarized as well as characteristics of the vibration distribution.

Keywords: transformer; axial vibration; multi-degree-of-freedom; displacement; acceleration

Citation: Gao, S.; Sun, L.; Tian, Y.; Liu, H. Research on the Distribution Characteristics of Transformer Axial Vibration under Short-Circuit Conditions Considering Damping Parameters. *Appl. Sci.* **2022**, *12*, 8443. https://doi.org/10.3390/app12178443

Academic Editors: Federico Barrero and Mario Bermúdez

Received: 31 May 2022
Accepted: 19 August 2022
Published: 24 August 2022

1. Introduction

Power transformer plays an extremely important role in the power system and is the nerve center of the power system, which undertakes the task of voltage and current conversion. Once it breaks down, it will affect a wide range and even lead to the paralysis of a power system [1,2]. About half of all transformer damages are due to insufficient short-circuit resistance of the transformer. The accumulation of axial deformation of a transformer due to short circuits is the primary cause of transformer short circuit damage accidents [3,4]. When a transformer is short-circuited, the maximum short-circuit current on the winding is 20–30 times the rated current [5]. Due to the action of the electromagnetic force, the huge electrodynamic force bends and deforms the winding in the axial direction. When the elastic deformation of winding transfers to plastic deformation, the transformer has irreversible structural failure [6,7].

Axial deformation of transformer winding caused by a short circuit can be attributed to the axial vibration of winding [8,9]. Euler established a mechanical model of a multi-DOF vibration system with mass particles such as stiffness spring connection in the 18th century [10], which has further evolved into the "spring-mass-damper" model in subsequent studies to describe the axial dynamic characteristics of winding [11,12]. So far, this model is still the mainstream analysis model for analyzing axial stability. Based on this model, many scholars have studied the axial vibration of winding through simulation and modeling. Liang Guangcheng et al. calculated the short-circuit electrodynamic force of a transformer by using a finite element method and short-circuit electrodynamic force calculation software, and carried out static and dynamic analysis and calculation, respectively [13]; Hyun-Mo Ahn et al. used the finite element method to calculate the short-circuit electrodynamic force acting on each disk of dry-type transformer winding in case of a short circuit by using a finite element method [14]; Hu Yiwei conducted modal analysis on the vibration signal corresponding to the axial vibration of a transformer under transient operating conditions during power failure by using an operating modal analysis method, and studied and analyzed the vibration characteristics of the transient process and the influence of loose winding pressing force on axial vibration [15]. These literatures studied the mechanical stability changes of winding caused by axial force. However, the influence of the distribution of axial vibration on the transformer cannot be determined. Therefore, some scholars have also conducted research on the distribution model of the axial vibration of the transformer winding.

Xu Yongming et al. calculated the short-circuit electrodynamic force distribution of transformer winding by using a three-dimensional finite element model [16]. Based on the field-circuit coupling method, Wang Xiwen et al. used the transient field analysis method to obtain the distribution of leakage magnetic field and electromagnetic force in the winding area and obtained the distribution law of the axial deformation of the transformer in the whole process in the transient structure field [17]; D. J. Allen believed when the transformer is short circuited, the axial displacement of winding presents a "V" shape distribution, that is, the axial displacement at the middle of the winding is greater than at the end of the winding [18]. Ji Shengchang et al. believed that axial vibration acceleration of winding presented an "M" shape distribution law, that is, the acceleration amplitude at the height of 1/4 and 3/4 of the winding is greater than at the end and middle of the winding [12,19]. However, in the process of calculation and solution, the damping parameters are ignored in the above literature, and only the ideal transformer without damping is considered, which is not universal and cannot be applied to a transformer in an actual situation.

Therefore, this paper firstly deduced the theoretical formula of the transformer axial deformation and built the axial theoretical model of the transformer winding under the condition of a short circuit with damping and multiple degrees of freedom. The accuracy of the model is verified by the frequency sweep, the axial distribution characteristics of the windings during the short-circuit impact of the transformer are obtained, and the influence of the real-time parameters of the axial deformation speed and acceleration on the axial stability is discussed. It provides theoretical support for the design and improvement of transformer windings and has a certain guiding significance.

2. Axial Vibration Model of Transformer Winding

Theoretical analysis and engineering test research show that only a few parameters, such as mass and its distribution, motion damping, and restoring force characteristics, play a leading role in vibration characteristics and the response of a vibration system. The winding of a large-scale power transformer is usually a disk structure, with the alternate distribution of disks and cushion blocks in the axial direction. High-, medium-, and low-voltage windings are stacked according to the number of turns. An alternating current is connected to the winding and generates an alternating magnetic field, which forms a magnetic circuit through the iron core. The current in the winding generates alternating electrodynamic force under the action of a magnetic field and acts on each disk, and the action of alternating current makes each disk vibrate (mainly vibration in the axial direction).

2.1. Equivalent Model of Axial Vibration of the Transformer

The lumped parameters of a linear vibration system are composed of mass, damping, and spring stiffness, which is usually called the mass-damping-spring model. The disk is regarded as a lumped mass; the insulating cushion block, gasket, and pressing plate between windings are regarded as elastic elements. The iron core and iron yoke are regarded as rigid bodies. The "mass-spring-damping" model is shown in Figure 1 [20].

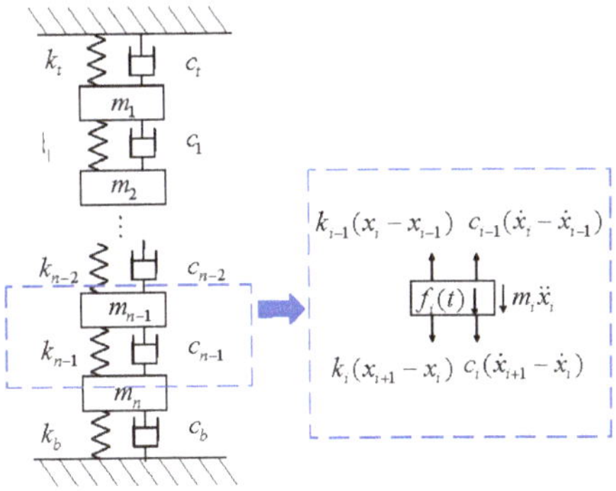

Figure 1. Winding axial mass-damping-spring model.

As is shown in Figure 1, the mass block m in the figure corresponds to the mass of a single-layer disk, and spring stiffness k_i and damping c_i correspond to the equivalent stiffness and damping of an insulating paper and insulating cushion block between two layers of the disk. According to the equivalent model and stress analysis of each disk, the dynamic equations of each disk can be written as follows:

$$\begin{cases} m_1\ddot{x}_1 + c_t\dot{x}_1 - c_1(\dot{x}_2 - \dot{x}_1) + k_t x_1 - k_1(x_2 - x_1) = f_1(t) \\ m_2\ddot{x}_2 + c_1(\dot{x}_2 - \dot{x}_1) - c_2(\dot{x}_3 - \dot{x}_2) + k_1(x_2 - x_1) - k_2(x_3 - x_2) = f_2(t) \\ \vdots \\ m_i\ddot{x}_i + c_{i-1}(\dot{x}_i - \dot{x}_{i-1}) - c_i(\dot{x}_{i+1} - \dot{x}_i) + k_{i-1}(x_i - x_{i-1}) - k_i(x_{i+1} - x_i) = f_i(t) \\ \vdots \\ m_n\ddot{x}_n + c_{n-1}(\dot{x}_n - \dot{x}_{i-1}) - c_b\dot{x}_n + k_{n-1}(x_n - x_{n-1}) - k_b x_n = f_n(t) \end{cases} \tag{1}$$

wherein, m_n is the mass of the n-th layer of disk; k_t and k_b are the equivalent stiffness coefficients between the upper and lower pressing plates, and between the first disk and the last disk; c_t and c_b are the equivalent damping coefficients between the upper and lower pressing plates and between the first disk and the last disk; k_n is the equivalent stiffness of the n-th disk; c_n is the equivalent stiffness of the n-th disk; x_n is the displacement of the n-th disk; and $f_n(t)$ is the electrodynamic force received by the n-th disk.

As shown in Table 1, the element definitions required for the solution of the axial multi-DOF vibration of the transformer winding.

Table 1. Definition of elements in the multi-DOF model.

Elements	Definition
m_n	the quality of the n-th layer of disk
k_t and k_b	equivalent stiffness between clamping rings and winding end
k_n	equivalent stiffness between the n-th disk and the $(n+1)$-th disk
c_t and c_b	equivalent viscous damping factors between clamping rings and winding end
c_n	equivalent viscous damping factors between the n-th disk and the $(n+1)$-th disk
$x_n(t)$	displacement of n-th disk
$\dot{x}_n(t)$	the speed of the n-th disk
$\ddot{x}_n(t)$	acceleration of the n-th disk
$q_n(t)$	displacement of n-th disk in canonical modal coordinates
$\dot{q}_n(t)$	the speed of n-th disk in canonical modal coordinates
$\ddot{q}_n(t)$	acceleration of n-th disk in canonical modal coordinates
ζ_r	r-th order damping ratio
Q_r	the magnitude of the electric power received by each disk
ω_{nr}	the r-order component of the natural frequency of the n-th mode shape
Ω	the frequency of the electric power received by each disk
φ_r	phase angle of electric power received by each disk

For convenience, Equation (1) is written in matrix form

$$[m]\{\ddot{x}(t)\}+[c]\{\dot{x}(t)\}+[k]\{x(t)\}=\{f(t)\} \tag{2}$$

wherein, $[m]$ is the mass matrix of each disk, $[c]$ is the damping matrix of each disk, $[k]$ is the stiffness matrix of each disk, $\{\ddot{x}(t)\}$ is the vibration acceleration matrix of each disk, $\{\dot{x}(t)\}$ is the vibration velocity matrix of each disk, $\{x(t)\}$ is the vibration displacement matrix of each disk, and $\{f(t)\}$ is the electrodynamic force matrix received by each disk.

When the transformer is short-circuited, only the steady-state part of the short-circuit current is considered,

$$I=I_m\cos(\Omega t+\varphi_0) \tag{3}$$

wherein, I_m is the amplitude of the AC component of the short-circuit current, and φ_0 is the initial phase.

Then the electromagnetic force acting on the winding is

$$F=\frac{1}{2}pI_m^2[1+\cos(2\Omega t+\varphi_0)] \tag{4}$$

wherein, p is the electromagnetic force coefficient.

It can be seen that the value of the electromagnetic force frequency on the winding is twice the current frequency.

The characteristic equation of the system is as follows

$$\left|[k]-\omega_n{}^2[m]\right|=0 \tag{5}$$

wherein, ω_n is the natural frequency of vibration mode in n-th order.

Substitute the r-order natural frequency successively into the eigenvalue problem equation of the system

$$[k]\{u\}=\omega_n{}^2[m]\{u\} \tag{6}$$

wherein, $\{u\}$ is the modal vector matrix.

Manually specify any one of the modal vectors, such as $\{u_1(r)\} = \{1\}$, to solve all other r-order modal vectors of the system

$$\{u^{(r)}\}=[u_1{}^{(r)},u_2{}^{(r)},\cdots,u_n{}^{(r)}]^T \tag{7}$$

wherein, $r = 1, 2, 3, \cdots, n$

Transform the physical coordinates of the system by using normal modal matrix $[u_N]$

$$\{q_N(t)\}=[u_N]^{-1}\{x(t)\} \tag{8}$$

wherein, $\{q_N(t)\}$ is the displacement matrix converted to normal modal coordinates.

Transform the vibration equation of a general multi-DOF system into a normal modal equation

$$\{\ddot{q}_N(t)\}+[2\zeta_r\omega_{\mathrm{nr}}]\{\dot{q}_N(t)\}+[\omega_{nr}{}^2]\{q_N(t)\}=\{N(t)\} \tag{9}$$

wherein, $\{\ddot{q}_N(t)\}$ is the acceleration matrix of each disk under normal modal coordinates, ζ_r is the r-order damping ratio, and $\{N(t)\}$ is the electrodynamic force matrix received by each disk under normal modal coordinates.

Therefore, the solution of Equation (9) can be regarded as the solution of n second-order partial differential equations with different coefficients but the same variables,

$$\begin{cases} q_{Nr}(t)=Q_r\cos(\Omega t-\varphi_r) \\ \dot{q}_{Nr}(t)=-Q_r\Omega\sin(\Omega t-\varphi_r) \\ \ddot{q}_{Nr}(t)=-Q_r\Omega^2\cos(\Omega t-\varphi_r) \end{cases} \tag{10}$$

$$Q_r = \frac{N_r}{\sqrt{\left(\omega_{nr}{}^2 - \Omega^2\right)^2 - \left(2\zeta_r\omega_{nr}\Omega\right)^2}} \tag{11}$$

$$\varphi_r = \arctan\frac{2\zeta_r\Omega/\omega_{nr}}{1-\left(\Omega/\omega_{nr}\right)^2} \tag{12}$$

wherein, Q_r is the amplitude of electrodynamic force received by each disk, φ_r is the phase angle of electrodynamic force received by each disk, and Ω is the frequency of the electric power received by each wire cake.

The displacement equation, velocity equation, and acceleration equation can be obtained by transforming it from a modal space coordinate system to a physical coordinate system.

2.2. Parameter Matrix Acquisition

The axial section model of a three-phase transformer is shown in Figure 2. Generally, the disk-like structure is adopted in large and super large transformers with a high voltage of 110 kV and above, that is, the windings are in the radial direction, and the wire turns are continuously wound into a disk shape, and several disks are stacked into windings. The model parameter settings are completely based on the actual parameters of the real transformer.

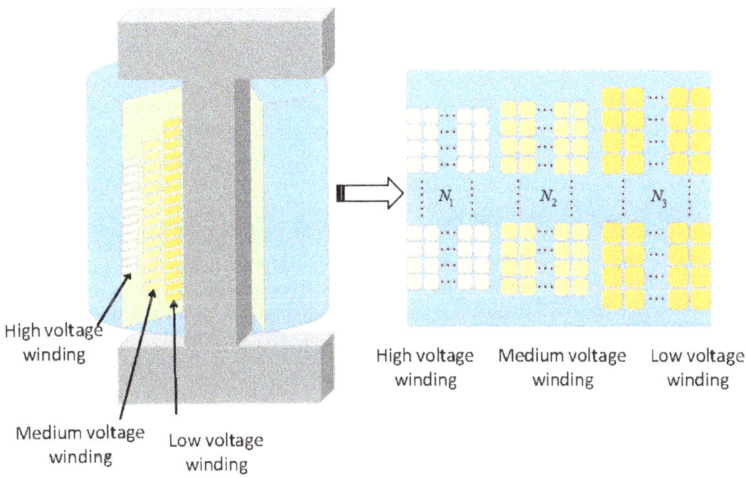

Figure 2. Transformer axial section model.

In the axial direction, the upper and lower pressure plates are fixed at the ends of the windings, and the spacers are stacked between the turns of the coil. The support stiffness mainly depends on the spacers, which can be expressed as the following formula:

$$k = \frac{abE}{h} \tag{13}$$

wherein, E is the elastic modulus of the cushion block, which is 50 MPa here; a is the width of the cushion block; b is the height of the conductor in the radial direction; and h is the distance between the iron core and the inner diameter of the coil.

To analyze the overall vibration, each disk is regarded as a coil as a whole. The calculation formula of disk quality is as follows:

$$m = \frac{LnS\rho}{1000N} \tag{14}$$

where m is the mass of the disk, L is the total length of a single-phase conductor, n is the number of turns of a single disk, S is the cross-sectional area of the conductor, ρ is the density of the conductor, and N is the number of turns of winding. Thus, the mass of each winding disk is obtained.

According to Rayleigh damping (proportional damping), the damping matrix

$$[c] = \alpha[m] + \beta[k] \tag{15}$$

where α and β are constants.

The damping ratio can be obtained

$$\zeta_r = \frac{\alpha + \beta{\omega_{nr}}^2}{2\omega_{nr}} \tag{16}$$

where ω_{nr} is the r-th order natural frequency.

3. Model Calculation Based on Actual Parameters of True Transformer

Combined with the theoretical formula in the second section and the winding parameters of a power transformer, the simulation calculation of high-voltage winding under a short circuit is carried out through MATLAB (R2016a, MathWorks, Natick, MA, USA).

3.1. Model-Specific Parameters

The specific parameters of transformer winding of a 31.5 MVA/110 kV three-phase three winding power transformer used in MATLAB simulation in this paper are shown in Table 2.

Table 2. Transformer parameters.

Parameters	Value	Unit
Conductor density	8.9	ρ/[g/cm^3]
Turns of high voltage winding N_H	647	[-]
Turns of medium voltage winding N_M	226	[-]
Turns of low voltage winding N_L	107	[-]
Cross-sectional area of high voltage winding S_H	61.44	[mm^2]
Cross-sectional area of medium voltage winding S_M	182.75	[mm^2]
Cross-sectional area of low voltage winding S_L	296.64	[mm^2]
Total length of high voltage winding L_H	2594	[m]
Total length of medium voltage winding L_M	700	[-]
Total length of low voltage winding L_L	245.5	[m]
Elastic modulus E	50	[MPa]
Cushion block width a_H	50	[mm]
Cushion block width a_M	40	[mm]
Cushion block width a_L	30	[mm]
Axial height of conductor b_H	11.8	[mm]
Axial height of conductor b_M	14	[mm]
Axial height of conductor b_L	6	[mm]

3.2. Natural Frequencies and Mode Shapes

By solving the characteristic Equation (5), it can be concluded that the eigenvalue and eigenvector correspond to the natural frequency and vibration mode vector of the axial winding vibration model, respectively.

Figure 3 shows the natural frequencies of each order of high-voltage winding, which are arranged in ascending order. As shown, the frequency values, which are equal to $\omega_{nr} = \sqrt{K_r/M_r}$, change linearly and change with the change of the square root of the ratio of stiffness to mass.

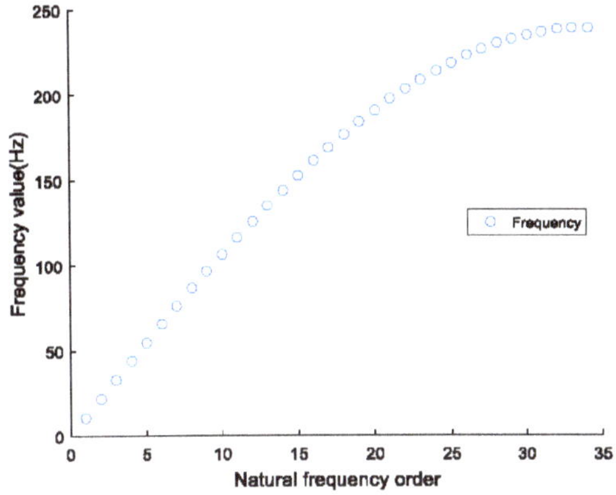

Figure 3. Natural frequencies of each order.

Figure 4 shows the typical modal shape diagram of the axial vibration of a winding with 34 disks by solving the system characteristic Equation (5), where the number of modal vectors is equal to the number of disks. It can be concluded from the figure that the lower the natural frequency, the greater the oscillation period of the disk, that is, the greater the impact on the system. Therefore, only the first four order modal vectors having the greatest impact on the system are taken in Figure 3. The first-order vibration mode shows the windings vibrate in the same direction in the radial direction and is distributed in a "V" shape; the second-order vibration mode shows the windings can be divided into halves from the middle part of the windings in the radial direction, and the upper half and lower half vibrate in the opposite direction, with an "N" distribution; the third-order and fourth-order modes show the higher the number of waves, the higher the distribution of waves in the fourth-order mode.

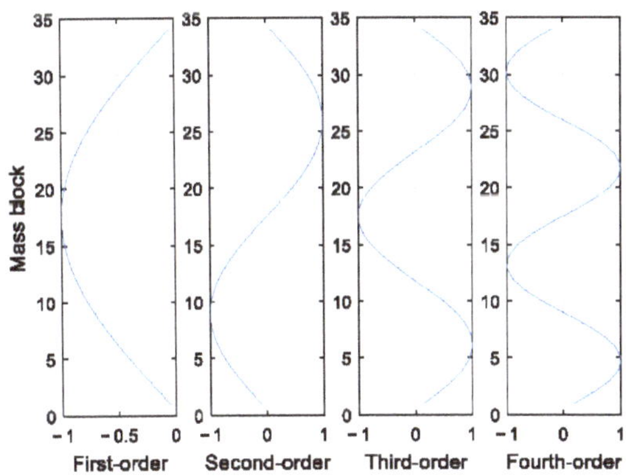

Figure 4. The first four modes.

3.3. Displacement and Acceleration Variation Law of Disks at Different Frequencies

Figure 5 shows winding displacement changes in steady-state condition (t = 0.15 s) at 0–250 Hz. In the low-frequency region, the displacement is large and the amplitude is very high, distributed in an 'arch' shape, while in the high-frequency region, the displacement amplitude is small, but fluctuation is increasing. The theory in the previous section proves the zero nodes of high-order vibration mode increase, but the amplitude change decreases; obviously, the displacement change in Figure 5 conforms to the above theory. To better observe the displacement distribution at different frequencies, take the displacement change at 0–40 Hz as a separate observation. As shown in Figure 5a, at 0–20 Hz, the displacement is distributed in a 'U' shape on the winding, which is the same as the first vibration mode. With an increase of frequency from 20 Hz to 40 Hz, the displacement shape gradually turns into multiple "arches" from the 'U' shape, but the displacement peak decreases gradually. As shown in Figure 5b, in the winding displacement distribution at 40–100 Hz, the displacement fluctuation amplitude decreases consistently, the maximum displacement amplitude gradually approaches the head and end of the winding, fluctuation in the middle part tends to be gentle, and fluctuation in the middle part is almost zero with an increase of frequency. The winding displacement distribution gradually changes from "M" shape distribution to "three-arch" and "four-arch" shapes, which is also consistent with the distribution of vibration modes from third-order vibration modes to high-order vibration modes. As shown in Figure 5c, as frequency increases from 100 Hz to 250 Hz, displacement amplitude continues to decrease in a terraced manner, gradually close to zero, and displacement fluctuation also slowly decreases to zero.

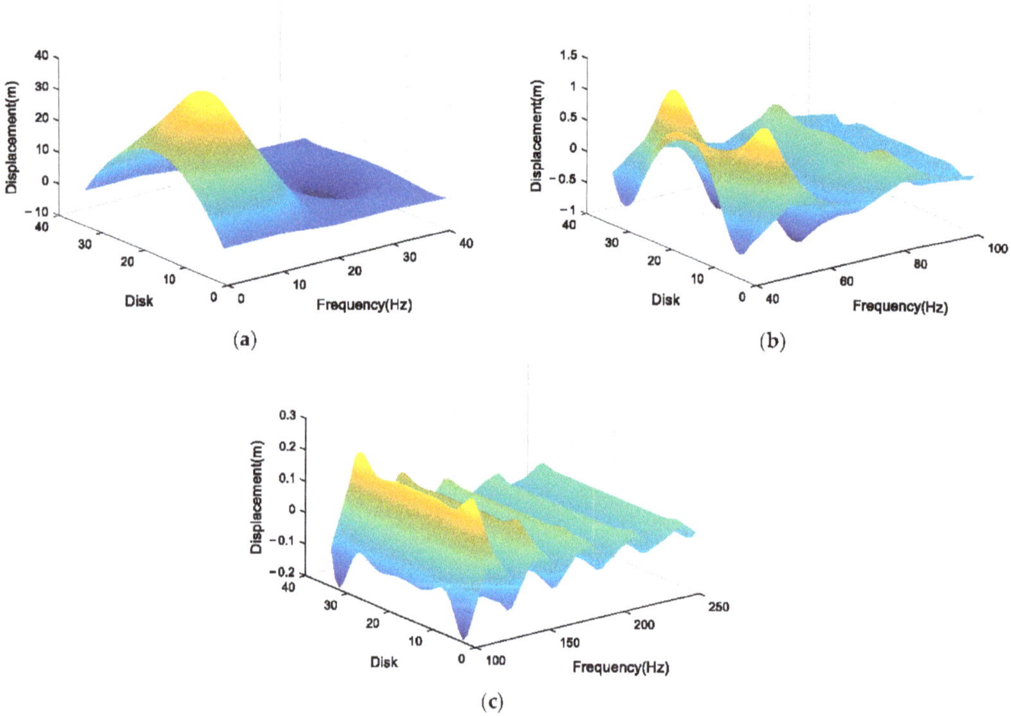

Figure 5. Displacement of each disk at 0–250 Hz. (**a**) Displacement of each disk at 0–40 Hz; (**b**) 40–100 Hz; and (**c**) 100–250 Hz.

Figure 6 shows the winding acceleration changes at 0–250 Hz. In the low-frequency region, the acceleration is large and the amplitude is high, distributed by the above vibration mode law, while in the high-frequency region, the acceleration amplitude changes little and gently.

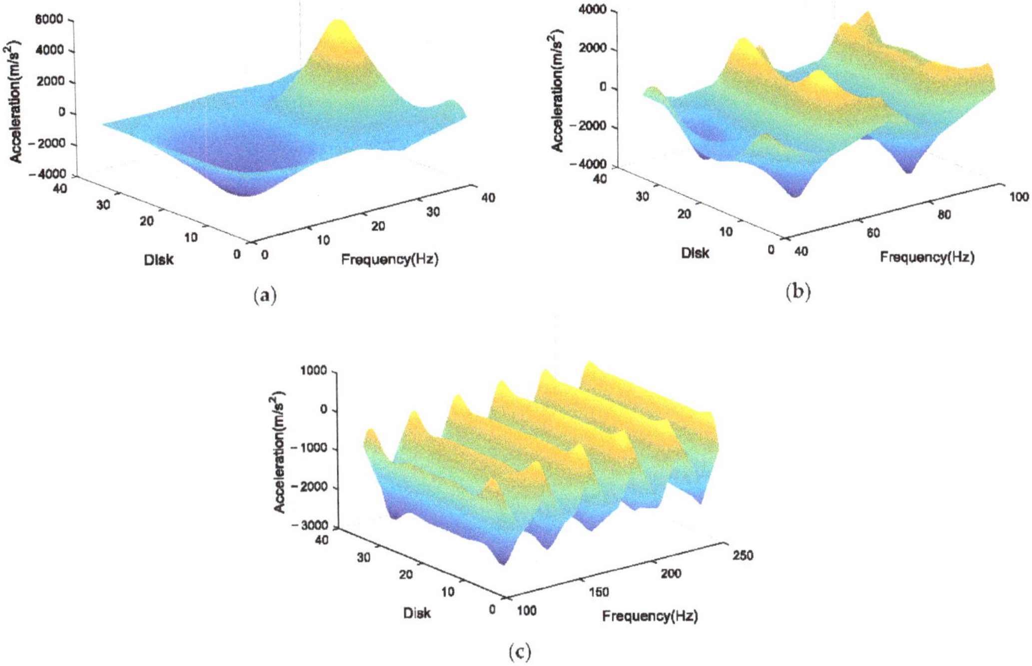

Figure 6. Acceleration of each disk at 0–250 Hz. (**a**) Acceleration of Each disk at 0–40 Hz; (**b**) 40–100 Hz; and (**c**) 100–250 Hz.

Moreover, all disks have the same acceleration distribution as above. As shown in Figure 6a, at 0–20 Hz, vibration acceleration is distributed in a 'U' shape on the winding, consistent with the distribution law of the first vibration mode on the winding. As the frequency increases from 20 Hz to 40 Hz, the displacement shape changes from 'U' to 'double-arch' and then to 'M', consistent with the distribution law from the first-order vibration mode to the third-order vibration mode. Different from the displacement, the peak acceleration in each disk increases gradually with an increase in applied frequency.

As the conclusion of the above analysis, as shown in Figure 6b, the acceleration distribution of each disk is consistent with the shape change law of the vibration mode, while the amplitude does not continue to decay, but changes around a fixed value, and peak acceleration still appears near both ends of the winding.

On the contrary, as shown in Figure 6c, as the frequency increases from 100 Hz to 250 Hz, the acceleration of each disk at each resonance frequency remains at a fixed value except at both ends, and the peak value appears closer and closer to both ends of the winding.

3.4. Displacement and Acceleration Variation Law of Disks at Power Frequency

Assuming that the initial displacement and initial speed of each disk coil are zero, Period T = 2 × pi/ω, when the short-circuit current frequency is equal to the power frequency current, it can be seen from Equation (4) that the short-circuit electromotive force frequency is equal to twice the frequency of the power frequency current, that is, 100 Hz; the external short-circuit current frequency is 50 Hz, that is, at this time, the short-circuit electrodynamic frequency of the disk on the winding is 100 Hz.

In the initial stage of system vibration, transient vibration and steady vibration coexist, and the total vibration of the system is the synthetic vibration of two simple harmonic vibrations with different frequencies. In the damped system, transient vibration is quickly attenuated due to damping, and finally, only the steady vibration is retained. Therefore, the steady motion of damped forced vibration in steady vibration reflects the motion law of a damped forced vibration system. This paper focuses on the transformation and law of displacement and acceleration of the windings when they vibrate axially under short-circuit conditions, so only the steady vibration is studied with the transient vibration omitted in the simulation analysis.

Figure 7 shows the change of the displacement of the top layer of the wire cake with time during a short circuit. When the time is at 0.15, the transient amplitude of the short-circuit current tends to zero, and the steady-state vibration displacement amplitude is at the maximum moment. It can better observe the displacement change of each disk in the axial direction, so the time t = 0.15 s is taken.

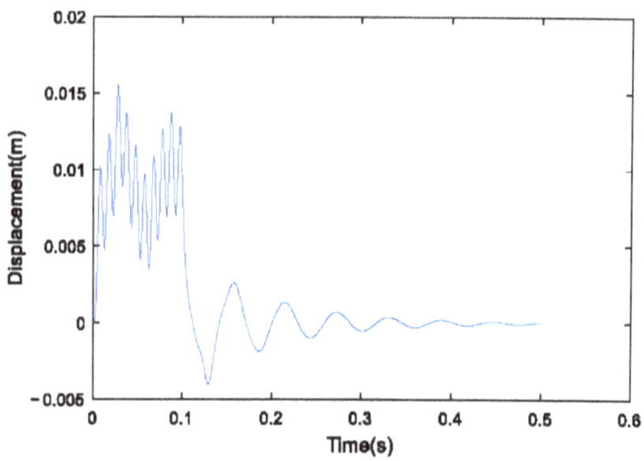

Figure 7. Displacement response of the top disk.

When t = 0.15 s and the damping ratio takes different values, the displacement change at this time is shown in Figure 8. The axial vibration displacement waveform of the winding is in a 'valley' shape, which is roughly consistent with the vibration mode law of the eighth order vibration mode, which is just around the frequency of 100 Hz, so it is believed the radial displacement distribution of the winding conforms to the theoretical calculation. The displacement amplitude reaches the peak at one-seventh of both ends, the displacements at the top and the end are close to zero, and the middle disk displacement is distributed in an "M" shape. Therefore, the deformation near both ends of the winding is large, and small in the middle part. As the damping ratio increases, the displacement oscillation amplitude of each disk gradually decreases.

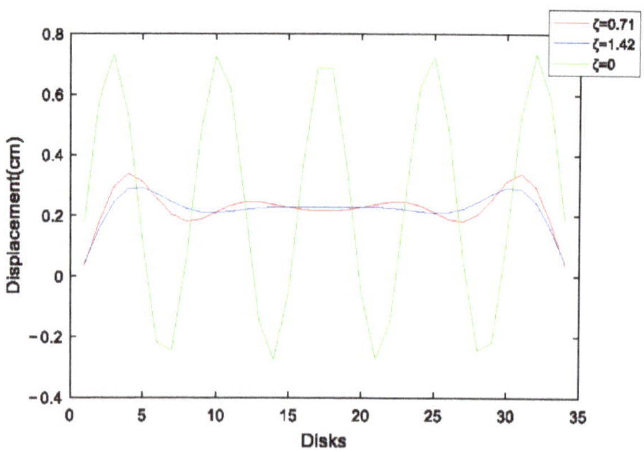

Figure 8. Displacement of each disk under different damping ratios.

When t = 0.15 s, the acceleration change of each winding disk at this time is shown in Figure 9. Similar to the displacement waveform of each winding disk, the radial vibration acceleration waveform of the winding is in the shape of a 'valley', which is also roughly consistent with the vibration mode law of the eighth order mode. The acceleration near both ends of the windings reaches a maximum value, about twice the acceleration of the middle part. With the increase of the damping ratio, the acceleration oscillation amplitude of each disk gradually decreases.

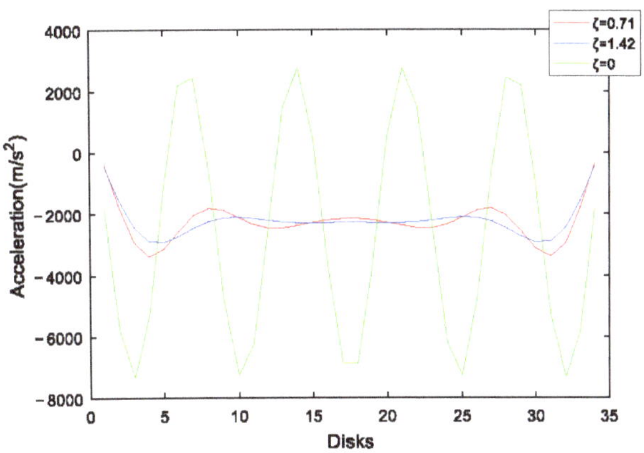

Figure 9. Acceleration of each disk under different damping ratios.

It can be concluded that when the power transformer is in a short circuit, the winding acts as a damped multi-DOF system and the short-circuit electrodynamic force of the winding is 100 Hz. The displacement distribution law of each disk is consistent with the displacement distribution law of the eighth vibration mode. The displacement is the largest near both ends of the winding, and the middle part is relatively gentle. The displacements of the top and end disks are close to zero. The acceleration distribution law of each disk is also consistent with the eighth order vibration mode distribution law, with the peak appearing near both ends of the winding, about twice the acceleration of the middle part, while the acceleration of the disk at the top and end of the winding being relatively small. The amplitude and vibration law of winding vibration are related to the physical properties, current, and frequency of the system itself, rather than the initial conditions.

4. Results

In this paper, the distribution law of winding axial deformation in the process of transformer short circuit is studied; a multi-DOF winding "spring-mass-damping" model considering damping is established; and the variation laws of axial displacement and acceleration when the transformer with different damping ratios is short circuited at different frequencies is analyzed, which provide a new idea for the design of winding to a certain extent, further improving the short-circuit-bearing capacity of a transformer. The main conclusions are as follows:

(1) In this paper, an axial vibration model of the transformer winding, considering the system damping parameters, is built. Compared with the previous V-type and M-type mode shapes, the mode shape structure calculated in this paper is more detailed and refined. At the same time, the displacement and acceleration distributions of the model at different frequencies are illustrated by sweeping the frequency.

(2) Under short-circuit conditions, the acceleration and displacement of the winding are distributed in a "VMV" type, and the displacement and acceleration reach peak values at 1/8 of the upper end and 1/8 of the lower end at the same time, and the displacement and acceleration at both ends of the winding are approaching to zero; in

addition, the numerical distribution of displacement and acceleration in the middle of the winding is relatively flat. With the increase of the damping ratio, the vibration amplitude shows a decaying trend.

(3) According to the calculation model, suggestions for the operation and maintenance of the transformer can be put forward. The positions of the sensors for measuring the vibration acceleration can be located at the two ends, the middle, one quarter, and one-eighth of the winding. In actual operation, attention should be paid to the vibration of the upper and lower ends of the winding and the deformation of the winding. When designing the structure of the transformer winding, it is especially necessary to improve the anti-deformation and stretching ability near both ends of the winding.

In future research, we can further expand the following aspects based on the research in this paper:

(1) Comparative analysis with the actual transformer short-circuit test to verify whether the theory of axial vibration of transformer windings is insufficient.
(2) Further analyze the influence of different short-circuit conditions or transformer overload on the axial vibration.
(3) Considering the response of transformer winding radial vibration under short-circuit conditions, comprehensive axial vibration transformer winding vibration characteristics reflect the transformer winding vibration characteristics more comprehensively.

Author Contributions: Conceptualization, S.G. and L.S.; methodology, S.G., L.S. and Y.T.; validation, L.S. and H.L.; formal analysis, S.G.; investigation, Y.T.; resources, S.G., L.S. and Y.T.; data curation, H.L.; writing—original draft preparation, S.G.; writing—review and editing, Y.T.; visualization, H.L.; supervision, L.S.; project administration, S.G. All authors have read and agreed to the published version of the manuscript.

Funding: This research was funded by Science and Technology Project of State Grid Hebei Electric Power, Fund No. Kj2019-060 and the Natural Science Foundation of Hebei Province, Fund No. E2021521004.

Institutional Review Board Statement: Not applicable.

Informed Consent Statement: Not applicable.

Data Availability Statement: Not applicable.

Conflicts of Interest: The authors declare no conflict of interest.

References

1. Secic, A.; Krpan, M.; Kuzle, I. Vibro-Acoustic Methods in the Condition Assessment of Power Transformers: A Survey. *IEEE Access* **2019**, *7*, 83915–83931. [CrossRef]
2. Yun, Z.; Shaoyi, L.; Xiaobo, H. Analysis of a 220kV transformer sudden short-circuit fault. *Electr. Technol.* **2018**, *19*, 94–98.
3. Zhu, Y.; Ji, S.; Zhang, F.; Liu, Y.; Dong, H.; Cui, Z.; Wu, W. Vibration mechanism and influence factors in power transformers. *J. Xi'an Jiaotong Univ.* **2015**, *49*, 115–125.
4. Mengyun, W. Fault statistics and analysis of transformers with 110 kV or higher voltage in 2004. *Electr. Equip.* **2005**, *6*, 35–41.
5. Tenbohlen, S.; Vahidi, F.; Jagers, J.; Gebauer, J. A Worldwide Transformer Reliability Survey. In Proceedings of the VDE High Voltage Technology 2016; ETG-Symposium, Berlin, Germany, 14–16 November 2016; pp. 1–6.
6. Bagheri, M.; Zollanvari, A.; Nezhivenko, S. Transformer fault condition prognosis using vibration signals over cloud environment. *IEEE Access* **2018**, *6*, 9862–9874. [CrossRef]
7. Yichun, L. Calculation and Stability Analysis of Transformer Winding Short-Circuit Electrodynamic Force. Master's Thesis, Harbin University of Science and Technology, Harbin, China, 2016.
8. Xue, W.; Tao, W. Research on dynamic analysis of transformer winding deformation based on magnetic-structural direct coupling. *Power Syst. Prot. Control* **2017**, *45*, 24–30.
9. Paoletti, G.J.; Herman, G. Monitoring of electrical equipment failure indicators and zero-planned outages: Past, present and future maintenance practices. In Proceedings of the Pulp and Paper Industry Conference, Charlotte, NC, USA, 23–27 June 2013; pp. 1–10.
10. Wenbo, B.; Quan, B.; Haiyan, L. *Vibration Mechanics Foundation and MATLAB Application*; Tsinghua University Press: Beijing, China, 2015; pp. 1–2.
11. Brincker, R.; Ventura, C. *Introduction to Operational Modal Analysis*, 1st ed.; Wiley: West Sussex, UK, 2015.

12. Zhou, H.; Hong, K.; Huang, H.; Zhou, J. Transformer winding fault detection by vibration analysis methods. *Appl. Acoust.* **2016**, *114*, 136–146. [CrossRef]
13. Zhenguang, L.; Chuncheng, W. Research on radial stability of transformer low-voltage windings using large-displacement geometric nonlinear theory. *Transformer* **2008**, *39*, 1–5.
14. Ahn, H.M.; Oh, Y.H.; Kim, J.K.; Song, J.S.; Hahn, S.C. Experimental verification and finite element analysis of short-circuit electromagnetic force for dry-type transformer. *IEEE Trans. Magn.* **2012**, *48*, 819–822. [CrossRef]
15. Yiwei, H. Research on Transformer Winding Fault Diagnosis Method Based on Spatial Vibration Distribution and Working Condition Modal Analysis. Master's Thesis, Zhejiang University, Hangzhou, China, 2020.
16. Yongming, X.; Rong, G.; Hongda, Z. Calculation of short-circuit electrodynamic force of power transformer windings. *J. Electr. Mach. Control.* **2014**, *18*, 36–42.
17. Xiwen, W.; Yongteng, J.; Yongchao, Z.; Yan, L.; Dongxue, L.; Qiang, M. Calculation and Analysis of Dynamic Process of Short-Circuit Winding Deformation in Transformer. *Transformers* **2021**, *58*, 1–6+77.
18. Allan, D.; Sharpley, W. The short circuit performance of trans-formers-a contribution to the alternative to direct testing. In Proceedings of the International Conference on Large High Voltage Electric Systems, Paris, France, 27 August–4 September 1980.
19. Shengchang, J.; Fan, Z.; Guochao, Q.; Yeye, Z.; Hongkui, D.; Dexu, Z. Axial vibration characteristics and influencing factors of transformer windings under steady-state conditions. *High Volt. Technol.* **2016**, *42*, 3178–3187.
20. Hong, K.; Huang, H.; Zhou, J. Winding condition assessment of power transformers based on vibration correlation. *IEEE Trans. Power Del.* **2017**, *32*, 1031–1038. [CrossRef]

applied sciences

MDPI

Article

Fault Diagnosis Method of Intelligent Substation Protection System Based on Gradient Boosting Decision Tree

Wei Ding, Qing Chen *, Yuzhan Dong and Ning Shao

Key Laboratory of Power System Intelligent Dispatch and Control of Ministry of Education, Shandong University, Jinan 250061, China

* Correspondence: qchen@sdu.edu.cn

Abstract: In order to improve the efficiency of the devices' fault diagnosis of the protection systems of intelligent substation, a fault diagnosis method based on a gradient boosting decision tree (GBDT) was proposed. Using the integrated alarm information, the device self-checking information, the link information of generic object-oriented substation event (GOOSE) and sampled value (SV) and the sampling value information generated during the fault of the protection system, the fault feature information set is constructed. According to different fault characteristics, the protection system faults are classified into simple faults and complex faults to improve the diagnosis efficiency. Using GBDT training rules, a fault diagnosis model of protection system based on GBDT is established and fault diagnosis steps are given. This study takes a 110 kV intelligent substation in southern China as an example, to verify the effectiveness and accuracy of the proposed fault diagnosis method, and compared it with the existing methods in terms of the accuracy. The diagnostic accuracy in the case of false alarms and the case of multiple faults are verified. The results show that the method can meet the practical engineering application.

Keywords: protection system; fault diagnosis; complex faults; GBDT

Citation: Ding, W.; Chen, Q.; Dong, Y.; Shao, N. Fault Diagnosis Method of Intelligent Substation Protection System Based on Gradient Boosting Decision Tree. *Appl. Sci.* **2022**, *12*, 8989. https://doi.org/10.3390/app12188989

Academic Editors: Federico Barrero and Mario Bermúdez

Received: 28 July 2022
Accepted: 5 September 2022
Published: 7 September 2022

Publisher's Note: MDPI stays neutral with regard to jurisdictional claims in published maps and institutional affiliations.

1. Introduction

The protection system is an important part of the smart substation, which mainly includes the merging unit, intelligent terminal, and protection device. It is of great significance for the safe operation of the whole power grid to detect the faults or abnormal conditions in the power system, so as to send an alarm signal, or directly isolate and remove the faulty part. Therefore, it is necessary to ensure the reliable operation of the protection system itself. Even if it fails, it should also quickly eliminate the fault and ensure its support for the normal operation of the power grid to the maximum extent. With the rapid development of smart grid technology, intelligent substations or digitally transformed smart stations have gradually become popular. Substations have the new features of intelligent equipment, data networking, and overall station informatization [1]. Compared with the traditional substation, the fault types and fault characteristics of its protection system have changed a lot, and it is necessary to continuously develop new fault diagnosis methods on the basis of traditional fault diagnosis methods [2]. Improvements in data acquisition, storage, and analysis in intelligent substations provide us with new ideas for developing new fault diagnosis methods. The function of the secondary system in the substation is more complete and the amount of monitoring information is larger, for example, the operation event record of the secondary system, the main station information, the key information of the protection device, the integrated alarm information, the network operation information, the device self-checking information, etc., which provide solid data support for the condition assessment and fault diagnosis of the substation. Reference [3] analyzed the operation behavior of the secondary system of intelligent substation, and sorted out 18 core evaluation sub-items, including sampling accuracy, internal environment, self-checking state, protection start, input consistency, real-time input and output, operation time of the whole group, port flow, fault-free time, correct action, correct control, station-control layer communication, process-layer communication, port function, port performance, on-time receiving, and on-time output. A Big Data mining technology for intelligent substations was proposed by Reference [4]. By deeply mining the integrated alarm information, the device self-checking information, the link information of GOOSE and SV, the sampling value information provided data support for fault diagnosis.

With regard to the research on the fault diagnosis of the protection system of the intelligent substation, Reference [5] proposed a method for evaluating the performance of secondary equipment in smart substations based on availability, dependability, and capability (ADC). The accuracy of its evaluation needs to be improved. Reference [6] proposed an online monitoring and fault diagnosis method of the secondary circuit of relay protection based on multi-parameter information. Through the monitoring and analysis of the SV, GOOSE, and manufacturing message specification (MMS) messages, the online state monitoring method, abnormal sampling value, switch monitoring, and abnormal alarm strategy for relay protection devices was proposed. At the same time, the typical alarm information of the protection device and secondary circuit when faults occur was collected, analyzed, and uploaded to construct a set of online monitoring and fault diagnosis systems for a secondary circuit of relay protection. References [7–10] proposed a method of locating the secondary equipment fault based on the substation configuration description (SCD) of the intelligent substation. However, the actual workflow of intelligent substations is highly dependent on the configuration tools of integrators and manufacturers, and the difference in configuration tools leads to a poor standardization of the files. Therefore, there are still many shortcomings in the method of locating faults simply using SCD files. With the research in machine learning and deep learning algorithms, their application in fault diagnosis is increasingly employed [11]. Reference [12] proposed a research method for the fault location of the secondary device in intelligent substations based on deep learning. According to the device self-checking information, a fault location model for the secondary device based on a recurrent neural network (RNN) was established and the fault location steps were given; however, the data source used had certain limitations, and due to the limitations of the time and accuracy of the algorithm, this method needs to be improved. Reference [13] proposed an intelligent state assessment of the protection systems based on random forest algorithm, but the prediction accuracy and robustness need to be improved, and the requirements for the parameters are high.

The above method can basically meet the needs of fault diagnosis, but there are still various problems in its practical application. In summary, the current research on the protection system of intelligent substations still needs to face these problems: There are many types of faults in the protection system, but the correlation between the fault characteristics is weak; the complex components of the protection system equipment and the connection relationship between the different devices cause a large amount of data to be generated when a fault occurs [14]. Conventional methods cannot efficiently and quickly analyze the massive multi-dimensional data; since the fault-feature information may be distorted and lost during the acquisition process, the results obtained by the conventional method fluctuate with the confidence of the fault feature information; In addition, the accuracy of the algorithm also needs to be improved.

In order to solve the problems of data source, data processing, fault diagnosis logic, diagnosis method, and fault accuracy rate faced in fault diagnosis, a new fault diagnosis method of an intelligent substation protection system based on a gradient boosting decision tree is proposed. The GBDT algorithm is a supervised ensemble learning method. Through the continuous iteration of the weak prediction model composed of decision trees, the strong prediction model is trained with the goal of minimizing the prediction errors of the previous round. It has extremely high accuracy and a fast convergence speed. Taking the protection system merging unit, intelligent terminal, and protection device as the main body of fault diagnosis, this method used the integrated alarm information, device self-checking information, link information of GOOSE and SV, and the sampling value information as the judgment basis to form the fault feature information set. According to the historical fault feature data and maintenance records, the faults of the protection system are divided into simple faults and complex faults. At the same time, the gradient boosting decision tree (GBDT) intelligent algorithm is used as a diagnostic tool, and the fault diagnosis process of the protection system is proposed to realize the diagnosis of complex faults of the protection system. The effectiveness of the method proposed in this paper is verified by example analysis.

2. Fault Type and Fault Feature Information of the Protection System

2.1. Classification of Protection System Fault Types

When the fault diagnosis of the protection system is carried out, the fault types are properly classified, which can ensure the accuracy of judgment, reduce the amount of calculation in the process of fault diagnosis, reduce the amount of computer resources, and improve the response speed and convergence speed [12].

By analyzing the alarm information, self-checking information, sampling value information, and fault maintenance-record data of the device of the protection system, we can divide the faults into two categories. One is the simple fault, that is, there is an obvious mapping relationship between the fault type and the fault feature information. After the fault occurs, the fault type can be simply deduced according to the fault feature information. For example, if the fault feature information is "Power failure alarm of merging unit", it can be directly deduced that the fault is "Power module fault of merging unit". Another type is the complex fault, which means that the mapping relationship between the fault type and fault feature information is weak, and cannot be directly deduced by simple reasoning of fault feature information. An intelligent algorithm is needed for the reasoning and diagnosis.

According to the equipment manual, fault data, and fault characteristics, the high-frequency faults are classified as shown in Tables 1 and 2.

Table 1. Simple fault types and corresponding fault feature information.

Number	Fault Types	Fault Feature Information
1	Power module fault of merging unit	Power failure alarm of merging unit
2	The RX1 receive Bus fault of the merge unit	4–8 Bus MU frame Loss/4–8 Bus check error
3	GPS timing signal of merging unit not accessed	B01 synchronous anomaly alarm of merging unit
4	Device board card configuration and specific engineering design drawings do not match	Board card configuration error
5	Error in merging unit memory check	Merging unit memory checking error
6	Incorrect configuration of send text for merge unit	Incorrect configuration of send text for merge unit
7	Errors in HTM data exchange between merging unit board boards	B01 HTM error of merging unit
8	Device stall error of merging unit	BO1 report BO3 Stall error of merging unit
9	B11 input power loss of merging unit	B11 optocoupler power loss alarm of merging unit
10	BO3 sampling plate anomaly of merging unit	BO3 sampling plate anomaly of merging unit
11	The sampling of BO3 dual AD in merging unit is inconsistent	The sampling of BO3 dual AD in merging unit is inconsistent
12	GOOSE-A network storm of merging unit	GOOSE-A network storm of merging unit
13	The sampling voltage of merging unit BO2 drops below the set value	BO2 sampling power drop of merging unit
14	Error in intelligent terminal memory check	Intelligent terminal memory checking error
15	BO1_A network link N of intelligent terminal disconnected	BO1_A network link N of intelligent terminal disconnected

Table 1. *Cont.*

Number	Fault Types	Fault Feature Information
16	BO1_GOOSE-A network storm of intelligent terminal	BO1_GOOSE-A network storm alarm of intelligent terminal
17	BO1_GOOSE configuration error of intelligent terminal	BO1_GOOSE configuration error of intelligent terminal
18	BO2_A network link N of intelligent terminal disconnected	BO2_A network link N of intelligent terminal disconnected
19	GPS timing signal of intelligent terminal not accessed	B01 synchronous anomaly alarm of intelligent terminal
20	Error in HTM data exchange between intelligent terminal device boards	B01 HTM error of intelligent terminal
21	Device stall fault of intelligent terminal	BO1 report BO2 stall error of intelligent terminal
22	The power supply of the intelligent terminal AD chip exceeds the normal range	BO3 sampling power supply anomaly of intelligent terminal
23	B09 input power loss of intelligent terminal	B09 optocoupler power loss alarm of intelligent terminal
24	Abnormal opening of intelligent terminal outlet power supply	Bus QD Signal Anomaly
25	GOOSE text configuration error of intelligent terminal	Link N configuration error
26	GPS input signal of intelligent terminal is lost or abnormal	GPS clock abnormal
27	Abnormal timing of the protection device	Abnormal timing of the protection device
28	Spring device of protection device does not store energy	No energy storage alarm for spring device of protection device
29	CT disconnection of protection device	CT disconnection alarm
30	PT disconnection of protection device	PT disconnection alarm
31	Synchronous voltage circuit disconnection	Abnormal synchronous voltage

Table 2. Complex fault types and corresponding main fault feature information.

Number	Fault Types	Fault Feature Information
f_1	Main DSP module failure of merging unit	Sampling anomaly of merging unit, Synchronization anomaly of merging unit, SV alarm of merging unit/protection device, Protection locking, etc.
f_2	Sampling DSP module failure of merging unit	
f_3	I/O module fault of merging unit (Merging unit ↔ Switch)	SV alarm of protection device, SV interruption of protection device/merging unit/measurement and control device, GOOSE interruption of merging unit, SV/GOOSE total alarm, Protection locking, etc.
f_4	I/O module fault of merging unit (Bus merging unit → Line merging unit)	
f_5	I/O plug – in fault of protection device (Protection device ↔ Intelligent terminal)	SV/GOOSE alarm of protection device, GOOSE alarm of intelligent terminal, GOOSE interruption of protection device, Protection locking, Reclosing device locking, etc.
f_6	I/O plug – in fault of protection device (Switch → Protection device)	
f_7	I/O plug – in fault of protection device (Merging unit → Protection device)	
f_8	Longitudinal channel fault	Channel differential exit, Longitudinal channel abnormal, In/out communication interruption of protection device, etc.
f_9	CPU plug-in fault of protection device	Protection device parameter error, CPU X exception, Memory self-check error, Protection locking, etc.
f_{10}	I/O board fault of intelligent terminal (Intelligent terminal ↔ Switch)	GOOSE interruption of merging unit/measurement and control device/intelligent terminal/protection device, GOOSE alarm of merging unit/intelligent terminal/protection device, etc.
f_{11}	I/O board fault of intelligent terminal (Protection device ↔ Intelligent terminal)	
f_{12}	GOOSE plug-in fault of protection device	SV/GOOSE alarm of protection device, GOOSE alarm of intelligent terminal, SV/GOOSE interruption of protection device, Protection locking, Reclosing lock, etc.

Set the complex fault set as Formula (1):

$$F = \{f_1, f_2, \ldots, f_{12}\} \tag{1}$$

In Formula (1), $f_1 - f_{12}$, respectively, represent the 12 faults in Table 2.

For the simple faults in Table 1, the expert system can be used for fault diagnosis according to the fault feature information summarized in Table 1. For space reasons, the performance of the method proposed in this paper is explored based on the faults in the complex fault set F. In addition, with the development of intelligent substations and the improvement in field complexity, the fault set F will be further expanded, and the method proposed in this paper is still applicable to subsequent faults.

2.2. Fault Feature Information of the Protection System

Based on the complex fault types of the protection system summarized in Table 2, this paper selects four features of integrated alarm information, device self-checking information, link information of GOOSE and SV, and sampling value information as the feature information of fault diagnosis, which can comprehensively reflect the change in feature quantity caused by the fault of the protection system [15].

The main function of the integrated alarm information is to reflect whether the protection system fails. If a fault occurs, the equipment will issue alarm information and upload it to the monitoring terminal, which can be used as one of the bases for equipment fault diagnosis while realizing fault warning.

Device self-checking is an important function of an intelligent protection system. When any abnormality occurs in the operation process of the device, the device will record the abnormal information through the event-recording function for the operator to query.

GOOSE (Generic Object-Oriented Substation Event) is equivalent to the DC control and signal cables in traditional substations, which transmit control instructions and signals. It mainly includes a switch/knife switch position, control switch position, abnormal/alarm signal, blocking signal, etc. SV (Sampled Value) is equivalent to the secondary AC cable in the traditional substation, which transmits the sampled instantaneous values of voltage and current, including the instantaneous value of voltage and current on the secondary side of the transformer. The link information of GOOSE and SV are important indicators to indicate whether the information links between the protection system equipment and between the equipment and the monitoring terminal work normally, reflecting the link connection state of the equipment.

The sampling value information is the sampling value of three-phase voltage and current transmitted by two channels, which can reflect whether the voltage and current-sampling function of the protection system are normal.

The protect system fault feature information, as shown in Table 3:

2.3. Fault Feature Information Set of the Protection System

According to the fault feature information in Table 3, the fault feature information set is established to provide data support for the subsequent fault diagnosis of the protection system of an intelligent substation.

The integrated alarm information set A_i of the protection system of an intelligent substation in the i-th fault event is established as shown in Formula (2):

$$A_i = \{a_1, a_2, a_3, \ldots, a_{11}\} \tag{2}$$

a_1–a_{11} in the above formula are the 11 kinds of fault feature information contained in the integrated alarm information in Table 3. When the monitoring host receives the alarm information, the element at the corresponding position is set to 1, otherwise it is set to 0.

Table 3. Protect system fault feature information.

Type of Fault Feature Information	Fault Feature Information
Integrated alarm information	Protection locking, Abnormal alarm of merging unit, Abnormal alarm of intelligent terminal, Abnormal alarm of protection device, SV total alarm, GOOSE total alarm, SV alarm of merging unit, SV alarm of protection device, GOOSE alarm of merging unit, GOOSE alarm of protection device, GOOSE alarm of intelligent terminal
Link information of GOOSE and SV	GOOSE interruption of merging unit, GOOSE interruption of protection device, GOOSE interruption of intelligent terminal, SV interruption of merging unit, SV interruption of protection device, SV interruption of measurement and control device
Device self-checking information	Sampling anomaly of merging unit, Synchronization anomaly of merging unit, Board card configuration error, Merging unit memory checking error, Incorrect configuration of send text for merge unit, B01 HTM error of merging unit, BO1 report, BO3 Stall error of merging unit, BO3 sampling plate anomaly of merging unit, Sampling of BO3 dual AD in merging unit is inconsistent, GOOSE-A network storm of merging unit, BO2 sampling power drop of merging unit, Intelligent terminal, Memory-checking error, B09 optocoupler power loss of intelligent terminal, BO1_A network link N of intelligent terminal disconnected, BO1_GOOSE-A network storm alarm of intelligent terminal, BO1_GOOSE configuration error of intelligent terminal, BO2_A network link N of intelligent terminal disconnected, B01 synchronous anomaly alarm of intelligent terminal, B01 HTM error of intelligent terminal, BO1 report BO2 stall error of intelligent terminal, BO3 sampling power supply anomaly of intelligent terminal, Channel differential exit, Longitudinal channel abnormal, Protection device parameter error, Abnormal timing of protection device, In/out communication interruption of protection device, Abnormal synchronous voltage, CPU X exception, Memory self-check error, 4–8 Bus MU frame loss, 4–8 Bus check error
Sampling value information	Double-channel voltage sampling value, Double-channel current sampling value

The link information of the GOOSE and SV set I_i of the protection system of an intelligent substation in the i-th fault event is established as shown in Formula (3):

$$I_i = \{i_1, i_2, i_3, \ldots, i_6\} \tag{3}$$

i_1–i_6 in the above formula are the 6 kinds of fault feature information contained in the link information of GOOSE and SV in Table 3. When the secondary monitoring system receives the alarm information, the element at the corresponding position is set to 1, otherwise it is set to 0.

The device self-checking information set C_i of the protection system of an intelligent substation in the *i*-th fault event is established as shown in Formulas (4)–(7):

$$C_i = \{C_{MU}, C_P, C_{IT}\} \tag{4}$$

$$C_{MU} = \{C_{MU1}, C_{MU2}, \ldots, C_{MUa}\} \tag{5}$$

$$C_P = \{C_{P1}, C_{P2}, \ldots, C_{Pb}\} \tag{6}$$

$$C_I = \{C_{I1}, C_{I2}, \ldots, C_{Ic}\} \tag{7}$$

In the above Formula (4), C_i contains the device self-checking information in Table 3, and it is divided into three parts: merging unit self-checking information C_{MU}, protection device self-checking information C_P, and intelligent terminal self-checking information C_{IT}, where Formulas (5)–(7) subscripts a, b, and c represent the number of these three types of device in the protection system of an intelligent substation. When the secondary monitoring system receives the alarm information, the element in the corresponding position is set to 1, otherwise, it is set to 0.

The sampling value information set S_i of the protection system of an intelligent substation in the *i*-th fault event is established as shown in Formula (3).

$$\begin{cases} S_i = \{M_1, M_2\} \\ M_1 = \{I_{A1}, I_{B1}, I_{C1}, U_{A1}, U_{B1}, U_{C1}\} \\ M_2 = \{I_{A2}, I_{B2}, I_{C2}, U_{A2}, U_{B2}, U_{C2}\} \end{cases} \tag{8}$$

In the above Formula (8), M_1 and M_2 represent the three-phase voltage and current sampling values in Channel 1 and Channel 2, respectively. *I* and *U* represent the three-phase current and voltage values of the dual channel.

To make the sample data of the different units comparable, improve the convergence speed of the model, and improve the accuracy of the model, the sampling value information is preprocessed by the Min-Max method, and the original value S_i in the dataset is mapped to the value $S_i{}'$ in the interval [0,1]. The conversion formula is shown in Formula (9):

$$S_i{}' = \frac{S - S_{min}}{S_{max} - S_{min}} \tag{9}$$

In the Formula (9), S_{max} and S_{min} are the maximum and minimum values of the sampled values, respectively.

3. Fault Diagnosis of the Protection System Based on Gradient Boosting Decision Tree (GBDT)

The gradient boosting decision tree intelligent algorithm belongs to the ensemble algorithm, which has a good processing ability for discrete data and is very prominent in dealing with small sample data. Gradient boosting is the core idea and step of this intelligent algorithm for the classification task. When it carries out 'multi-classification' work, it is based on 'two classifications' and adopts the idea of 'one positive class, multiple negative classes'. The training process of the gradient boosting decision tree intelligent algorithm is the main work of this section, which mainly includes selecting the optimal value of the learning rate and the number of iterations, and finally gives the fault diagnosis process of the protection system based on the gradient boosting decision tree.

3.1. *Principle and Training Steps of Gradient Boosting Decision Tree*

GBDT intelligent algorithm belongs to the ensemble learning algorithm. The ensemble learning algorithm is a hot topic in the field of engineering applications. It is a method to improve the learning ability through the combination of multiple weak learners [16]. Compared with conventional methods, it has a good performance in terms of accuracy and generalization ability. Bagging and Boosting algorithms are two typical ensemble learning algorithms. The schematics are shown in Figures 1 and 2.

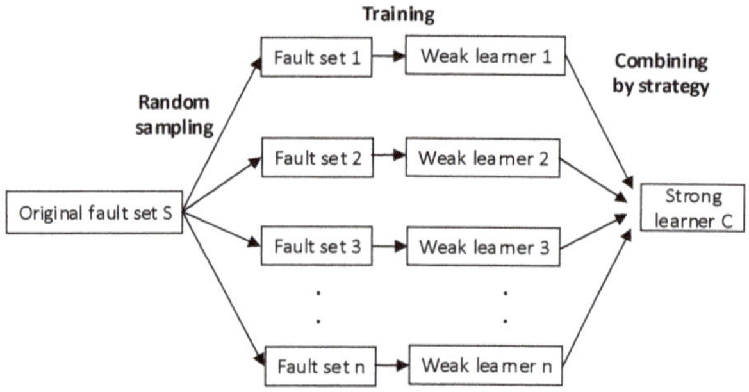

Figure 1. General schematic diagram of bagging algorithm.

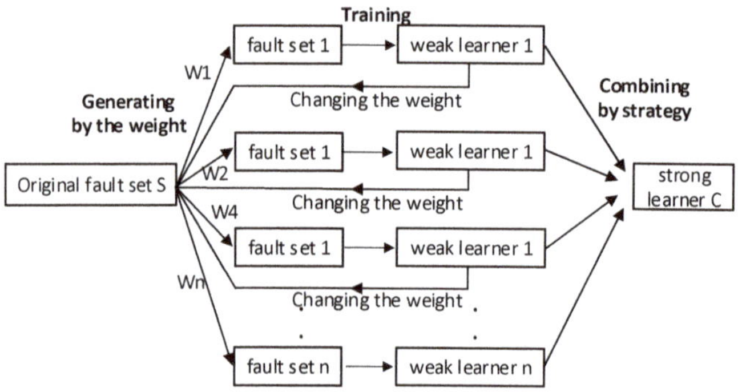

Figure 2. General schematic diagram of boosting algorithm.

The bagging algorithm generates n training sample sets from the total sample library according to the random sampling method with playback. Each sample set trains a weak learner and uses the sample set to train n weak learners. The weak learners run in parallel. According to different combination strategies, n weak learners are combined to generate strong learners. The boosting algorithm is an inherited algorithm, in which the weak learners operate in a serial manner. The data weight in the training set of each iteration is changed by the learning results of the weak learners. The learning results are fitted according to the residuals, and then n weak learners are combined according to different combinations to generate strong learners [17].

The gradient boosting decision tree (GBDT) intelligent algorithm is one of the most widely used boosting algorithms in the engineering field, and it combines the sampling idea of the bagging algorithm, allowing sampling samples and features to increase the independence between weak learners. The GBDT intelligent algorithm does not change the sample weight in the iteration process, but continuously learns the negative gradient of the loss function, generates multiple new weak learners, and combines multiple weak learners into strong learners. Compared with the traditional machine learning algorithm, the GBDT intelligent algorithm can achieve higher accuracy in many of the application scenarios, and has a faster operation speed, stronger generalization ability, and lower requirements for parameter adjustment.

Regardless of whether the GBDT intelligent algorithm performs regression tasks or classification tasks, its core idea is "gradient boosting", and the negative gradient of the loss function in the iterative process is shown in Formula (10):

$$\widetilde{y}_i = -\frac{\partial L(y_i, F(x_i))}{\partial F(x_i)},\ i = 1,2,\ldots,\mathrm{K} \tag{10}$$

In the above formula, $\widetilde{y}_i$ is the negative gradient of the current loss function, namely the fitting target of the next iteration, $L(y_i, F(x_i))$ is the current loss function, y_i is the learning target value of the current weak learner, $F(x_i)$ is the output value of the current weak learner, x is the input variable (refers to the fault feature information in this paper), K is the number of training samples, and i is the current training sample.

GBDT multi-classification is an organic combination of the GBDT binary classifier. In the training process, the idea of 'one positive class, multiple negative classes' is adopted. There are 12 kinds of fault types in the fault set. In the training process of a single sample, when the sample is the i-th type of fault ($i = 1, 2, \ldots, 12$), it is assumed that the fault type of this sample is from 1 to 12, and each time it is assumed that the other 11 types of fault are unified as the negative samples of this sample, and 12 binary classifiers are trained to generate independent classification. Then, according to the real fault type of the sample, only the output result of one binary classifier (assuming that the i-th class is a positive class) is correct, and the rest are errors, then the final classification result is the fault type corresponding to this binary classifier. The schematic diagram of the GBDT multi-classification algorithm is shown in Figure 3.

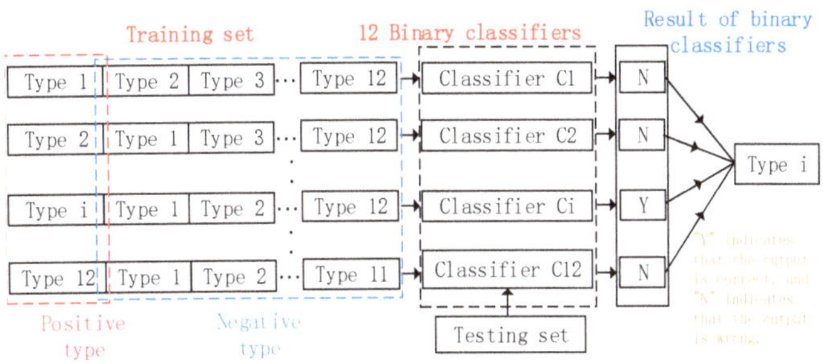

Figure 3. Schematic diagram of GBDT multi-classification algorithm.

In this paper, when conducting 12-classification training for complex faults in the protection systems of intelligent substations, the fault feature information of the protection system in Table 3 is used as a variable, and the complex fault types of the protection system in Table 2 are used as the fitting target. The training process is as follows:

Step 1 Select the fault sample i in the training set. The fault feature information set of this sample is = $X_i = \{A_i, I_i, C_i, S_i{}'\}$, and the fault type is f_1—"Main DSP module failure of merging unit" in Table 2. Then, the true classification label (probability) of the fault sample in the 12 binary classifiers is the fitting target y = (1, 0, 0, 0, 0, 0, 0, 0, 0, 0, 0, 0). Because the fault sample belongs to the first fault f_1, '1' is used to indicate that it belongs to the fault, and '0' is used to indicate that it does not belong to the other 11 faults, forming the input consisting of the fault feature information set and the fitting target: $(X_i,1)(X_i,0)(X_i,0) \ldots (X_i,0)$;

Step 2 Input the input into 12 weak classifiers and obtain the output results: $F_1(X_i)$, $F_2(X_i)$, $F_3(X_i), \ldots, F_{12}(X_i)$;

Step 3 Convert the output result into probability, as shown in Formula (11):

$$p_n(X) = \frac{\exp(F_n(X_i))}{\sum_{k=1}^{12} \exp(F_n(X_i))},\ n = 1,\ 2,\ 3,\ \ldots,\ 12 \tag{11}$$

Step 4 Calculate the loss function and solve the negative gradient of the loss function. The loss function formula is shown in Formula (12):

$$L(y, p(X)) = -\sum_{n=1}^{12} y_n \log p_n(X),\ n = 1,\ 2,\ 3,\ \ldots,\ 12 \tag{12}$$

The negative gradient formula of the loss function is shown in Formula (13):

$$\widetilde{y}_n = -\frac{\partial L(y, p(X))}{\partial F_n(X_i)}, \ n = 1, 2, 3, \ldots, 12 \quad (13)$$

Step 5 Generate a new fitting target, namely $(X_i, \widetilde{y}_1)(X_i, \widetilde{y}_2)(X_i, \widetilde{y}_3) \ldots (X_i, \widetilde{y}_{12})$, and repeat steps 2 to 5.

Iterate M times according to the above steps, and after generating M weak learners, one training is completed. The training set contains 12 kinds of complex faults in the fault set and has several corresponding training samples. After each sample of the training set is trained once, the GBDT 12 classification model with a high accuracy can be obtained. It should be noted that the values of the number of iterations M and the learning rate σ require multiple verifications, and this process is described in detail in Section 3.2.

3.2. Training of Fault Diagnosis Model Based on Gradient Lifting Tree

The GBDT model is trained by using the fault information of the protection system of a typical 110 KV intelligent substation in southern China. Figure 4 shows the information topology between the devices of the protection system of intelligent substation. The protection system includes a line merging unit, a line protection device, a Bus protection device, and an intelligent terminal. The GOOSE/SV message-receiving form between devices is shown in Table 4.

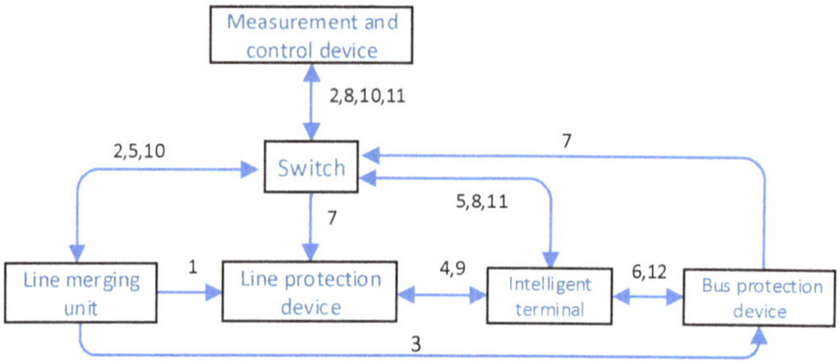

Figure 4. Information topology structure of intelligent substation.

Table 4. Interval information flow in intelligent substations.

Message Number	Transmit Port	Receive Port	Form of Message Transmission
1	Line merging unit	Line protection device	Point-to-point SV Communication
2	Line merging unit	Measurement and control device	Networking SV Communication
3	Line merging unit	Bus protection device	Point-to-point SV Communication
4	Intelligent terminal	Line protection device	Point-to-point GOOSE Communication
5	Intelligent terminal	Line merging unit	Networking GOOSE Communication
6	Bus protection device	Intelligent terminal	Point-to-point GOOSE Communication
7	Bus protection device	Line protection device	Networking GOOSE Communication
8	Intelligent terminal	Measurement and control device	Networking GOOSE Communication
9	Line protection device	Intelligent terminal	Point-to-point GOOSE Communication
10	Measurement and control device	Line merging unit	Networking GOOSE Communication
11	Measurement and control device	Intelligent terminal	Networking GOOSE Communication
12	Intelligent terminal	Bus protection device	Point-to-point GOOSE Communication

Select 4200 actual fault samples of this intelligent substation, and the distribution of fault types in the samples is shown in Table 5.

Table 5. Distribution of fault samples.

Fault Types	Number of Samples	Number of Training Samples	Number of Test Samples
f_1	351	316	35
f_2	371	334	37
f_3	347	312	35
f_4	344	310	34
f_5.	345	310	35
f_6	353	318	35
f_7	352	317	35
f_8	348	313	35
f_9	346	311	35
f_{10}	354	319	35
f_{11}	340	306	34
f_{12}	349	314	35

The GBDT fault diagnosis model is trained by using the fault sample set of the protection system, with 75% of the samples as the training set and 25% as the test set. Use the training set to train the model according to the steps in Section 3.1. Taking the diagnostic accuracy of the test sample set as the optimization index, the model is optimized by adjusting the learning rate σ and the number of iterations M, because these two parameters have the greatest impact on the accuracy of the model. The training results are shown in Table 6.

Table 6. Diagnostic accuracy of GBDT under different parameters.

M \ σ	0.05	0.1	0.3	0.5	0.7	1
10	0.97619	0.98333	0.98095	0.98095	0.97857	0.97381
30	0.97857	0.99048	0.98333	0.98095	0.97857	0.97142
50	0.97619	0.98809	0.98571	0.98333	0.97857	0.97381
70	0.97619	0.98809	0.98809	0.98333	0.98095	0.97619
100	0.97619	0.98571	0.98571	0.98571	0.98095	0.97857

It can be seen from Table 6 that the accuracy of the GBDT model for fault diagnosis of the protection system is quite high, and when the number of iterations is 30 and the learning rate is 0.1, the accuracy is the highest, reaching 99.048%. The specific diagnosis results of the test set samples at this time are shown in Table 7.

Table 7. Diagnostic results of test set at σ = 0.1, M = 30.

	f_1	f_2	f_3	f_4	f_5	f_6	f_7	f_8	f_9	f_{10}	f_{11}	f_{12}
Number of samples	35	37	35	34	35	35	35	35	35	35	34	35
Number of correct diagnoses	35	37	34	33	35	34	34	35	35	35	34	35

Compared with the existing research methods, such as recurrent neural network (RNN) [12] and random forest algorithm (RF) [13], the fault diagnosis accuracy under the same dataset is shown in Table 8:

Table 8. Comparison of diagnostic accuracy of different models.

Model	Diagnostic Accuracy
GBDT	0.9905
[12]	0.9871
[13]	0.9609

It can be seen from Table 8 that the GBDT algorithm has the highest accuracy compared with the other two algorithms when dealing with the same dataset due to its excellent performance on small sample sets, and GBDT has fewer iterations in training, faster-running speed, and training process.

To explore the influence of the number of samples in the training set on the accuracy of the model and compare the accuracy of the three methods, according to the distribution ratio of fault samples in Table 5, the number of samples in the training set is changed for training, and the test results are shown in Figure 5

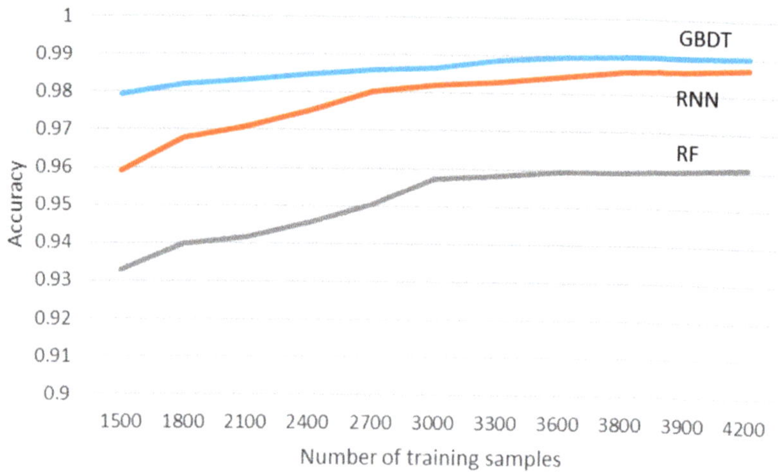

Figure 5. Accuracy of models with different training set samples.

Figure 5 shows that when the number of samples in the training set reaches 3800, the accuracy of the model reaches 99%, and with the increase in the number of samples in the training set, the accuracy of the model does not improve much. Therefore, in practical application, higher accuracy can be achieved when the number of samples in the training set reaches 3800.

3.3. Fault Diagnosis Process of the Protection System

Based on the above content, the fault diagnosis process of the protection system of an intelligent substation based on the gradient boosting decision tree is constructed, as shown in Figure 6.

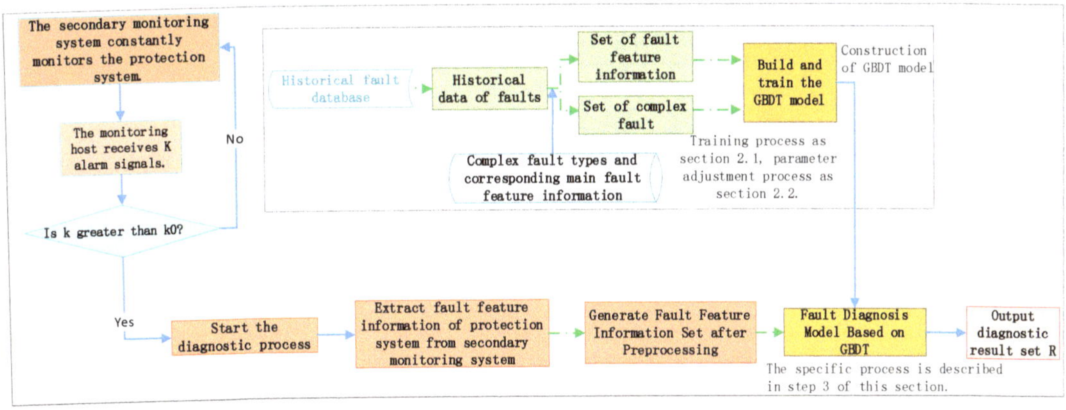

Figure 6. Fault diagnosis process of the protection system.

The specific steps are:

Step 1 To avoid the false start of the diagnosis process, set the minimum number k_0 of alarm messages within 30 s after receiving the first alarm message. When the number of alarm messages received by the secondary monitoring system within the specified time is greater than or equal to k_0, the fault diagnosis of the protection system of this intelligent substation is triggered. The intelligence and integration of the secondary system make it produce a lot of alarm messages when a fault occurs. When the maintenance personnel repair the equipment incorrectly or the equipment is disturbed by environmental factors, alarm information is also generated, but the alarm information is single and small in number. In this case, the fault diagnosis of the protection system should not be started. To avoid the false start of diagnosis, according to the actual fault data analysis and field experience of the intelligent substation, within 30 s after the first alarm message appears, whether the number of alarm messages received by the secondary equipment monitoring system is greater than or equal to k_0 is the trigger diagnosis condition, and set $k_0 = 3$;

Step 2 If the number of alarm messages is greater than k_0, the fault feature information of the protection system of the intelligent substation is extracted to form a set of fault feature information. The feature information in Table 3 is collected, including the integrated alarm information, link information of GOOSE and SV, device self-checking information, and sampling value information from the secondary monitoring system. After data processing, the fault feature information set $X = \{A, I, C, S'\}$ in Section 2.3 is generated, which prepares the data for the fault diagnosis of the protection system of this intelligent substation;

Step 3 Input the processed fault feature information set $X = \{A, I, C, S'\}$ into the fault diagnosis system based on GBDT for diagnosis. The specific process is: Input X into the binary classifiers in the GBDT model, respectively, and calculate the probability $P_{f_i}(i = 1, 2, \ldots)$ that this fault belongs to each complex fault. The one with the highest probability determines that the fault belongs to this type of complex fault and outputs the diagnostic result set $R = \{f_1, f_2, \ldots\} = \{1, 0, \ldots\}$ (Suppose the fault is f_1).

4. Case Analysis

To verify the effectiveness of the fault diagnosis method based on the gradient boosting decision tree proposed in this paper, some fault cases are selected from the historical fault data of the intelligent substation described in Section 3.2 for analysis. According to the three special cases of complete fault feature information, false alarm, and multiple faults, the method in this paper is used for fault diagnosis. Finally, the effectiveness and general applicability of the method are proved.

4.1. Cases with Complete Fault Feature Information

Taking f_4-"I/O module fault of merging unit (Bus merging unit → Line merging unit)" as an example, when the fault occurs, the fault feature information is shown in Table 9:

Table 9. Fault feature information of f_4.

Fault Type	Fault Feature Information
I/O module fault of merging unit (Bus merging unit → Line merging unit)	Abnormal alarm of merging unit SV total alarm GOOSE total alarm SV interruption of merging unit GOOSE interruption of merging unit SV interruption of protection device SV alarm of protection device GOOSE alarm of intelligent terminal Protection locking

Establish the integrated alarm information set A as shown in (14):

$$A = \{a_1, a_2, a_3, \ldots, a_{16}\} = \{1, 1, 1, 0, 1, \ldots, 0\} \tag{14}$$

Establish the link information of GOOSE and SV set I as shown in (15):

$$I = \{I_1, I_2, I_3, \ldots, I_9\} = \{1, 0, 0, 1, 1, \ldots, 0\} \tag{15}$$

Establish device self-checking information set I as shown in (16):

$$\begin{cases} C = \{C_{MU}, C_P, C_{IT}\} \\ C_{MU_{ZT}} = \{0, 0, \ldots, 0\} \\ C_{P_{ZT}} = \{0, 0, \ldots, 0\} \\ C_{IT_{ZT}} = \{0, 0, \ldots, 0\} \end{cases} \tag{16}$$

The voltage and current sampling values are shown in (17):

$$\begin{cases} S = \{M_1, M_2\} \\ M_1 = \{I_{A1}, I_{B1}, I_{C1}, U_{A1}, U_{B1}, U_{C1}\} \\ M_2 = \{I_{A2}, I_{B2}, I_{C2}, U_{A2}, U_{B2}, U_{C2}\} \end{cases} = \begin{cases} S = \{M_1, M_2\} \\ M_1 = \begin{Bmatrix} 3.023, 3.028, 3.016, \\ 57.488, 57.558, 57.549 \end{Bmatrix} \\ M_2 = \begin{Bmatrix} 2.969, 2.958, 2.894, \\ 57.402, 58.218, 56.473 \end{Bmatrix} \end{cases} \tag{17}$$

After preprocessing by the Min-Max method, the sampling data are transformed into (18):

$$\begin{cases} S' = \{M_1', M_2'\} \\ M_1' = \begin{Bmatrix} 0.9781, 0.9768, 0.9729, \\ 0.9810, 0.9822, 0.9821 \end{Bmatrix} \\ M_2' = \begin{Bmatrix} 0.9577, 0.9542, 0.9335, \\ 0.9796, 0.9935, 0.9637 \end{Bmatrix} \end{cases} \tag{18}$$

After the fault occurs, a total of 9 alarm messages are received within 30 s after receiving the alarm message of "abnormal alarm of the merging unit ", which is greater than 3, so the fault diagnosis of the protection system of the intelligent substation is triggered. The fault feature set $X = \{A, I, C, S'\}$ is constructed by calling fault feature information and processing data, as shown in Formulas (14)–(18). Input into the GBDT-based fault diagnosis system for diagnosis, input *X* into the 12 binary classifiers in the GBDT algorithm, respectively, and calculate the probability P as shown in Table 10.

Table 10. The output probability of the binary classifier under f_4 fault.

	f_1	f_2	f_3	f_4	f_5	f_6	f_7	f_8	f_9	f_{10}	f_{11}	f_{12}
P	0.02	0.09	0.41	0.96	0.03	0.04	0.11	0.09	0.14	0.14	0.11	0.04

It can be seen that this fault is most likely to be f_4, and the output diagnostic result set R $= \{f_1, f_2, f_3, f_4, \ldots\} = \{0, 0, 0, 1, \ldots\}$, then it is considered that this fault belongs to the "I/O module fault of merging unit (Bus merging unit → Line merging unit)", so the diagnosis is correct.

To verify whether the method in this paper has a generally high accuracy under the condition of complete fault feature information, 1125 groups of samples that have not been used were selected from historical fault samples. The fault diagnosis is carried out by this method. The steps are shown in the above, and the distribution of fault samples and diagnosis results are shown in Table 11.

Table 11. Fault sample distribution and diagnosis results when fault feature information is complete.

	f_1	f_2	f_3	f_4	f_5	f_6	f_7	f_8	f_9	f_{10}	f_{11}	f_{12}
Number of samples	97	89	96	95	97	88	96	96	89	94	95	93
Number of correct diagnoses for GBDT	96	86	96	94	95	87	96	96	89	94	95	93

The final accuracy rate is 99.29%, which has high accuracy under the condition of complete fault feature information.

4.2. Cases of False Alarms in Fault Feature Information

The fault feature information of the protection system of intelligent substation may be misreported or missed, especially the integrated alarm information, the link information of GOOSE and SV, and the device self-checking information. The sampling value information is obtained by the dual-channel method, and the reliability is high, with almost no false positives or omissions. To improve the diagnostic accuracy in the case of falsely reported fault feature information, this paper extends the training set by adding an appropriate proportion of noise information in the feature information of the fault sample-set data to enhance the generalization ability of the model and reduce the influence of interference information on the final diagnosis. This process was completed in the training process of Section 3.2. The verification shows that when the proportion of noise in each feature information reaches 5%, the generalization ability of the model can meet the requirements.

To verify the reliability of the method proposed in this paper when the information is falsely reported, randomly select a piece of information from the fault feature information *A* or *C* or *I* of the case in Section 4.1 to invert (the original "1" is set to "0" or the original "0" set to "1") to simulate the situation that the fault feature information is falsely reported. In this case, the "SV alarm of protection device" is set to "0" to simulate information missed, and the "Sampling anomaly of merging unit" is set to "1" to simulate information misreported. The diagnosis steps are the same as those in Section 4.1, and the output result set R $= \{f_1, f_2, f_3, f_4, \ldots, f_{12}\} = \{0, 0, 0, 1, \ldots, 0\}$. The fault is diagnosed as the fault of "I/O module fault of merging unit (Bus merging unit → Line merging unit)", and the diagnosis is correct.

To verify whether the proposed method has a generally high accuracy in the case of false alarms of fault feature information, 120 sets of fault data are selected from the test samples in Section 4.1. These 120 sets of data can obtain the correct diagnosis results through the fault diagnosis of the protection system of an intelligent substation based on GBDT. In each sample of fault feature information, one, two, or three information samples are randomly selected to invert, respectively, to simulate the situation that the fault feature information has one, two, or three false reports, and provides a comparison of the diagnostic results of GBDT with RNN and RF for one false report. According to the diagnosis process in Section 4.1, the 120 faults are re-diagnosed, and the results are shown in Tables 12–14.

Table 12. Fault sample distribution and diagnosis results when one false alarm occurs.

	f_1	f_2	f_3	f_4	f_5	f_6	f_7	f_8	f_9	f_{10}	f_{11}	f_{12}
Number of samples	10	10	10	10	10	10	10	10	10	10	10	10
Number of correct diagnoses for GBDT	9	10	10	9	10	10	10	10	10	9	10	10
Number of correct diagnoses for RNN	7	8	9	9	9	10	9	9	8	7	9	10
Number of correct diagnoses for RF	5	6	9	9	10	10	9	7	9	6	9	9

Table 13. Fault sample distribution and diagnosis results when two false alarms occur.

	f_1	f_2	f_3	f_4	f_5	f_6	f_7	f_8	f_9	f_{10}	f_{11}	f_{12}
Number of samples	10	10	10	10	10	10	10	10	10	10	10	10
Number of correct diagnoses for GBDT	9	8	10	9	9	9	10	10	9	9	10	10

Table 14. Fault sample distribution and diagnosis results when three false alarms occur.

	f_1	f_2	f_3	f_4	f_5	f_6	f_7	f_8	f_9	f_{10}	f_{11}	f_{12}
Number of samples	10	10	10	10	10	10	10	10	10	10	10	10
Number of correct diagnoses for GBDT	8	8	9	8	9	7	10	9	9	7	10	10

It can be seen from Tables 12–14 that when the fault feature information is falsely reported, the GBDT model has a very strong anti-overfitting ability. When there are one and two false alarms in the fault feature information, the diagnostic accuracy is high, which is 97% and 92%, respectively. When the false alarm information reaches three, the diagnostic accuracy is 84%. Therefore, this method has a high accuracy when the number of false alarms is less than or equal to three. With the increasing number of false alarms, the lack of feature information will seriously affect the diagnosis of GBDT.

4.3. Case of Multiple Faults

Due to the high integration of the device of the protection system, the fault of a component in the merging unit, protection device, and intelligent terminal is likely to cause other component faults in the current device. To verify the accuracy of the GBDT diagnostic model under multiple faults, the multiple fault types shown in Table 15 are considered.

Table 15. Type of multiple fault.

Number	Complex Faults Included
f_{13}	Main DSP module failure of merging unit Sampling DSP module failure of merging unit
f_{14}	I/O module fault of merging unit (Merging unit ↔ Switch) I/O module fault of merging unit (Bus merging unit → Line merging unit)
f_{15}	I/O plug − in fault of protection device (Protection device ↔ Intelligent terminal) I/O plug − in fault of protection device (Switch → Protection device) I/O plug − in fault of protection device (Merging unit → Protection device)
f_{16}	I/O board fault of intelligent terminal (Intelligent terminal ↔ Switch) I/O board fault of intelligent terminal (Protection device ↔ Intelligent terminal)

Four types of fault samples in Table 15 are selected from the sample set for training based on the GBDT model formed in Section 3.2. The sample distribution and diagnosis results are shown in Table 16.

Table 16. Multiple fault sample distribution and diagnosis results.

Fault Type	Number of Training Samples	Number of Test Samples	Number of Correct Diagnoses for Test Samples
f_{13}	75	24	23
f_{14}	79	25	24
f_{15}	74	24	20
f_{16}	80	25	23

It can be seen from Table 16 that the diagnostic accuracy of this method for multiple faults is as high as 91.8367%. Considering the low probability of multiple faults in the actual operation environment, this method has a reliable fault diagnosis ability.

Taking f_{13}—"Main DSP module failure of merging unit, Sampling DSP module failure of merging unit" as an example, when the fault occurs, the fault feature information is shown in Table 17.

Table 17. Fault feature information of this case.

Fault Type	Fault Feature Information
Main DSP module failure of merging unitSampling DSP module failure of merging unit	Sampling anomaly of merging unit, Synchronization anomaly of merging unit, SV alarm of merging unit/protection device, Protection locking, etc.

The diagnosis steps are the same as those in Section 4.1, and the output result set $R = \{f_1, f_2, f_3, f_4, \ldots, f_{13}, f_{14}, f_{15}, f_{16}\} = \{0, 0, 0, 0, \ldots, 1, 0, 0, 0\}$. The fault is diagnosed as the fault of "Main DSP module failure of merging unit, Sampling DSP module failure of merging unit", and the diagnosis is correct.

5. Conclusions

This paper sorts out the common faults of the protection system and proposes simple fault types and corresponding fault feature information and complex fault types and corresponding main fault feature information. The integrated alarm information, link information of GOOSE and SV, device self-checking information, and sampling value information that can be used as fault feature information of the protection system of an intelligent substation are sorted out to form a set of fault feature information. The model parameter adjustment of GBDT is completed according to the fault data. The fault diagnosis model of the protection system of an intelligent substation based on GBDT is studied and verified.

The method proposed in this paper has a high diagnostic accuracy and stronger generalization ability and is more suitable for processing the fault feature data of the protection system of the intelligent substation. The calculation example shows that the overall accuracy of the method proposed in this paper can reach 99.0476%. Compared with the existing methods based on recurrent neural networks and random forest algorithms, the method proposed in this paper has a higher fault diagnosis accuracy. In the case of one false alarm in the fault feature information data, the accuracy rate of the proposed method can reach 97%. In the case of two false alarms in the fault feature information data, the accuracy rate of the proposed method can reach 92%. In the case of three false alarms in the fault feature information data, the accuracy rate of the proposed method can reach 84%. In multiple fault diagnosis, the accuracy of the proposed method is 91.8367%. Through the above analysis, it can be concluded that the method proposed in this paper gives full play to the high accuracy and anti-overfitting ability of the GBDT algorithm when dealing with device faults in the protection system. Compared with the RNN and RF algorithms, this method is more convenient to adjust algorithm parameters in addition to higher accuracy. Compared with the existing methods, this method also performs very well when faced with bad data (false alarms of fault information, multiple faults). In conclusion, the method proposed in this paper can play a better role in practical applications.

Author Contributions: Conceptualization, W.D. and Q.C.; data curation, W.D. and Y.D.; formal analysis, W.D. and Q.C.; funding acquisition, Q.C.; methodology, W.D., Q.C., and N.S.; project administration, Q.C.; software, W.D.; validation, W.D. and Q.C.; visualization, W.D. and N.S.; writing—original draft, W.D.; writing—review and editing, W.D. and Q.C. All authors have read and agreed to the published version of the manuscript.

Funding: This work was supported by the National Natural Science Foundation of China (No. 5187070349).

Institutional Review Board Statement: Not applicable.

Informed Consent Statement: Not applicable.

Data Availability Statement: The data presented in this study are available in the article.

Conflicts of Interest: The authors declare no conflict of interest. The funders had no role in the design of the study; in the collection, analyses, or interpretation of data; in the writing of the manuscript or in the decision to publish the results.

References

1. Liu, Z.; Zhang, Q. Study on the development mode of national power grid of China. *Proc. CSEE* **2013**, *33*, 1–10.
2. Wang, D.; Li, Z.; He, F. Design of Integrative Information Platform for Smart Substations. *Power Syst. Technol.* **2010**, *10*, 20–25.
3. Li, Y.; Huo, G.; Che, Y. Research on secondary system on-line monitoring and evaluation in smart substation. *Power Syst. Prot. Control.* **2016**, *44*, 66–70. [CrossRef]
4. Wang, L.; Chen, Q.; Gao, H. Framework of Fault Trace for Smart Substation Based on Big Data Mining Technology. *Autom. Electr. Power Syst.* **2018**, *42*, 84–91. [CrossRef]
5. Tao, W.; Wang, Y.; Li, J.; Ye, Y.; Zhang, X.; Chen, X. ADC Based Effectiveness Evaluation of Secondary Equipment in Smart Substation. *Autom. Electr. Power Syst.* **2016**, *23*, 118–124.
6. Ye, Y.; Sun, Y.; Huang, T. Online state detection and fault diagnosis technology of relay protection secondary circuits in smart substation. *Power Syst. Prot. Control.* **2017**, *44*, 148–153. [CrossRef]
7. Zhang, Q.; Jia, H.; Ye, H. Design and application of virtual secondary circuit monitoring in smart substation. *Power Syst. Prot. Control.* **2015**, *43*, 123–128.
8. Gao, X.; Yang, Y.; Jiang, J. Analysis of secondary circuit monitoring methods based on SCD. *Power Syst. Prot. Control.* **2014**, *42*, 149–154.
9. Hu, D.; Wo, J. Virtual Circuit System of Smart Substations Based on IEC 61850. *Autom. Electr. Power Syst.* **2010**, *34*, 78–82.
10. Cao, H.; Gao, X.; Yang, Y. Analysis of online monitoring and intelligent diagnosis based on the full model SCD secondary system. *Power Syst. Prot. Control.* **2016**, *44*, 136–141. [CrossRef]
11. Lin, X.; Ke, S.; Li, Z. A fault diagnosis of electric transmission lines using modular neural networks. *IEEE Trans. Power Deliv.* **2010**, *25*, 1268–1274. [CrossRef]
12. Ren, B.; Zheng, Y.; Wang, Y.; Sheng, S.; Li, J.; Zhang, H.; Zheng, C. Fault Location of Secondary Equipment in Smart Substation Based on Deep Learning. *Power Syst. Technol.* **2021**, *45*, 713–721. [CrossRef]
13. Zhang, L.; Wang, G.; Cao, L.; Dai, Z.; Kou, B. Intelligent state evaluation and early warning of protection system based on GAN model and random forest algorithm. *J. Electr. Power Sci. Technol.* **2012**, *6*, 104–112. [CrossRef]
14. Xu, C.; Zhuang, C.; Jiang, H. Technical research of secondary equipment's state monitoring in smart substation. *Power Syst. Prot. Control.* **2015**, *43*, 127–131.
15. Xie, M.; Wang, T.; Sun, Y. Research and application of classification of relay protection information in smart substations. *Electr. Power Inf. Commun. Technol.* **2014**, *12*, 15–18.
16. Xiang, W.; Ban, L.; Zhou, P. Online prediction and optimal control method for subsynchronous oscillation of wind power based on an interpretable surrogate model for machine learning. *Power Syst. Prot. Control.* **2021**, *16*, 67–75. [CrossRef]
17. Liao, W.; Guo, C.; Jin, Y.; Gong, X. Oil-immersed Transformer Fault Diagnosis Method Based on Four-stage Preprocessing and GBDT. *Power Syst. Technol.* **2019**, *43*, 2195–2203. [CrossRef]

applied sciences

MDPI

Article

Optimal Selection of the Diesel Generators Supplying a Ship Electric Power System

Panayiotis Michalopoulos [1], George J. Tsekouras [2], Fotios D. Kanellos [3,*] and John M. Prousalidis [1]

1 School of Naval Architecture and Marine Engineering, National Technical University of Athens, Heroon Polytechniou 9, 15780 Athens, Greece
2 Department of Electrical and Electronics Engineering, University of West Attica, 250 Thivon Str., 12241 Athens, Greece
3 School of Electrical and Computer Engineering, Technical University of Crete, University Campus, Akrotiri, 73100 Chania, Crete, Greece
* Correspondence: fkanellos@tuc.gr; Tel.: +30-2821037339

Featured Application: Evaluation of electric generation system during the ship design or selection process.

Abstract: It is very common for ships to have electric power systems comprised of generators of the same type. This uniformity allows for easier and lower-cost maintenance. The classic way to select these generators is primarily by power and secondarily by dimensions and acquisition cost. In this paper, a more comprehensive way to select them, using improved cost indicators, is proposed. These take into account many factors that have a significant impact in the life-cycle cost of the equipment. A realistic and detailed profile of the ship's electric load spanning a full year of her operation is also developed to allow for a solution that is tailor-made to a specific case. The method used is highly iterative. All combinations of genset quantities and capacities are individually considered to populate a power plant, taking into account the existing redundancy requirements. For each of these and for every time interval in the load profile, the engine consumption is Lagrange-optimized to determine the most efficient combination to run the generators and the resulting cost. The operating cost throughout the year is thus derived. In this way, the method can lead to optimal results as large data sets regarding ship operation and her power system's technical characteristics can be utilized. This intense calculation process is greatly accelerated using memorization techniques. The reliability cost of the current power plant is also considered along with other cost factors, such as flat annual cost, maintenance, and personnel. The acquisition and installation cost are also included, after being distributed in annuities for various durations and interest rates. The results provide valuable insight into the total cost from every aspect and present the optimum generator selection for minimal expenditure and maximum return of investment. This methodology may be used to enhance the current power-plant design processes and provide investors with more feasible alternatives, as it takes into consideration a multitude of technical and operational characteristics of the examined ship power system.

Keywords: power-plant design; generator selection; consumption; Lagrange optimization; load profile; reliability cost; return on investment

Citation: Michalopoulos, P.; Tsekouras, G.J.; Kanellos, F.D.; Prousalidis, J.M. Optimal Selection of the Diesel Generators Supplying a Ship Electric Power System. *Appl. Sci.* **2022**, *12*, 10463. https://doi.org/10.3390/app122010463

Academic Editors: Federico Barrero and Mario Bermúdez

Received: 20 September 2022
Accepted: 13 October 2022
Published: 17 October 2022

1. Introduction

The shipping industry is ever growing and today numbers more than 63,000 commercial ships worldwide. Each year, more than 2000 new ships are built in the world [1], while the global shipbuilding industry market is expected to exceed $195 billion by 2030 [2]. In this context, the cost related to building, acquiring and operating a ship is a major concern to investors, but also has a significant impact in the world economy.

One aspect of ship design is its electric power system. This is usually overshadowed by the propulsion plant and thus overlooked in the decision-making process. However, if properly examined, it can turn out to be a substantial financial concern, especially when the requirements for electric power are increased, such as in large container ships, or even more, as technology moves towards electric propulsion.

With this goal in mind, researchers have proposed many innovative hybrid multi-energy plants [3] that include such renewable sources as photovoltaics [4,5], wind turbines [6], fuel cells [7], and batteries or supercapacitors for energy accumulation [8,9].

For classic power plants with diesel generator sets, the common and easiest way to operate them is by sharing the load proportionally among them and adding or removing generators to the grid when the load reaches certain thresholds. This simplistic management scheme allows for little efficiency improvement.

On the other hand, several techniques have been presented to achieve performance optimization and efficiency increase in a ship energy efficiency management plant (SEEMP) [10,11]. These involve sophisticated load management and distribution [12,13], smart grids and microgrids [10,14], multiagent systems [15,16], distributed power management [17] and other methodologies, even exotic ones using quantum computing [18]. According to a complicated but also efficient approach, the load distribution on the gensets is optimized according to their fuel consumption curves [19], leading to notable fuel savings.

However, little has been discussed on the selection process of the gensets. A classic ship power plant is typically designed using the following steps. First, the number of generators is determined, usually based on reservation or redundancy requirements. Afterwards, the nominal power of the generators is calculated so that the maximum total load and the maximum critical load can be adequately supplied according to the reservation and redundancy requirements. Finally, the manufacturer and the exact type of the generators is determined, based on financial criteria, usually purchase price and average fuel consumption. At this point, if the cost seems too high, the design process is restarted in a spiral fashion and all the parameters are redetermined until an acceptable outcome is eventually reached, as in Figure 1. This may be satisfactory, but optimality is not guaranteed. Furthermore, the whole process is mostly empirical and thus not efficient.

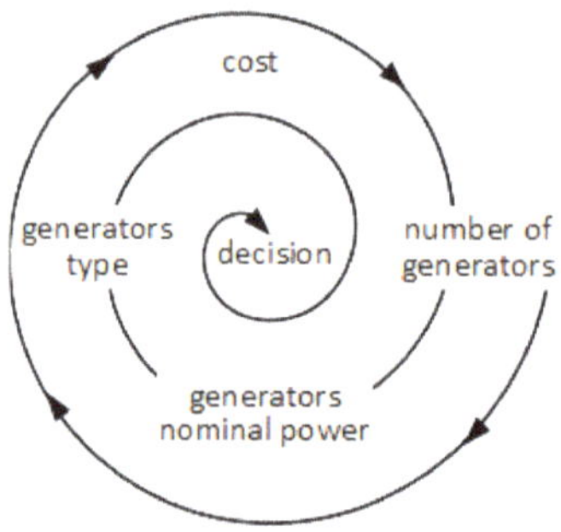

Figure 1. Typical genset selection process.

The goal of this paper is to present a new method that will definitely produce the optimum result in very little computation time and with little effort. It is noted that there is no best solution fitting all cases. On the contrary, each problem has unique requirements and constraints necessitating particular handling. To this end, the proposed method uses as inputs a detailed load profile of a real passenger ship based on real data, and all upcoming calculations are performed in this realistic context.

Additionally, the cost of gensets is much more than just their acquisition price and their average fuel consumption. Therefore, several parameters are also contemplated, among them the cost of installation, maintenance and payroll of the crew members assigned to it and detailed and optimized fuel and lubricating oil consumption. The reliability of the installation is also considered and the cost it entails. This way, the true life-cycle cost of the installation is estimated.

Furthermore, insight is provided allowing financiers to preview various interest rates and number of annuities combinations in order to select the most suitable return on investment (ROI) scheme.

In Section 2, the methodology followed is described in detail. In Section 3, the applied computational speed improvement technique is described. In Section 4, a representative case study and the results obtained are provided, while discussion on the presented work and results are given in Section 5.

2. Methodology

2.1. Overview

An auxiliary graphical overview of the calculation process is shown in Figure 2 and described in detail in the following:

Load profile creation:

Preliminarily, a detailed load profile of the ship is drafted to become the frame in which all calculations will be based upon. More details are provided in Section 2.2.

Genset pool:

A pool of diesel engine generators and their specifications is formed to combine and populate the ship's power plant. More details are provided in Section 2.3.

Algorithm main loop:

A loop begins by selecting from the pool one genset type after the other.

Genset installed capacity:

Their installed capacity is determined so that they are sufficient to supply the maximum load of the ship, taking into account any redundancy requirements. More details are provided in Section 2.4. Note that all generators in the power plant are assumed to be of the same type. If their quantity is excessive (i.e., >18), the current generator type is rejected and the loop continues with the next iteration and type selection.

Operating cost estimation:

For every time interval throughout the load profile, the Lagrange optimization method is used to establish the genset combination that will supply this particular load with the smallest fuel and lubricating oil consumption. This produces the lowest operating cost and System Marginal Cost (SMC) for each time period. More details are provided in Section 2.5. All operating costs are summed to produce the total operating cost throughout the year for the particular genset type.

Reliability cost estimation:

The Capacity Outage Probability Table (COPT) of the selected power plant is estimated. For every time interval throughout the load profile, this is used to evaluate the expected loss of load energy (LOLE) separately for each type of load and load conditions of each time interval in the profile. Afterwards, these are summed up to produce the total LOLE for each load type throughout the year. More details are provided in Section 2.6.

For every time interval in the load profile, the above SMC and LOLE values are used to calculate the total cost of power loss, for the whole year, for the selected generator type.

Initial cost estimation:

The initial cost includes acquisition and installation of the genset and it is broken down to annuities for a range of years and for a range of interest rates. More details are provided in Section 2.7.

Total cost estimation:

The flat cost related for maintenance and payroll is estimated. Then, this is added to the aforementioned operating cost and initial cost to form the total annual cost. The reliability cost is also added separately, providing the total annual cost with reliability considerations. These are both calculated for the range of annuities and interest rates mentioned above and for the current generator type. More details are provided in Section 2.9. Afterwards, the loop continues with the next genset type selection.

Optimal genset selection:

After the loop completes and all genset types are evaluated, the least expensive is selected and the total annual cost of the plant with and without reliability considerations is displayed, for the given range of annuities and interest rates.

The whole process is illustrated below. Subsequently, each individual aspect is more thoroughly discussed.

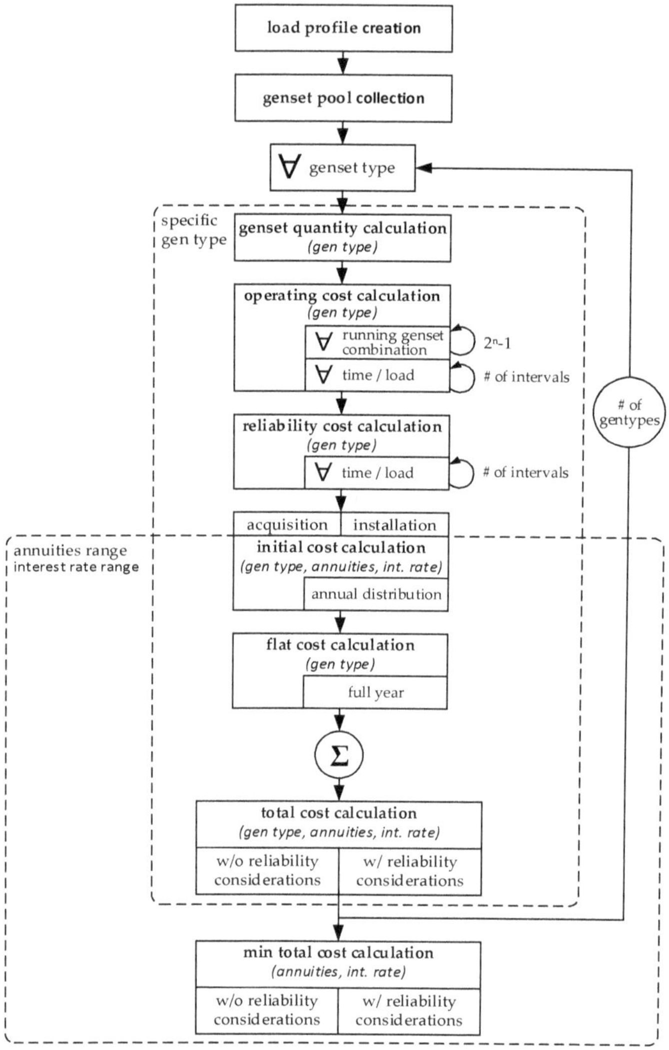

Figure 2. Calculation process overview.

2.2. Load Profile

The electric load of a ship varies greatly versus time and is very specific to her condition and performed operations. For example, the load of a ferry is much greater when she is underway filled with passengers than when she is at port with only a skeleton crew. Moreover, as the ship's schedule is usually predetermined, a load profile can be drafted with sufficient accuracy.

On the other hand, the load requirements of each ship are very distinctive and vary greatly, not only among different types and sizes but also among similar ships with different operating schedules. For example, a ship will have a different load signature when she is mostly underway and has only brief port time than when she is on a daily short-cruise routine. Therefore, it makes sense for a generator selection process to be shaped around the specific load requirements of the ship. For the purposes of this study, as well as for further research, a complete profile of the electric load of a passenger ship has been created. It is based on actual data from a real ship and it spans the range of a full operational period with relatively high resolution.

A ship is a complex structure like a small autonomous mobile city, containing a large variety of equipment. These extend from propulsion and energy production to air-conditioning, galleys and other hotel facilities. As such, the electric load associated with each of them may be characterized as more or less significant. In general, the total load $P_{load}(t_j)$ for every time t_j can be divided into K parts $P_{load-k}(t_j)$, first being the least and K-th being the most significant. In this paper, it is divided into inessential (P_{load-1}), essential (P_{load-2}), and critical parts (P_{load-3}).

$$P_{load}(t_j) = \sum_{k=1}^{K} P_{load-k}(t_j) \tag{1}$$

Inessential load refers to equipment that may become unavailable for a long time without any significant effect on the ship's operation, the performance of her crew or the living conditions of her passengers. This can be air-conditioning, hot water and lighting in living quarters, etc.

Essential load refers to equipment that when unavailable has a significant impact on the ship's operation, the performance of her crew and the living conditions of her passengers. This can be ventilation and lighting in compartments with running machinery, transfer pumps, air compressors, etc.

Finally, critical load refers to equipment that when unavailable seriously affects the safety of the ship and all those onboard. This can be auxiliaries necessary for running the gensets, propulsion and navigation (when the ship is underway), firefighting, damage control, etc.

2.3. Generator Specifications

In order to provide applicable results for this process, more than 30 actual generator sets, from several manufacturers, were studied and used to determine the optimal one (Table 1). The specifications in Table 1 were collected or derived from their datasheets.

The most characteristic information of a generator is its nominal power P_{nom}. This is provided along with its minimum and maximum power P_{min} and P_{max}, respectively. These are the limits of the equipment outside which operation is not permitted.

$$P_{min} < P_{nom} < P_{max} \tag{2}$$

For the reliability calculations, the probability of a genset not being available, also known as the forced outage rate (FOR) [20], was used.

The fuel type, fuel consumption and lubricating oil consumption were used to estimate the operating cost of the engine, while its physical characteristics and its acquisition price were used to estimate the installation cost.

It is noted that engines have additional restrictions and costs in their operation e.g., minimum running time and a minimum time between shutting down and starting up. There is also a maximum power increase/decrease rate and a starting cost. This information can be considered in future work.

Table 1. Generator data collected.

Specification
Electrical
Nominal power Minimum power percentage Maximum power percentage P_{max} Forced outage rate (FOR)
Mechanical
Fuel type Fuel consumption curve Lubricating oil consumption Weight Length Width Height
Cost
Acquisition cost Maintenance cost

2.4. Power Requirements

This paper assumes that all generators used in a single power plant are of the same type. Therefore, the maximum load can be supplied by n^* gensets of nominal power P_{nom} each, as shown in (3).

$$n* = \left\lceil \frac{\max\limits_{\forall t_j} P_{load}(t_j)}{P_{nom}} \right\rceil \tag{3}$$

This number is adequate for the ship's needs, if no redundancy is required, or if there is an extra emergency generator to take up all critical loads. However, if no extra emergency generator exists and the power plant is to withstand the failure of a single genset, then $n^* + 1$ generators will be required. Similarly, if the whole compartment may fail, then $2{\cdot}n^*$ gensets are required in a different location. This is summarized in Table 2.

Table 2. Number of generators required.

Redundancy	Generator Quantity n
none or emergency generator	n^*
1 generator	$n^* + 1$
full: 1 power compartment	$2{\cdot}n^*$

2.5. Operating Cost

The amount of fuel consumed by an engine is a function of its power output or load. As the power increases, so does the consumption versus time (see the fuel consumption curve in Figure 3). However, the consumption versus power and time (i.e., energy), also called specific consumption, reveals the existence of a point of optimal operation (see the corresponding specific fuel consumption curve in Figure 3).

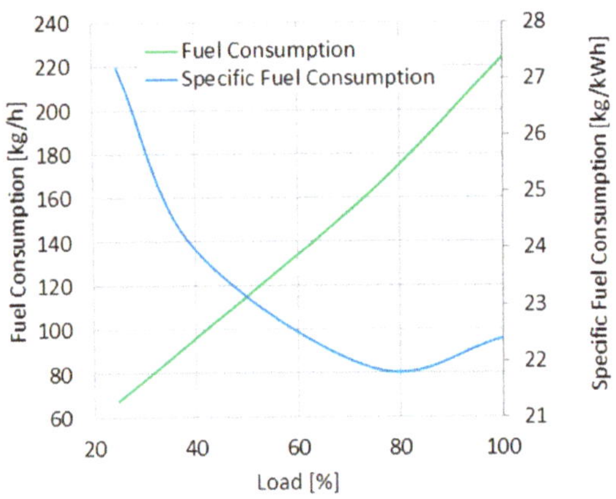

Figure 3. Fuel consumption curves.

The fuel consumption cost F_{fuel} may accurately be approximated by a second- or third-degree polynomial function (4) of the electric power P_m produced, with coefficients derived from its fuel consumption curve, or specific fuel consumption curve, provided by the manufacturer or actually measured.

$$F_{fuel}(P_m) = a + b \cdot P_m + c \cdot {P_m}^2 + d \cdot {P_m}^3 \tag{4}$$

In this paper, the approximation was calculated using a second-degree polynomial; therefore, F_{fuel} became:

$$F_{fuel}(P_m) = a + b \cdot P_m + c \cdot {P_m}^2 \tag{5}$$

On the other hand, it is specified that the lubricating oil consumption may accurately be approximated as proportional to the electric power P_m produced.

$$F_{lub}(P_m) = e \cdot P_m \tag{6}$$

Therefore, the total operational cost became:

$$F_{operation}(P_m) = a + (b + e) \cdot P_m + c \cdot {P_m}^2 \tag{7}$$

It is common practice to share the load equally among the running generators. This is efficient when all generators are of the same type and have the exact same consumption curve. However, this is never reality, since even generators of the same type will have significant differences in their consumption curves, due to their running hours, maintenance history, mechanical ware, etc. These curves can be obtained by taking periodic measurements. It has been proven that taking into account these differences and distributing the load using optimization methods, allows for extra fuel savings [19].

The quantity n of the generators required has been established above. Assuming, for the sake of generality that each one is different, there are $2^n - 1$ possible combinations $B_{combination}$ of them running. For every one $A_{operation-v}$ of them and for a particular time period t_j, the load requirements $P_{load}(t_j)$ were distributed in each running generator m producing power $P_m(t_j)$ with operating cost $F_{operation-m}(P_m(t_j))$. This distribution was optimized using the Lagrange method [19,21], because of its suitability to solve optimization problems that are constrained with equalities and/or inequalities. As such, that the total operating cost $F_{operation-A_{operation-v}}$ for this case became minimal (8) under the constraints (9) and (10).

$$F_{operation-A_{operation-v}}(t_j) = \min \sum_{m \in A_{operation-v}} F_{operation-m}(P_m(t_j)) \quad (8)$$

$$P_{load}(t_j) = \sum_{m \in A_{operation-v}} P_m(t_j) \quad (9)$$

$$m \in A_{operation-v} : P_{\min-m} \leq P_m(t_j) \leq P_{\max-m} \quad (10)$$

The system marginal cost $SMC_{operation-v}(t_j)$ was also calculated:

$$SMC_{operation-v}(t_j) = \frac{\partial F_{operation-m}(P_m(t_j))}{\partial P_m}, \forall m \in A_{operation-v} \quad (11)$$

Out of all combinations $B_{combination}$, the most efficient was selected, as in (12), and the total cost due to fuel and lubricating oil consumption throughout the year (i.e., N_T time intervals) was calculated, as in (13).

$$F_{operation}(t_j) = \min_{\forall v \in B_{combination}} F_{operation-A_{operation-v}}(t_j) \quad (12)$$

$$Cost_{operation} = \sum_{j=1}^{N_T} F_{operation}(t_j) \cdot \Delta t_j \quad (13)$$

The whole process is illustrated in Figure 4.

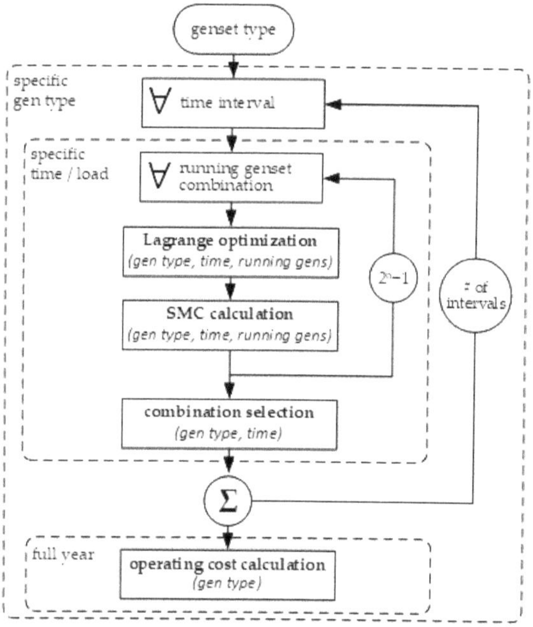

Figure 4. Operating cost calculation process.

If all the engines populating the power plant have identical behavior, the optimization process may be simplified using equal distribution. However, the algorithm uses optimization to address different duty cycles of the gensets and any future expansion of this work.

2.6. Reliability Cost

The reliability of a system is a factor of paramount importance. However, most of the time, industrial systems use rather simplistic and crude redundancy techniques to achieve the required reliability levels.

A more innovative and detailed way is using the COPT of the power plant. This is formulated, for N_P amount of generator combinations each with power-outage probability p_i, during an amount of N_T time intervals each with duration Δt_i. From this, the expected Loss Of Load Power (LOLP) is derived. This is the amount of time the available power $P_{available\_power\_i}$ is not sufficient to supply the ship's load $P_{load}(t_j)$, thus leading to a power outage, expressed here using the step function $u()$.

$$LOLP = \sum_{j=1}^{N_T}\sum_{i=1}^{N_P} p_i \cdot \Delta t_j \cdot u\left(P_{load}(t_j) - P_{available\_power-i}\right) \quad (14)$$

Similarly, the expected LOLE is derived, showing the amount of active energy not supplied to the load for the same time period and is expressed here using the ramp function $r()$.

$$LOLE = \sum_{j=1}^{N_T}\sum_{i=1}^{N_P} p_i \cdot \Delta t_j \cdot r\left(P_{load}(t_j) - P_{available\_power-i}\right) \quad (15)$$

This is also equal to:

$$LOLE = \sum_{j=1}^{N_T}\sum_{i=1}^{N_P} p_i \cdot \Delta t_j \cdot \left(P_{load}(t_j) - P_{available\_power-i}\right) \cdot u\left(P_{load}(t_j) - P_{available\_power-i}\right) \quad (16)$$

Furthermore, the LOLE can be individually expressed for each load category as:

$$LOLE_k = \sum_{j=1}^{N_T}\sum_{i=1}^{N_P} p_i \cdot \Delta t_j \cdot r\left(\sum_{b=n}^{k} P_{load-b}(t_j) - P_{available\_power-i}\right) \quad (17)$$

One way to calculate the cost of LOLE is by assuming a constant cost per load category $Cost_{loss\_energy-k}$, as seen in:

$$Cost_{LOLE} = \sum_{k=1}^{K} LOLE_k \cdot Cost_{loss\_energy-k} \quad (18)$$

A more innovative way is by assuming a cost proportional to the SMC calculated earlier:

$$Cost_{LOLE} = \sum_{j=1}^{N_T}\sum_{k=1}^{K} LOLE_k(t_j) \cdot SMC(t_j) \cdot Factor\_Cost_{loss\_energy-k} \quad (19)$$

2.7. Initial Cost

The first type of cost that comes to mind is the initial cost $F_{initial-total}$ of the generators. This is usually limited to their purchase price A_m, provided by the vendors.

However, when building a ship, there is an additional cost resulting from the space allocated for the generators and its impact on the ship's size. This is estimated as a fraction of the total ship cost C, which in turn is approximated using semiempirical relations like the following, where a and b are constants and DWT is the DeadWeight Tonnage [22].

$$C = \alpha \cdot DWT^b \quad (20)$$

A more detailed way to approach this is by considering the area E_m and the volume V_m occupied by the generator and also its mass M_m, along with their associated unit costs $Cost_{Area}$, $Cost_{Volume}$ and $Cost_{Mass}$, respectively, as shown below:

$$F_{installation-area} = \sum_{m=1}^{n} (E_m \cdot Cost_{Area}) \quad (21)$$

$$F_{installation-volume} = \sum_{m=1}^{n} (V_m \cdot Cost_{Volume}) \quad (22)$$

$$F_{installation-mass} = \sum_{m=1}^{n} (M_m \cdot Cost_{Mass}) \quad (23)$$

Therefore, the total installation and initial costs become:

$$F_{installation-total} = \sum_{m=1}^{n} (E_m \cdot Cost_{Area} + V_m \cdot Cost_{Volume} + M_m \cdot Cost_{Mass}) \quad (24)$$

$$F_{initial-total} = \sum_{m=1}^{n} (A_m + E_m \cdot Cost_{Area} + V_m \cdot Cost_{Volume} + M_m \cdot Cost_{Mass}) \quad (25)$$

As the operating period of the ship is set to one year, all costs need to refer to this. In order for the initial cost to be projected to the total annual cost, the investment scheme must be examined. For an interest rate i_{cap} and a number of T_{per} annuities, the Capital Recovery Factor (CRF) [23] becomes:

$$CRF(i_{cap}, T_{per}) = \frac{i_{cap} \cdot (1 + i_{cap})^{T_{per}}}{(1 + i_{cap})^{T_{per}} - 1} \quad (26)$$

Therefore, the annual cost for the total recovery of the investment, or equivalent initial cost $F_{initial-eq}$, becomes:

$$F_{initial-eq}(i_{cap}, T_{per}) = F_{initial-total} \cdot CRF(i_{cap}, T_{per}) \quad (27)$$

2.8. Flat Cost

No machinery may be left running unattended and without adequate maintenance. There is additional cost associated with this: the spare parts and the consumables used. This kind of work also requires specialized crew members, devoting a major portion of their time. As a consequence, their payroll was also included. This flat cost F_{flat}, has been statistically approximated as cost per calendar hour $Cost_{flat-m}$ for the m-th generator and for a whole year became:

$$F_{flat} = 8760 \cdot \sum_{m=1}^{n} Cost_{flat-m} \quad (28)$$

2.9. Total Cost

Taking into account all the above, the equivalent annual cost of the electric power generating equipment is the following:

$$F_{total}(i_{cap}, T_{per}) = F_{initial-eq}(i_{cap}, T_{per}) + F_{flat} + Cost_{operation} \quad (29)$$

If reliability considerations are also taken into account, the equivalent annual cost becomes:

$$F_{total\_LOLE}(i_{cap}, T_{per}) = F_{initial-eq}(i_{cap}, T_{per}) + F_{flat} + Cost_{operation} + Cost_{LOLE} \quad (30)$$

3. Computational Speed Improvement

Performing the above calculations proved to be a very computationally intensive task, even for modern computers, requiring several hours to complete. The major cause of delay was the Lagrange optimization and its repetition for every combination of running gensets, as well as for every time interval in the load profile, as previously seen in Figure 4.

The classic method of proportional load distribution is trivial and thus much faster. However, it has none of the efficiency benefits provided by the otherwise-rigorous Lagrange optimization. Achieving improved generator efficiency and fuel savings outweighed the convenience and speed of the classic method. Moreover, it is a well-established and documented method [24], especially for load distribution among thermal engines [25,26].

Dynamic programming could also be used, but it seemed more complex and less efficient, as it is a multilayer method that would be better suited to solve time-dependent problems [27].

To alleviate the speed concern, the memorization technique, shown in Figure 5 and described next, was also applied.

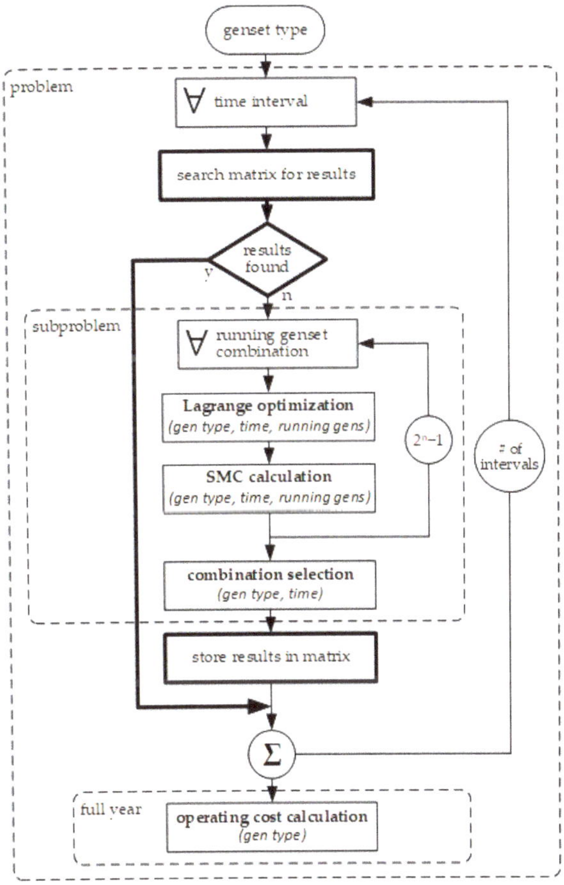

Figure 5. Speed improvement modification.

If the load profile has a duration of m months and a resolution of s samples per hour, then it will contain q intervals, where:

$$q = m \cdot 30 \times 24 \cdot s \tag{31}$$

Consequently, a profile of one year with a resolution of 30 min contains 365·24·2 = 17,520 intervals. On the other hand, the quantity n of generators populating the power plant, as determined in Section 2.4, can be quite high. Depending on the nominal power of a genset type and the redundancy and load requirements of a certain interval, the combinations of running generators can be as much as $2^n - 1$.

Therefore, for an average n and for g different genset types, the Lagrange optimization code is executed on average l times, where:

$$\bar{l} = q \cdot (2^{\bar{n}} - 1) \cdot g \tag{32}$$

This amount can easily be in the order of several million, hence the large total execution time.

Then again, it is apparent that for the same generator type and the same total load, the optimization outcome is the same. If the calculation of the operating cost is the problem, then the Lagrange optimization section, with all its repetitions, is the subproblem. Due to the uniformity of the load profile, many load conditions are the same; therefore, an overlapping of subproblems exists. This is a strong indication that running time can be reduced [28].

According to the memorization technique, an empty matrix is created for storing all optimization (i.e., subproblem) results. Any time such a calculation is required, the code quickly checks the matrix for an existing solution. If one is found, meaning that this particular optimization was performed before, the results are retrieved and the detailed calculation is bypassed.

This approach achieved a computational time reduction of more than 300 times and the running time of the code was reduced from several hours to less than a minute.

4. Case Study

As a case study, the above method was applied to a real passenger ship. To populate her power plant and to come up with tangible results, an extended data base comprising the functional parameters from several real diesel generators was used. Of course, many different scenarios can also be tested and numerical data better, may easily be applied.

4.1. Load Profile

The ship performs the same routine every year. Its load profile was formed to span this time period with a resolution of 30 min. In detail, she completes an 8-hour cruise every weekday, as shown in Table 3.

Table 3. Ship's weekday routine.

Status	Duration
at port	8.5 h
preparation for departure	1 h
underway	8 h
preparation for arrival	1 h
at port	5.5 h
Total:	24 h

Weekends are holidays and only maintenance takes place. The crew also has 4 weeks of holidays every year. The total electrical load is therefore drafted as in Figure 6.

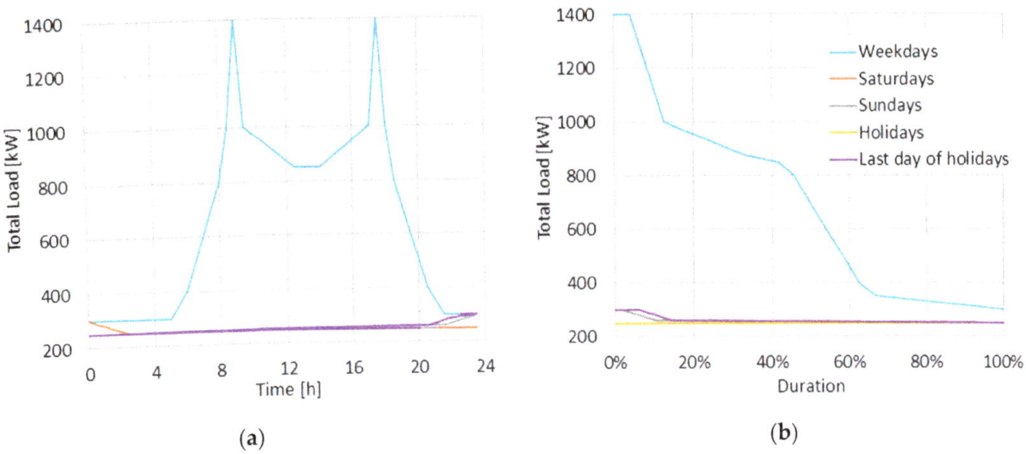

Figure 6. Load profile of ship: (**a**) versus time; (**b**) load duration curve.

As mentioned above, the total load is distinguished in critical, essential and inessential load. The critical load was measured and approximated as follows in Table 4.

Table 4. Critical load approximation.

Status	Critical Load
at port	20 kW
preparation for departure	100 kW
underway	100 kW
preparation for arrival	100 kW

Noncritical load was divided into essential and inessential, as follows in Table 5.

Table 5. Essential and inessential load division.

Status	Essential Load	Inessential Load	
at port	70%	30%	of noncritical
preparation for departure	60%	40%	of noncritical
underway	60%	40%	of noncritical
preparation for arrival	60%	40%	of noncritical

For reliability purposes, the cost of losing essential load was estimated at 100 times more that of losing inessential load. Similarly, the cost of losing critical load was estimated to be 100 times even higher, as shown below in Table 6.

Table 6. Relative reliability cost.

Load Type	Relative Cost
inessential	1
essential	100
critical	10,000

4.2. Generator Data: Electrical

The generators examined [29–37] covered an area of nominal power from 30 to 2250 kW. The whole range, along with their respective allowable limits of minimum and maximum power, may be seen in Figure 7. A common FOR equal to 0.0113 was used.

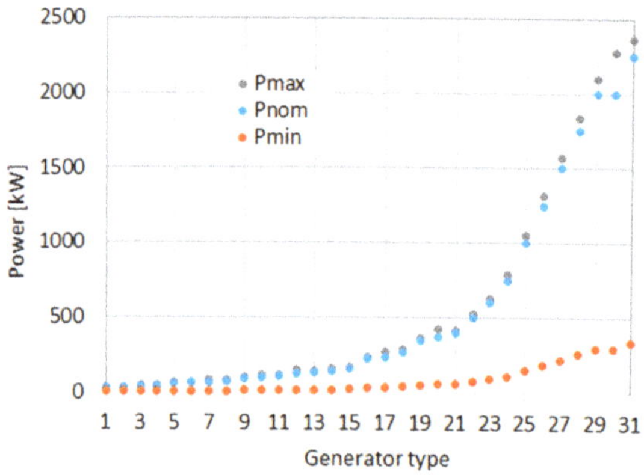

Figure 7. Power nominal, min, max.

4.3. Generator Data: Mechanical

The equipment runs on light fuel (i.e., marine diesel) with a cost of 0.40 €/kg. Its fuel consumption was approximated by a second-degree polynomial (with coefficients *a*, *b* and *c*) versus its power output, as seen in Figure 8.

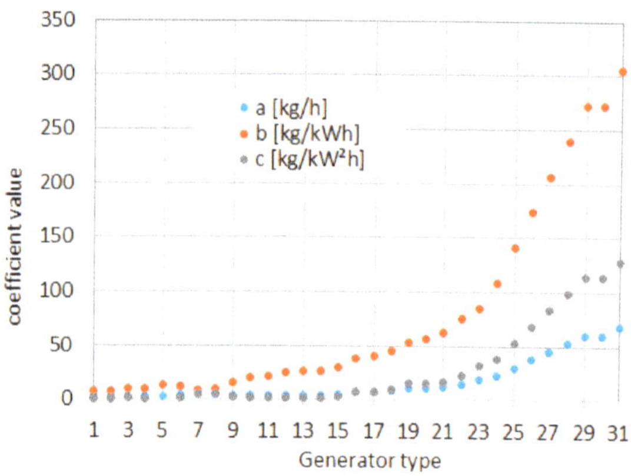

Figure 8. Fuel consumption coefficients.

The lubricating oil consumption cost was found to be proportional to the output power and was approximated in all cases as 0.006 €/kWh. The dimensions and the weight of the engines are shown in Figure 9.

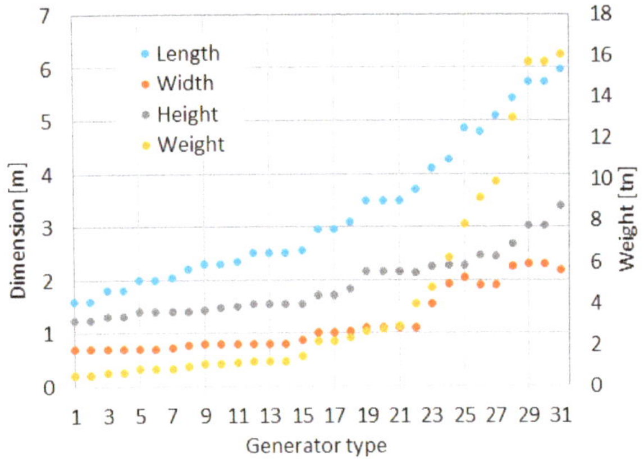

Figure 9. Dimensions and weight.

4.4. Generator Data: Cost

The acquisition and the maintenance cost of each genset are shown in Figure 10.

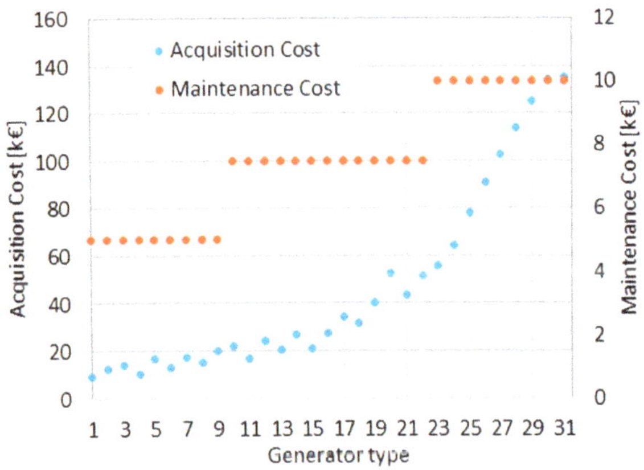

Figure 10. Acquisition cost and maintenance cost.

The unit costs of installation due to area, volume and weight used were the following, as seen in Table 7.

Table 7. Installation unit costs.

Installation Cost Type	Unit Cost
due to area	0 €/m^2
due to volume	463 €/m^3
due to weight	0 €/kg

The complete set of data can be found in Table A1 in the Appendix A.

4.5. Results

Assuming that no redundancy ($n = n^*$) is required, the most efficient combination turned out to be one engine of 1500 kW nominal power when reliability was not considered. On the other hand, the most efficient combination turned out to be three engines of 500 kW nominal power each when reliability was considered, as seen in Figure 11.

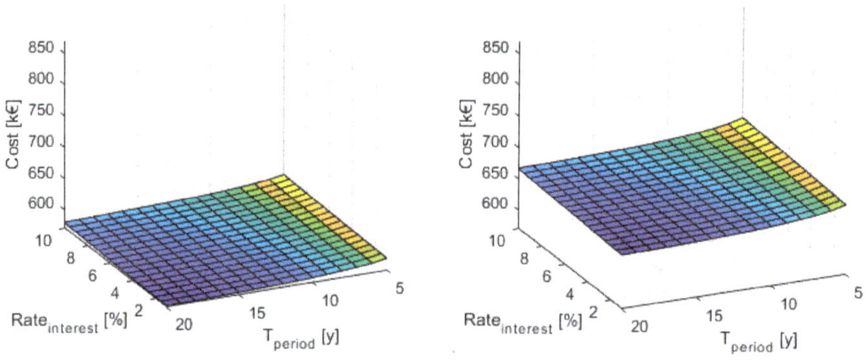

(a) (b)

Figure 11. Minimum annual total cost for the best solution for power plant without redundancy: (**a**) without reliability considerations; (**b**) minimum annual total cost for the best solution with reliability considerations.

Assuming that redundancy of a whole power compartment ($n = 2{\cdot}n^*$) is required, the most efficient combination turned out to be two engines of 1500 kW nominal power each when reliability of the ship power system was not considered. On the other hand, the most efficient combination turned out to be four engines of 750 kW nominal power each when reliability of the ship power system was considered, as seen in Figure 12.

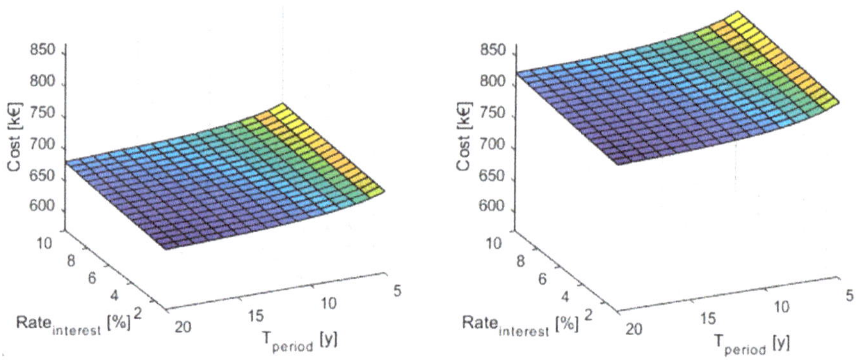

(a) (b)

Figure 12. Minimum annual total cost for the best solution for power plant with full redundancy: (**a**) without reliability considerations; (**b**) minimum annual total cost for the best solution with reliability considerations.

The results can be summarized as follows in Table 8.

Table 8. Results summary.

	No Redundancy		Full Redundancy	
	w/o Reliability	**w/ Reliability**	**w/o Reliability**	**w/Reliability**
Selected power plant	1 × 1500 kW	3 × 500 kW	2 × 1500 kW	4 × 750 kW
duty cycle 1 engine	100%	64%	100%	70%
duty cycle 2 engines	-	33%	0%	30%
duty cycle 3 engines	-	3%	-	0%
duty cycle 4 engines	-	-	-	0%
min optimal annual cost *(20 annuities with 1% interest rate)*	€572,000	€656,000	€666,000	€804,000
max optimal annual cost *(5 annuities with 10% interest rate)*	€596,000	€691,000	€713,000	€865,000

5. Discussion

As observed in the examined designs, a ship with 1400 kW maximum load requirement can be sufficiently supplied by a single 1500 kW generator, assuming that no redundancy and reliability considerations exist.

When reliability begins to matter, one might expect a solution of two 750 kW engines. However, the proposed combination was three 500 kW engines. Although the total power supply capability remained the same, the larger number of engines is obviously more reliable.

Despite the fact that in the first case, the large engine ran most of the time at a load less than 30% of its nominal value, it was still more economical than the combination of the second case, which probably ran more efficiently per engine.

Next, when full redundancy became a requirement, as was expected, the scheme of the first case (without reliability) doubled, even though again only one generator was running at any certain time.

The same did not occur when both full redundancy and reliability were required, and the scheme of the second case was not doubled like before. Instead, four engines with 750 kW nominal power were selected as more efficient. This configuration is seen in many types of ships. Again, although the total power in both plants with full redundancy was the same, the cost of the reliable one was higher.

6. Conclusions

In this paper, a novel method was introduced to facilitate the selection process of the generators in a ship power plant. It uses many parameters related to all aspects of the life-cycle cost of the engines and to the actual operating routine of the ship; however, computational time is significantly low. This way, the designers can have a complete idea of the cost involved in their selection and its return on investment.

This method may be used for different operation scenarios simply by changing the numerical data. It can also be used for applications other than shipping, since industrial installations have similar needs. Even more exotic applications may also benefit from this, by calibrating the indicators used here, or simply adding new ones.

An idea for future work could be performing a sensitivity analysis to determine how much each the examined factors affect the outcome.

Another probably useful addition might be the consideration of the minimum running time, the minimum time between shutting down and starting up, the power-increase rate, and the starting cost.

Finally, it might prove advantageous to expand this method by testing combinations of different gensets and possible exploitation of renewable energy onboard. This way, ships with shaft generators and electric propulsion, but also terrestrial power factories, may be examined.

Author Contributions: Conceptualization, P.M., G.J.T., F.D.K. and J.M.P.; methodology, P.M. and G.J.T.; software, P.M. and G.J.T.; validation, P.M., G.J.T. and F.D.K.; formal analysis, P.M. and G.J.T.; investigation, P.M. and G.J.T.; resources, P.M., G.J.T., F.D.K. and J.M.P.; data curation, P.M. and G.J.T.; writing—original draft preparation, P.M.; writing—review and editing, P.M., G.J.T. and F.D.K.; supervision, J.M.P.; project administration, J.M.P. All authors have read and agreed to the published version of the manuscript.

Funding: This research received no external funding.

Institutional Review Board Statement: Not applicable.

Informed Consent Statement: Not applicable.

Data Availability Statement: Not applicable.

Conflicts of Interest: The authors declare no conflict of interest.

Appendix A

The diesel generator data used are shown in detail below.

Table A1. Genset data.

#	P_{nom}	Fuel Consumption Coefficients			Weight	Dimensions			Price	Flat Cost
		a	b	c		L	W	H		
	[kW]	[kg/h]	[kg/kW h]	[kg/kW^2 h]	[tn]	[m]	[m]	[m]	[k€]	[€/h]
1	30.0	2.3621	6.8028	0.0000	0.55	1.60	0.70	1.25	9.5	5.0
2	42.0	2.7085	9.1712	1.0078	0.55	1.60	0.70	1.25	12.5	5.0
3	62.0	2.6823	13.1900	−0.2520	0.64	1.80	0.70	1.30	14.4	5.0
4	72.0	4.6770	8.9067	4.7872	0.66	1.80	0.70	1.30	10.6	5.0
5	92.0	3.9904	15.1600	2.8723	0.80	2.00	0.70	1.40	16.7	5.0
6	105.0	3.4487	19.7160	1.2598	0.81	2.00	0.70	1.40	13.2	5.0
7	131.0	3.7951	24.9560	1.2598	0.84	2.04	0.71	1.40	17.5	5.0
8	141.0	4.0156	26.4930	1.7637	0.96	2.20	0.77	1.40	15.4	5.0
9	238.0	6.8343	39.6330	7.5587	1.05	2.30	0.80	1.43	20.3	5.0
10	370.0	11.0550	56.7160	15.6210	1.10	2.30	0.80	1.48	22.2	7.5
11	2000.0	60.4700	272.2400	113.3800	1.11	2.34	0.80	1.49	16.7	7.5
12	30.0	2.3621	6.8028	0.0000	1.16	2.50	0.80	1.55	24.1	7.5
13	45.0	2.6377	9.9838	0.6299	1.18	2.50	0.80	1.55	20.6	7.5
14	65.0	3.2807	11.9050	1.2598	1.18	2.50	0.80	1.55	26.7	7.5
15	80.0	4.8974	9.8389	5.2911	1.43	2.57	0.87	1.55	21.1	7.5
16	110.0	3.5117	20.7230	1.2598	2.14	2.96	1.00	1.72	27.3	7.5
17	140.0	3.9893	26.3720	1.6797	2.18	2.96	1.00	1.72	34.3	7.5
18	160.0	4.4722	29.1260	3.0235	2.36	3.10	1.03	1.83	31.8	7.5
19	230.0	6.7713	37.9200	7.5587	2.67	3.50	1.10	2.16	40.2	7.5
20	275.0	7.9130	44.8170	9.4484	2.75	3.50	1.10	2.16	52.6	7.5
21	350.0	10.7080	53.1630	15.1170	2.86	3.50	1.10	2.16	43.7	7.5
22	400.0	11.5740	62.0440	16.3770	3.96	3.70	1.10	2.14	51.7	7.5
23	500.0	14.6450	74.9570	22.6760	4.79	4.11	1.54	2.25	55.4	10.0
24	600.0	18.6610	84.4690	31.4950	6.19	4.28	1.91	2.28	64.4	10.0
25	750.0	22.2040	107.9600	37.7940	7.85	4.86	2.05	2.28	78.2	10.0
26	1000.0	29.9200	140.7200	52.9110	9.08	4.79	1.90	2.45	90.8	10.0
27	1250.0	37.4790	173.7200	68.0280	9.91	5.10	1.90	2.44	102.7	10.0
28	1500.0	45.1950	206.4800	83.1460	12.95	5.42	2.24	2.68	113.9	10.0
29	1750.0	52.7530	239.4900	98.2630	15.70	5.73	2.30	3.02	124.6	10.0
30	2000.0	60.4700	272.2400	113.3800	15.70	5.73	2.30	3.02	133.8	10.0
31	2250.0	68.0280	305.2500	128.5000	16.07	5.97	2.18	3.40	134.9	10.0

P_{min} = 15% P_{nom}, P_{max} = 105% P_{nom}, FOR = 0.0113, fuel type = light fuel, lubricating oil consumption cost = 0.006 €/kWh.

References

1. Shipping Fleet Statistics: 2021—GOV.UK. Available online: https://www.gov.uk/government/statistics/shipping-fleet-statistics-2021/shipping-fleet-statistics-2021--2 (accessed on 14 September 2022).
2. Shipbuilding Market Size, Share, Report, Analysis, Growth—2030. Available online: https://www.alliedmarketresearch.com/shipbuilding-market-A08511 (accessed on 14 September 2022).
3. Tsekouras, G.J.; Kanellos, F.D.; Prousalidis, J. Simplified Method for the Assessment of Ship Electric Power Systems Operation Cost Reduction from Energy Storage and Renewable Energy Sources Integration. *IET Electr. Syst. Transp.* **2015**, *5*, 61–69. [CrossRef]
4. Tsekouras, G.J.; Kanellos, F.D. Optimal Operation of Ship Electrical Power System with Energy Storage System and Photovoltaics: Analysis and Application. *WSEAS Trans. Power Syst.* **2013**, *8*, 145–155.
5. Sun, Y.; Yan, X.; Yuan, C.; Tang, X.; Malekian, R.; Guo, C.; Li, Z. The Application of Hybrid Photovoltaic System on the Ocean-Going Ship: Engineering Practice and Experimental Research. *J. Mar. Eng. Technol.* **2019**, *18*, 56–66. [CrossRef]
6. Huang, X.; Sun, S. Application of Wind Power Generation Technology in Ships. In Proceedings of the 2022 IEEE Asia-Pacific Conference on Image Processing, Electronics and Computers, IPEC 2022, Dalian, China, 14–16 April 2022; pp. 1591–1593.
7. Chen, X.; Guo, Y. Research on Energy Management Strategy of Fuel Cell-Battery Hybrid Power Ship. In Proceedings of the SPIE—The International Society for Optical Engineering, Guangzhou, China, 10–12 December 2021; Volume 12164.
8. Chen, H.; Zhang, Z.; Guan, C.; Gao, H. Optimization of Sizing and Frequency Control in Battery/Supercapacitor Hybrid Energy Storage System for Fuel Cell Ship. *Energy* **2020**, *197*, 117285. [CrossRef]
9. Jeong, B.; Jeon, H.; Kim, S.; Kim, J.; Zhou, P. Evaluation of the Lifecycle Environmental Benefits of Full Battery Powered Ships: Comparative Analysis of Marine Diesel and Electricity. *J. Mar. Sci. Eng.* **2020**, *8*, 580. [CrossRef]
10. Kanellos, F.D.; Tsekouras, G.J.; Prousalidis, J. Onboard DC Grid Employing Smart Grid Technology: Challenges, State of the Art and Future Prospects. *IET Electr. Syst. Transp.* **2015**, *5*, 1–11. [CrossRef]
11. Ballou, P.J. Ship Energy Efficiency Management Requires a Total Solution Approach. *Mar. Technol. Soc. J.* **2013**, *47*, 83–95. [CrossRef]
12. Sudhoff, S.D. Currents of Change. *IEEE Power Energy Mag.* **2011**, *9*, 30–37. [CrossRef]
13. Feng, X.; Zourntos, T.; Butler-Purry, K.L.; Mashayekh, S. Dynamic Load Management for NG IPS Ships. In Proceedings of the IEEE PES General Meeting, PES 2010, Minneapoli, MI, USA, 25–29 July 2010.
14. Fang, S.; Xu, Y.; Wen, S.; Zhao, T.; Wang, H.; Liu, L. Data-Driven Robust Coordination of Generation and Demand-Side in Photovoltaic Integrated All-Electric Ship Microgrids. *IEEE Trans. Power Syst.* **2020**, *35*, 1783–1795. [CrossRef]
15. Feng, X.; Butler-Purry, K.L.; Zourntos, T. Real-Time Electric Load Management for DC Zonal All-Electric Ship Power Systems. *Electr. Power Syst. Res.* **2018**, *154*, 503–514. [CrossRef]
16. Zeng, Y.; Zhang, Q.; Liu, Y.; Zhuang, X.; Lv, X.; Wang, H. An Improved Distributed Secondary Control Strategy for Battery Storage System in DC Shipboard Microgrid. *IEEE Trans. Ind. Appl.* **2022**, *58*, 4062–4075. [CrossRef]
17. Tang, D.; Wang, H. Energy Management Strategies for Hybrid Power Systems Considering Dynamic Characteristics of Power Sources. *IEEE Access* **2021**, *9*, 158796–158807. [CrossRef]
18. Si, Y.; Wang, R.; Zhang, S.; Zhou, W.; Lin, A.; Zeng, G. Configuration Optimization and Energy Management of Hybrid Energy System for Marine Using Quantum Computing. *Energy* **2022**, *253*, 124131. [CrossRef]
19. Michalopoulos, P.; Kanellos, F.D.; Tsekouras, G.J.; Prousalidis, J.M. A Method for Optimal Operation of Complex Ship Power Systems Employing Shaft Electric Machines. *IEEE Trans. Transp. Electrif.* **2016**, *2*, 547–557. [CrossRef]
20. Nasioulas, E.C.; Tsekouras, G.J.; Kanellos, F.D. Bottom-up Reliability Analysis of a Base Load Diesel Engine Driven Electric Power Unit. *WSEAS Trans. Power Syst.* **2014**, *9*, 327–340.
21. Bertsekas, D.P. *Constrained Optimization and Lagrange Multiplier Methods*, 1st ed.; Athena Scientific: Nashua, NH, USA, 1996; ISBN 978-1-886529-04-5.
22. Bureau of Transport Economics. *An Estimate of Operating Costs for Bulk, Ro-Ro and Container Ships*; Australian Government Publishing Service: Canberra, Australia, 1982; ISBN 978-0-642-01774-1.
23. Jäger, L. *Handbook of Capital Recovery (CR) Factors*; Books on Demand GmbH: Norderstedt, Germany, 2021; ISBN 978-3-7534-5672-0.
24. Momoh, J.A. *Electric Power System Applications of Optimization*; Marcel Dekker Inc.: New York, NY, USA, 2001; ISBN 0-8247-9105-3.
25. Saccomanno, F. *Electric Power Systems: Analysis and Control*, 1st ed.; Wiley-IEEE Press: Piscataway, NJ, USA; Hoboken, NJ, USA, 2003; ISBN 978-0-471-23439-5.
26. Tsekouras, G.J.; Kanellos, F.D.; Tsirekis, C.D.; Mastorakis, N.E. Optimal operation of thermal electric power production system without transmission losses: An alternative solution using Artificial Neural Networks based on external penalty functions. In Proceedings of the 12th WSEAS International Conference on Artificial Intelligence, Knowledge Engineering and Databases (AIKED 2013), Cambridge, UK, 22–24 February 2013.
27. Chow, G.C. *Dynamic Economics: Optimization by the Lagrange Method*; Oxford University Press: Oxford, UK, 2011; p. 248, ISBN 978-0-19-985503-2.
28. Karumanchi, N. *Algorithm Design Techniques: Recursion, Backtracking, Greedy, Divide and Conquer, and Dynamic Programming*; CareerMonk Publications: Madinaguda, India, 2018; ISBN 978-81-932452-5-5.
29. Approximate Diesel Generator Fuel Consumption Chart. Available online: https://www.generatorsource.com/Diesel_Fuel_Consumption.aspx (accessed on 14 September 2022).

30. Argyriou SA Generator Sets—Industrial. Available online: https://www.argyriousa.com.gr/ell/product/ILEKTROPARAGOGA-ZEYGI-Viomixanika (accessed on 14 September 2022).
31. Wartsila. LNG Shipping Solutions. Wartsila, 2017. Available online: https://cdn.wartsila.com/docs/default-source/oil-gas-documents/brochure-lng-shipping-solutions.pdf?sfvrsn=12 (accessed on 12 October 2022).
32. Dual Fuel Engines. Available online: https://www.wartsila.com/marine/products/engines-and-generating-sets/dual-fuel-engines (accessed on 12 October 2022).
33. Wartsila. Wärtsilä 50DF Product Guide. Wartsila, 2018. Available online: https://cdn.wartsila.com/docs/default-source/product-files/engines/df-engine/product-guide-o-e-w50df.pdf?sfvrsn=9 (accessed on 12 October 2022).
34. MAN Diesel & Turbo, MAN L51/60DF; MAN Diesel & Turbo. Available online: https://mandieselturbo.com/docs/default-source/shopwaredocuments/man-51-60df0a0c6de48eda40e380a07ede431aa5bd.pdf?sfvrsn=1 (accessed on 12 October 2022).
35. CIMAC. *Guidelines for the Lubrication of Medium Speed Diesel Engines*; CIMAC: Solihull, UK, 2008.
36. GenSets Closed Type 1500 rpm. Available online: https://www.genitries.gr/product-category/gennitries-ilektroparagoga-zeygi/iz-kleistoy-typoy/ (accessed on 14 September 2022).
37. Marine Generators. Available online: https://www.zenoro.com/marine-generators/ (accessed on 14 September 2022).

applied sciences

MDPI

Review

Energy Harvesting Methods for Transmission Lines: A Comprehensive Review

Jordi-Roger Riba *, Manuel Moreno-Eguilaz and Santiago Bogarra

Electrical Engineering Department, Universitat Politècnica de Catalunya, 08222 Terrassa, Spain
* Correspondence: jordi.riba-ruiz@upc.edu; Tel.: +34-937398365

Abstract: Humanity faces important challenges concerning the optimal use, security, and availability of energy systems, particularly electrical power systems and transmission lines. In this context, data-driven predictive maintenance plans make it possible to increase the safety, stability, reliability, and availability of electrical power systems. In contrast, strategies such as dynamic line rating (DLR) make it possible to optimize the use of power lines. However, these approaches require developing monitoring plans based on acquiring electrical data in real-time using different types of wireless sensors placed in strategic locations. Due to the specific conditions of the transmission lines, e.g., high electric and magnetic fields, this a challenging problem, aggravated by the harsh outdoor environments where power lines are built. Such sensors must also incorporate an energy harvesting (EH) unit that supplies the necessary electronics. Therefore, the EH unit plays a key role, so when designing such electronic systems, care must be taken to select the most suitable EH technology, which is currently evolving rapidly. This work reviews and analyzes the state-of-the-art technology for EH focused on transmission lines, as it is an area with enormous potential for expansion. In addition to recent advances, it also discusses the research needs and challenges that need to be addressed. Despite the importance of this topic, there is still much to investigate, as this area is still in its infancy. Although EH systems for transmission lines are reviewed, many other applications could potentially benefit from introducing wireless sensors with EH capabilities, such as power transformers, distribution switches, or low- and medium-voltage power lines, among others.

Keywords: batteryless; distributed sensors; electric power systems; energy harvesting; predictive maintenance; transmission lines; ultra-low power

Citation: Riba, J.-R.; Moreno-Eguilaz, M.; Bogarra, S. Energy Harvesting Methods for Transmission Lines: A Comprehensive Review. *Appl. Sci.* **2022**, *12*, 10699. https://doi.org/10.3390/app122110699

Academic Editors: Federico Barrero and Mario Bermúdez

Received: 29 September 2022
Accepted: 20 October 2022
Published: 22 October 2022

1. Introduction

Energy Harvesting (EH) refers to the process by which energy from ambient or other sources is converted into electrical energy to supply autonomous devices [1] as wireless sensors and to improve their effective lifetime and capability [2]. Although many of the existing EH systems generate limited power, on the order of $\mu W/cm^2$ to mW/cm^2 [3,4], EH is gaining popularity due to the development of very low-power sensors and wireless communication systems. Self and ubiquitous energy generation capability characterize EH, so it avoids battery replacement in many applications, allowing electronic sensors to be deployed in hostile or inaccessible places. EH also contributes to reducing carbon footprints, as in many cases, electrochemical batteries can be replaced by EH units in autonomous systems [5]. Although EH technologies often do not completely eliminate the use of storage batteries, they can maximize the duration of their use [6]. Ambient energy harvesting, e.g., wind, thermal, solar, vibration, or radio frequency (RF) [2] promises low-cost, small form factor, and an endless lifetime of low-power electronic wireless sensor nodes (WSNs) by minimizing or eliminating battery use, replacement, and related maintenance costs [7]. The finite lifetime of WNSs severely limits their ability to collect data because they are energy-constrained, as WSNs require energy to supply the sensors and transmit their data to external data collectors or gateways [2,8]. Energy Harvesting Strategies Address

WSN's limited lifetime challenge. However, different challenges related to the stochastic nature of the incoming energy [9] and the influence of environmental conditions have to be addressed to ensure the optimal performance of WSNs [10]. WSNs can be connected, forming a wireless sensor network comprising a network of numerous WSNs using a multi-hop communication arrangement that allows the expansion of the coverage area compared to a single WS while requiring very low power for operation [11].

After the digital revolutions due to the introduction of the computer and the Internet, the Internet of Things (IoT) is considered to be the third revolution, offering many potential advantages to smart grids, such as fault detection and prediction or disaster prevention and prediction, which are complex problems that transmission system operators have to deal with. Therefore, IoT technology can potentially improve power transmission stability, availability, and reliability [12]. Electric power lines are the arteries of today's power systems; therefore, their efficiency, reliability, and stability profoundly impact daily life and the national economy. Transmission lines may operate in remote areas, making routine inspections complex and expensive. Environmental factors such as wind, rain, ice, or extreme temperatures affect the operation of transmission lines. Although the rate of failures in transmission systems is lower than in distribution systems, the first tends to affect more customers and generate higher outage costs. Therefore, the use of self-powered IoT sensors for real-time monitoring of transmission lines, as shown in Figure 1 [13], has great potential to solve the problems mentioned above [12].

Figure 1. Schematic of wireless monitoring of transmission lines.

Electrical grid reliability is a major challenge for utility companies, so solutions based on grid monitoring, diagnosis, and control are receiving special attention [14]. In countries with an important growth of renewable energy sources, grid balancing and congestion is an issue [15,16]. The safe and reliable operation of transmission lines must take into account the thermal rating to ensure that conductors do not exceed the maximum allowable temperature [17]. Today, dynamic line rating (DLR) approaches are gaining popularity because they offer a solution to the congestion problem since they are based on utilizing the maximum capacity of transmission lines. To this end, DLR approaches are based on measuring weather and line variables in order to dynamically determine the maximum allowable line current to ensure that the line operates within safe operation limits [18]. These solutions also warrant an increased return on investment due to more efficient asset utilization [11]. However, due to the geographical spread of power systems, monitoring power lines and assets outside the substation is challenging due to the harsh outdoor weather conditions, the vast distances to be covered, and the power requirements of the required electronic sensing devices [11].

In general, in indoor applications, there is no urgent need for EH devices due to easy access to a power supply and accessibility of sensor nodes, which reduces maintenance costs, so the need for EH systems is more typical of harsh outdoor applications [19]. Developing EH systems for high-voltage applications remains a big challenge, so most sensor nodes are still battery-powered, being inconvenient due to periodic replacements and unfavorable maintenance-free operations. As a result, many studies are focusing on self-powered wireless sensors [20].

The basic architecture of a WSN includes the sensor or sensors that convert the measured variables (temperature, current, voltage, ...) into electric signals, the power management module, the energy storage unit [21], the energy harvesting, and the communications module, which are represented in Figure 2.

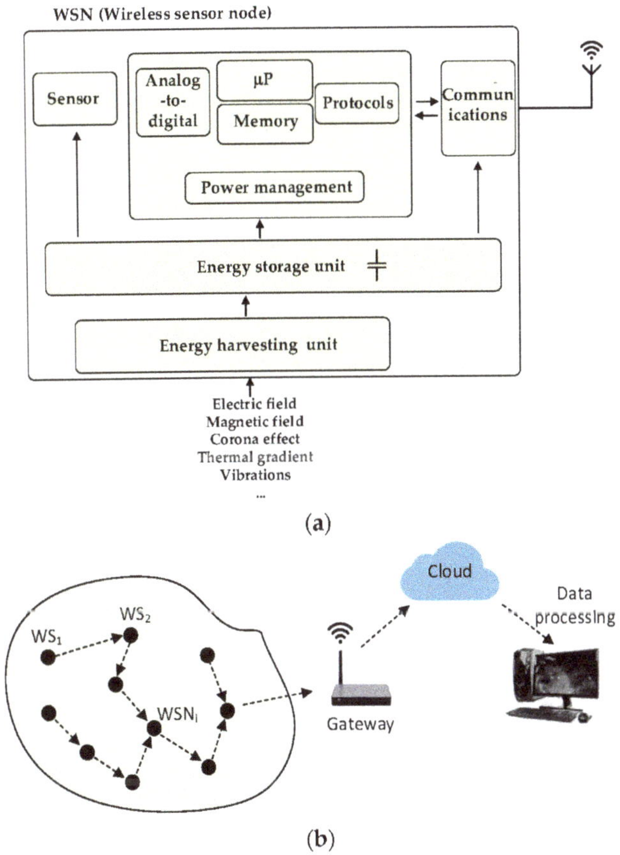

Figure 2. (**a**) Basic structure of a wireless sensor node (WSN). (**b**) Structure of a wireless sensor network, adapted from [22].

This work reviews and analyzes the state-of-the-art and recent progress in EH systems for WSNs intended for transmission lines while identifying challenges and research needs. This topic has a huge impact and expansion potential, but this area is still in its infancy due to the specificities and complexities of transmission lines. The information condensed in this review paper can also be applied to other applications, such as power transformers, distribution switches, or medium- and low-voltage power lines. The knowledge presented in this work has been compiled from the latest technical and scientific literature, including journal papers, review articles, conference works, doctoral theses, technical reports, and white papers.

The rest of the article is organized as follows. Section 2 reviews ultra-low-power energy harvesting technology for wireless sensor networks. Section 3 reviews the state-of-the art batteryless energy harvesting systems. Section 4 reviews the state-of-the-art of energy harvesting methods for transmission lines. Section 5 discusses the main characteristics of the analyzed energy harvesting methods. Section 6 identifies the challenges and research needs related to this topic, and finally, Section 7 concludes this review paper.

2. Ultra-Low-Power Technology for Energy Harvesting Wireless Sensor Networks

Although there is no exact definition of ultra-low-power (ULP) in the field of wireless sensors, it usually refers to designs that consume, on average, in the order of fractions of μW to fractions of mW. Most EH sources fall within the ULP category because they can generate as few as tens of μW [23]. Batteryless WSNs require the harvesting of energy from ambient sources in their application volume. However, due to the small volume of EH systems, very little ambient energy is available, requiring ULP circuits. In [24], it is stated that micro-power EH from indoor small-size solar cells, piezoelectric, thermoelectric (gradient temperature 1–10 °C), or RF Wi-Fi harvesters can scavenge power at ULP levels (some μW). However, power at μW level is not enough to supply IoT using commercial integrated circuits, so in addition to increasing the amount of energy harvested, reducing system power consumption to ULP levels is also required [24]. For example, according to [25], WSNs typically require 1–5 μW in standby mode, 0.5–1 mW in active mode, and 50 mW while transmitting. Similar values are provided in [21].

ULP technology aims to extend the lifetime of WSNs through significant energy savings [26]. The basic diagram of a ULP WSN is the same as shown in Figure 2a. Since WSNs are constrained in terms of processing capacity, memory and energy, and power usage, energy management is one of the challenges that must be addressed for successful WSN deployment [27]. In particular, energy-efficient wireless communication modules play a critical role in developing battery-powered and batteryless WSNs. Different approaches can be applied to address these issues at both the physical and network levels without sacrificing important metrics, such as transmission range, latency, or immunity to interference [28]. By combining energy-efficient wireless protocols with low-power communication architectures, operation below 10^{-4} W can be achieved [28]. Communication modules typically operate discontinuously at low duty cycles, hibernating between consecutive active states to save energy [19]. This strategy, based on periodically alternating between sleep and active modes, allows a large reduction in the average power of the power-hungry radio [29]. In [23], it is reported that by combining different techniques, such as short frame sizes, optional and synchronous receptions, asynchronous transmissions, and a light architecture, it is possible to reduce power consumption by 1 to 2 orders of magnitude compared to Bluetooth Low Energy (BLE).

3. Batteryless Energy Harvesting Systems

Although many EH systems use supplementary batteries, they have a finite lifetime, require regular charging or replacement, and produce electronic waste [30]. Due to IoT devices' fast deployment, millions of batteries probably need to be replaced daily [28], so battery replacement is impractical and has significant environmental impacts [3]. By reducing power consumption and voltage supply, WSNs could be self-powered, i.e., supplied directly from the EH source without using batteries [28,31]. The use of batteryless WSNs is directly related to the application of ULP devices and approaches [31].

Due to recent advances in low-power sensors, when combined with communication modules that operate in a discontinuous mode at low duty cycles hibernating between consecutive active states, they can, in some cases, be supplied directly from an independent source without using any battery [19]. Batteryless systems can use a variety of power sources, including vibration, solar power, temperature gradient, or wireless power transfer, among others, making them very attractive for applications where any intervention is challenging. In many cases, the energy collected by the EH unit is stored in small capacitors [3]. Due to the often stochastic nature of harvested energy and the restricted energy storage, associated WSNs typically operate in an intermittent on-off pattern [32]. However, in the case of power lines, this problem can be partially overcome due to the "predictability" of the energy flow, which allows better planning of the energy harvesting.

Instead of batteries, capacitors can be used to store energy, being smaller than batteries and offering lifetimes of more than a decade since they withstand a larger number of charge-discharge cycles [3,33]. However, the use of storage capacitors presents some disadvantages and challenges related to the stochastic nature of the energy transferred from the source to the energy harvester [34] and energy storage capacity, especially when using small capacitors. The WSN lacks a constant power supply so that power failure can last any duration, from seconds to hours. Further, depending on the WSN's mode of operation, the amount of power consumed can vary considerably, for example, when switching from sleep to transmission modes or during the sensing period, causing voltage changes that affect the energy stored in the capacitor. Therefore, for effective use of batteryless WSNs, applications based on such devices must have the ability to handle this intermittent behavior [32].

Another possibility is to use supercapacitors, which deliver and accept charge much faster than batteries while tolerating many more charge-discharge cycles than rechargeable batteries [35], but having much lower energy densities and exhibiting higher self-discharge currents, so once fully charged, compared to batteries they discharge at a much faster rate. Therefore, supercapacitors are used in applications that require many rapid charge-discharge cycles rather than compact, long-term energy storage. Small supercapacitors are used in WSNs because they can boot and charge batteryless WSNs in a matter of seconds when the energy is available. However, the use of small supercapacitors does not allow the WSN to perform energy-intensive operations. On the other hand, using large capacitors implies long charging times, so for start-up, the node must wait for the capacitor to reach a minimum voltage level. Moreover, the energy stored in the supercapacitor is quickly dissipated as the energy scavenged is not enough to maintain a minimum voltage across the supercapacitor, so a design trade-off is required for irregularly powered batteryless systems. However, according to [36], this topic is not well-studied in the technical literature.

Communication technologies play a critical role in deploying reliable batteryless WSNs, as they have relatively high power requirements compared to the accompanying sensors. Different low-energy communication technologies have been applied in batteryless WSNs, such as Bluetooth Low Energy (BLE) [3] or LoRaWAN, probably the most popular technology within low-power wide-area networks (LPWAN) [32].

4. State-of-the-Art Energy Harvesting Methods for Transmission Lines

EH circuits are designed to collect energy from ambient sources to provide a stable supply to the other modules [37]. Continuous energy supply is an important research topic for IoT and WSNs applications [7] because the power supply is a bottleneck for developing autonomous transmission line monitoring systems [38]. Monitoring and maintenance of transmission lines can be performed by line robots or by using line-mounted distributed wireless sensors, which typically use batteries as the main power source, but as explained at the expense of environmental impacts and maintenance costs [38], and presenting significant maintenance difficulties in remote areas [39].

This section reviews the EH methods that can be effectively applied to transmission lines. However, some EH technologies are not included in this section, such as wind collectors that rely on moving parts due to insulation issues and sensor costs [38], because they cannot be effectively applied in power transmission lines.

4.1. Solar Energy Harvesting

Solar EH is a mature technology that is being applied in many areas. Solar energy is harnessed using a photovoltaic (PV) system that converts sunlight into useful electric power. For example, in [40], a temperature monitoring system for overhead transmission lines using two solar panels fixed at the top of the tower was analyzed.

The main disadvantage of solar panels is that, due to the stochastic nature of weather conditions, a stable energy supply cannot be guaranteed [41] because, in many places, it is possible to reliably generate solar power for less than 8 h a day [42]. Further, weather conditions, dust, sand, dirt, ice, or snow can significantly impact the availability of solar energy, increasing maintenance costs [43]. Since a constant electrical supply is not feasible [44], combining solar EH with an ultracapacitor or a rechargeable battery is necessary to provide a stable supply to the sensor system [11,45]. Due to the drawbacks mentioned above, recent developments point toward using hybrid solutions that combine solar EH with other sources, such as electromagnetic or RF. For example, in [41,46], a harvester combining solar and magnetic field EH is proposed for WSNs intended for smart grid applications, where the magnetic field harvester is based on a multi-turn coil surrounding a magnetic core. In [47], a hybrid solar-RF energy harvesting system combining a solar panel and an antenna is proposed.

4.2. Vibration Energy Harvesting

Vibration energy harvesting transforms mechanical vibrational energy into useful electric energy [21]. Therefore, mechanical vibrations in power lines or induced by them can be converted into electrical energy. Different energy extraction methods from vibrations are possible, such as piezoelectric [1,5,48,49], inductive [26], or electrostatic [25,50,51] approaches.

Piezoelectric EH is based on the use of piezoelectric materials, which generate electric charges, and thus voltage, when the material is subjected to mechanical stress [25]. Piezoelectric harvesters can be designed and manufactured at the scale of microelectromechanical systems (MEMS) to take advantage of ambient vibrations with a frequency usually less than 200 Hz [52]. In [53], a piezoelectric energy harvester was presented using a magnetic metal as the tip mass that vibrates due to the action of the AC magnetic field generated by a power conductor. The piezoelectric ceramic element is composed of $Ba_{0.85}Ca_{0.15}Ti_{0.90}Zr_{0.10}O_3$ + CuO 0.3 wt%, which can produce an instantaneous maximum power of 8.2 mW in a weak magnetic field of 250 μT, and exhibits an energy density of 107.9 mW/cm^3. In [54], a device for harvesting the energy of a galloping conductor was proposed, using a swinging piezoelectric cantilever beam collision structure and a pendulum with a swinging ball, whose motion was caused by the low-frequency vibration of the galloping conductor. Similarly, in [55], a linear Halbach array mounted on the free end of a piezoelectric cantilever beam is used to harvest the energy from the AC magnetic field generated by a power line. The presented results show that this harvester can generate 897 μW across a 212 kOhm resistor when mounted on a two-wire cable carrying opposite currents of 10 A. In [56], a cantilever (piezoelectric) beam was used together with a miniaturized permanent magnet attached to the tip, which vibrates due to the interaction with the magnetic field of the AC line, as shown in Figure 3a. In [57], a nickel–zinc ferrite ring-type core, with a piezoelectric ceramic toroidal solenoid wound around it, was presented. The nickel-zinc ferrite ring was required to convert the 50 Hz circumferential magnetic field into a 50 Hz circumferential strain, transmitted to the piezoelectric layer, generating a voltage.

Inductive EH from vibrations is based on the voltage generated due to the relative motion of a permanent magnet and a harvesting coil [26,58], as shown in Figure 3b. The amplitude of the induced voltage in the coil depends on the amplitude of the beam vibration, the number of turns, the line frequency, and the flux gradient. To increase the magnitude of the flux gradient and the flux density, [59] proposes the use of an array of tiny and powerful permanent magnets with opposite polarities attached to the vibrating beam.

The electrostatic method (see Figure 3c) is based on the capacitance change of a variable capacitor or electret due to relative motion between the two plates [11,25,50,60], although different methods can be combined to improve performance [61]. The main drawback of electrostatic EH is related to the need for an external voltage source during operation [25].

The energy produced by these harvesters increases with their volume [50]. In [51], an electrostatic energy harvester that produces 45 μW for an acceleration of 0.6 m/s^2 at 50 Hz with a volume of 150 mm^3 is presented.

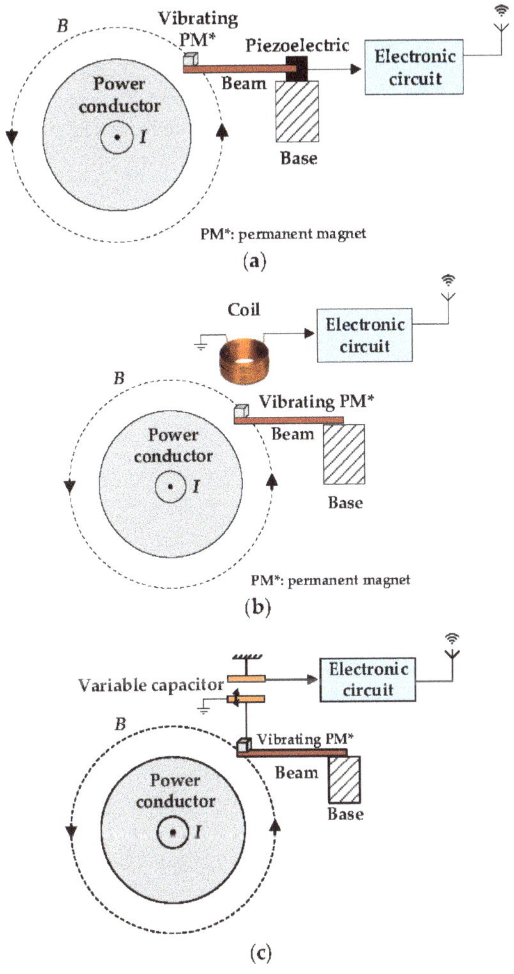

Figure 3. Hybrid magnetic–vibration EH using a permanent vibrating magnet due to the action of the AC magnetic field of the power line. (**a**) Piezoelectric energy harvesting adapted from [62]. (**b**) Inductive energy harvesting adapted from [63]. (**c**) Electrostatic energy harvesting.

Other methods, such as airflow energy harvesting, are not considered in this section due to the incompatibility of using rotating elements in power lines.

4.3. Thermoelectric Energy Harvesting

The action of an electric current heats any conductor due to the ohmic resistance, which produces a temperature gradient between the conductor and the environment [4] so that objects at different temperatures allow energy to be harvested through heat transfer [11]. Thermoelectric energy harvesting techniques are typically based on thermoelectric generators (TEGs); that is, devices that convert the temperature difference between their two sides into electrical energy [5,64], according to the Seebeck, Peltier, and Thomson effects. Thermoelectric energy harvesting converts the temperature difference between the environment and the power line into electric energy [2]. However, the efficiency of thermal EH is governed by the Carnot cycle, so for small temperature gradients, low efficiencies are expected [11]. In addition, similar to solar and wind, thermoelectric EH often cannot generate stable electric power since it depends on environmental variables [14], and in the case of power lines, it depends on the intensity of the line current, which largely determines the temperature of the conductor. When applied to transmission lines, the wide temperature swings of both ambient and line conductors make it necessary to take special care and apply specific energy management approaches to ensure reliable and stable thermoelectric generation [38]. In the case of low temperature gradients, thermoelectric energy harvesting is increasingly applied to supply wearable devices [65]. According to [65], a 900 mm^2 TEG allows the extraction of between 5 and 50 µW with a few °C temperature difference, whereas in [66], it is shown that a 4-layer TEG with 5000 thermocouples can generate up to 200 µW with a temperature gradient of approximately 8 °C.

It is possible to harvest the energy of the heat flux between the conductor and the environment by wrapping a thermoelectric generator around it. A heat-transmitting paste is usually placed between the conductor and the thermoelectric generator to maximize the heat flux. A heat exchanger is often attached to the cold outer surface of the thermoelectric generator to maximize the temperature difference between the cold and hot surfaces of the thermoelectric generator [4].

In DC systems, due to the lack of a time-varying variable such as the electric or magnetic field, the development of EH systems becomes more difficult. Thermoelectric EH is especially appealing for DC systems since Joule losses are characteristic of both AC and DC systems [4]. Since AC transmission lines are much more common than DC transmission lines, few works apply thermoelectric generators for EH in DC transmission systems. In [67], thermoelectric EH was applied to supply a WSN that includes several sensors (temperature, current, and voltage drop) mounted on a cylindrical busbar for a substation connector. It was shown that the temperature gradient is a key parameter to determine the output power of the thermoelectric generator, which generated 155 µW for a temperature difference between the busbar and the ambient of 19 °C.

4.4. Magnetic Field Energy Harvesting

Magnetic field EH harvesting has been extensively discussed in the literature. The magnetic field generated by a line conductor is directly proportional to the intensity of the current and decreases with distance from the conductor as,

$$B = \frac{\mu_0 I}{2\pi r} \tag{1}$$

I being the intensity of the current, μ_0 the magnetic permittivity of free air, and r the distance from the measuring point and the center of the conductor.

From (1), it is evident that the magnetic field strength of power lines is substantial only in their close proximity, which limits the possible locations of EHs close to AC conductors. Another limitation is that energy harvesters are based on transformer action, so usually, they have wound coils, with their internal resistance limiting the extraction of energy, so a highly efficient circuit is required [68]. For efficient use, they need to be clamped around the conductor, thus limiting their application because, in some cases, it is not practical [11].

Current transformers are typically used for this purpose (see Figure 4), although power availability is highly dependent on the current level of the transmission line, so a minimum current flow is required [14].

Figure 4 shows the basic principle for energy harvesting from the magnetic field generated by power lines. Similar designs are found in numerous scientific works [59,69–74].

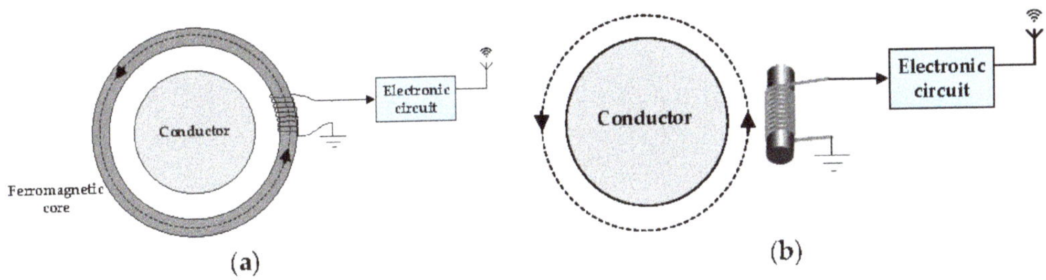

Figure 4. Magnetic field energy harvesting concept in power conductors, adapted from [75]. (**a**) Clamped topology. (**b**) Clampless topology.

In [39], it was shown that using a multi-winding up-conversion current transformer configuration is able to scavenge energy with a current as low as 1 A, thus providing enough power to supply wireless sensors. In [68], two inductors were placed between two parallel wires in opposite directions, carrying 8.4 A each, showing that depending on the configuration, up to 850 μW can be generated. In [76], it was shown that by using nano circular cut crystalline cores, high power densities could be achieved, around 100 mW/cm^3, 50 times higher than using conventional cores. In [77], the material role of magnetic toroidal cores was evaluated, showing that nanocrystalline alloy cores increase the power density about four times compared to ferrite cores. In [78], the role of core size was analyzed, showing that it is essential to optimize the power output of the harvester.

A miniature wound multicore structure made of mu-metal was proposed in [79], showing that it is able to deliver 10 mW to a 50 Ohm load. A magnetic field EH device based on a double-ring core was proposed in [80], showing that 32.8 mW can be harvested when the line current is as low as 10 A. In [11], a wound flux concentrator core made of silicon steel stuck to the conductor was proposed, which allows the concentration of the nearby flux very efficiently. The proposed design was shown to generate up to 257 mW when 1000 A flowed through the conductor. The design proposed in [11] does not have to be clamped around the conductor, which limits practical applications. In [81], an annular cored current transformer with high power density was presented, showing that it can generate 350 mW for a line current of 10 A. In [75], a toroidal core design methodology considering the saturation effect was proposed.

4.5. Electric-Field Energy Harvesting

The electric field generated by a transmission line is independent of the current level since it only depends on the applied voltage. Therefore, electric field energy harvesting is the only method that can allow effective EH at any time the line is energized, even when not carrying any current. It makes electric-field EH the most feasible option to energize sensors from the point of view of predictability, availability, and controllability [14]. The idea of harvesting energy from the electric field is not new [11]. Electric-field energy harvesting does not depend on environmental variables, unlike many conventional harvesting methods [82], so it provides a more durable and reliable operation since it allows operation with any conductive material to which voltage is applied, making it ideal for applications requiring a certain quality of service. Since the frequency and voltage of transmission lines are tightly regulated, the electric field they generate is stable, allowing predictable amounts of energy to be harvested due to the constant rate of power harvested [14]. Electric-field EH is well suited for high-voltage transmission lines due to the strong associated electric fields, although different works have shown that it is also feasible in low-voltage applications [83,84] using low-power electronics and switches.

Any energized conductor generates a radial electric field. In the case of alternating current (AC) lines, the time-varying electric field produces a displacement current that can charge a nearby capacitor so that the energy stored in this capacitor E_C can be expressed as,

$$E_C = \frac{1}{2}CV^2 \quad (2)$$

C being the capacitance, and V the voltage accumulated between the armatures of the capacitor terminals.

As shown in Figure 5, the EH unit includes diodes to rectify the generated voltage and prevent the scavenged energy from back feeding, capacitors or supercapacitors, and a controlled switch to regulate energy usage. The switch allows automatic charging of the capacitor (position 1) when the voltage is below a specific value and connects the capacitor to the load (position 2) when the stored energy is high enough for transmission. Therefore, it Is essential to use highly efficient rectifiers, microcontrollers, and regulators to optimize the overall efficiency of EH [14].

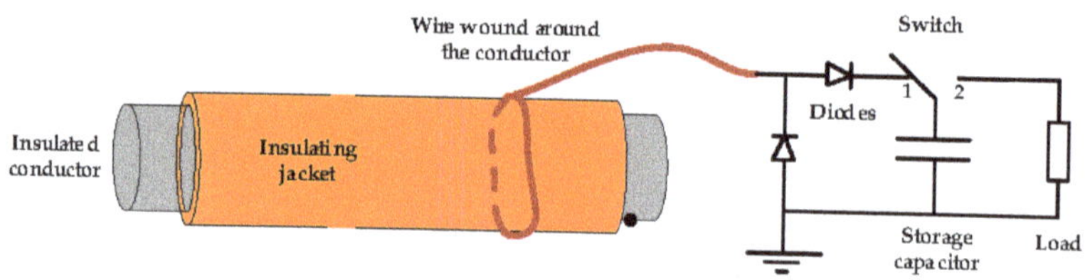

Figure 5. Electric-field energy harvesting using a wire wound around a conductor adapted [85].

In [85], a wire wound around an insulated single-phase 3-wire 220 V cable was used to harvest energy from the stray electric field, generating 680 nW on average. A sketch of this EH system is shown in Figure 5.

In [14,86], an EH system is proposed using a dielectric layer and a conducting sheath wrapped around the conductor that generates a stray capacitance, which is used to harvest energy from the electric field generated by the conductor. A multi-layer structure is also possible, as shown in Figure 6. A similar approach, shown in Figure 6a, was applied in [84] using a 220 V power line as a reference, showing an average power extracted of about 47 µW. In [87], using a copper sheet wrapped around a 230 V power line, the authors harvested 367.5 µW. In [88], a similar EH method was applied using a power line insulator, the authors stating that a continuous power of up to 17 mW can be extracted from a 12.7 kV medium voltage power line. Similar circuits can be applied using metallic plates instead of wrapping a conductive sheath around the conductor [85,89]. In [90], it has been shown that 2.5 µW can be extracted from a 120 V power line using a metallic plate. A similar approach is proposed in [91], showing that a displacement current of fractions of mA can be induced.

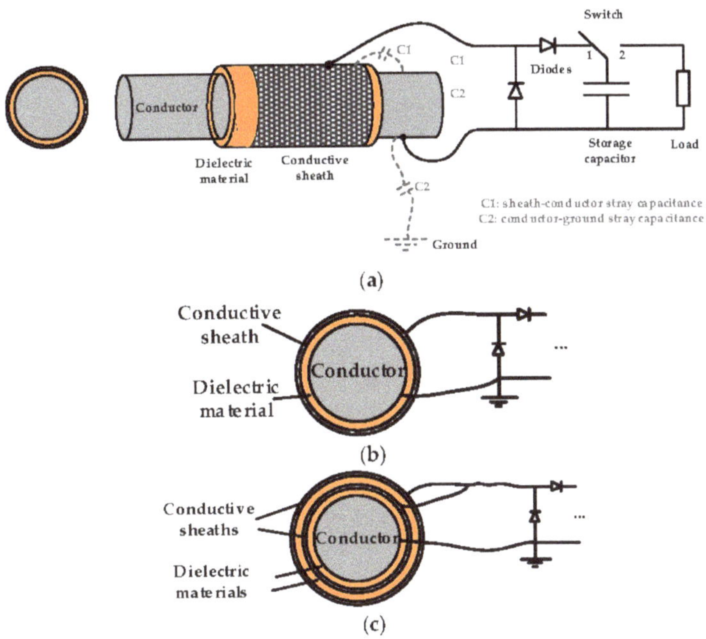

Figure 6. Electric-field energy harvesting concept in power conductors, adapted from [92]. (**a**) Schematics of the energy harvester. (**b**) Single-layer structure. (**c**) Two-layer structure.

4.6. Radio-Frequency (RF) Energy Harvesting

RF enables energy harvesting from RF sources such as base transceiver stations (BTS), TV towers, Wi-Fi, radio broadcast [93], or from dedicated RF transmitters [36] through the use of receiving antennas. These signals are then converted to DC power to supply sensor nodes [14]. RF power transmitters are broadly classified as dedicated units [94] and ambient RF power sources [95]. Ambient RF energy is found in a wide range of frequency bands, mainly in the MHz and GHz ranges [96]. Due to the widespread installation of communication systems, RF energy harvesting has become a widely analyzed technology, although it is still a challenge for low power regions [7]. RF energy harvesting can provide a regular energy source in the presence of strong RF signals due to dedicated RF power transmitters.

Unlike wind and solar energy, local weather conditions do not affect RF sources [7]. RF EH presents notable features such as predictability, reliability, or the possibility of simultaneous power supply to different WSNs [96]. Additionally, RF signals can be found almost everywhere, 24 h a day, offering low-cost implementation, compactness, and a wireless nature [97,98].

The major drawback is related to the amount of power that can be harvested [99] due to the decay of the radiated power with the inverse square relationship with the distance to the radiation source [100]. The power density near powerful transmitters is relatively low, in the order of tens to hundreds nW/cm^2 [95,101] to some few µW/cm^2 [102–104], so power-dedicated transmitters are often required to supply WSns using inductive, capacitive, or radiative tag readers [11]. In addition, the input signal may not be permanently available due to random service usage or the duty cycle (mobile phones or Wi-Fi routers) or the size of the antenna required (AM radio waves) [105,106]. Therefore, automatic frequency tuning could be useful to maximize the output power [104].

As shown in Figure 7, a conventional RF EH circuit includes an antenna to convert the incoming RF signal into a voltage difference or vice versa, a rectifier to convert the high-frequency signal to DC, a voltage multiplier that produces a higher DC output voltage, and an energy storage device, such as a battery or a supercapacitor [96].

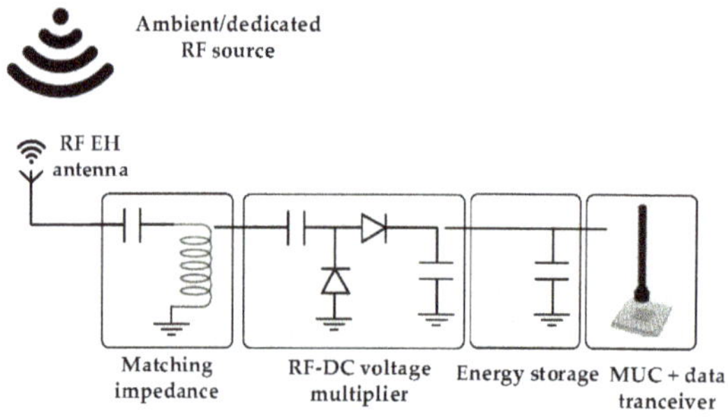

Figure 7. Basic structure of an RF EH node adapted from [107].

In [108], a BLE device was combined with a wireless RF energy harvester equipped with a 50 mF capacitor, charged by a GSM mobile, while in [109], a WSN prototype using an RF EH system from Wi-Fi transmissions was proposed. In [110], an ultra-high-frequency (UHF) RF energy harvesting solution was proposed to supply the electronics to monitor power lines. A planar monopole antenna was used for better resonator frequency performance. In [102], a multi-hop displacement of an array of magnetic antennas was proposed to harvest ambient energy for power cord monitoring. A far-field RF EH approach was also proposed for the same purpose.

Batteryless operation of RF-powered WSNs can be achieved by integrating the sensor into a radio-frequency identification (RFID) tag using an external RFID transmitter receiver [111]. Batteryless RF sensors can receive a steady energy flow from a nearby RFID reader during the entire measurement cycle, sufficient to supply the sensing and communication tasks [36].

Wake-up radio (WuR) is an energy-saving technique being applied in WSNs based on saving power by transiting into a low-power sleeping mode and turning off their main radios [112]. Devices implementing WuR usually maintain a low-power Wake-up Receiver waiting for any incoming Wake-up Signal (WuS). They can also sometimes include a wake-up transmitter capable of sending WuSs. The WuS wakes up the receiver device, which connects its main radio, so it is ready to communicate with other nearby devices [113]. WuR power consumption falls in the μW domain, but the IoT device often consumes several mW [114], although fractions of mW can be achieved in some cases [28]. Different works have integrated the capabilities of WuR technology into BLE devices [115–117]. In [118], a passive RF WuR device for a batteryless WSN is proposed.

4.7. Corona Energy Harvesting

Electric-field and magnetic-field EH are appropriate for AC power systems, but they are not suitable for DC systems due to the lack of an alternating field. Therefore, in DC systems, other strategies are required. Corona energy harvesting overcomes this difficulty.

Corona is a type of electric discharge that occurs in gas insulation systems under a highly inhomogeneous AC or DC electric field [119,120]. It is generated at a lower voltage than required for complete breakdown [121]. Due to its ionization effect, the corona generates a space charge area, producing a flow of ions perceived as current pulses. In [122], it was shown that in the case of a single conductor energized at 100 kV-DC, the corona ion flow density is approximately 200 $\mu A/m^2$.

In [20], an EH approach based on the current induced by corona discharges is proposed based on using the corona current pulses generated by a corona electrode and a simple energy-harvesting circuit, resulting in a stable output of approximately 10 mW in a 22 kV-DC system, which is sufficient for the discontinuous operation of most sensors.

Figure 8 shows the main characteristics of a corona discharge taken by the authors of this work.

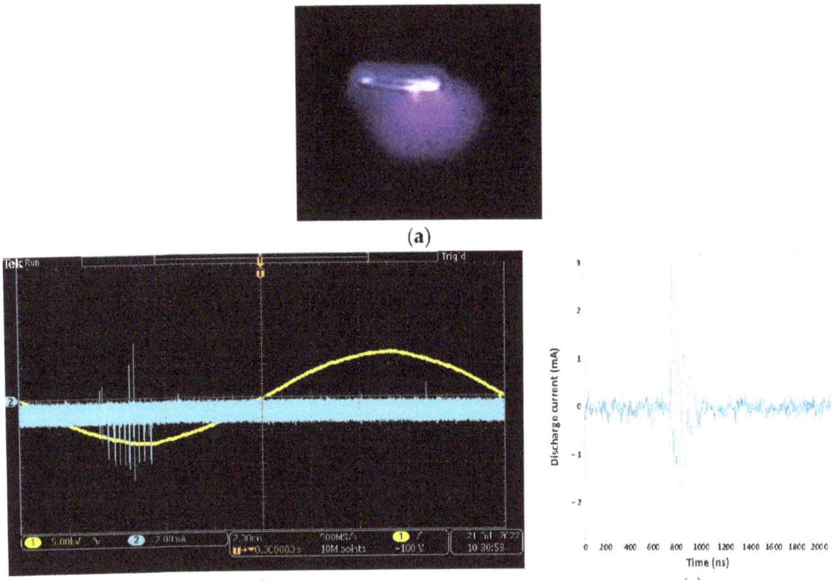

Figure 8. Corona discharges in a needle-plate geometry (needle diameter = 0.8 mm, needle tip-to-ground distance = 70 mm, 60 kPa, 3.6 kV, 50 Hz). (**a**) Photograph of the corona discharge in the tip of the needle. (**b**) Voltage (yellow) and current (blue) waveforms during the corona discharges. (**c**) Detail of a corona discharge as seen in the electrical current pattern.

5. Summary of EH Methods for Power Lines

This section summarizes the state-of-the-art EH systems for transmission lines found in the technical literature, shown in Figure 9.

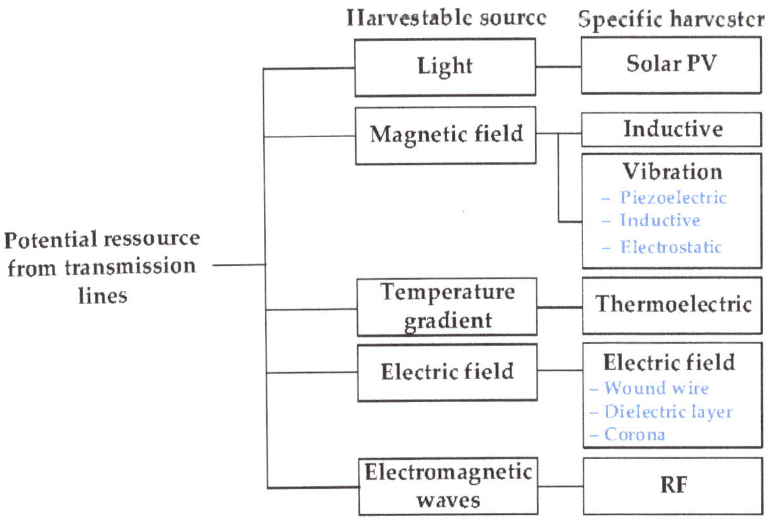

Figure 9. State-of-the-art EH systems for transmission lines.

Table 1 summarizes the main characteristics of the EH methods for power lines reviewed in this section.

Table 1. EH methods comparison for power lines.

EH Type	Power Density	Advantages	Disadvantages	References
Solar	<100 mW/cm^2	For AC and DC grids and high output power	No constant supply and regular maintenance	[14]
Vibration piezoelectric	10–200 μW/cm^3	No external supply required and high voltage output	Only for AC systems & Moving elements	[123]
Vibration inductive	1–2 μW/cm^3	No external supply required	Only for AC systems & Moving elements & low voltage output	[123]
Thermoelectric	50 μW/cm^2 (ΔT = 5 °C)	For AC and DC grids	Requires current flow and a heat sink	[14]
Magnetic field	280 μW/cm^2	Easy to implement in AC grids	Difficult in DC grids and requires current flow	[14]
Electric field	170 μW/cm^2	Available under no load conditions	Difficult in DC grids	[124]
RF	0.0002–1 μW/cm^2	Abundant in urban areas For AC and DC grids	Low power density and scarce in remote areas	[14]
Corona	Not available	For AC and DC grids	Risk of overcurrent & requires more research	[20]

The results presented in Table 1 clearly show that solar is the EH system with the highest power density, and it can be applied to both AC and DC transmission lines, although this method presents two important drawbacks related to periodic maintenance requirements and the impossibility of a constant supply. Other EH methods compatible with AC and DC lines are thermoelectric, RF and corona. RF is probably the EH system with the lowest power density. Solar, thermoelectric, electric field, and RF EH allow for generating energy even when the line is not carrying any current, i.e., under no load conditions. The results show that there is not a clear winner or universal solution, so the most suitable EH system depends on the characteristics of each particular application. These characteristics include the geographical location (it defines dust, sand, dirt, ice. or snow conditions, solar resource, or ambient RF energy), the intensity of the current flowing in the line, or the line voltage, among others.

It is worth noting that WSNs in transmission lines are often enclosed inside a metallic box that acts as a Faraday cage to block the intense electric field. In such cases, the common ground of the electronic board is connected to the metallic box, which, in turn, is in contact with the line conductor.

6. Identified Challenges and Research Needs

WSNs are often supplied at low power and are energy constrained, resulting in low processing capabilities, reduced power storage, and limited operational lifetime, although they offer multifunctional abilities. WSNs randomly deactivate when their storage elements are drained, thus leading to reduced duty cycles and reliability or short-range transmission [125]. All of these limitations contribute to quickly draining their stored energy when operating for extended periods [22]. Therefore, different critical issues around EH circuits have to be addressed, which are listed below in no specific order.

- WSNs must support hybrid EH configurations to extend communication frequency and range as well as onboard features. This area requires more research and development [124].
- Batteryless systems require a true self-starting operation; that is, to start their operation from any amount of ambient power without needing any special arrangement [124].

- Further developments in low-loss rectifiers, low-power micro-controllers, mV-range DC/DC converters, and highly efficient regulators are required to maximize system efficiency, thereby increasing communication reliability and system lifetime [14].
- Efficiency and integration are key points for a widespread application of WSNs, specifically batteryless, so the harvesters, interface circuits, and sensors could be integrated into the same package and even into the same semiconductor component. These advances could lead to dramatic reductions in the size and costs of WSNs [124].
- The use of improved materials with higher efficiency, such as better materials for solar cells, thermoelectric generators, piezoelectric cantilevers, magnetic components, or capacitors with loess leakage [124].
- The minimum voltage and power levels at which the system can harvest energy must be reduced to the minimum levels possible to extract all available energy [24].
- WSN devices operate a large portion of their time in sleep mode, which largely determines total power requirements, so to increase their lifespan, technologies that minimize power consumption in sleep mode need to be further optimized [24].
- The entire system can only be turned on when the voltage on the storage capacitor is beyond a threshold level, so the system's lifetime can be significantly extended by lowering the operating voltage [24].
- Due to the energy requirements, energy-efficient wireless communication modules play a critical role in increasing the lifetime of WSNs. Therefore, to optimize power transmission efficiency, ultra-low-power (ULP) transceivers are needed, along with the development of simpler modulation techniques and optimized signal waveforms and bandwidth [14].
- Multi-layer harvesting architectures allow more compact energy harvesters to be developed, which reduces storage needs, so much research is needed in this area [14].
- Data collected by WSNs must be transmitted securely and protected from cyber-attacks. Therefore, future developments need to consider this aspect [104] since power grids are strategic and critical facilities.

7. Conclusions

Due to the continuous increase in energy consumption worldwide, power grids in general, and transmission lines in particular, face important challenges related to their optimal use, availability, and security aspects. Predictive maintenance plans that are based on network monitoring approaches are known to be useful in increasing the stability, reliability, availability, and security of power grids. Transmission lines are found everywhere, sometimes in remote and inaccessible areas, where routine inspections are complex and expensive. For this, monitoring strategies based on wireless sensors placed in strategic places that acquire and send data in rea time to a control center becomes essential. To maximize their lifetime, such sensors must incorporate an energy harvesting (EH) unit to supply all electronic components. Due to the specificities of transmission lines, this work has reviewed and analyzed the state-of-the-art of energy harvesting strategies for transmission lines in depth, an area with enormous potential for expansion. EH methods applicable to transmission lines include solar, thermoelectric, magnetic field (inductive, and different types of vibrations induced by the magnetic field), electric field (wound wire, dielectric layer, and corona), and RF. From the state-of-the-art, it has been shown that among the different existing energy harvesting methods, there is not a universal solution since the most suitable EH system depends on the characteristics of each particular application, characteristics such as the geographical location (it defines dust, sand, dirt, ice, or snow conditions, solar resource, or available ambient RF energy), the intensity of the current flowing in the line or the line voltage among others. This work has also identified the challenges and research that needs to be addressed for a successful application of this technology because, despite the importance of this topic, there is still a long way to go as this area is still in its early stages. There is a wide range of possible applications that could potentially benefit from the introduction of such energy harvesting systems, including power transformers, distribution switches, or medium- and low- voltage power lines, among others.

Author Contributions: Conceptualization, J.-R.R.; methodology, J.-R.R. and S.B.; formal analysis, J.-R.R., M.M.-E. and S.B.; investigation, J.-R.R.; writing—original draft preparation, J.-R.R.; writing—review and editing, J.-R.R. and M.M.-E.; funding acquisition, J.-R.R. All authors have read and agreed to the published version of the manuscript.

Funding: This research was funded by Ministerio de Ciencia e Innovación de España, grant number PID2020-114240RB-I00 and by the Generalitat de Catalunya, grant number 2017 SGR 967.

Institutional Review Board Statement: Not applicable.

Informed Consent Statement: Not applicable.

Conflicts of Interest: The authors declare no conflict of interest.

References

1. Eunice Akin-Ponnle, A.; Borges Carvalho, N.; Poniszewska-Maranda, A.; Maranda, W. Energy Harvesting Mechanisms in a Smart City—A Review. *Smart Cities* **2021**, *4*, 476–498. [CrossRef]
2. Sudevalayam, S.; Kulkarni, P. Energy Harvesting Sensor Nodes: Survey and Implications. *IEEE Commun. Surv. Tutor.* **2011**, *13*, 443–461. [CrossRef]
3. Sultania, A.K.; Delgado, C.; Famaey, J. Enabling Low-Latency Bluetooth Low Energy on Energy Harvesting Batteryless Devices Using Wake-Up Radios. *Sensors* **2020**, *20*, 5196. [CrossRef] [PubMed]
4. Schlechter, T. Energy Harvesting Approaches for Wireless Sensor Nodes in High Voltage Direct Current Systems. In Proceedings of the IEEE 5th World Forum on Internet of Things (WF-IoT), Limerick, Ireland, 15–18 April 2019; pp. 243–246. [CrossRef]
5. Prajwal, K.T.; Manickavasagam, K.; Suresh, R. A review on vibration energy harvesting technologies: Analysis and technologies. *Eur. Phys. J. Spéc. Top.* **2022**, *231*, 1359–1371. [CrossRef]
6. Liu, Y.; Khanbareh, H.; Halim, M.A.; Feeney, A.; Zhang, X.; Heidari, H.; Ghannam, R.; Correspondence, R.; Ghannam, J. Piezoelectric energy harvesting for self-powered wearable upper limb applications. *Nano Sel.* **2021**, *2*, 1459–1479. [CrossRef]
7. Verma, G.; Sharma, V. A Novel RF Energy Harvester for Event-Based Environmental Monitoring in Wireless Sensor Networks. *IEEE Internet Things J.* **2022**, *9*, 3189–3203. [CrossRef]
8. Ang, K.L.-M.; Seng, J.K.P.; Zungeru, A.M. Optimizing Energy Consumption for Big Data Collection in Large-Scale Wireless Sensor Networks with Mobile Collectors. *IEEE Syst. J.* **2017**, *12*, 616–626. [CrossRef]
9. Ijemaru, G.; Ang, K.; Seng, J. Mobile Collectors for Opportunistic Internet of Things in Smart City Environment with Wireless Power Transfer. *Electronics* **2021**, *10*, 697. [CrossRef]
10. Hong, Y.-W.P.; Hsu, T.-C.; Chennakesavula, P. Wireless Power Transfer for Distributed Estimation in Wireless Passive Sensor Networks. *IEEE Trans. Signal Process.* **2016**, *64*, 5382–5395. [CrossRef]
11. Moghe, R.; Yang, Y.; Lambert, F.; Divan, D. A Scoping Study of Electric and Magnetic Field Energy Harvesting for Wireless Sensor Networks in Power System Applications. In Proceedings of the 2009 IEEE Energy Conversion Congress and Exposition, San Jose, CA, USA, 20–24 September 2009; pp. 3550–3557.
12. Chen, X.; Sun, L.; Zhu, H.; Zhen, Y.; Chen, H. Application of Internet of Things in Power-Line Monitoring. In Proceedings of the 2012 International Conference on Cyber-Enabled Distributed Computing and Knowledge Discovery, Sanya, China, 10–12 October 2012; pp. 423–426. [CrossRef]
13. Qin, X.; Wu, G.; Lei, J.; Fan, F.; Ye, X.; Mei, Q. A Novel Method of Autonomous Inspection for Transmission Line based on Cable Inspection Robot LiDAR Data. *Sensors* **2018**, *18*, 596. [CrossRef]
14. Cetinkaya, O.; Akan, O.B. Electric-Field Energy Harvesting in Wireless Networks. *IEEE Wirel. Commun.* **2017**, *24*, 34–41. [CrossRef]
15. Zafeiropoulou, M.; Mentis, I.; Sijakovic, N.; Terzic, A.; Fotis, G.; Maris, T.I.; Vita, V.; Zoulias, E.; Ristic, V.; Ekonomou, L. Forecasting Transmission and Distribution System Flexibility Needs for Severe Weather Condition Resilience and Outage Management. *Appl. Sci.* **2022**, *12*, 7334. [CrossRef]
16. Sijakovic, N.; Terzic, A.; Fotis, G.; Mentis, I.; Zafeiropoulou, M.; Maris, T.I.; Zoulias, E.; Elias, C.; Ristic, V.; Vita, V. Active System Management Approach for Flexibility Services to the Greek Transmission and Distribution System. *Energies* **2022**, *15*, 6134. [CrossRef]
17. Zainuddin, N.M.; Rahman, M.S.A.; Kadir, M.Z.A.A.; Ali, N.H.N.; Ali, Z.; Osman, M.; Mansor, M.; Ariffin, A.M.; Nor, S.F.M.; Nasir, N.A.F.M. Review of Thermal Stress and Condition Monitoring Technologies for Overhead Transmission Lines: Issues and Challenges. *IEEE Access* **2020**, *8*, 120053–120081. [CrossRef]
18. Liu, Y.; Riba, J.-R.; Moreno-Eguilaz, M.; Sanllehí, J. Analysis of a Smart Sensor Based Solution for Smart Grids Real-Time Dynamic Thermal Line Rating. *Sensors* **2021**, *21*, 7388. [CrossRef]
19. Wang, Z.; Hu, J.; Han, J.; Zhao, G.; He, J.; Wang, S.X. A Novel High-Performance Energy Harvester Based on Nonlinear Resonance for Scavenging Power-Frequency Magnetic Energy. *IEEE Trans. Ind. Electron.* **2017**, *64*, 6556–6564. [CrossRef]

20. Shi, Y.; Cui, X.; Qi, L.; Zhang, X.; Li, X.; Shen, H. A Novel Energy Harvesting Method for Online Monitoring Sensors in HVDC Overhead Line. *IEEE Trans. Ind. Electron.* **2022**, *70*, 2139–2143. [CrossRef]
21. Kamruzzaman, S.M.; Fernando, X.; Jaseemuddin, M. Energy Harvesting Wireless Sensors for Smart Cities. In Proceedings of the 2017 IEEE Canada International Humanitarian Technology Conference (IHTC), Toronto, ON, Canada, 21–22 July 2017; pp. 218–222. [CrossRef]
22. Ijemaru, G.K.; Ang, K.L.-M.; Seng, J.K. Wireless power transfer and energy harvesting in distributed sensor networks: Survey, opportunities, and challenges. *Int. J. Distrib. Sens. Netw.* **2022**, *18*, 15501477211067740. [CrossRef]
23. Moreno-Cruz, F.; Toral-López, V.; Escobar-Molero, A.; Ruíz, V.; Rivadeneyra, A.; Morales, D. *treNch*: Ultra-Low Power Wireless Communication Protocol for IoT and Energy Harvesting. *Sensors* **2020**, *20*, 6156. [CrossRef] [PubMed]
24. Shafiee, N.; Tewari, S.; Calhoun, B.; Shrivastava, A. Infrastructure Circuits for Lifetime Improvement of Ultra-Low Power IoT Devices. *IEEE Trans. Circuits Syst. I Regul. Pap.* **2017**, *64*, 2598–2610. [CrossRef]
25. Khan, F.U.; Qadir, M.U. State-of-the-art in vibration-based electrostatic energy harvesting. *J. Micromechanics Microengineering* **2016**, *26*, 103001. [CrossRef]
26. Mazunga, F.; Nechibvute, A. Ultra-low power techniques in energy harvesting wireless sensor networks: Recent advances and issues. *Sci. Afr.* **2021**, *11*, e00720. [CrossRef]
27. Mallick, C.; Satpathy, S. Challenges and Design Goals of Wireless Sensor Networks: A Sate-of-the-art Review. *Int. J. Comput. Appl.* **2018**, *179*, 42–47. [CrossRef]
28. Abdelatty, O.; Chen, X.; Alghaihab, A.; Wentzloff, D. Bluetooth Communication Leveraging Ultra-Low Power Radio Design. *J. Sens. Actuator Netw.* **2021**, *10*, 31. [CrossRef]
29. Loo, M.H.-W.; Ramiah, H.; Lei, K.-M.; Lim, C.C.; Lai, N.S.; Mak, P.-I.; Martins, R.P. Fully-Integrated Timers for Ultra-Low-Power Internet-of-Things Nodes—Fundamentals and Design Techniques. *IEEE Access* **2022**, *10*, 65936–65950. [CrossRef]
30. Jiang, C.; Li, X.; Lian, S.W.M.; Ying, Y.; Ho, J.S.; Ping, J. Wireless Technologies for Energy Harvesting and Transmission for Ambient Self-Powered Systems. *ACS Nano* **2021**, *15*, 9328–9354. [CrossRef]
31. Vougioukas, G.; Dimitriou, A.; Bletsas, A.; Sahalos, J. Practical Energy Harvesting for Batteryless Ambient Backscatter Sensors. *Electronics* **2018**, *7*, 95. [CrossRef]
32. Sabovic, A.; Delgado, C.; Subotic, D.; Jooris, B.; De Poorter, E.; Famaey, J. Energy-Aware Sensing on Battery-Less LoRaWAN Devices with Energy Harvesting. *Electronics* **2020**, *9*, 904. [CrossRef]
33. Delgado, C.; Sanz, J.M.; Famaey, J. On the Feasibility of Battery-Less LoRaWAN Communications Using Energy Harvesting. In Proceedings of the 2019 IEEE Global Communications Conference (GLOBECOM), Waikoloa, HI, USA, 9–13 December 2019. [CrossRef]
34. Altinel, D.; Kurt, G.K. Modeling of Multiple Energy Sources for Hybrid Energy Harvesting IoT Systems. *IEEE Internet Things J.* **2019**, *6*, 10846–10854. [CrossRef]
35. Häggström, F.; Delsing, J. IoT Energy Storage—A Forecast. *Energy Harvest. Syst.* **2018**, *5*, 43–51. [CrossRef]
36. Munir, B.; Dyo, V. On the Impact of Mobility on Battery-Less RF Energy Harvesting System Performance. *Sensors* **2018**, *18*, 3597. [CrossRef] [PubMed]
37. Yang, C.-C.; Pandey, R.; Tu, T.-Y.; Cheng, Y.-P.; Chao, P.C.-P. An efficient energy harvesting circuit for batteryless IoT devices. *Microsyst. Technol.* **2019**, *26*, 195–207. [CrossRef]
38. Guo, F.; Hayat, H.; Wang, J. Energy harvesting devices for high voltage transmission line monitoring. In Proceedings of the 2011 IEEE Power and Energy Society General Meeting, Detroit, MI, USA, 24–28 July 2011; pp. 1–8. [CrossRef]
39. Li, P.; Wen, Y.; Zhang, Z.; Pan, S. A High-Efficiency Management Circuit Using Multiwinding Upconversion Current Transformer for Power-Line Energy Harvesting. *IEEE Trans. Ind. Electron.* **2015**, *62*, 6327–6335. [CrossRef]
40. Bernauer, C.; Bohme, H.; Grossmann, S.; Hinrichsen, V.; Kornhuber, S.; Markalous, S.; Darmstadt, T.; Muhr, M.; Strehl, T.; Teminova, R. Temperature Measurement on Overhead Transmission Lines (OHTL) Utilizing Surface Acoustic Wave (SAW) Sensors. In *19th International Conference on Electricity Distribution*; CIRED, Ed.; CIRED: Vienna, Austria, 2007; pp. 1–4.
41. Yildiz, H.U.; Gungor, V.C.; Tavli, B. A Hybrid Energy Harvesting Framework for Energy Efficiency in Wireless Sensor Networks Based Smart Grid Applications. In Proceedings of the 2018 17th Annual Mediterranean Ad Hoc Networking Workshop (Med-Hoc-Net), Capri, Italy, 20–22 June 2018; pp. 1–6.
42. Takshi, A.; Aljafari, B.; Kareri, T.; Stefanakos, E. A Critical Review on the Voltage Requirement in Hybrid Cells with Solar Energy Harvesting and Energy Storage Capability. *Batter. Supercaps* **2020**, *4*, 252–267. [CrossRef]
43. Zhao, X.; Keutel, T.; Baldauf, M.; Kanoun, O. Energy harvesting for a wireless-monitoring system of overhead high-voltage power lines. *IET Gener. Transm. Distrib.* **2013**, *7*, 101–107. [CrossRef]
44. Dondi, D.; Bertacchini, A.; Brunelli, D.; Larcher, L.; Benini, L. Modeling and Optimization of a Solar Energy Harvester System for Self-Powered Wireless Sensor Networks. *IEEE Trans. Ind. Electron.* **2008**, *55*, 2759–2766. [CrossRef]
45. Antony, S.M.; Indu, S.; Pandey, R. An efficient solar energy harvesting system for wireless sensor network nodes. *J. Inf. Optim. Sci.* **2020**, *41*, 39–50. [CrossRef]
46. Strzalkowski, B.; Mo, K. Solving Isolation- and Power Supply Problems for Current Monitoring in High Voltage Power Line Application. In Proceedings of the PCIM Europe 2018, International Exhibition and Conference for Power Electronics, Intelligent Motion, Renewable Energy and Energy Management, Nuremberg, Germany, 5–7 June 2018; 2018; pp. 1295–1297.

47. Wang, T.; Huang, K.; Liu, W.; Hou, J.; Zhang, Z. A hybrid solar-RF energy harvesting system based on tree-shaped antenna array. *Int. J. RF Microw. Comput. Eng.* **2022**, *32*, e23301. [CrossRef]
48. Yang, C.C.; Bin Rohimi, R.I.; Hanif, N.H.H.M.; Ibrahim, S.N. Non-intrusive Energy Harvesting from Vibration of Air Conditioning Condenser Unit Utilizing Piezoelectric Sensors. In Proceedings of the 8th International Conference on Computer and Communication Engineering (ICCCE), Kuala Lumpur, Malaysia, 22–23 June 2021; pp. 126–130. [CrossRef]
49. Khaligh, A.; Zeng, P.; Zheng, C. Kinetic Energy Harvesting Using Piezoelectric and Electromagnetic Technologies—State of the Art. *IEEE Trans. Ind. Electron.* **2010**, *57*, 850–860. [CrossRef]
50. Shaikh, F.K.; Zeadally, S. Energy harvesting in wireless sensor networks: A comprehensive review. *Renew. Sustain. Energy Rev.* **2016**, *55*, 1041–1054. [CrossRef]
51. Beeby, S.P.; Torah, R.N.; Tudor, M.J.; Glynne-Jones, P.; O'Donnell, T.; Saha, C.R.; Roy, S. A micro electromagnetic generator for vibration energy harvesting. *J. Micromech. Microeng.* **2007**, *17*, 1257–1265. [CrossRef]
52. Kanno, I. Piezoelectric MEMS for energy harvesting. *J. Phys. Conf. Ser.* **2015**, *660*, 012001. [CrossRef]
53. Wang, Q.; Kim, K.-B.; Woo, S.; Sung, T. Magnetic Field Energy Harvesting with a Lead-Free Piezoelectric High Energy Conversion Material. *Energies* **2021**, *14*, 1346. [CrossRef]
54. Gao, S.; Zeng, X.; Tao, B.; Ke, T.; Feng, S.; Chen, Y.; Zhou, J.; Lan, W. Self-powered sensing of power transmission lines galloping based on piezoelectric energy harvesting. *Int. J. Electr. Power Energy Syst.* **2023**, *144*, 108607. [CrossRef]
55. He, W.; Li, P.; Wen, Y.; Zhang, J.; Lu, C.; Yang, A. Energy harvesting from electric power lines employing the Halbach arrays. *Rev. Sci. Instrum.* **2013**, *84*, 105004. [CrossRef]
56. Abasian, A.; Tabesh, A.; Nezhad, A.Z.; Rezaei-Hosseinabadi, N. Design Optimization of an Energy Harvesting Platform for Self-Powered Wireless Devices in Monitoring of AC Power Lines. *IEEE Trans. Power Electron.* **2018**, *33*, 10308–10316. [CrossRef]
57. Meng, Q.; Huo, F.; Teng, S.; Ding, Z.; Gu, T.; Cao, C. An Autonomous Vibration-Sensing System for Power Transmission Lines Monitoring. *J. Phys. Conf. Ser.* **2021**, *2095*, 012014. [CrossRef]
58. Xiao, H.; Peng, H.; Yuan, J. A Coil Connection Switching Strategy for Maximum Power Delivery in Electromagnetic Vibration Energy Harvesting System. In Proceedings of the 2021 IEEE Energy Conversion Congress and Exposition (ECCE), Vancouver, BC, Canada, 10–14 October 2021; pp. 5867–5872. [CrossRef]
59. Hosseinimehr, T.; Tabesh, A. Magnetic Field Energy Harvesting from AC Lines for Powering Wireless Sensor Nodes in Smart Grids. *IEEE Trans. Ind. Electron.* **2016**, *63*, 4947–4954. [CrossRef]
60. Li, M.; Luo, A.; Luo, W.; Wang, F. Recent Progress on Mechanical Optimization of MEMS Electret-Based Electrostatic Vibration Energy Harvesters. *J. Microelectromechanical Syst.* **2022**, *31*, 726–740. [CrossRef]
61. Chen, S.-M.; Hu, J.-H. Experimental Study of a Hybrid Vibration Energy Harvesting Mechanism. In Proceedings of the 2011 Symposium on Piezoelectricity, Acoustic Waves and Device Applications (SPAWDA), Shenzhen, China, 9–11 December 2011; pp. 56–59. [CrossRef]
62. Wang, Q.; Kim, K.-B.; Woo, S.-B.; Song, Y.; Sung, T.-H. A Magneto-Mechanical Piezoelectric Energy Harvester Designed to Scavenge AC Magnetic Field from Thermal Power Plant with Power-Line Cables. *Energies* **2021**, *14*, 2387. [CrossRef]
63. Muscat, A.; Bhattacharya, S.; Zhu, Y. Electromagnetic Vibrational Energy Harvesters: A Review. *Sensors* **2022**, *22*, 5555. [CrossRef] [PubMed]
64. Becker, T.; Kluge, M.; Schalk, J.; Otterpohl, T.; Hilleringmann, U. Power Management for Thermal Energy Harvesting in Aircrafts. In Proceedings of the IEEE Sensors Conference, Lecce, Italy, 26–29 October 2008; pp. 681–684. [CrossRef]
65. Proto, A.; Bibbo, D.; Cerny, M.; Vala, D.; Kasik, V.; Peter, L.; Conforto, S.; Schmid, M.; Penhaker, M. Thermal Energy Harvesting on the Bodily Surfaces of Arms and Legs through a Wearable Thermo-Electric Generator. *Sensors* **2018**, *18*, 1927. [CrossRef] [PubMed]
66. Torfs, T.; Leonov, V.; Van Hoof, C.; Gyselinckx, B. Body-Heat Powered Autonomous Pulse Oximeter. In Proceedings of the 5th IEEE Sensors, Daegu, Korea, 22–25 October 2006; pp. 427–430. [CrossRef]
67. Kadechkar, A.; Riba, J.-R.; Moreno-Eguilaz, M.; Perez, J. SmartConnector: A Self-Powered IoT Solution to Ease Predictive Maintenance in Substations. *IEEE Sens. J.* **2020**, *20*, 11632–11641. [CrossRef]
68. Gupta, V.; Kandhalu, A.; Rajkumar, R. Energy Harvesting from Electromagnetic Energy Radiating from AC Power Lines. In Proceedings of the 6th Workshop on Hot Topics in Embedded Networked Sensors, Killarney, Ireland, 28–29 June 2010. [CrossRef]
69. Wright, S.W.; Kiziroglou, M.E.; Spasic, S.; Radosevic, N.; Yeatman, E.M. Inductive Energy Harvesting from Current-Carrying Structures. *IEEE Sens. Lett.* **2019**, *3*, 1–4. [CrossRef]
70. Toh, T.T.; Wright, S.; Kiziroglou, M.; Mueller, J.; Sessinghaus, M.; Yeatman, E.; Mitcheson, P.D. Inductive energy harvesting from variable frequency and amplitude aircraft power lines. *J. Phys. Conf. Ser.* **2014**, *557*, 012095. [CrossRef]
71. Tashiro, K.; Wakiwaka, H.; Inoue, S.-I.; Uchiyama, Y. Energy Harvesting of Magnetic Power-Line Noise. *IEEE Trans. Magn.* **2011**, *47*, 4441–4444. [CrossRef]
72. Wang, H.; Shi, G.; Han, C. A Free-Standing Electromagnetic Energy Harvester for Condition Monitoring in Smart Grid. *Wirel. Power Transf.* **2021**, *2021*, 6685308. [CrossRef]
73. Kabakulak, M.; Arslan, S. An Electromagnetic Energy Harvester for Wireless Sensors from Power Lines: Modeling and Experiment Verification. *GAZI Univ. J. Sci.* **2020**, *34*, 786–806. [CrossRef]
74. White, R.M.; Nguyen, D.-S.; Wu, Z.; Wright, P.K. Atmospheric Sensors and Energy Harvesters on Overhead Power Lines. *Sensors* **2018**, *18*, 114. [CrossRef]

75. Park, B.; Kim, N.; Park, J.; Kim, K.; Koo, J.; Park, H.; Ahn, S. Optimization design of toroidal core for magnetic energy harvesting near power line by considering saturation effect. *AIP Adv.* **2018**, *8*, 056728. [CrossRef]
76. Najafi, S.A.A.; Ali, A.A.; Sozer, Y.; De Abreu-Garcia, A. Energy Harvesting from Overhead Transmission Line Magnetic Fields. In Proceedings of the 2018 IEEE Energy Conversion Congress and Exposition (ECCE), Portland, OR, USA, 23–27 September 2018; pp. 7075–7082. [CrossRef]
77. dos Santos, M.P.; Vieira, D.A.; Rodriguez, Y.P.; de Souza, C.P.; de Moraes, T.O.; Freire, R.C. Energy Harvesting Using Magnetic Induction Considering Different Core Materials. In Proceedings of the 2014 IEEE International Instrumentation and Measurement Technology Conference (I2MTC) Proceedings, Montevideo, Uruguay, 12–15 May 2014; pp. 942–944.
78. Gaikwad, A.; Kulkarni, S. Evaluation of dimensional effect on electromagnetic energy harvesting. *Procedia Comput. Sci.* **2018**, *143*, 58–65. [CrossRef]
79. Bhuiyan, R.H.; Dougal, R.A.; Ali, M. A Miniature Energy Harvesting Device for Wireless Sensors in Electric Power System. *IEEE Sens. J.* **2010**, *10*, 1249–1258. [CrossRef]
80. Zhang, J.; Tian, X.; Li, J.; Yan, D. A Novel Electromagnetic Energy Harvester Based on Double-Ring Core for Power Line Energy Harvesting. *J. Circuits Syst. Comput.* **2020**, *29*, 2050265. [CrossRef]
81. Liu, Y.; Xie, X.; Hu, Y.; Qian, Y.; Sheng, G.; Jiang, X.; Liu, Y. A novel high-density power energy harvesting methodology for transmission line online monitoring devices. *Rev. Sci. Instrum.* **2016**, *87*, 075119. [CrossRef] [PubMed]
82. Zangl, H.; Bretterklieber, T.; Brasseur, G. Energy Harvesting for Online Condition Monitoring of High Voltage Overhead Power Lines. In Proceedings of the 2008 IEEE Instrumentation and Measurement Technology Conference, Victoria, BC, Canada, 12–15 May 2008; pp. 1364–1369. [CrossRef]
83. Kim, H.; Choi, D.; Gong, S.; Park, K. Stray electric field energy harvesting technology using MEMS switch from insulated AC power lines. *Electron. Lett.* **2014**, *50*, 1236–1238. [CrossRef]
84. Chang, K.-S.; Kang, S.-M.; Park, K.-J.; Shin, S.-H.; Kim, H.-S.; Kim, H.-S. Electric Field Energy Harvesting Powered Wireless Sensors for Smart Grid. *J. Electr. Eng. Technol.* **2012**, *7*, 75–80. [CrossRef]
85. Kang, S.; Yang, S.; Kim, H. Non-intrusive voltage measurement of ac power lines for smart grid system based on electric field energy harvesting. *Electron. Lett.* **2017**, *53*, 181–183. [CrossRef]
86. Zhao, X.; Keutel, T.; Baldauf, M.; Kanoun, O. Energy Harvesting for Overhead Power Line Monitoring. In Proceedings of the International Multi-Conference on Systems, Signals & Devices, Chemnitz, Germany, 20–23 March 2012. [CrossRef]
87. Zhou, J.; Zhang, J.; Xu, C.; Fang, L.; Wang, Y.; Zhuang, Y.; Jia, T.; Yi, H.; Han, C. On the Improvement of Electric Field Energy Harvesting from Domestic Power Lines. *AEU-Int. J. Electron. Commun.* **2022**, *155*, 154349. [CrossRef]
88. Rodríguez, J.C.; Holmes, D.G.; McGrath, B.P.; Teixeira, C. Energy Harvesting from Medium Voltage Electric Fields Using Pulsed Flyback Conversion. In Proceedings of the 2016 IEEE 8th International Power Electronics and Motion Control Conference (IPEMC-ECCE Asia), Hefei, China, 22–26 May 2016; pp. 3591–3598. [CrossRef]
89. Gulati, M.; Parizi, F.S.; Whitmire, E.; Gupta, S.; Ram, S.S.; Singh, A.; Patel, S.N. CapHarvester: A Stick-On Capacitive Energy Harvester Using Stray Electric Field from AC Power Lines. *Proc. ACM Interact. Mob. Wearable Ubiquitous Technol.* **2018**, *2*, 1–20. [CrossRef]
90. Khan, M.R.; Islam, M.A.; Rana, M.M.; Haque, T.; Joy, S.I.I. A Circuit Model for Energy Harvesting from Fringing Electric Fields for Mobile Wearable Device Applications. *Energies* **2021**, *14*, 7016. [CrossRef]
91. Moser, M.J.; Bretterklieber, T.; Zangl, H.; Brasseur, G. Strong and Weak Electric Field Interfering: Capacitive Icing Detection and Capacitive Energy Harvesting on a 220-kV High-Voltage Overhead Power Line. *IEEE Trans. Ind. Electron.* **2011**, *58*, 2597–2604. [CrossRef]
92. Menéndez, O.; Romero, L.; Cheein, F.A. Serial Switch Only Rectifier as a Power Conditioning Circuit for Electric Field Energy Harvesting. *Energies* **2020**, *13*, 5279. [CrossRef]
93. Alneyadi, F.; Alkaabi, M.; Alketbi, S.; Hajraf, S.; Ramzan, R. 2.4 GHz WLAN RF Energy Harvester for Passive Indoor Sensor Nodes. In Proceedings of the 2014 IEEE International Conference on Semiconductor Electronics (ICSE2014), Kuala Lumpur, Malaysia, 27–29 August 2014; pp. 471–474. [CrossRef]
94. Mouapi, A.; Hakem, N.; Delisle, G.Y. A new approach to design of RF energy harvesting system to enslave wireless sensor networks. *ICT Express* **2018**, *4*, 228–233. [CrossRef]
95. Pinuela, M.; Mitcheson, P.D.; Lucyszyn, S. Ambient RF Energy Harvesting in Urban and Semi-Urban Environments. *IEEE Trans. Microw. Theory Tech.* **2013**, *61*, 2715–2726. [CrossRef]
96. Ibrahim, H.H.; Singh, M.J.; Al-Bawri, S.S.; Ibrahim, S.K.; Islam, M.T.; Alzamil, A.; Islam, S. Radio Frequency Energy Harvesting Technologies: A Comprehensive Review on Designing, Methodologies, and Potential Applications. *Sensors* **2022**, *22*, 4144. [CrossRef]
97. Kamalinejad, P.; Mahapatra, C.; Sheng, Z.; Mirabbasi, S.; Leung, V.C.M.; Guan, Y.L. Wireless energy harvesting for the Internet of Things. *IEEE Commun. Mag.* **2015**, *53*, 102–108. [CrossRef]
98. Lu, X.; Wang, P.; Niyato, D.; Kim, D.I.; Han, Z. Wireless Networks with RF Energy Harvesting: A Contemporary Survey. *IEEE Commun. Surv. Tutor.* **2015**, *17*, 757–789. [CrossRef]
99. Cruz, F.M.; Molero, A.E.; Castillo, E.; Becherer, M.; Rivadeneyra, A.; Morales, D.P. Why Use RF Energy Harvesting in Smart Grids. In Proceedings of the IEEE 23rd International Workshop on Computer Aided Modeling and Design of Communication Links and Networks (CAMAD), Barcelona, Spain, 17–19 September 2018. [CrossRef]

100. Nintanavongsa, P. A Survey on RF Energy Harvesting: Circuits and Protocols. *Energy Procedia* **2014**, *56*, 414–422. [CrossRef]
101. Paradiso, J.; Starner, T. Energy Scavenging for Mobile and Wireless Electronics. *IEEE Pervasive Comput.* **2005**, *4*, 18–27. [CrossRef]
102. Kawahara, Y.; Wei, W.; Narusue, Y.; Shigeta, R.; Asami, T.; Tentzeris, M. Virtualizing Power Cords by Wireless Power Transmission and Energy Harvesting. In Proceedings of the 2013 IEEE Radio and Wireless Symposium, Austin, TX, USA, 20–23 January 2013; pp. 37–39. [CrossRef]
103. Sojan, S.; Kulkarni, R. A Comprehensive Review of Energy Harvesting Techniques and its Potential Applications. *Int. J. Comput. Appl.* **2016**, *139*, 14–19. [CrossRef]
104. Jayasinghe, D.H.G.A.; Jayakody, D.N.K. Survey on Energy Harvesting in Smart Grid Networks. In *International Research Conference of SLTC 2020—IRC 2020*; Sri Lanka Technological Campus: Colombo, Sri Lanka, 2020; pp. 1–7.
105. Liu, X.; Sanchez-Sinencio, E. A Highly Efficient Ultralow Photovoltaic Power Harvesting System with MPPT for Internet of Things Smart Nodes. *IEEE Trans. Very Large Scale Integr. (VLSI) Syst.* **2015**, *23*, 3065–3075. [CrossRef]
106. Jayakumar, H.; Lee, K.; Lee, W.S.; Raha, A.; Kim, Y.; Raghunathan, V. Powering the Internet of Things. In Proceedings of the 2014 international Symposium Low Power Electronics and Design (ISLPED), La Jolla, CA, USA, 11–13 August 2014; pp. 375–380. [CrossRef]
107. Muhammad, S.; Tiang, J.J.; Wong, S.K.; Iqbal, A.; Alibakhshikenari, M.; Limiti, E. Compact Rectifier Circuit Design for Harvesting GSM/900 Ambient Energy. *Electronics* **2020**, *9*, 1614. [CrossRef]
108. Sanislav, T.; Zeadally, S.; Mois, G.D.; Folea, S.C. Wireless energy harvesting: Empirical results and practical considerations for Internet of Things. *J. Netw. Comput. Appl.* **2018**, *121*, 149–158. [CrossRef]
109. Tran, V.H.; Misra, A.; Xiong, J.; Balan, R.K. WiWear: Wearable Sensing via Directional WiFi Energy Harvesting. In Proceedings of the 2019 IEEE International Conference on Pervasive Computing and Communications (PerCom), Kyoto, Japan, 11–15 March 2019. [CrossRef]
110. Mahamat, Y.; Hussain, S.R.R.; Eroglu, A. Far-Field RF Energy Harvesting System for Distribution Power Lines. In Proceedings of the 2016 IEEE/ACES International Conference on Wireless Information Technology and Systems (ICWITS) and Applied Computational Electromagnetics (ACES), Honolulu, HI, USA, 13–18 March 2016; Institute of Electrical and Electronics Engineers Inc.: Piscataway, NJ, USA, 2016; pp. 1–2.
111. Li, P.; Long, Z.; Yang, Z. RF Energy Harvesting for Batteryless and Maintenance-Free Condition Monitoring of Railway Tracks. *IEEE Internet Things J.* **2021**, *8*, 3512–3523. [CrossRef]
112. Oller, J.; Demirkol, I.; Casademont, J.; Paradells, J.; Gamm, G.U.; Reindl, L. Has Time Come to Switch from Duty-Cycled MAC Protocols to Wake-Up Radio for Wireless Sensor Networks? *IEEE/ACM Trans. Netw.* **2016**, *24*, 674–687. [CrossRef]
113. Caballe, M.C.; Auge, A.C.; Aspas, J.P. Wake-Up Radio: An Enabler of Wireless Convergence. *IEEE Access* **2021**, *9*, 3784–3797. [CrossRef]
114. Mafi, Y.; Amirhosseini, F.; Hosseini, S.A.; Azari, A.; Masoudi, M.; Vaezi, M. Ultra-Low-Power IoT Communications: A Novel Address Decoding Approach for Wake-Up Receivers. *IEEE Trans. Green Commun. Netw.* **2022**, *6*, 1107–1121. [CrossRef]
115. Ding, M.; Zhang, P.; Lu, C.; Zhang, Y.; Traferro, S.; van Schaik, G.-J.; Liu, Y.-H.; Huijts, J.; Bachmann, C.; Dolmans, G.; et al. A 2.4 GHz BLE-Compliant Fully-Integrated Wakeup Receiver for Latency-Critical IoT Applications Using a 2-dimensional Wakeup Pattern in 90 nm CMOS. In Proceedings of the 2017 IEEE Radio Frequency Integrated Circuits Symposium (RFIC), Honolulu, HI, USA, 4–6 June 2017; pp. 168–171. [CrossRef]
116. Abdelhamid, M.R.; Paidimarri, A.; Chandrakasan, A.P. A −80 dBm BLE-Compliant, FSK Wake-Up Receiver with System and Within-Bit Dutycycling for Scalable Power and Latency. In Proceedings of the 2018 IEEE Custom Integrated Circuits Conference (CICC), San Diego, CA, USA, 8–11 April 2018; pp. 1–4. [CrossRef]
117. Wang, P.-H.P.; Mercier, P.P. 28.2 A 220 μW −85 dBm Sensitivity BLE-Compliant Wake-up Receiver Achieving −60 dB SIR via Single-Die Multi- Channel FBAR-Based Filtering and a 4-Dimentional Wake-Up Signature. In Proceedings of the 2019 IEEE International Solid- State Circuits Conference—(ISSCC), San Francisco, CA, USA, 17–21 February 2019; pp. 440–442. [CrossRef]
118. Liu, Q.; IJntema, W.; Drif, A.; Pawełczak, P.; Zuniga, M. BEH: Indoor Batteryless BLE Beacons using RF Energy Harvesting for Internet of Things. *arXiv* **2019**, arXiv:1911.03381. [CrossRef]
119. Borghei, M.; Ghassemi, M. Separation and Classification of Corona Discharges Under Low Pressures Based on Deep Learning Method. *IEEE Trans. Dielectr. Electr. Insul.* **2022**, *29*, 319–326. [CrossRef]
120. Riba, J.-R. Application of Image Sensors to Detect and Locate Electrical Discharges: A Review. *Sensors* **2022**, *22*, 5886. [CrossRef] [PubMed]
121. Anis, H.; Srivastava, K.D. Pre-Breakdown Discharges in Highly Non-uniform Fields in Relation to Gas-insulated Systems. *IEEE Trans. Electr. Insul.* **1982**, *2*, 131–142. [CrossRef]
122. Zhou, X.; Cui, X.; Lu, T.; Fang, C.; Zhen, Y. Spatial Distribution of Ion Current Around HVDC Bundle Conductors. *IEEE Trans. Power Deliv.* **2012**, *27*, 380–390. [CrossRef]
123. Muhtaroğlu, A. Micro-scale Energy Harvesting for Batteryless Information Technologies. In *Lecture Notes in Energy*; Springer: Berlin, Germany, 2017; Volume 37, pp. 63–85. [CrossRef]

124. Pavana, H.; Deshpande, R. Energy Harvesting Techniques for Monitoring Devices in Smart Grid Application. In Proceedings of the 2020 Third International Conference on Advances in Electronics, Computers and Communications (ICAECC), Bengaluru, India, 11–12 December 2020. [CrossRef]
125. Madhja, A.; Nikoletseas, S.; Raptis, T.P. Distributed wireless power transfer in sensor networks with multiple Mobile Chargers. *Comput. Netw.* **2015**, *80*, 89–108. [CrossRef]

applied sciences

Article

Frequency Stability Analysis of a Low Inertia Power System with Interactions among Power Electronics Interfaced Generators with Frequency Response Capabilities

Lucio Radaelli [1] and Sergio Martinez [2,*]

1 Politecnico di Milano, 20133 Milan, Italy
2 Department of Electrical Engineering, Escuela Técnica Superior de Ingenieros Industriales, Universidad Politécnica de Madrid, 28040 Madrid, Spain
* Correspondence: sergio.martinez@upm.es

Abstract: One of the main actions required to face and limit global warming is the substitution of conventional fossil-fueled electrical generators with renewable ones. Thus, it becomes fundamental to create non-dispatchable renewable generators able to provide services for power system stabilization that nowadays are delivered by conventional ones. Particularly, renewable generators are usually connected to the electrical power system through power electronic converters lacking natural responses to frequency variations. This challenges conventional frequency control methods that are based on synchronous generators' capabilities, particularly in systems with high levels of non-synchronous generation. Solutions based on advanced controls that allow renewable generators to participate in frequency control are the subject of current research efforts worldwide. This paper contributes to these efforts by studying the benefits of introducing Power Reserve Control in photovoltaic generators and Extended Optimal Power Point Tracking control in wind generators to provide frequency control in low inertia power systems and the interactions between them. The tests and the simulations, prove that these kinds of controls help in stabilizing the system frequency thanks to the cooperative action of both types of renewable generators.

Keywords: primary frequency control; frequency stability; LFC scheme; renewable generators; power curtailment; inertial control

Citation: Radaelli, L.; Martinez, S. Frequency Stability Analysis of a Low Inertia Power System with Interactions among Power Electronics Interfaced Generators with Frequency Response Capabilities. *Appl. Sci.* **2022**, *12*, 11126. https://doi.org/10.3390/app122111126

Academic Editors: Federico Barrero and Mario Bermúdez

Received: 1 October 2022
Accepted: 27 October 2022
Published: 2 November 2022

1. Introduction

In recent decades, decarbonization targets have been set in order to face and limit global warming. On a timeline, the last milestone of this challenge has been set with the Paris Agreement in 2015, which established the goal of keeping global warming well below 2 °C, and possibly under 1.5 °C, with respect to the pre-industrial level [1,2]. In this scenario, one of the most important layers of intervention is the energy sector, which in 2019 accounted for 41.8% of the CO_2 emissions in the world, consequently representing their major contributor [3]. An outlined solution to reduce its impact is to increase renewable energies utilization. In this sense, for example, the EU set the target of covering 32% of its energy consumptions with renewable energies by 2030 [4].

These ambitious targets are inducing deep changes in the energy mix, leading to equally difficult technological challenges. In particular, renewable generators are displacing conventional fossil-fueled ones, leading to a variety of issues mainly related to their non-dispatchability and inability in providing ancillary services.

A potential future scenario is presented in report [5]. It offers an immediate idea of the modification that will hit the electric system in less than thirty years: if the world will adopt measures to comply with the Paris agreement ("REmap Case"), the share of renewable energies in the power generation sector could reach 86%, while the sector of variable renewables such as solar and wind would be about 60%, with an annual addition of capacity, respectively, up to 360 GW/yr for photovoltaic plants and up to 240 GW/yr for wind plants in 2050.

Starting from this estimation, it is clear how important it will be in the future to create renewable generators able to provide the services that nowadays are the responsibility of traditional ones. According to this, regulators and System Operators (SO) are starting to ask for some of these services also to renewable plants in the updated versions of grid codes to face the system developing and keep ensuring its safety and reliability [6]. For variable renewable sources, the implementation of techniques to meet these requirements is asking huge research efforts due to their nature. In fact, they are used to be connected with the grid through power electronic devices, which have no rotational parts and are, therefore, unable to contribute to the total inertia of the power system.

Inertia is a fundamental concept while analyzing a power system. Indeed, transient stability and frequency stability are strongly related to the kinetic energy present in the system, whose amount in turn is correlated to the inertia constant. A reduction in the total inertia constant is reflected in a stability decrease [7].

Focusing on the frequency stability problem, inertia constant impact can be appreciated in Equation (1) [7]:

$$df/dt = 0.5\, f_0\, (P_{gen} - P_{load})/(S_{tot}\, H_{tot}), \tag{1}$$

where f is the frequency of the system, f_0 is the nominal frequency, P_{gen} is the total generated power in a certain time instant and P_{load} the total consumed one, S_{tot} is the sum of the rated power of all the generators, and H_{tot} is the equivalent inertia constant of the system as a whole. It is evident that if H_{tot} decreases, given a certain active power unbalance between generation and load, the frequency variation will be larger.

To contain the frequency variations and keep the frequency as close as possible to its nominal value, it is thus necessary to create certain levels of reserve available at different timescales.

The spread of variable renewable energies (VRE) mentioned above impacts on the reserve problem in more than one way: the uncertainty of the primary source availability, and so of the power production, requires additional reserve in the system; and the participation in the system of the conventional units that are able to provide reserves reduces, and so the stability of the system.

More specifically, the operation of electrical power systems requires frequency to remain tightly close to its rated value. Traditional frequency control strategies are based on synchronous generators and their ability to modify the mechanical power from their prime movers to compensate for frequency deviations. These are caused by real-time mismatches between the mechanical power input and the electrical power output (plus losses) of synchronous generators. The inertia of rotational masses of each turbine-generator set offers a natural temporal buffer for these mismatches that limits the rate of change of frequency (RoCoF). As stated, renewable generators are commonly connected to the power system through power converters, which lack an inherent inertial response to frequency deviations. This increases frequency deviations and RoCoF values as a growing share of renewable generators displace synchronous ones. Thus, the main objective of this work is to assess the ability of renewable generators to contribute to frequency control in power systems where synchronous generators are being displaced by power electronics interfaced ones.

This work is focused on the first seconds immediately after an unbalance, in which the primary reserve activates to contain as much as possible the initial oscillation of frequency after a perturbation. In particular, the work faces the issue of providing primary frequency control with renewable generators as photovoltaic modules and wind turbines, which are traditionally not technically able to perform it, trying to understand how to implement it and whether the performances are acceptable.

Before going through the developed model, it is extremely important to offer an overview of the methods and algorithms proposed in the literature for the power reserve control of photovoltaic systems. This bibliographic analysis is very useful, since each one of the methods can have some issues and some advantages, which can be compared to the ones of the model developed in this work, understanding its effectiveness.

For what concerns the photovoltaic (PV) technology, two main types of solutions have been reported in the literature for complying with future requirements in terms of the modulation of active power as a response to a frequency deviation: the use of energy storage systems (ESSs) or the implementation of innovative control schemes for power curtailment [8]. The adoption of ESSs for primary frequency control can potentially solve the issues in highly renewable penetrated systems, as the small island ones or the power systems of the future. The biggest advantage with respect to power curtailment methods is that no energy is wasted. Indeed, to keep a certain reserve level always available, renewable generators have to exploit just a certain amount of the producible energy. If the remuneration and the incentives for the participation in primary frequency control are not enough, the profitability of renewable plants decreases. In that case, adopting an ESS integrated with the generators would allow a more flexible utilization without reducing the energy exploited. Another winning aspect of some ESSs is that they are able to adapt the power with rapid ramp rates, which is one of the key targets for the future primary frequency control [9]. On the other hand, the ESS solution presents also some drawbacks, mainly related to investment cost and complexity. In fact, the ESS solution requires an additional investment to purchase the storage system itself. Furthermore, in the case of batteries as ESSs, their lifetime is limited [8]. These negative aspects can lead to an increase in the cost of the PV electricity produced, which is in contrast to the expectations of cost reduction in the next decade [10]. It is also important to state that power curtailment and storage are not necessarily mutually exclusive solutions, since, in some situations, combining the two is the best option from the economic point of view [11].

In this work, the impact of the second solution is analyzed, which allows a simpler control and lower investment costs. If the choice is to modify the control algorithm of a photovoltaic module, three main functionalities may be implemented [12]: Power Limiting Control (PLC), Power Ramp-Rate Control (PRRC), and Power Reserve Control (PRC), that are conceptually described in the following.

1.1. Power Limiting Control (PLC)

The concept presented in the following is the basis for all the power curtailment methods. It is based on the imposition of a certain level of active power required from the plant (P_{res}), that has to be reached by modifying the voltage applied to the PV panels (v_{PV}*), taking into account the power-voltage curve, as shown in Figure 1.

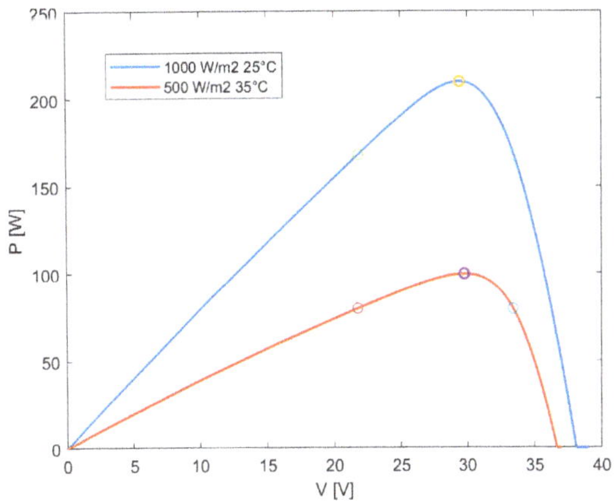

Figure 1. Power-voltage curve of a PV module for two different ambient conditions, showing the maximum power point (MPP) and the possible reference operating points for a 20% reserve.

It is evident that, given a certain P_{res}, it is fundamental to characterize the irradiance (G) and the temperature (T) in order to establish the shape of the power curve. In this sense, this raises another problem: the choice of the method to obtain the power curve. Different solutions are available, for example, the direct measurement of G and T, a non-linear least-square curve fitting or using solar forecasting methods. The choice represents a trade-off between cost and accuracy, due to the fact that the most accurate methods are also the most expensive to be implemented. Once the actual current-voltage curve is determined, all the inputs required for the algorithm are available:

- If $P_{res} \geq P_{MPP} \rightarrow$ MPP operation: the PV system will operate at its maximum power point, so $v_{PV}{}^* = v_{MPP}$ and $P_{PV} = P_{MPP}$.
- If $P_{res} < P_{MPP} \rightarrow$ curtailed operation: the PV system will operate at a voltage $v_{PV}{}^* = v$ and $P_{PV} = P_{res}$.

It can be noticed that, in the curtailed case, two voltage levels are possible for the same P_{res}, respectively, on the left and on the right of the MPP: a comparison between both options is provided in the following.

1.2. Power Ramp-Rate Control (PRRC)

The aim of this strategy is to smooth the active power output oscillations of the PV system. In particular, the algorithm imposes a maximum rate of variation of the active power output when variations of irradiance occur. In this case, the criterion to curtail the power does not come from the absolute value of the active power output but from its change rate with time [12]. The PV power ramp-rate can be calculated as in Equation (2):

$$R_r(t) = dP_{pv}/dt \tag{2}$$

If $R_r(t)$ is higher than the limit value required by the system operator, the voltage $v_{PV}{}^*$ is perturbed in order to reduce the change rate of the PV power to that value. This kind of control can be performed both by controlling the voltage (as just explained) or the power.

1.3. Power Reserve Control (PRC)

This control logic can be seen as a particular case of the PLC algorithm. In fact, instead of imposing a certain constant power output level, in the PRC algorithm the PV output is regulated as a percentage of the maximum available power in each time instant. This dynamically changing fraction of the MPP is called Power Reserve (ΔP). The output power is then obtained as:

$$P_{res} = P_{MPP} - \Delta P \tag{3}$$

Additionally, in this case, the challenge is to estimate the MPP, since it is one of the inputs necessary for the controller. The detailed explanation of this algorithm as well as of the determination of the power-voltage curve (and so of the MPP) is provided in Section 2, where the complete implementation of a model to control the active power output of photovoltaic modules in order to enable their capability to participate in the primary frequency control is described.

This work is focused on the third solution. To better organize the literature review regarding PRC, the differences between approaches are outlined under some key aspects: PV plant model, estimation output, controlled variable, MPP tracking method, and operational side of the power-voltage curve. The summary of reviewed articles is presented in Table 1.

Table 1. Summary of reviewed articles about Power Reserve Control of PV systems.

N°	Authors	Reference
1	Batzelis et al.	[13]
2	Sangwongwanich et al.	[14]
3	Li et al.	[15]
4	Hoke et al.	[16]
5	Riquelme et al.	[17]

The starting point of each work about the power reserve control of a photovoltaic grid-connected system is the model adopted for the plant itself. There are two main approaches: the most complete one is the two-stage model [18], which comprehends all the conversion stages between the modules and the grid. It includes the boost converter to step up the voltage to the DC-link value, the inverter, and an LCL filter. However, since the actual power output control is operated on the DC-DC converter, the single-stage model is the most commonly used in the literature, as represented in Figure 2, where the duty cycle command is obtained using a control module. The main differences between all the reviewed methods are found in this control module.

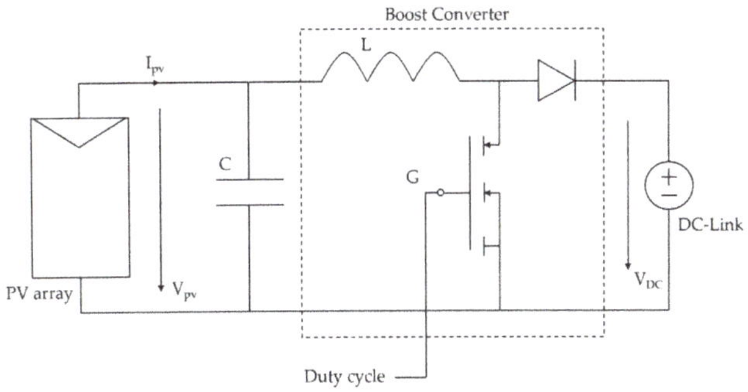

Figure 2. Single-stage grid-connected PV system.

The logical process of the control block is a sequence of some key operations. First of all, the maximum power point is estimated in real-time. This calculation can be performed through a variety of techniques. The approach proposed in [13] is probably the most widespread: the power-voltage curve is estimated from past measurements of power and voltage, in the current window close to the operating point, through the equation for the power-voltage curve shown in Section 2. The measurements are referred to the last temperature value recorded and fitted through a least-squares curve fitting method, whose output are the five characteristic parameters of the module, as presented in Section 2. Qualitatively, without entering in the detail of the mathematical formulation, which is not the focus of this paper, the least squares fitting consist in finding the power-voltage curve that best approximates the measurements, i.e., gives the lowest difference between the measurements and the estimated values. A similar approach is used in [14], but with a combination of a linear and a quadratic approximation. The interesting aspect of [14] is that, focusing on multi-string PV plants, one string called "master string" is operated at the MPP, but using the estimation of the MPP of the master as reference, while the others are in the curtailed mode. A non-linear least-squares fitting is also used in [17], proposing further and deeper observations about the possibility to follow the actual characteristics of the PV modules when temperature and irradiance change, and demonstrating that the non-linear least squares fitting works well also in changing ambient conditions if measurements close to the MPP or on its right are used. Since the current-voltage curves of PV modules are almost superposed for different temperatures on the left side, it is not possible to estimate the correct one if measurements are taken on that side. However, since the temperature dynamics are slower than the irradiance one, it is possible to operate with measurements on the left side, at least for short periods. A different method is used in [16]: the MPP is found using a polynomial function of the irradiance and the temperature, which in this case, differently from the previous ones, are measured. The coefficients for the polynomial relation are found by calculating the value of the module current (with Equation (7)) for all the expected irradiance, temperature, and voltage values, then numerically finding the power maximums and interpolating them with a linear regression. It is also worth stating that, in photovoltaic systems, another possibility to find the MPP are the Perturb and Observe methods [19], representing the traditional methods of estimation, which in this case are not useful since the knowledge of the entire power-voltage curve is necessary.

Once the MPP is known for the actual operating condition (so for an irradiance/temperature pair) the required operating point to deliver the curtailed active power output desired has to be calculated. The most used control variable is the PV voltage: the voltage level corresponding to the desired output power is calculated and then it is compared with the actual voltage applied to the PV system to produce a duty cycle signal by means of a proportional-integral (PI) controller. This approach is used in [14,16,17]. The same reasoning is adapted with the power as a reference variable in [13,15], where the desired value of power is compared with the actual value to produce a duty cycle signal for the boost converter.

It is also important to notice that, for a certain level of curtailed power desired, there are two options in terms of voltage, respectively, on the left and on the right side of the maximum power point in the power-voltage curve. The scientific literature does not agree on which is the best side of the power curve of the PV module to operate the power curtailment. All the existing algorithms work only on one side. Since the operating condition throughout the year can be very different, this can be limiting.

In fact, operating on the right side allows a bigger variation range of the power, but at the same time modest reductions in irradiance can lead to a drop in the power output. This behavior can be easily explained looking at the characteristic power-voltage curve shape: reasoning at fixed irradiance, a request of power reduction from the system operator can be easily satisfied with small variations of the operational voltage. This is intrinsic to the high slope of the right part of the power-voltage curve. On the contrary, in a context in which the irradiance variates a lot, if a reduction in irradiance occurs at constant voltage the power output would immediately reduce, and even become null.

Instead, on the left part of the curve the pros and the cons are shifted due to the lower slope of the curve: there is a moderate insensitivity to the irradiance, but at the same time the range of variation of power is very restricted.

From these observations, it can be immediately pointed out how the choice of a single side of the power-voltage curve for power curtailment can influence the actual capability of the plant to participate in frequency control. In this sense, it becomes crucial to develop an algorithm able to adapt the control strategy to the boundary conditions of the problem (i.e., meteorological conditions, level of power reserve demanded by the system operator) [17]. This is why the present work adopts the approach proposed in [17], to be able to work on both sides of the power-voltage curve, exploiting the advantages of the operation on the two sides as a function of the situation.

Together with the photovoltaic technology, wind turbines are certainly the most exploited solution to drive the energy sector through the decarbonization. For photovoltaic systems, the algorithms to control the active power output through the power reserve control are relatively similar to each other in the logical sequence, but for wind turbines this does not happen. Indeed, there are different techniques to provide inertial control and, in particular, primary frequency control. The large variety of solutions is well described in [20]. Primary frequency control with wind turbines is an autonomous response to a change in system frequency that emulates the governor control of synchronous generators. More precisely, when a change in frequency is detected, the wind turbine modulates its power output to respond to this change and help to balance the system. Figure 3 summarizes different categories of methods for inertial control and primary frequency control in wind power plants.

First of all, it is important to offer an overview of the different types of wind turbines present in the market. In general, there are fixed speed wind turbines, semi-fixed speed wind turbines, and variable speed wind turbines (VSWT) [20]. Fixed and semi-fixed speed turbines have an inertia constant but are smaller with respect to the conventional synchronous generator ones, due to the lower coupling of induction generators with the grid frequency. This makes their inertial response lower and slower compared to the conventional generation [20]. On the other hand, as already mentioned, variable speed wind turbines (which are dominant in the market due to their higher efficiency) are decoupled from grid frequency due to the presence of the power electronics converter, so their contribution to the system inertia is null. However, it is possible to achieve a short-term inertial response from VSWTs by introducing an additional control loop in their control system. This control modifies the operation, allowing the turbine to exploit the kinetic energy of its rotating mass to modulate the electrical power output of the system and so to contribute to participation in the inertial control.

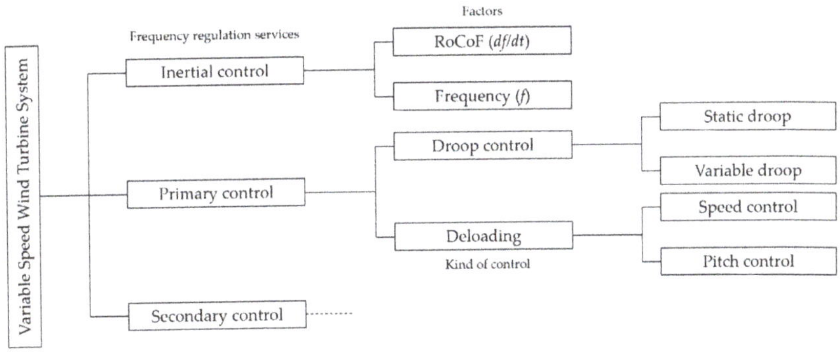

Figure 3. Active power frequency control research lines in wind-based power plants.

The functioning principle is simple: if the control detects a frequency variation, a signal is sent to the turbine speed governor or torque governor, temporarily modifying the power setup according to the magnitude of the frequency variation and enabling the possibility to exploit an amount of the kinetic energy of the rotor. For example, if an increase in power output is required to balance a frequency decay, the kinetic energy stored in the rotor is temporarily released. In this case, an unbalance between the electrical power injected into the grid and the mechanical power extracted from the wind is created, causing a deceleration of the rotor (i.e., a decrease in kinetic energy). This slowing down reduces the lift, discharging aerodynamically the blades, and increasing the stall risk at the bottom sections of the blade. To avoid stalling, the overproduction period has to be followed by an underproduction period in which the rotational speed of the rotor is increased [20].

There are different models for the inertial control described in the literature, but they can be conceptually divided into two main categories: a "simple" inertial control, which takes into account just the absolute value of the frequency deviation Δf or the frequency derivative df/dt, or an inertial-droop control, which takes into account both. The first category is presented for example in [21,22], while the second is adopted in [23,24]. The inertial-droop approach allows for a very fast inertial control as well as an active power support more extended in time. An extended explanation of this kind of control is given in Section 2, since is the one adopted in the present work. The droop concept for VSWTs is analogous to the synchronous generators one. There are two main types of droop control: fixed or variable droop value. The second is more accurate in responding to frequency deviations, since it depends on wind and power reserve conditions [25] or on the RoCoF [24,26]. Indeed, it ensures improved system stability and avoids reserve exhaustion. The drawbacks of this kind of control are intrinsic to the physical behavior of the rotor during frequency oscillations. In fact, when the rotor slows down, it has always to respect aero-mechanical limits and avoid stall, keeping a safety margin; consequently, the control is limited, especially for low wind speeds. Furthermore, while the operational life in normal operation has been extensively tested at industrial levels, the operation with inertial control enabled has not, leading to the prescription of a practical utilization of this kind of control in only a few situations by the grid codes [20]. Another issue is related to the fact that, to re-accelerate to optimal speed, the power released to the grid has to be lower than the one coming from the wind. This implies that, for a certain time interval, the power output of the turbine is lowered to recover kinetic energy, delaying frequency recovery and, in some situations, requiring an extra number of reserves to prevent a "double-dip" in system frequency, increasing the risk of triggering protective relays at the substation level and causing blackouts [27].

In normal operation, wind turbines follow the MPP to extract the maximum energy available from the wind, achieving the highest aerodynamic efficiency possible. However, if an underfrequency event happens (i.e., the load increases its power), with such an operation there is no margin to respond to the frequency variation. On the other hand, through de-loading, the operation is shifted from the maximum power point to a sub-optimal one, keeping a power margin to face that event [28]. To work in a de-loaded condition, it is sufficient to reduce the coefficient of performance of the turbine (C_p), which is a function of the tip-speed ratio (λ) and of the pitch angle (β). The tip-speed ratio is proportional to the rotor rotational speed (ω_r). Consequently, it immediately derives the two possibilities to de-load the wind turbine: working in a rotational overspeed condition (as in [29]) or through a pitching control. Usually, rotor overspeeding is applied when the rotor speed is low, while for a rotor speed equal or higher than the rated one the pitch control technique is used [30]. The de-loading of wind turbines has some limitations that historically confined it to very few practical applications. First of all, similarly to what happens in the power curtailment of photovoltaic plants, the de-loading leads to a reduction in the annual capacity factor of the turbine. At the same time, overspeeding can reduce the turbine's life. To determine the power set-point, accurate wind speed measurements are important, otherwise incorrect or rough estimations can highly affect power output and turbine life [20].

2. Model for a Small Island Power System with Renewable Generators Participating in the Primary Frequency Control

After an overview of the techniques that allow the considered renewable generators to participate in the inertial and primary frequency control, it is important to go through the model developed in this work.

The negative effects on stability due to a massive presence of variable renewable sources are more severe in small and isolated power systems, due to a lack of interconnections with other systems and to their limited inertia. For this reason, many studies have assessed the impact on the frequency stability in small island power systems [29,31]. In this work, a generic small island power system has been considered, i.e., the most challenging situation, to evaluate the coordinated effect on frequency stability of photovoltaic modules equipped with a power reserve control and wind turbines with inertial control enabled. The models described in the following have been developed using the Matlab/Simulink software, which allows for easily the observation and study of a large variety of scenarios.

This section starts with an overview of the complete model to offer a general introduction to the problem and it continues with the detailed description of the different subsystems of the model.

2.1. Assumptions of the Model

Before starting to describe the model, it is important to state the fundamental assumptions on which it is based, which are necessary to understand the work and its main objectives.

The most important observation regards the power system representation: in primary frequency control studies, i.e., in the time frame from 1 to 100 s, it is a common practice to schematize it using the Load Frequency Control (LFC) approach. In fact, in such problems, the time constants of the electromechanical variables are much smaller than the ones representing the dynamics of the mechanical variables of synchronous generation prime movers [32]. Furthermore, the time constants of the wind turbines and photovoltaic converters are very small, justifying this approach [24]. Due to this reasoning, the equivalent inertia constant of the system is the key parameter to obtain the dynamic frequency characteristics [33], together with the load damping constant, which describes the sensitivity of load power to the frequency variations [32]. This allows for the representation of the entire system response just through a transfer function block, making the problem at the same time easier to be modeled (the system is not modeled node by node) and more general. A similar kind of reasoning is related to the generation blocks. Indeed, since the only aim of this work is to observe the inertial and the primary frequency responses, each single generating unit is not represented, all of the same type are aggregated in equivalent generating units. This is valid in particular for the variable speed wind turbines and for photovoltaics. The same principle is applied to the load, which is schematized as a unique block. The main advantages of this approach are the simplicity and the versatility of the results to a lot of practical cases, while the main drawback is the losing of detail.

To account for the penetration of the different generating sources in the energy system, a participation factor (p_i) is used, which represents the fraction of the total power of the system which is generated by the i-th source.

Other assumptions, which are quite usual in this kind of study, are related to the ambient conditions. In the simulations, the effects of variable wind speed, variable irradiance, and ambient temperature are considered. At the same time, the non-uniform distribution of the wind resources among different wind turbines is neglected, since they are considered as an aggregate; the same happens for photovoltaic modules, which work under the assumption of uniform radiation, and shading is neglected.

2.2. Overview of the General Model

The upper layer of the model of the system under study is shown in Figure 4. It represents a small isolated power system, comprehending three kind of generating plants: a reheated steam turbine, wind turbines and photovoltaic units. It has to be specified that the total generated power (that in equilibrium condition equals the load power), as well as the power of the three generating blocks, vary with the magnitude of the participation factors. In fact, in this kind of study, the importance is related to the relative weight of the three generating blocks rather than to the absolute magnitude of the system, since it is based on per unit variations in a reference common basis. This concept is particularly relevant for simulations at different participation factors, as the ones presented in Section 4.

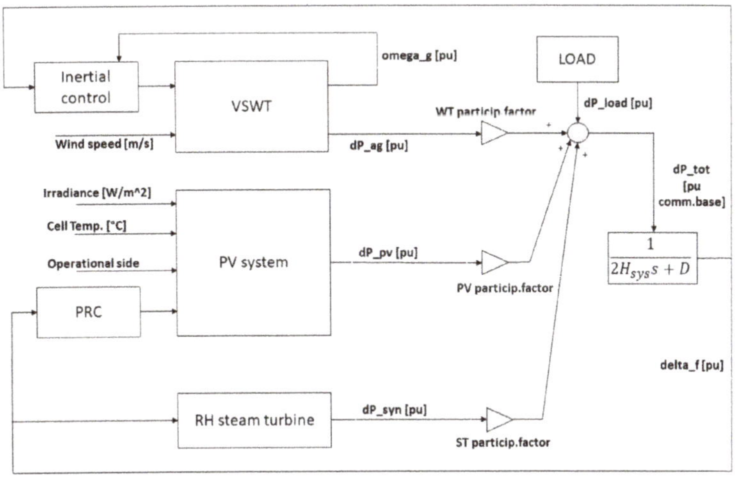

Figure 4. General layout of the model.

Before going deeper into the characteristics of the different subsystems present in the model, it can be useful to understand the main connections between them, to facilitate their study. Each of the three generation blocks produces a signal of variation of power in per unit with respect to the steady state value (dP_i), which is the power corresponding to the initial condition:

$$P_{i,0}; \quad \Sigma_i (P_{i,0}) = P_{load,0} \tag{4}$$

Indeed, all the simulations are initialized to have a dP_i [pu] = dP_{load} [pu] = 0, corresponding to a frequency value equal to the rated one, which is 50 Hz (i.e., $\Delta f = 0$).

The three signals of variation of power of the generating blocks are initially in per unit with respect to the generator nominal power, so they require to be weighted to obtain the equivalent variation on a common basis (the sum of the nominal powers of the three generators blocks): each of them passes through a gain block in which it is multiplied by its participation factor, as in Equation (5). The dP_{load}, on the contrary, is already defined in per unit common basis.

$$dP_{i[pu\_common\ basis]} = p_i\ dP_{i[pu]} \tag{5}$$

For simplicity, from now on, "common basis" is omitted in the equations. Then, the four dP signals are summed up to obtain the instantaneous power unbalance of the entire system:

$$dP_{tot[pu]} = \Sigma_i (dP_{i[pu]}) - dP_{load[pu]} \tag{6}$$

A non-negative power unbalance leads, through a transfer function, to a frequency deviation from the nominal value. The transfer function describes the frequency response behavior of the system. The meaning is captured by two parameters: H_{eq} is the equivalent inertia constant of the synchronous generator present in the system, and D is the damping constant of the load, which represents the sensitivity of the load power to the frequency variations. The function output is the signal representing the frequency variation in per unit with respect to the nominal value, which is also one of the metrics to evaluate the system behavior in the analysis presented in Section 4. The frequency variation signal is also the main input to the controllers of the three generators. Indeed, the synchronous one is traditionally able to respond to a change in system frequency by adapting its power output, but in this study also photovoltaics and wind turbines are, thanks to the presence of the power reserve control block (PRC) and the inertial control block, respectively.

After this brief introduction, a more detailed analysis of each block is required, to understand the modeling of the generators as well as the functioning of the above-mentioned controllers.

2.3. Photovoltaic Generators Model

One of the three generating blocks represents the photovoltaic modules present in the system. The modeling of this block is quite challenging, requiring a lot of different literature sources as well as customized solutions to improve its functioning. Figure 5 gives an overview of the model.

The model is divided into two subsystems: the upper one contains the electrical model of a photovoltaic string, while the lower block represents the control system of the PV string modified to implement the power reserve control. Observing the inputs, some observations can be performed:

- Irradiance and cell temperature are the two ambient conditions that mainly influence the performance of the photovoltaic modules. They are assumed to be measured in real time: this is a strong assumption, which implies higher costs with respect to other solutions such as non-linear least squares fitting methods. This assumption comes from the fact that the aim of this work is centered on the control logic, not on the determination of the irradiance and cell temperature.
- The reserve signal is produced by another subsystem that is presented in the following. It is sensible to frequency variation: if the frequency is at its nominal value, the reserve level is fixed at a certain value imposed by the steady state conditions. When the frequency oscillates, the reserve level is adapted automatically to face this variation. Of course, this signal is required by the control system to calculate the operating PV string voltage corresponding to a curtailed operation.

- The last external input, indicated with *Side*, represents the desired operating side of the power-voltage curve for implementing the reserve control. In this work, the innovative approach proposed in [17] has been adopted, which allows, depending on the requirements, to operate on both sides of the power curve, by simply switching this input value.

The control system receives these four inputs, as well as the measurement of the actual voltage applied to the PV string, and calculates the maximum power point, the corresponding voltage, the reference voltage to operate in the desired curtailed condition and, above all, a duty cycle signal for the boost converter necessary to impose that desired operation to the PV string.

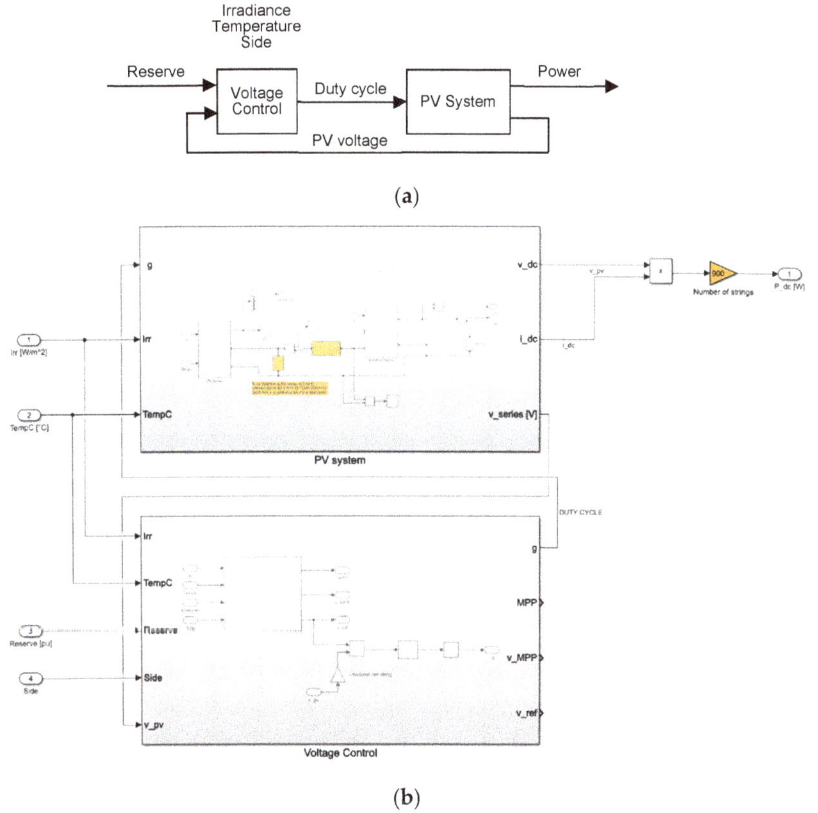

Figure 5. Photovoltaic system model: (**a**) General flowchart showing the electrical model (left block) and the voltage control (right block); (**b**) Simulink implementation, including the electrical model (upper subsystem) and the voltage control (bottom subsystem).

The general diagram of the voltage control subsystem is shown in Figure 6.

This subsystem is mainly composed of a function and a PID controller. The function is the most important element of the PV system, since it contains almost all the features that characterize the PV model proposed in this work. For this reason, it is appropriate to explain how the function works to obtain v_{ref}, i.e., the control variable of the PRC.

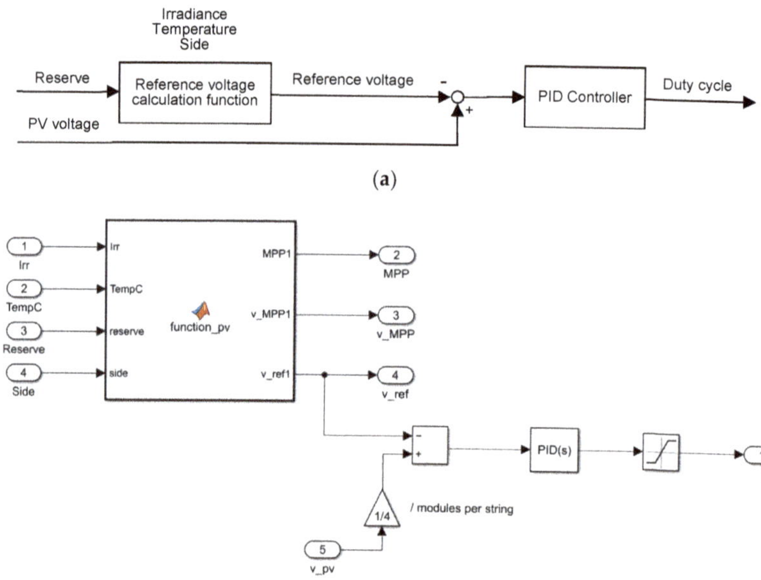

Figure 6. Voltage control subsystem: (**a**) General flowchart; (**b**) Simulink implementation.

The first aim of the function is the determination in real-time of the power-voltage characteristic of the module. As already observed in Figure 1, there is a strong dependence of the shape of the power-voltage curve on the environmental conditions. This means that the model has to properly account for this fact. Two of the characteristic parameters present in the datasheet of all the photovoltaic modules, the open-circuit voltage (V_{oc}) and the short-circuit current (I_{sc}), participate in solving the equations and are impacted by irradiance and temperature. This means that, to be precise, they have to be adapted to the actual environmental conditions. In the literature, different approaches are proposed to do so, which are reviewed in [34]. In this work, the following correlations have been used:

$$I_{sc} = I_{sc,ref}\, G/G_{ref}\, [1 + \alpha_{sc}\, (T_{cell} - T_{cell,ref})] \tag{7}$$

$$\begin{aligned} V_{oc} = V_{oc,ref}\, [1 + \beta_{oc}\, (T_{cell} - T_{cell,ref})] + 54.68511{\cdot}10^{-3} \log(G/G_{ref}) + \\ 5.973869{\cdot}10^{-3} \log(G/G_{ref})^2 + 761.6178{\cdot}10^{-6} \log(G/G_{ref})^3 \end{aligned} \tag{8}$$

The *Single-Diode Model* (SDM) of the photovoltaic cell [35] adopted in this work is characterized by five parameters: $I_{ph,cell}$ (also called I_L) is the photocurrent, $I_{s,cell}$ is the saturation current of the diode (also called I_0), n is the diode ideality factor, and $R_{s,cell}$ and $R_{sh,cell}$ are the series and shunt resistances, respectively. The electrical variables of interest are the PV cell terminal voltage ($V_{pv,cell}$) and current ($I_{pv,cell}$) [35]. The parameters solve the implicit equation describing the current-voltage characteristic of the cell (which is extendable to the module):

$$I = I_L - I_0 \left\{ \exp\left(\frac{V + R_s\, I}{n\, V_{th}} \right) - 1 \right\} - \frac{V + R_s\, I}{R_{sh}}, \tag{9}$$

where V_{th} is the thermal voltage. In principle, all the parameters necessary to solve the equation can be calculated numerically as in [35]. On the other hand, to estimate I_L and I_0, Equations (10) and (11) can be used, respectively [35]:

$$I_L \approx \frac{R_s + R_{sh}}{R_{sh}} I_{sc}, \tag{10}$$

$$I_0 \approx \frac{(R_s + R_{sh}) I_{sc} - V_{oc}}{R_{sh}} \exp\left(\frac{-V_{oc}}{n\, V_{th}}\right). \tag{11}$$

Once the current is known for each voltage between 0 and the actual V_{oc}, the power curve is easily obtained as the product between it and the corresponding voltage. Once the power-voltage characteristic is known, the maximum power point can be easily extracted, as well as the corresponding voltage level. Indeed, the MPP is found as the maximum of the power-voltage curve. In the last part of the function, the reference voltage for the power reserve control is found. In particular, knowing the desired reserve amount, it is possible to calculate the corresponding power. Once this is known, an operational side of the power curve is imposed through an external signal, determining the reference voltage (v_{ref}) thanks to a simple minimization problem and finding the voltage vector element on the desired side, which corresponds to the calculated curtailed power.

The reference voltage signal is then compared with the actual voltage applied to the module. The associated PI controller acts on the duty cycle signal to minimize the difference. Indeed, the output signal of the voltage control subsystem is the duty cycle, which is imposed to the DC-DC boost converter in order to set the module (the series) voltage to the desired value.

The second subsystem forming the photovoltaic system block is the one containing its electrical model, which is in practice the object of the control subsystem's operation. It is basically composed of a PV array in parallel with a capacitor and a boost converter. The DC-DC converter connects these elements to a DC-link, whose voltage for the aims of this work is assumed to be fixed to 230 V, as common practice in the literature when studying the power reserve control of a PV system [17]. In practice, the single-stage PV model already mentioned (as shown in Figure 2) has been used. The corresponding Simulink implementation is the subsystem diagram presented in Figure 7.

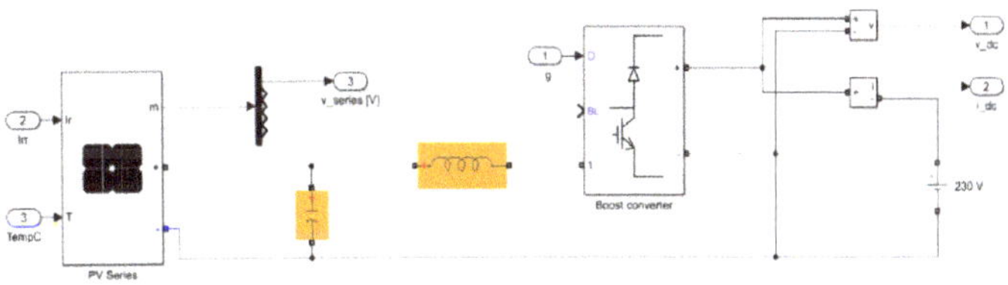

Figure 7. PV electrical subsystem diagram.

To fully describe the photovoltaic system, it is necessary to specify another aspect: how the adaptation of the active power output happens in response to a frequency change. This is implemented partially in the voltage control block, but to close the loop, it is important to notice how the *reserve* input signal is generated. The reserve adaptation block (PRC in Figure 4) receives the per unit frequency deviation as an input. The reserve is adapted through a droop characteristic so that the output variable is changed proportionally to the input one. In this case, the reserve is increased if the frequency increases, i.e., the power produced is reduced by working away from the MPP.

It is fundamental to set a proper droop value. To do so, extensive tests have been performed. On the one hand, a too high droop constant means that the reserve varies very slightly, providing insufficient help to the primary frequency control. On the other hand, a too low droop constant leads to a reserve depletion or to an extremely low power production. As a compromise, a value of 0.05 has been used, which is identical to the droop constant of the synchronous generators presented in the following of this work.

The rest of constants used can be found in the Appendix A.

2.4. Variable Speed Wind Turbines Model

There are many different solutions when designing a wind turbine. For example, there are several types of generators that can be coupled to the rotor. In this work, a variable-speed wind turbine using a DFIG (*doubly fed induction generator*) with pitch control enabled has been chosen as representative.

The block diagram of the model is presented in Figure 8 [36]. This layout is formed by five subsystems interacting with each other. Following a conceptual path from the wind speed to the output power: a predefined Simulink wind turbine block, which receives the wind speed, the low-speed shaft rotational speed, and the pitch angle signals and calculates the mechanical torque applied to that shaft; this torque signal is compared in the mechanical system (which represents the gearbox mechanisms) with the torque applied to the DFIG generator and the rotational speeds of the two shafts are calculated through a first-order dynamics represented by a transfer function; the high-speed shaft signal is used in the speed governor block to calculate the new power setup, compared in the same block with the actual power to calculate the reference generator torque; in the DFIG, simply modeled through another first-order dynamics, the reference torque is imposed to the generator; and a pitch control block adapts the pitch angle, if necessary, as a function of the high-speed shaft rotational speed, to limit the output power to the maximum value for wind speeds higher than the nominal one.

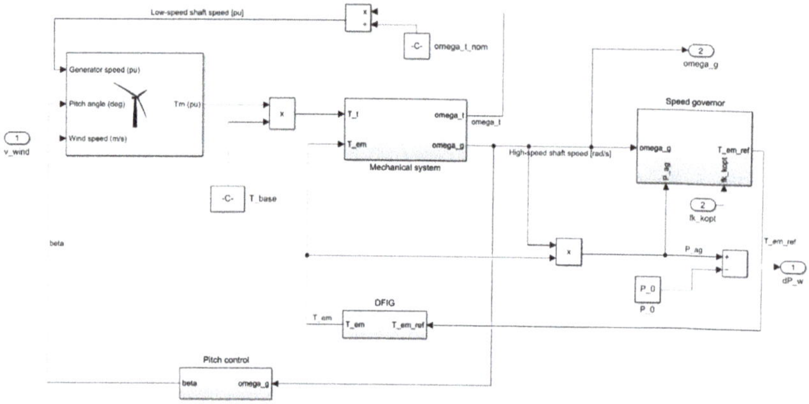

Figure 8. Block diagram of a VSWT model.

The inertia control is grafted in this model inside the speed-governor block. In particular, a multiplication factor $f_{K,opt}$ (which is calculated in the inertial control block) called *signal for the optimization zone* is used to shift the operation of the wind turbine from the MPPT curve to other curves by exploiting the kinetic energy of the rotor (Figure 9). This method is called the Extended OPPT Method [24]. The control function of the method is shown in Equation (12). As can be observed, $f_{K,opt}$ depends both on the frequency deviation and on the time derivative of the frequency. Indeed, the effectiveness of this virtual inertia algorithm depends both on the magnitude of the frequency disturbance (through Δf_s and df_s/dt) and on the initial kinetic energy stored in the rotor before the disturbance.

$$f_{k,opt} = [\omega_{g0[pu]}/(\omega_{g0[pu]} + k_{vir}\ \Delta f_{s[pu]})]^3 - W_{vir}\ K_0/(K_{opt}\ \omega_{g0[pu]}{}^3)\, f_{s[pu]}\ df_{s[pu]}/dt, \tag{12}$$

where:

$$K_0 = 2\ \omega_{g0[pu]}/\omega_{s0[pu]}\ k_{vir}\ H_{WT}, \tag{13}$$

and $\omega_{s0[pu]}$ is the pre-disturbance rotational speed corresponding to the system frequency, 1 p.u.

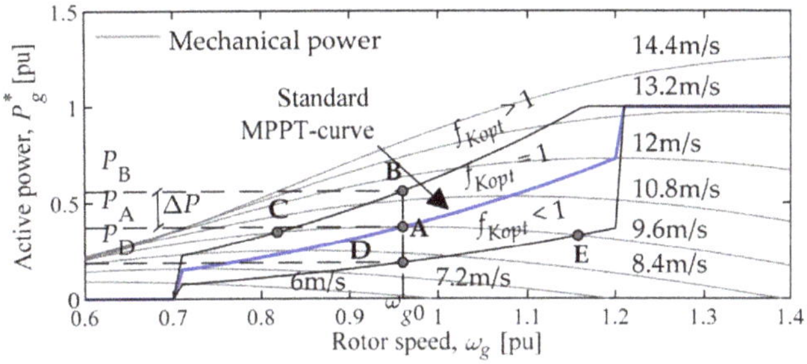

Figure 9. Extended OPPT Method [24].

The inertia control block simply implements these equations in a block diagram, providing the $f_{K,opt}$ signal as the final output, which, as already mentioned, is used to shift the operating point of the turbine.

Looking at the P-ω chart in Figure 9, it is possible to understand the functioning of the OPPT algorithm. Under the hypothesis of constant wind speed and normal operating conditions, the generator will deliver a power P_A associated with a rotational speed ω_{g0}. If an underfrequency event happens, caused for example by a mismatch between the injected and the consumed power in the system, an increase in the power output of the turbine will be required to reduce the frequency oscillation as much as possible. In particular, the non-null values of Δf_s and df_s/dt will activate the inertial response emulation of the wind turbine, with a consequent shift of the MPPT curve upwards (A $\rightarrow$ B). The higher the severity of the underfrequency event (and so the magnitude of Δf_s and df_s/dt), the higher will be the value assumed by P_B. In this situation, the imbalance between the active power extracted from the electric generator and the mechanical power delivered at the turbine shaft will cause a deceleration of the rotating masses of the turbine, testifying to the exploitation of the turbine kinetic energy. This deceleration moves the operational point to the left (B $\rightarrow$ C), reaching a new equilibrium condition when P_C coincides with P_A; this is a situation that corresponds to a balance between the mechanical power delivered by the turbine and the active power extracted by the generator (P_C). When the support is not necessary anymore, the system returns to the initial state (C $\rightarrow$ A), preparing it for future requirements [24].

Putting together these two renewable generators with a reheated steam turbine (modeled as in [32]), the scheme proposed before in Figure 4 is obtained, which is the one on which the following simulations have been performed.

3. Simulations Summary and Metrics to Evaluate the Results

To be as clear as possible, after showing the model layout, this section lists the simulations performed, with particular attention on detailing the aim of the different simulation categories and the metrics that will be used in next section to evaluate the results.

It is important to remember which situation this model is describing: a small island power system characterized by the presence of three equivalent generating units (PV, wind, and synchronous generator) and an equivalent load, which can be seen as a grouping of all the smaller loads present in the island. This situation, in fact, is the one in which a mismatch between the active power produced and consumed can have more serious effects on frequency stability, due to the small size of the power system, its lack of interconnections due to the remote geographic position, as well as the high penetration of renewable sources that is usually observed on the islands.

The simulations can be divided into some categories, as a function of the perturbation introduced with respect to a common reference condition. It is thus fundamental to describe this reference condition before going on. The common reference condition that has been chosen from now on will be indicated with "*base case*" and is characterized as follows:

- Participation factors of the three equivalent generating units: p_{wind} = 0.2; $p_{synchronous}$ = 0.7; p_{pv} = 0.1 (30% of renewable penetration on nominal power basis).

- H_{eq} = 2.88 s [24]; D = 1 [32].
- Incoming wind speed at t = 0 s: 9.6 m/s for constant wind simulations; 11.457 m/s for real wind profile simulations.
- Irradiance at t = 0 s: 1000 W/m^2.
- Cell temperature: 25 °C.
- Constant load power requirement.

In each simulation, a modification with respect to this base case is introduced to study different aspects linked to the frequency stability. As stated before, the simulations can be grouped as a function of the perturbation introduced. In particular, the situations compiled in Table 2 have been studied.

Table 2. Simulation groups and aims.

Simulation Group	Aim
Load active power step up/down	Observing the frequency response of the system after a sudden variation of the load power for different step magnitudes
Small sinusoidal oscillations + step of load active power	Observing the frequency response of the system to small and continuous variations of the load, i.e., similar to the real behavior, and to a step if a part of the reserve is already in use to balance the small variations
Variable environmental conditions (real wind speed profile, irradiance variations)	Understanding how the system responds to environmental condition changes, as well as to a sudden load power variation in such a situation
Variable participation factors	Understanding what is the impact of an increasing penetration of renewable sources on the system's frequency stability

Furthermore, to have other points of view on the system, each simulation is performed on four different versions of the complete system:

- Primary frequency control provided only by the synchronous generators (wind turbines without virtual inertia control, PV without power reserve control).
- Primary frequency control provided by synchronous generators and wind turbines (Extended OPPT Method enabled).
- Primary frequency control provided by synchronous generators and photovoltaic strings (Power Reserve Control enabled).
- All the generators participate in the primary frequency control (from now on it is denoted as *complete FC*).

The first scenario is representing the current situation in most of the power systems, since the renewables are mostly operated in their MPP condition. The fourth scenario, on the other hand, could represent a future situation in which renewables are flexible and able to help synchronous generators (or displace them) in balancing the frequency of a power system. This captures the interaction between the various controllers, the help that the renewables controlled with the updated logics can provide to the system, and the impact of each technology in frequency stabilization.

The final summary of the simulations performed on the model is reported in Tables 3 and 4. A number has been used to designate each of them, so that the results can be commented in a clearer way.

As is probably noticed, this work does not analyze the case of a generator loss. Indeed, this kind of model is not capable to catch that kind of event, excluding a loss of a part of the photovoltaic generation. This is intrinsic to the model developed for the system, in which the generators have to be intended as an "equivalent" of all the similar generating units present in the system. With some modifications, it is just possible to model the loss of the entire fleet of wind or synchronous generators. For PV, since a gain block is included at the output of the subsystem, it could be used to model a partial loss of generation.

Table 3. Summary of Simulation 1–8.

N° Simulation	Features
1	Step up P_{load} +1%
2	Step down P_{load} −1%
3	Step down P_{load} −2%
4	Step down P_{load} −10%
5	Small sinusoidal variation of P_{load} (amplitude 5‰) + step down P_{load} −10%
6	Real wind profile
7	Real wind profile + step down P_{load} −10%
8	Steep irradiance ramp up—ramp down

Table 4. Summary of Simulation 9–17.

N° Simulation	p_{wind}	$p_{synchronous}$	p_{pv}
9	0.3	0.6	0.1
10	0.4	0.5	0.1
11	0.5	0.4	0.1
12	0.2	0.6	0.2
13	0.2	0.5	0.3
14	0.2	0.4	0.4
15	0.3	0.5	0.2
16	0.4	0.3	0.3
17	0.4	0.2	0.4

Before presenting the simulation results, it is important to define quantitative instruments that can help in interpreting them. Indeed, the next section includes both qualitative analysis on the results and quantitative observations to catch some important concepts. It is first of all required to report a definition of primary frequency adequacy. As stated in [37], a primary frequency control is adequate if it is able to ensure an uninterrupted delivery of electricity after a sudden imbalance between generation and demand, i.e., if it is capable of arresting and stabilizing the frequency deviation without exceeding the limits imposed by grid codes.

In this work, the following metrics are used:

- Frequency nadir [Hz]: it is a direct measure of the primary frequency control adequacy. It is calculated as the maximum/minimum value of frequency deviation occurring after an active power imbalance. The closer it is to the nominal frequency, the better.
- RoCoF [Hz/s]: it is the time derivative of the system's frequency. The smaller it is, the better for the system.
- *Nadir-based frequency response*: it expresses how good the primary frequency control has performed in stabilizing the frequency after a perturbation. It can be calculated as [37]:

$$\text{Nadir-based frequency response} = \Delta P_{perturbation} / \Delta f_{nadir}. \tag{14}$$

4. Simulation Results

Once the model has been presented in detail, it is now possible to finally match the modeling part with the conceptual part. Indeed, the previous sections have not addressed with quantitative and qualitative considerations the core of this work. It is interesting to approach the problem about the primary frequency control with some major questions: How the increasing penetration of renewable generators affects it? Is the possibility of including them in the provision of this ancillary service effectively helping the system and encouraging the displacement of conventional ones?

This chapter is trying to answer these questions both qualitatively, commenting some graphs, and quantitatively, thanks to the metrics proposed earlier in Section 3. For the sake of clarity, the results are divided and commented on by groups of simulations.

4.1. Load Power Step Up/Down

The first and simpler test to analyze the frequency response of a dynamic system is obtained by observing its behavior after a step perturbation. In this case, the step is an instantaneous change of the load power (P_{load}). This situation is extremely simplified with respect to what happens during the normal operation of a power system, but nevertheless this test is really important to take a first look at the effectiveness of the ability of the primary frequency control in stabilizing the frequency.

In the small island system that is described here through an LFC approach, a sudden step change of the load power is more likely to happen as a reduction. In other words, it is more common that the load decreases suddenly (for example due to a disconnection of an industrial user) than the vice versa. This is why the maximum amplitude tested for the steps in the reduction is higher than the one in the step up of the load power.

The first plot reported regards Simulation 1, in which the load suddenly increases its power by 1%. Figure 10 shows the frequency oscillation of the system due to the consequent imbalance between generation power and load power.

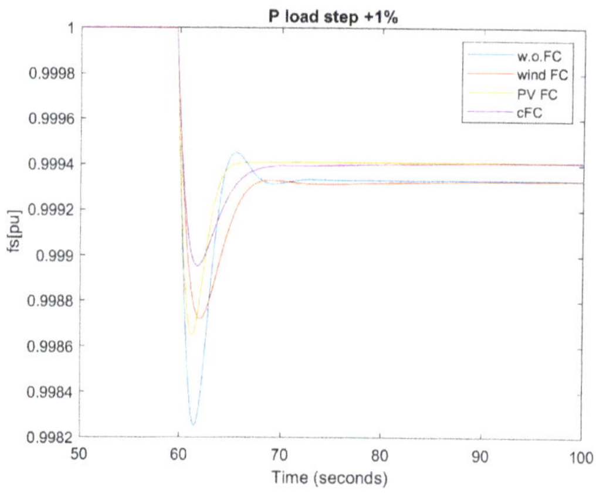

Figure 10. Simulation 1 frequency trend.

First of all, it can be observed that an increase in the load power causes a decrease in the frequency. In Figure 10, the frequency nadir can be easily individuated as the minimum of the frequency in the time plot. Observing it in the four scenarios: as expected, in the absence of frequency response capabilities from the renewable sources (w.o.FC), it is lower than in the other three scenarios. Indeed, it is equal to 0.9983 pu (49.915 Hz) against the 0.9990 pu (49.95 Hz) in case all the generators participate in the primary control (cFC). It is evident how the intervention of wind and photovoltaic generators in stabilizing the frequency is helping the system, since they reduce the magnitude of the nadir and, after a few instants, they also reduce the oscillations of the system's frequency and the stabilization time. The superposition of their effects lead to the behavior in purple, which is one of the scenarios in *complete FC*, i.e., all the generators participate in the primary frequency control.

Of course, since this work focuses just on the primary (and inertial) frequency control, the final value of the frequency is not the nominal one, that requires also the actions of the secondary and tertiary control to be restored.

Looking at the RoCoF plot in Figure 11, it can be seen that the PV system produces oscillations that increase the RoCoF in proximity of the step. This is a negative fact, which would require an additional PRRC to be solved, that reflects also on the *complete FC* scenario. After the peak of the RoCoF, anyway, the scenario for which it is higher is the one without frequency control. This allows for the understanding that the PRC for the PV system is capable of containing the frequency and the RoCoF in the instants after a sudden variation in the load power. On the other hand, the wind virtual inertia operates very well both during and after the load power step.

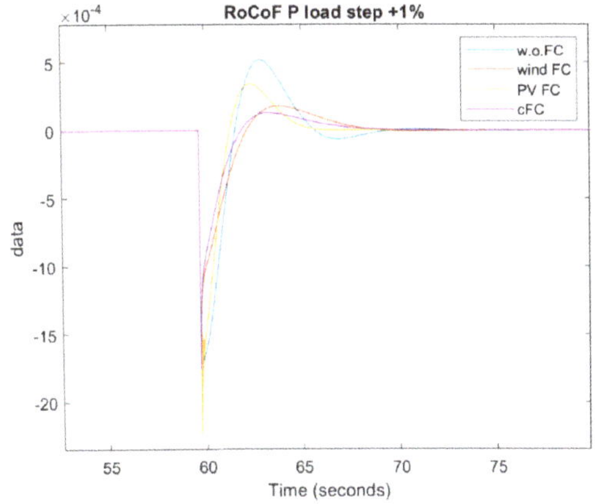

Figure 11. Simulation 1 RoCoF.

Remarkably similar results are obtained from Simulation 2 as shown in Figure 12: a negative step of load power of the same amplitude produces the same behavior of the frequency but mirrored to an increase in frequency. The nadir is also simply mirrored upwards with respect to Simulation 1: in the scenario with the renewables with disabled frequency control, it is equal to 1.0017 pu, while for the *complete FC* it is equal to 1.0010 pu.

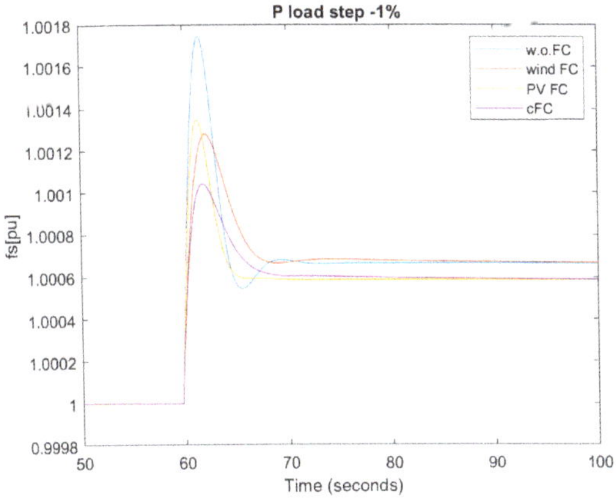

Figure 12. Simulation 2 frequency trend.

In the next simulations, larger steps are introduced. Due to the similarity already discussed between the first two, they are expected to produce a frequency waveform with the same shape but a different amplitude (dilated by a factor of 2 and 10, respectively). Looking at the results for Simulation 3 shown in Figure 13, this expectation is confirmed. This behavior is caused by the constancy of the load damping constant and of the inertia of the synchronous generator, which do not depend on the step amplitude.

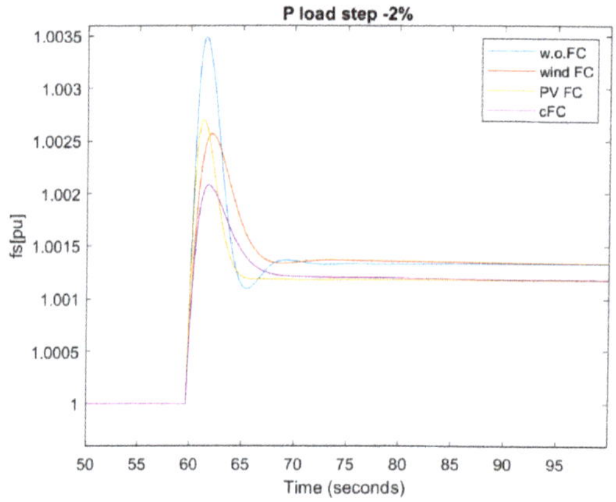

Figure 13. Simulation 3 frequency trend.

It can be also observed that the time required to stabilize the frequency is practically independent on the step amplitude, since in all the simulations the new steady state condition after the load power step is reached almost at the same time instant.

These kind of results are coherent also with the plots for Simulation 4 shown in Figure 14. In this case, such a large loss of load could be associated with the disconnection of an industrial user, and it is evident how large the frequency variation associated with this loss is. The nadir reached in this extreme case is equal to 1.0175 pu (50.875 Hz), which is not compliant with many grid codes, for the "traditional" renewables control scenario. In the case of the updated control, on the other hand, the nadir is equal to 1.0106 pu (50.53 Hz), which also exceeds the upper frequency limit but is lower than the previous, highlighting the huge benefit introduced.

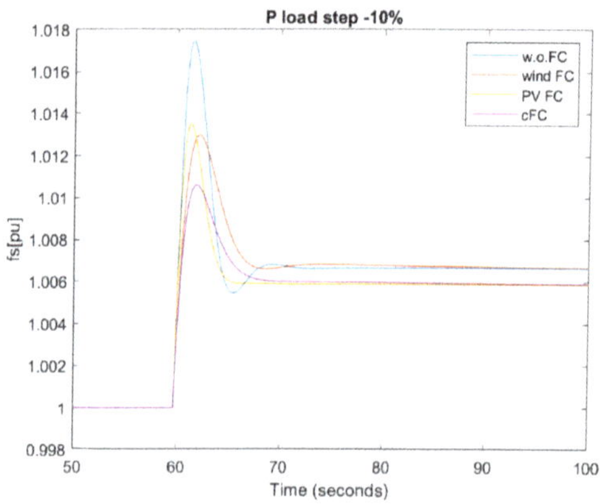

Figure 14. Simulation 4 frequency trend.

It is very interesting to observe how the different generators contribute to the frequency control by adapting their power, as shown in Figure 15.

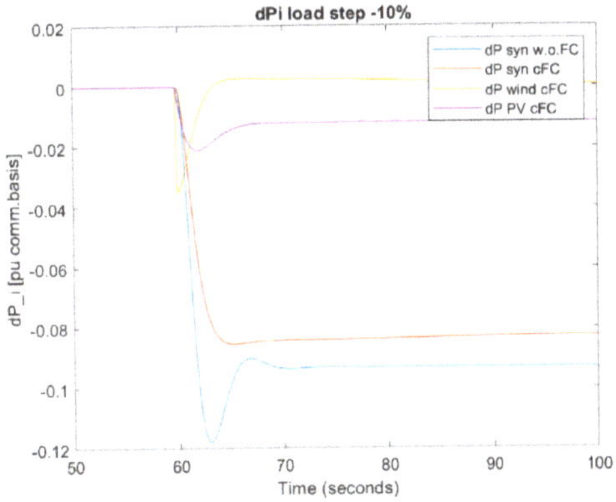

Figure 15. Generator power variations in Simulation 4.

It is clearly visible how the two renewables reduce their power to face a decrease in load. This helps the synchronous generator in containing the frequency increase: it can be seen that the steam turbine power reduction in the case of the *complete FC* is less intense than that in the case of the frequency control operated just by themselves. The stabilization effect is also emphasized by the reduction in the oscillations after the nadir, that further facilitates the intervention of the synchronous generator.

Considering the third metric (nadir-based frequency response), it has been shown with these numerical results that it is practically the same in all the four simulations for what concerns the conventional scenario, while in the case of the *complete FC* it is a little bit lower in increasing the step magnitude.

4.2. Small Sinusoidal Oscillations and Step down Load Active Power

In the previous case, before the perturbation, the reserve was entirely available to stabilize the frequency due to its load power constancy. This is far from the reality of a power system: the load continuously oscillates, entailing that a part of the reserve is already involved in the stabilization of the frequency. This means that if a sudden perturbation of the load power happens the system has less reserve available to face the variation.

The aim of this second type of simulations is to address how the continuous and small variability of the load affects the primary frequency control and its capability to face a step in load power, as well as to understand if the designed control for renewables is able to catch and follow these small load oscillations. The load power oscillations introduced in this simulation have an amplitude of 0.005 pu. The resulting frequency of Simulation 5 is presented in Figure 16.

It can be observed that the system with a *complete FC* needs some seconds to adapt to the sinusoidal variation of the load, but then it is able to follow it without problems. It is also possible to appreciate the huge reduction in the frequency nadir with respect to the base scenario, which is a fact that testifies one more time to the effectiveness of the updated control of the renewable generators. After the frequency peak, both systems are able to follow the sinusoidal oscillations of the load, but of course in the *complete FC* scenario the stabilization of the frequency is more evident and requires less time to be reached.

This simulation is particularly important for the validation of the innovative control of the renewables, since it is closer to the normal behavior of the load power.

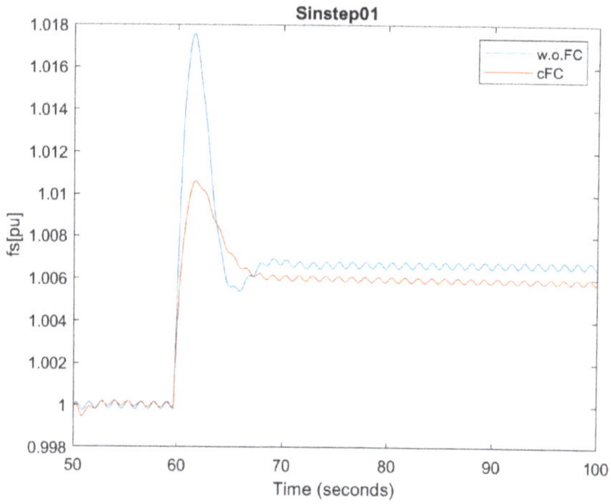

Figure 16. Simulation 5 frequency trend.

4.3. Variable Environmental Conditions

Two of the major drawbacks of renewables are the difficult predictability and the high variability of their primary resources, i.e., the wind and solar irradiance. In the field of primary frequency control, which takes place in the time window of a few seconds after a disturbance, this variability affects their response, especially for what concerns the wind energy. If the sun variations are fast but occur less frequently, the wind variations are huge and occur continuously. In this sense, the most critical situation is represented by windy days with some small clouds, in which both the wind and the irradiance are very variable.

The aim of these tests is to understand how much the variability of the renewable source affects primary frequency control if the renewables are rigid and they do not participate in it and, on the other hand, in the case they are able to modulate their power output to smooth their power variations. Furthermore, as for the second group, a part of the reserve will be already occupied to balance the natural variability of the resource, so, in case of a perturbation of the load, the reserve level could be less than the nominal amount.

The first analysis regards the most critical of the two resources: wind. In Simulation 6 and 7, the system is tested in the four scenarios with a real high variable wind profile. The wind profile has been downloaded from a DTU database of measurements in [38] and has the shape depicted in Figure 17.

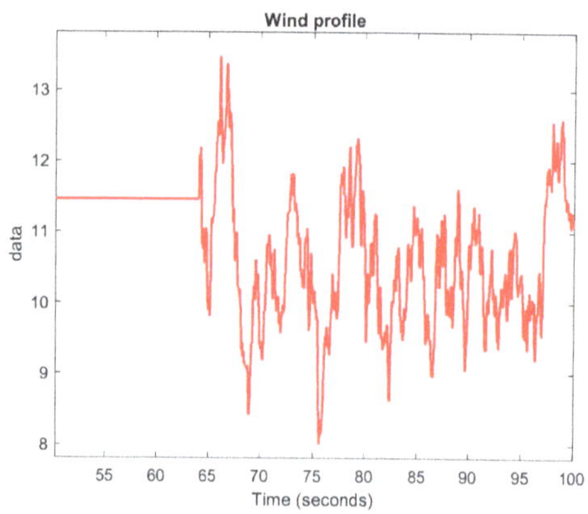

Figure 17. Real wind profile used in the simulations.

The profile has been used as an input for the wind turbine system, which is not anymore free to adapt its output only to balance the frequency: now it also has to face a variable availability of the wind resource. This means that a part of the reserve could be already involved in the balancing of the frequency consequent to a wind speed variation, reducing the capability to counter a step of load power.

Simulation 6 studies the effect of the variable wind profile on the frequency. In particular, it is interesting to observe the response in the case of the wind turbine with virtual inertia control enabled compared with the base case without frequency response capabilities. The results are shown in Figure 18. In this simulation, the variability of the wind has been introduced after 64 s, so that its effects are more visible starting from a steady state condition.

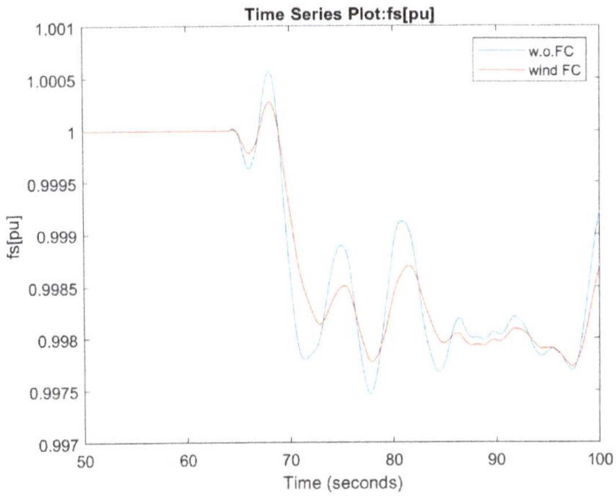

Figure 18. Simulation 6 frequency trend.

In the base case without frequency control enabled, the wind turbines modulate their power to always produce at their maximum. This means that they do not detect that their behavior is causing the system's frequency to oscillate. It is possible to observe, indeed, that the frequency has a number of peaks in correspondence with the peaks of the wind speed. This trend will stress the system, requiring an intense control action from the synchronous generators.

On the other hand, if the wind turbines are able to exploit their kinetic energy using the Extended OPPT method, it is evident how they smooth their power variation, producing lower peaks and stressing the power system less.

Looking at the largest nadir reached in the variable wind period, in the first scenario it is equal to 0.9975 pu (49.875 Hz), while in the updated scenario is 0.9977 pu (49.885 Hz). The reduction in the nadir is not so pronounced between the two, but the benefits consequent to the introduction of the Extended OPPT Method are evident considering the trend of frequency, since the amplitude of each oscillation is lowered. This fact is confirmed by the plot of the RoCoF, which highlights how the rate of variation of frequency is slower and smoother in the second scenario (Figure 19).

For what concerns the irradiance variation, Simulation 8 has been inspired by the dynamic EN50530 test, which is recognized as one of the techniques to evaluate the MPP tracking effectiveness [19]. It consists of a sequence of ramping up, constancy, and ramping-down the solar irradiance hitting the module. In this case, it has been adapted to model the shading of PV modules consequent to the passage of a small cloud. As can be seen in Figure 20, it consists of a steep ramping down (100 $W/m^2/s$) from the nominal irradiance of 1000 W/m^2 to 300 W/m^2. Then, for few seconds, the irradiance remains low due to the shading and then it increases with the same steepness again to its nominal value.

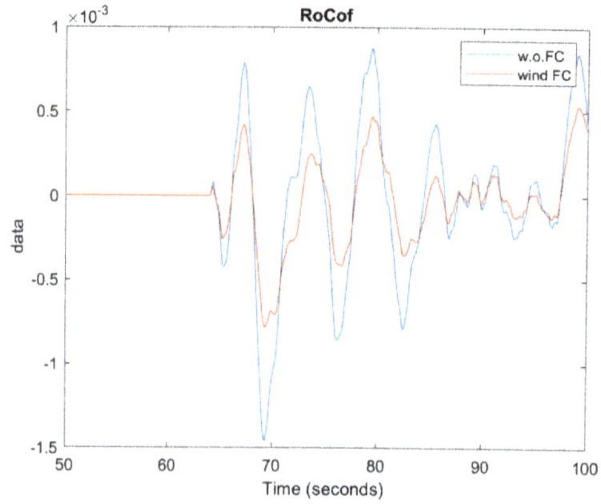

Figure 19. Simulation 6 RoCoF.

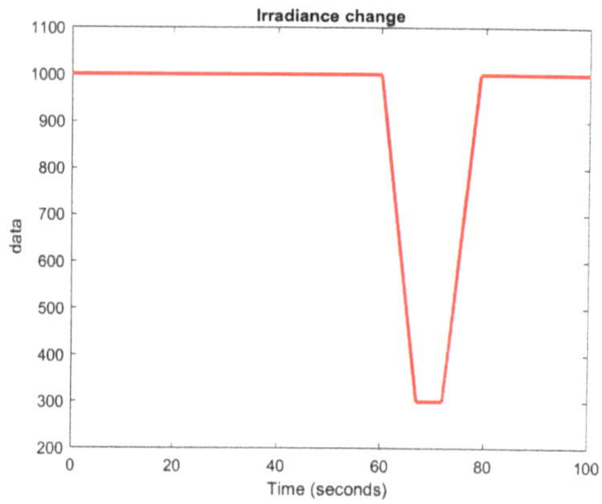

Figure 20. Simulation 8 irradiance profile.

The system with renewables participating in primary frequency control also shows variable irradiance to be more stable in terms of frequency, as depicted in Figure 21.

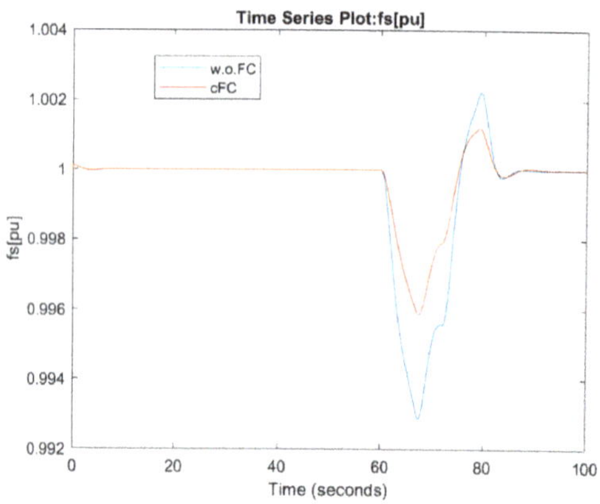

Figure 21. Simulation 8 frequency trend.

4.4. Increasing Renewables Penetration

The last scenario studied in this work is interesting in a long-term perspective. As explained in the introduction, in the future, a large spread of renewable sources is expected that will displace the conventional ones. In these simulations, the effect of such an increase on the frequency stability is addressed, both in the case of the absence of frequency response capability from renewables and in the presence of the advanced controls considered in this work.

It is important to remember the expected result of the tests in this first scenario: the increase in renewable penetration, due to their unpredictability and non-dispatchability, requires an increase in the reserve level, which at the same time displaces the conventional plants that are responsible for providing primary frequency controls and contributing to systems' inertias. Due to these facts, the magnitude of the frequency nadir consequent to a perturbation is expected to increase as the penetration of renewables increases. The aim of the advanced controls, on the other hand, is to allow the substitution of the conventional fleet of generators with renewable ones without losing the ability to control the frequency. This is why, in the second case, a smaller frequency nadir and frequency oscillations are expected.

When adapting the participation factors, it is important to take into account that the displacement of synchronous generators causes a reduction in the system's equivalent inertia. In this work, the reduction is proportional to the synchronous generator participation factor. This means, for example, that if the participation factor of the equivalent steam turbine passes from 0.7 to 0.35, the equivalent inertia constant will be halved.

The first group of simulations regards an increase in VSWT penetration, whose results are shown in Figure 22.

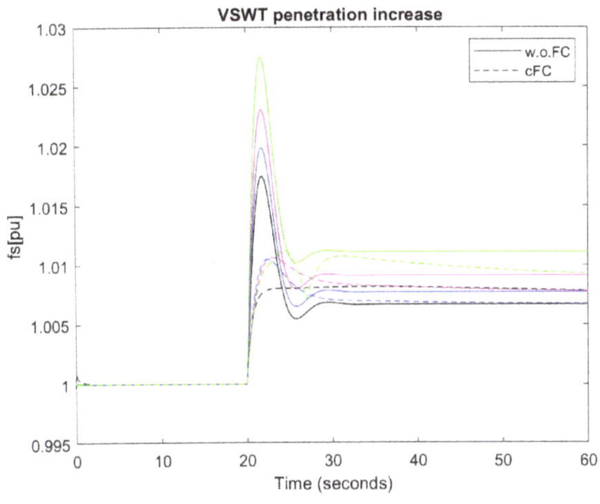

Figure 22. VSWT penetration increase.

The solid lines represent the frequency variation consequent to a negative load power step of −10% in the scenario of primary frequency control provided just with synchronous generators, while the dashed lines are resulting from the scenario with a *complete FC*. Taking a first look at the plot, the negative effect of the increasing penetration of renewables if they are not able to participate in the primary frequency control is immediately clear. Indeed, for the same step of load power, the nadir increases from 1.0174 pu in the base case (20% wind, 10% PV) to 1.0275 pu in the most extreme case (50% wind, 10% PV). Furthermore, the difference in terms of the nadir between one curve and the other is not equal: increasing the wind penetration, the increase in the nadir is larger and larger from one curve to the next one. This happens because there is not just an effect due to the reduction in the synchronous generator power variation weight, but also of the system's inertia.

Once stated thus, it can be seen how large the benefit is associated with the introduction of the frequency response capability in renewables: the nadir is much lower than in the first scenario and, as the penetration of VSWT increases, the mismatch between the two scenarios is way more evident. Unlike what happens in the first scenario, the nadir difference between the various curves in the second scenario reduces as the wind turbine penetration increases, meaning that the provision of this service from renewables (in this case, it prevails over the effect of the wind turbines inertial control since their weight is dominant) is able to "cover" the reduction in the system's inertia, thus improving frequency stability.

The same kind of observations are discernable from the other simulations, which regard an increase in PV penetration (Figure 23).

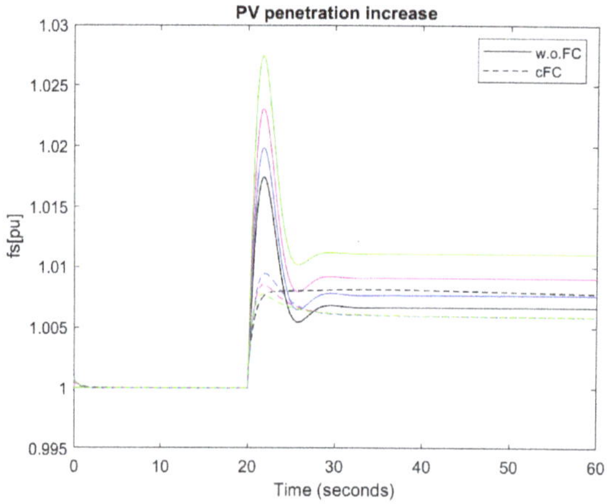

Figure 23. PV penetration increase.

In this case, the benefit is even larger, since, after a certain value of p_{pv}, the nadir even reduces as the penetration of the photovoltaic increases. For example, in the most extreme case (40% PV, 20% wind), the nadir is lower than in the less PV-penetrated systems (10% PV, 20% PV, and 30% PV).

In a future scenario, it is not difficult to imagine power systems that are even largely penetrated by renewables. This case is studied in Simulation 17, with 80% of renewable participation. The correspondent results are shown in Figure 24.

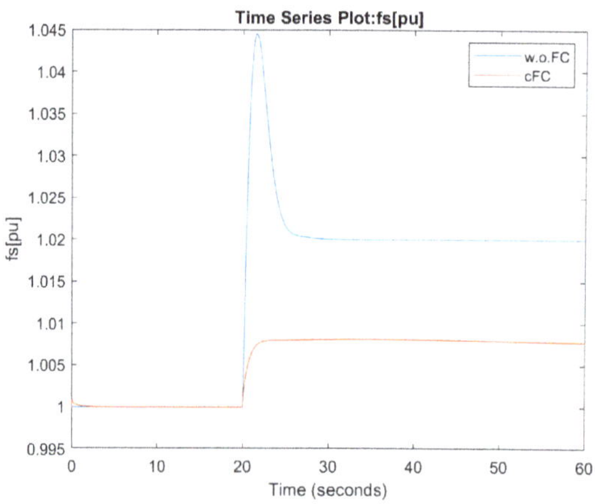

Figure 24. Simulation 17 frequency trend.

It is evident that the benefit was obtained using the synergy between the Extended OPPT Method and the Power Reserve Control on the frequency stability. With such a renewable penetration, the system's inertia is really small (−72% with respect to the base case) and, at the same time, a variation of power of the synchronous generator has just a small influence on the overall active power balance, accounting just for 20% of the per unit common basis power. The consequence of a drop in load power is potentially critical: a frequency nadir of 1.0446 pu corresponds to 52.23 Hz, which is an incredibly large frequency deviation.

On the other hand, the introduction of renewable updated controls and their coordination allows for a limiting of the nadir to 1.0082 pu (50.41 Hz), which is a perfectly acceptable value even facing an exceptionally strong perturbation event. By looking at the results of both simulations (11 and 15), in which, respectively, wind and PV penetration are equal to 40% as in Simulation 17, it can be observed that, if taken singularly, the frequency nadir is more accentuated, while in 17 the trend of the frequency is very flat. This fact enhances the idea that the two controls do not just act individually but that they support each other in the frequency control task due to their different response rates and they integrate very well with each other (Figure 25), obtaining a very smooth frequency trend and a satisfying frequency containment and stabilization.

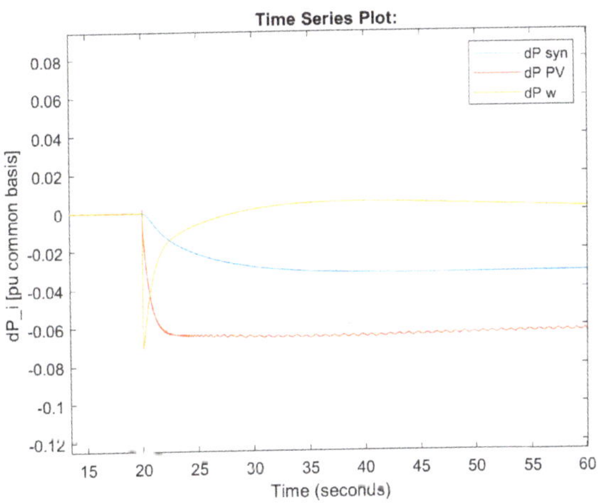

Figure 25. Simulation 18 power variations.

Finally, it is very interesting to observe the effect of this control in a more dynamic situation. In particular, combining Simulation 7 (variable wind profile and load power step −10%) with Simulation 17 (40% wind, 40% PV), the frequency profile shown in Figure 26 is obtained.

It can be observed that the variability of the wind, as already demonstrated, has an increasing effect on the frequency instability if the penetration of the PV and wind turbine increases. If the system's inertia decreases, indeed, the system frequency is more sensible to the perturbation, including the step introduced at time 60 s. In the case of an updated control of renewables, when they constitute the large majority of the generation fleet, it is evident how good they control the frequency in cases of variable wind profiles, smoothing almost all the peaks produced.

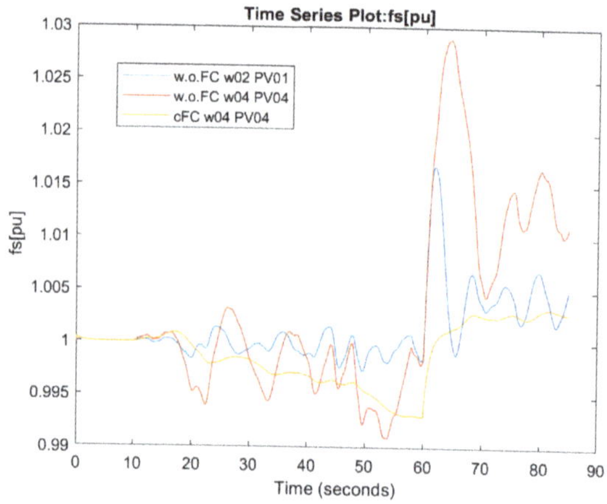

Figure 26. Simulation 7 for different scenarios frequency trend.

5. Conclusions and Future Developments Proposals

After the detailed analysis of the results of the simulations of the developed model shown in Section 4, it is possible to take stock of the work that has been carried out.

Looking ahead to a future in which renewables will be increasingly relevant in the energy mix, two strategies have been implemented to ensure they are able to help (or substitute) conventional generators in keeping the power system stable. In particular, Power Reserve Control and the Extended OPPT Method have been tested in this work, respectively, aiming at allowing photovoltaic modules and variable-speed wind turbines to participate in primary frequency control. Indeed, the challenge of using non-programmable renewable generators to provide the services that nowadays are delivered by fossil-fueled plants will become increasingly central, as the renewable presence increases to a more significant level.

In this sense, the results obtained in this work, especially when both controls coexist, are encouraging. Indeed, even if the Power Reserve Control introduces a very small increase in the RoCoF in the very few instants after a strong and sudden perturbation, the frequency nadir is reduced with respect to the case in which the frequency is not supported by renewables, especially in a future scenario in which they are massively present in the power system. The tradeoff between RoCoF increases and the nadir reduces in a given power system depending on its characteristics and, particularly, on the settings of the respective relays.

The model and the results rely on some simplifications, which generalize them but at the same time are lacking in detail. For this reason, there is an open field for future developments of the present work in some areas.

One of the possible fields of improvement is the modeling of the photovoltaic system, in particular for what concerns shading conditions and inequalities in sun resource distribution among the modules. Indeed, this would allow for a deeper study of the capability of photovoltaic systems in supporting the system's frequency in more realistic situations.

Another possible improvement regards the modeling of the load. Indeed, it could be very interesting to better characterize its active power demand profile in case of a small island to cause the study to be even more realistic.

The last and more challenging proposal for future developments would be to model a small system node by node as the one studied in an overall way with the LFC approach in this work. This will completely fit it to a specific practical situation and ensure it is able to catch some details that are invisible with a general approach. At the same time, the adaptation to a very specific context removes one of the biggest advantages of the LFC approach, which is the generality and versatility of the developed model. Nevertheless, since the results obtained are promising, such improvements are interesting and can add cues to go deeper in the knowledge of this field.

Author Contributions: Conceptualization, S.M.; methodology, S.M.; software, L.R.; validation, L.R. and S.M.; formal analysis, L.R. and S.M.; investigation, L.R. and S.M.; resources, L.R. and S.M.; data curation, L.R.; writing—original draft preparation, L.R.; writing—review and editing, S.M.; visualization, L.R.; supervision, S.M.; project administration, S.M.; funding acquisition, S.M. All authors have read and agreed to the published version of the manuscript.

Funding: This research was funded by the Spanish national research agency Agencia Estatal de Investigación, grant number PID2019-108966RB-I00/AEI/10.13039/501100011033.

Institutional Review Board Statement: Not applicable.

Informed Consent Statement: Not applicable.

Data Availability Statement: Not applicable.

Conflicts of Interest: The authors declare no conflict of interest. The funders had no role in the design of the study; in the collection, analyses, or interpretation of data; in the writing of the manuscript; or in the decision to publish the results.

Appendix A. Values of Constants Used in the Model

This appendix compiles the values of the constants used in the model, organized as a function of the subsystems in which they are located.

- PV subsystem
  - PV module: Solartech Power SPM210P (four modules in series and variable number of arrays as a function of the participation factors)
  - $C_{capacitor}$ = 100 µF
  - $L_{inductor}$ = 10 mH
  - D-controlled boost converter with default Simulink settings
  - PI controller: $P = 0.03$; $I = 1$
  - Droop constant of active power reserve adaptation = 0.05
  - Initial reserve level = 0.2
- VSWT system
  - $P_{base} = P_{t,base} = P_{g,base} = 1.5$ MW, $v_{nom} = 12$ m/s, $\omega_{t,base} = 1.644$ rad/s, $\omega_{g,base} = 157.08$ rad/s, $f = 50$ Hz
  - Values for the model blocks taken from [36]
  - Speed governor PI controller: $P = 3$; $I = 80$
  - Pitch governor P controller: $P = 500$
  - Inertial control [24,39]: $K_{opt} = 0.4225$, $k_{vir} = 8$, $W_{vir} = 0.2$, $T_{wo} = 10$ s, $T_{lp} = 100$ ms, $H_{WT} = 5.29$ s
- Steam turbine with one reheat [24,32]
  - $H_{eq} = 2.88$ s
  - Droop constant = 0.05
  - $\tau_g = 0.2$
  - $F_{HP} = 0.3$
  - $T_{RH} = 5$ s
  - $T_{CH} = 0.3$ s

References

1. European Council. Available online: https://www.consilium.europa.eu/en/policies/climate-change/paris-agreement/ (accessed on 1 August 2022).
2. UNFCCC. Available online: https://unfccc.int/process-and-meetings/the-paris-agreement/the-paris-agreement (accessed on 1 August 2022).
3. IEA. Available online: https://www.iea.org/data-and-statistics/data-browser?country=WORLD&fuel=CO2%20emissions&indicator=CO2BySector (accessed on 2 August 2022).
4. European Energy Agency. Available online: https://www.eea.europa.eu/ims/share-of-energy-consumption-from (accessed on 2 August 2022).

5. IRENA. *Global Energy Transformation—A Roadmap to 2050*; International Renewable Energy Agency: Abu Dhabi, United Arab Emirates, 2019.
6. IRENA. *Grid Codes for Renewable Powered Systems*; International Renewable Energy Agency: Abu Dhabi, United Arab Emirates, 2022.
7. Hassan, F.; Bollen, M. *Integration of the Distributed Generation in the Power System*; IEEE Press: New York, NY, USA, 2011.
8. Riquelme-Dominguez, J.M.; De Paula Garcia-Lopez, F.; Martínez, S. Power Ramp-Rate Control via power regulation for storageless grid-connected photovoltaic systems. *Int. J. Electr. Power Energy Syst.* **2022**, *138*, 107848. [CrossRef]
9. Brivio, C.; Mandelli, S.; Merlo, M. Battery energy storage system for primary control reserve and energy arbitrage. *Sustain. Energy Grids Netw.* **2016**, *6*, 152–165. [CrossRef]
10. NREL. *On the Path to Sunshot: The Role of Advancements in Solar Photovoltaic Efficiency, Reliability, and Costs*; NREL: Golden, CO, USA, 2016.
11. Solomon, A.A.; Bogdanov, D.; Breyer, C. Curtailment-storage-penetration nexus in the energy transition. *Appl. Energy* **2019**, *235*, 1351–1368. [CrossRef]
12. Sangwongwanich, A.; Yang, Y.; Blaabjerg, F. Development of Flexible Active Power Control Strategies for Grid-Connected Photovoltaic Inverters by Modifying MPPT Algorithms. In Proceedings of the 2017 IEEE 3rd International Future Energy Electronics Conference and ECCE Asia, Kaohsiung, Taiwan, 3–7 June 2017.
13. Batzelis, E.I.; Kampitsis, G.E.; Papathanassiou, S.A. Power Reserves Control for PV Systems With Real-Time MPP Estimation via Curve Fitting. *IEEE Trans. Sustain. Energy* **2017**, *8*, 1269–1280. [CrossRef]
14. Sangwongwanich, A.; Yang, Y.; Blaabjerg, F.; Sera, D. Delta Power Control Strategy for Multi-String Grid-Connected PV Inverters. *IEEE Trans. Ind. Appl.* **2017**, *53*, 3862–3870. [CrossRef]
15. Li, X.; Wen, H.; Zhu, Y.; Jiang, L.; Hu, J.; Xiao, W. A Novel Sensorless Photovoltaic Power Reserve Control With Simple Real-Time MPP Estimation. *IEEE Trans. Power Electron.* **2019**, *34*, 7521–7531. [CrossRef]
16. Hoke, A.; Chakraborty, S.; Shirazi, M.; Muljadi, E. Rapid Active Power Control of Photovoltaic Systems for Grid Frequency Support. *IEEE J. Emerg. Sel. Top. Power Electron.* **2017**, *5*, 1154–1163. [CrossRef]
17. Riquelme-Dominguez, J.M.; Martinez, S. A Photovoltaic Power Curtailment Method for Operation on Both Sides of the Power-Voltage Curve. *Energies* **2020**, *13*, 3906. [CrossRef]
18. Nanou, S.; Papakonstantinou, A.; Papathanassiou, S. A generic model of two-stage grid-connected PV systems with primary frequency response and inertia emulation. *Electr. Power Syst. Res.* **2015**, *127*, 186–196. [CrossRef]
19. Riquelme-Dominguez, J.M.; Martínez, S. Systematic Evaluation of Photovoltaic MPPT Algorithms Using State-Space Models Under Different Dynamic Test Procedures. *IEEE Access* **2022**, *10*, 45772–45783. [CrossRef]
20. Aziz, A.; Oo, A.T.; Stojcevski, A. Frequency regulation capabilities in wind power plant. *Sustain. Energy Technol. Assess.* **2018**, *26*, 47–76. [CrossRef]
21. Clark, K.; Miller, N.W.; Sanchez-Gasca, J.J. *Modelling of GE Wind Turbine-Generators*; Tech. Rep.; Version 4.5; General Electric International: Schenectady, NY, USA, 2010.
22. Engelken, S.; Mendonca, A.; Fischer, M. Inertial response with improved variable recovery behaviour provided by type 4 WTs. *IET Renew. Power Gener. Spec.* **2017**, *11*, 195–201. [CrossRef]
23. Wu, Y.; Shu, W.; Hsieh, T.; Lee, T. Review of inertial control methods for DFIG-based wind turbines. *Int. J. Electr. Energy* **2015**, *3*, 174–178. [CrossRef]
24. Ochoa, D.; Martínez, S. Analytical Approach to Understanding the Effects of Implementing Fast-Frequency Response by Wind Turbines on the Short-Term Operation of Power Systems. *Energies* **2021**, *14*, 3660. [CrossRef]
25. Vidyanandan, K.V.; Senroy, N. Primary frequency regulation by deloaded wind turbines using variable droop. *IEEE Trans. Power Syst.* **2013**, *28*, 837–846. [CrossRef]
26. Hwang, M.; Muljadi, E.; Park, J.W.; Sorensen, P.; Kang, T.C. Dynamic droop-based inertial control of a doubly-fed induction generator. *IEEE Trans. Sustain. Energy* **2016**, *7*, 924–933. [CrossRef]
27. Fairley, P. Can Synthetic Inertia from Wind Power Stabilize Grids? *IEEE Spectrum* **2016**, *7*.
28. Fu, Y.; Zhang, X.; Hei, Y.; Wang, H. Active participation of variable speed wind turbine in inertial and primary frequency regulations. *Electr. Power Syst. Res.* **2017**, *147*, 174–184. [CrossRef]
29. Martínez-Lucas, G.; Sarasua, J.I.; Perez-Diaz, J.I.; Martínez, S.; Ochoa, D. Analysis of the Implementation of the Primary and/or Inertial Frequency Control in Variable Speed Wind Turbines in an Isolated Power System with High Renewable Penetration. Case Study: El Hierro Power System. *Electronics* **2020**, *9*, 901. [CrossRef]
30. Díaz-González, F.; Hau, M.; Sumper, A.; Bellmunt, O. Participation of wind power plants in system frequency control: Review of grid code requirements and control methods. *Renew. Sustain. Energy Rev.* **2014**, *34*, 551–564. [CrossRef]
31. O'Sullivan, J.; Rogers, A.; Flynn, D.; Smith, P.; Mullane, A.; O'Malley, M. Studying the maximum instantaneous non-synchronous generation in an island system—Frequency stability challenges in Ireland. *IEEE Trans. Power Syst.* **2014**, *29*, 2943–2951. [CrossRef]
32. Kundur, P. *Power System Stability and Control*; Electric Power Research Institute: Palo Alto, CA, USA, 1993.
33. Li, S.; Deng, C.; Shu, Z.; Huang, W.; Jun He, J.; You, Z. Equivalent inertial time constant of doubly fed induction generator considering synthetic inertial control. *J. Renew. Sustain. Energy* **2016**, *8*, 053304. [CrossRef]
34. Ibrahim, H.; Anani, N. Variations of PV module parameters with irradiance and temperature. *Energy Procedia* **2017**, *134*, 276–285. [CrossRef]

35. Carrero, C.; Ramírez, D.; Rodríguez, J.; Platero, C.A. Accurate and fast convergence method for parameter estimation of PV generators based on three main points of the I-V curve. *Renew. Energy* **2011**, *36*, 2972–2977. [CrossRef]
36. Ochoa, D.; Martínez, S. A Simplified Electro-Mechanical Model of a DFIG-based Wind Turbine for Primary Frequency Control Studies. *IEEE Lat. Am. Trans.* **2016**, *14*, 3614–3620. [CrossRef]
37. Eto, J.H. *Use of Frequency Response Metrics to Assess the Planning and Operating Requirements for Reliable Integration of Variable Renewable Generation*; Lawrence Berkeley National Laboratory: Berkeley, CA, USA, 2011; Available online: https://escholarship.org/uc/item/0kt109pn (accessed on 20 October 2022).
38. DTU Data: Database on Wind Characteristics. Available online: https://gitlab.windenergy.dtu.dk/fair-data/winddata-revamp/winddata-documentation/-/blob/master/readme.md/ (accessed on 20 October 2022).
39. Ochoa, D.; Martinez, S. Fast-Frequency Response Provided by DFIG-Wind Turbines and Its Impact on the Grid. *IEEE Trans. Power Syst.* **2017**, *32*, 4002–4011. [CrossRef]

applied sciences

MDPI

Article

Triple Phase Shift Control of Wireless Charging DAB LCC Resonant Converter for Unity Power Factor Operation with Optimized Rectifier AC Load Resistance

Yanni Cheong [1], Shuyu Cao [1], Ramasamy Thaiyal Naayagi [2,*] and Szesing Lee [2]

1 Singapore Institute of Technology, Electrical Power Engineering, 10 Dover Drive, Singapore 138683, Singapore
2 School of Electrical and Electronic Engineering, Newcastle University International, SIT@NYP, 172A Ang Mo Kio Avenue 8 #05-01, Singapore 567739, Singapore
* Correspondence: naayagi.ramasamy@ncl.ac.uk; Tel.: +65-6908-6069

Featured Application: The proposed triple phase shift (TPS) dual active bridge (DAB) resonant converter control scheme is suitable for high-efficiency electrical vehicle (EV) wireless charging applications.

Abstract: This paper presents a new triple phase shift (TPS) closed-loop control scheme of a dual active bridge (DAB) LCC resonant DC/DC converter to improve wireless charging power transfer efficiency. The primary side inverter phase shift angle regulates the battery charging current/voltage. The secondary side rectifier phase shift angle regulates the rectifier AC load resistance to match its optimized setting. The inverter-to-rectifier phase shift angle is set to achieve unity power factor operation of the DAB rectifier and inverter. The mathematical formulation of the TPS shift control is given for each phase shift angle. The analytical calculation, circuit simulation, and experimental test are carried out in a power scaled-down DAB LCC resonant wireless charging converter laboratory hardware setup to validate the proposed TPS close-loop control scheme. The PLECS circuit simulation shows that DAB LCC resonant SiC MOSFET operates at zero-voltage-switching (ZVS) with a unity power factor in emulated constant current (CC) mode battery charging. In constant voltage (CV) mode operation, one inverter/rectifier Leg does not operate at ZVS switching when Sic MOSFET is switched on near zero current. The experimental results show that the efficiency is greatly improved for CV mode charging with large DC load resistance connected if rectifier AC load resistance matching control is enabled. The measured efficiency matches well with the analytical calculation. The estimated efficiency improvement will be much more significant for EV applications in the kW power range with greater winding loss. The challenges and possible solutions to implement TPS PWM modulation in two separate inverter and rectifier control hardware are explained for future TPS control algorithm development in practical wireless charging products.

Keywords: wireless charging; dual active bridge (DAB); DC/DC converter; LCC resonant circuit; triple phase shift (TPS); constant current (CC) battery charging; constant voltage (CV) battery charging; zero-phase-angle (ZPA); zero-voltage switching (ZVS); and unity power factor operation

Citation: Cheong, Y.; Cao, S.; Naayagi, R.T.; Lee, S. Triple Phase Shift Control of Wireless Charging DAB LCC Resonant Converter for Unity Power Factor Operation with Optimized Rectifier AC Load Resistance. *Appl. Sci.* **2022**, *12*, 11871. https://doi.org/10.3390/app122211871

Academic Editors: Federico Barrero and Mario Bermúdez

Received: 28 September 2022
Accepted: 18 November 2022
Published: 21 November 2022

1. Introduction

Wireless power transfer has gained popularity over the years as this technology can charge electric vehicles (EVs) without requiring a power cable and a plug-in connector. In recent years, inductive power transfer by a DC/DC resonant converter has become an effective wireless charging method [1]. For EV wireless charging applications, the considerable variation of wireless transformer parameters exists practically due to the significant magnetic coupling factor change caused by air-gap distance variation, the misalignment between the transmitter and the receiver pads, and the receiver coil parameter vehicle-to-vehicle variation. For the basic series-series (S-S), series-parallel (S-P), parallel-series (P-S), and parallel-parallel (P-P) resonant circuit topologies, the resonant frequency is sensitive to the magnetic coupling factor variation and the receiver coil parameter variation. The high-order LCC resonant circuit is a more appealing topology for its robustness to the resonant circuit parameter variation and battery operation condition variation [2–4].

The resonant frequency selection is an important design decision for a wireless charging system. High resonant frequency improves the magnetic coupling and reduces the size of the wireless charging transformer. Therefore, increased resonant frequency reduces winding loss, core loss, and shield loss. The cost for high frequency is increased power device switching loss [5]. SiC MOSFET in the market's 600 V~1700 V voltage range opens the opportunity to implement high resonant frequency for EV wireless charging systems. Compared to SI MOSFET devices, SIC MOSFET has the advantage of much smaller and thermal-stable On-resistance Rds, which leads to low conduction loss. The intrinsic fast diode in SIC MOSFET has a much lower reverse recovery charge (Qrr) (low switching-off loss). SIC MOSFET is an ideal choice for wireless charging DC/DC resonant converter, which runs at hard-switching control with acceptable power device loss in low current at CV battery charging mode [5]. The switching loss of Sic MOSFET is estimated to reduce to around 50% with ZVS switching. The ZVS switching can be realized by introducing a phase shift between inverter/rectifier AC voltage and current so that the turn-off current at MOSFET switching ON transition is large enough to discharge the MOSFETs junction capacitors within dead time [6,7].

For a conventional unidirectional wireless charging system, the battery current or voltage feedback control is implemented in the transmitter side high-frequency DC/AC inverter, and the receiver side AC/DC converter is an uncontrolled diode rectifier bridge. The loss distribution of a unidirectional 8 kW wireless charging MOSFET DC/DC LCC resonant converter is estimated as 52.3% from transmitter and receiver coils, 16.7% from LCC compensation network, 11.9% from power MOSFETs, 16.8% from rectifier diodes, and 2.2% from output filter [2]. Similar power loss distributions are reported in [8] in CC mode charging operation. For wireless charging systems, the loosely coupled transmitter and receiver coils dissipate most of the power loss because of a relatively larger resonant current [2,8].

In LCC resonant compensation parameter design calculation, inverter AC circuit zero-phase-angle (ZPA) constraint is imposed to minimize reactive power flow into AC resonant circuit and maximize LCC resonant circuit efficiency [2,9–11] The resonant compensation circuit design with inverter ZPA operation constraint is critical for achieving high-efficiency operation of the wireless charger at CC charging mode with low battery equivalent resistance. However, resonate circuit efficiency significantly degrades in CV charging mode operation when battery equivalent load resistance is significantly increased [12–14].

Intensive research efforts are published to minimize the efficiency degradation caused by battery equivalent resistance variation in CV battery charging mode. This research effort can be generally classified as load impedance estimation and load impedance matching control, which requires modifying the rectifier side circuit and implementing dual side control of the DC/DC resonant converter. A simple solution is to add a post-regulation DC/DC converter after the diode rectifier to regulate the effective load impedance of the resonant circuit when the battery charging condition is changed. In [14], dynamic maximum-efficiency tracking control is proposed for S-S resonant converter wireless charging system by adding a pre-regulation DC/DC converter before the inverter on the transmitter side and a post-regulation DC/DC converter after the diode rectifier on the receiver side. On top of battery voltage/current/power regulation in inverter control, the duty cycles for pre-regulation DC/DC converter and post-regulation DC/DC converter are adjusted based on "perturb & observer" methods to minimize the measured efficiency. In [15], a post-regulation buck-boost converter is added after the diode rectifier bridge for the LCC resonant converter wireless charging system. The diode rectifier output DC load resistance is regulated around a reference load resistance value by duty cycle control of the buck-boost converter to improve efficiency in CV charging mode operation. The CC/CV/power regulation is realized in inverter control using model-based power/current calculation from the estimated inverter AC load resistance. In [16], a pre-regulation buck converter is added before the inverter. A semi-active rectifier replaces the diode bridge. The battery charging current is controlled by rectifier side phase shift PWM modulation with voltage pulse synchronized with measured receiver coil current. The inverter phase shift angle is kept at maximum value. The buck-boost duty cycle is adjusted to optimize the system efficiency based on the "perturb & observer" method. The drawback of this extra pre/post-DC/DC converter solution is that the efficiency improvement is compromised by the power loss generated by the additional power circuit. The wireless charging system also becomes more bulky, costly, and less reliable.

Recently, intensive research interests are raised in using DAB resonant converter control to improve the efficiency of the wireless charging system by simply replacing the diode bridge with an active rectifier bridge [7,17]. The advantage of DAB resonant converter topology is that it does not increase power device component count and opens the possibility to use low loss Sic MOSFET devices. For the DAB DC/DC resonant converter, it is crucial to ensure that both the inverter and rectifier work in ZPA conditions to minimize the reactive power input and output to the resonant circuit. The resonant compensation circuit design based on inverter ZPA constraint was originally developed for unidirectional DC/DC resonant with the diode rectifier. The underline assumption for unity power factor operation of the inverter is that the resonant circuit output has a purely resistive load. However, DAB resonant converter control can operate the active rectifier bridge in either a capacitive or inductive power factor by advancing or delaying the rectifier AC voltage relative to the current. It is essential that DAB DC/DC resonant converter control takes care of pure resistance AC load assumption and ensures that the DAB rectifier bridge operates at unity power factor for the DAB inverter bridge operating at ZPA condition. This restriction on DAB resonant converter control freedom has yet to be explained and discussed in publications.

In [7], the method to implement triple phase shift (TPS) control of the DAB LCC resonant converter wireless charging circuit is reported. The input/output current/voltage waveform relationship is utilized to derive the inverter phase shift angle βp and rectifier phase shift angle βs according to the load power demand. The inverter-to-rectifier phase shift angle is set to $\delta = 90^{\circ} + \Delta\delta$ with $\Delta\delta$ calculated analytically from the discharging current of the power electronic switch, the resonant circuit parameters, and the input/output DC voltage to maximize the ZVS operation range of the inverter and rectifier bridges. The robustness performance of this TPS control method is questionable because it is challenging to obtain accurate circuit parameters for the inverter and rectifier phase shift angle online calculation in a practical wireless charging system.

In [17], a phase-lock-loop (PLL) based double phase shift control scheme is proposed for a DAB S-S resonant converter wireless charging system. The inverter phase shift angle controls battery charging voltage/current. The rectifier phase shift angle controls the effective AC load resistance of the resonant circuit. A PLL algorithm was proposed to adjust the rectifier voltage window based on the zero-crossing detection of the receiver winding current to achieve closer to zero reactive power operation of the rectifier AC circuit. The PLL timing synchronization control algorithm implicitly adjusts the inverter-to-rectifier phase shift angle. The inverter ZPA can be realized with rectifier-side PLL timing synchronization control. It is hard to understand rectifier-side PLL auto-tuning results concerning inverter AC voltage and current because the inverter-to-rectifier phase shift angle for this DAB S-S resonant converter needs to be analyzed and evaluated. An explanation needs to be given about how the PLL algorithm starts to work in DC/DC converter power-on the startup process when there is current in the resonant circuit.

This paper presents a detailed TPS closed-loop control implementation method on DC/DC LCC resonant converter for the wireless charging system. The battery charging current/voltage feedback control generates the inverter phase shift angle. The rectifier phase shift angle is generated by LCC resonant circuit AC load resistance feedback control to improve the efficiency performance in CV charging operation. The theoretical DAB inverter-to-rectifier phase shift angle, which satisfies the LCC resonant compensation circuit ZPA design constraint for unity power factor operation of the inverter, is derived from the equivalent LCC resonant circuit model. The inverter-to-rectifier phase shift angle is set to $\delta = 90°$ for unity power factor operation of the DAB inverter and rectifier. The possible methods for TPS control implementation in decoupled inverter and rectifier control hardware are explained for future engineering development in wireless charging products. The simulation and measurement study results are discussed in the conclusion section.

2. DAB DC/DC Resonant Converter TPS Close Loop Control Scheme

2.1. DAB DC/DC Resonant Converter TPS PWM Modulation

DAB DC/DC resonant converter has a symmetrical circuit structure. It is suitable for the bi-directional operation of a wireless charging system at a single resonant frequency with high-power density and high efficiency.

Figure 1 shows the wireless charging DAB LCC resonant DC/DC converter circuit with transmitter side H-bridge DC/AC inverter and receiver side H-bridge AC/DC active rectifier. The function description of the symbols denoted in Figure 1 is given in Table 1. In this paper, R_{ac_out} is the study focus for rectifier side control and is denoted as "R_{ac}".

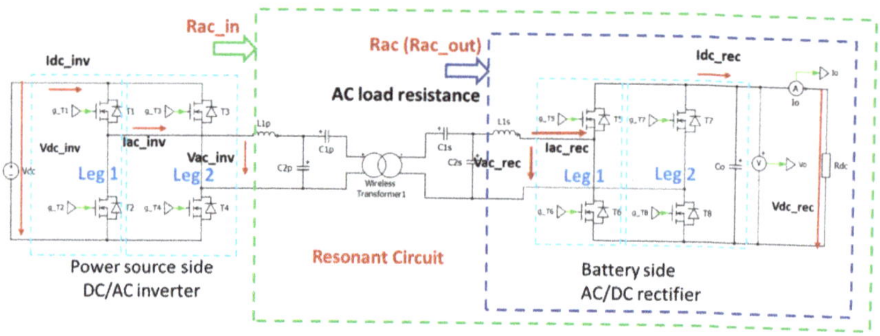

Figure 1. Wireless charging DAB LCC Resonant DC/DC resonant converter.

Table 1. DAB LCC resonant converter circuit notation definition.

Symbol	Function Description
C_{1p} and C_{1s}	LCC primary and secondary side series compensation capacitances
C_{2p} and C_{2s}	LCC primary and secondary side parallel compensation capacitances
L_{1p} and L_{1s}	LCC primary and secondary side compensation inductance
$V_{dc\_inv} = V_{dc}$	inverter DC input voltage
I_{dc_inv}	inverter DC input current
$V_{dc\_rec} = Vo$	rectifier DC output voltage
$I_{dc\_rec} = Io$	rectifier DC output current
R_{dc}	rectifier DC output side equivalent battery DC resistance
$R_{ac} = R_{ac\_out}$	rectifier AC side equivalent resistance = LCC resonant circuit load resistance
$R_{ac\_in} = R_{ac\_inv}$	Inverter AC side equivalent resistance = Inverter AC load resistance

Figure 2 illustrates the triangle PWM carrier signals, the PWM gating signals of the top switches of the inverter and rectifier Legs, and the resulting AC voltage waveforms of the inverter and rectifier for DAB TPS PWM modulation. The same color is used for signal notation and the signal itself.

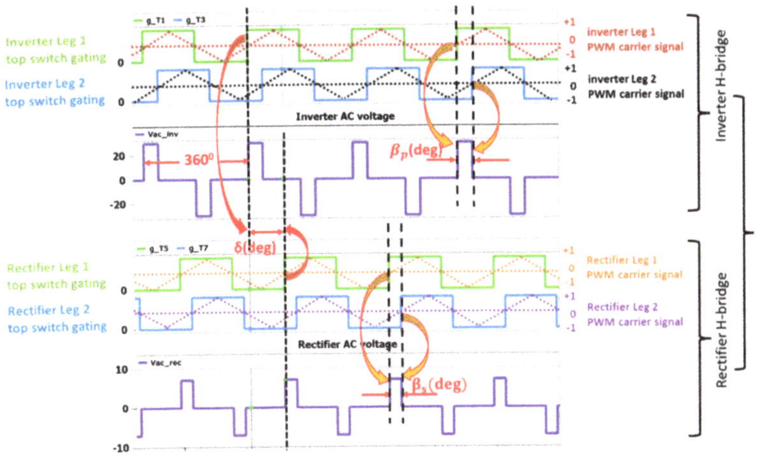

Figure 2. Inverter and rectifier AC voltage pulse generation from DAB TPS PWM modulation.

The PWM carrier signals generate the PWM gating signals. The gating signal for the bottom switch has the opposite polarity as the top switch of the same Leg. A turn-on delay is applied for all gating signals to generate a dead time to prevent short-through fault of the inverter and rectifier Leg. The inverter and rectifier AC voltage waveforms are generated by the inverter and rectifier PWM gating signals.

Table 2 listed out phase shift angles with their PWM carrier signal sources and the results shown in inverter/rectifier AC voltage waveforms. The phase shift time between two Legs is defined as the period between positive zero-crossings of the PWM carrier signals for the two Legs. The phase shift time is converted to phase shift angle in degree, with the ratio of one PWM cycle 360°.

Table 2. DAB Phase Shift Angle Generation and Results.

Phase shift angle (degree)	Symbol	Generation: Phase shift angle generated by the positive zero-crossing time (angle) difference of PWM carrier signals	Results: phase shift angle shown in DAB AC voltage waveforms
Inverter	"βp (deg)"	between inverter "Leg 1" and inverter "Leg 2"	pulse-width of inverter AC voltage
Rectifier	"βs (deg)"	between rectifier "Leg 1" and rectifier "Leg 2"	Pulse-width of rectifier AC voltage
Inverter-to-rectifier	"δ (deg)"	between inverter "Leg 1" and rectifier "Leg 1"	phase difference between inverter and rectifier AC voltages

The magnitude of the inverter/rectifier AC voltage pulse is the same as its corresponding DC voltage. The polarity of phase shift angle δ determines the power flow direction of the bidirectional DAB converter. For DAB LCC resonant converter, the optimized inverter-to-rectifier phase shift angle δ = 90° with inverter AC voltage leads the rectifier AC voltage to achieve unity power factor operation of the DAB converter for battery charging. The detailed derivation process will be given in Section 2.2.2.

In the real-time control algorithm implementation, the values of phase shift angle βp, βs, and δ are normalized over 3600 (one PWM switching period) as βp = βp (deg)/360, βs = βs (deg)/360, and δ = δ (deg)/360.

2.2. Mathematical Formulation of DAB TPS Control Method

2.2.1. Charging Power Regulation with Inverter Phase Shift Control

Using Fourier series expansion of the inverter AC pulsing voltage as explained in [18–20], the inverter steady state AC RMS voltage fundamental frequency component can be expressed as (1) as a function of the inverter phase shift angle βp and DC voltage. When phase shift angle βp = 0.5 (βp (deg) = 180°), the inverter voltage waveform becomes a square waveform with maximum inverter AC RMS voltage, as shown in (2).

$$\mathrm{v_{ac\_inv}} = \frac{2\sqrt{2}}{\pi}\sin(\pi\beta_{\mathrm{p}}) * \mathrm{V_{dc\_inv}} \tag{1}$$

$$\mathrm{v_{ac\_inv}}\left(\beta_{\mathrm{p}} = 0.5\right) = \frac{2\sqrt{2}}{\pi}\mathrm{V_{dc\_inv}} \tag{2}$$

The fundamental resonant frequency component of inverter and rectifier AC RMS voltage and AC RMS current can be decomposed into real and imaginary components in the phasor diagram. With power loss from switching devices ignored, the inverter's DC power equals its AC power.

The averaged DC input power can be expressed as a product of the inverter's AC RMS voltage, AC RMS current, and the inverter power factor, as shown in (3), where $(\varphi(PF)_{inv})$ is the power factor angle of the inverter AC output.

$$\mathrm{I_{dc\_inv}} * \mathrm{V_{dc\_inv}} = \mathrm{v_{ac\_inv}} * \mathrm{I_{ac\_inv}} * \cos(\varphi(PF)_{inv}) \tag{3}$$

Substituting (1) into (3), the averaged inverter DC input current can be calculated from (4). If the resonant compensation circuit is designed to operate the inverter at ZPA condition with $\varphi(PF)_{inv} = 0$, the relationship of the inverter averaged DC input current and its AC output current is simplified as (5).

$$\mathrm{I_{dc\_inv}} = \frac{2\sqrt{2}}{\pi}\sin\left(\pi\beta_{\mathrm{p}}\right) * \mathrm{I_{ac\_inv}} * \cos(\varphi(PF)_{inv}) \tag{4}$$

$$\mathrm{I_{dc\_inv}} = \frac{2\sqrt{2}}{\pi}\sin\left(\pi\beta_{\mathrm{p}}\right) * \mathrm{I}_{ac\_inv} \tag{5}$$

Equation (1) means that inverter output voltage can be regulated through inverter phase shift angle for a given fixed inverter DC input voltage. The inverter AC current reflects the battery charging current, the rectifier operation condition, and the resonant circuit operation condition.

Equations (4) and (5) show that DAB input DC current or power is decided by the inverter AC current loading and inverter shift angle. Equations (4) and (5) establish the foundation for battery charging power regulation in constant current (CC) mode or constant voltage (CV) mode using inverter phase shift control.

2.2.2. Resonant Circuit Load Resistance Regulation with Rectifier Phase Shift Control

Using Fourier series expansion on rectifier AC pulsing voltage generated by phase shift PWM modulation, the relationship of rectifier DC voltage and AC RMS voltage can be derived as (6). Applying rectifier input AC and output DC power are balanced with power device loss ignored, the rectifier AC input RMS current and averaged output DC current is derived as (7), where $\varphi(PF)_{rec}$ is rectifier AC input power factor angle.

$$\mathrm{v_{ac\_rec}} = \frac{2\sqrt{2}}{\pi}\sin(\pi\beta_{s}) * \mathrm{V_{dc\_rec}} \tag{6}$$

$$I_{dc\_rec} = \frac{2\sqrt{2}}{\pi}\sin(\pi\beta_s) * I_{ac\_rec} * \cos(\varphi(PF)_{rec}) \quad (7)$$

If the rectifier PWM gating signals are disabled, the body diodes of the rectifier H-bridge devices will work as a diode rectifier. In this case, βs = 0.5 (βs (deg) = 180°). The relationship of DC vs. AC RMS voltage/current of the diode bridge rectifier is further simplified as shown in (8) and (9).

$$v_{ac\_rec} = \frac{2\sqrt{2}}{\pi} * V_{dc\_rec} \quad (8)$$

$$I_{dc\_rec} = \frac{2\sqrt{2}}{\pi} * I_{ac\_rec} \quad (9)$$

For DAB resonant converter with rectifier phase shift control, the AC load resistance of the resonant circuit is the rectifier AC resistance (Rac), which can be derived as (10) from (6) and (7).

$$R_{ac} = \frac{v_{ac_{rec}}}{I_{ac_{rec}}} = \frac{\frac{2\sqrt{2}}{\pi}\sin(\pi\beta_s) * v_{ac_{rec}}}{\frac{I_{dc_{rec}}}{\left(\frac{2\sqrt{2}}{\pi}\sin(\pi\beta_s) * \cos(\varphi(PF)_{rec})\right)}} = \frac{8}{\pi^2}(\sin(\pi\beta_s))^2\cos(\varphi(PF)_{rec})\frac{V_{dc\_rec}}{I_{ac\_rec}} \quad (10)$$

For rectifier unity power factor operation, the relationship of rectifier AC and DC equivalent resistance is further simplified as (11), where $R_{dc} = V_{dc\_rec}/I_{dc\_rec}$ is the equivalent battery resistance.

$$R_{ac} = \frac{8}{\pi^2}(\sin(\pi\beta_s))^2\frac{V_{dc\_rec}}{I_{ac\_rec}} = \frac{8}{\pi^2}(\sin(\pi\beta_s))^2 R_{dc} \quad (11)$$

Figure 3 illustrates significant equivalent battery load resistance variation in a typical CC-CV mode battery charging operation. The battery CC or CV charging mode is decided by the battery state of charge (SOC) and the battery voltage level. When the battery voltage is much smaller than its nominal value, and the battery SOC is low, the maximum charging current is applied for CC mode charging. When the battery voltage is approaching its nominal value, and the battery SOC is approaching 100%, the charging current is greatly reduced, and the battery works in CV mode charging.

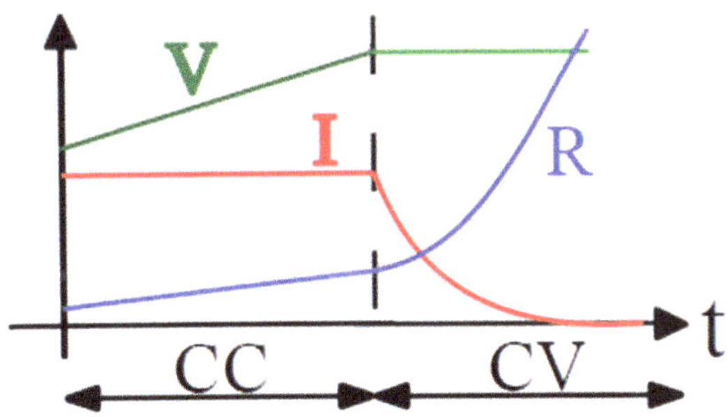

Figure 3. Battery equivalent resistance variation in CC-CV charging operation.

The load resistance of the DAB resonant converter is the equivalent battery resistance which varies significantly from CC to CV mode charging operation. For DAB resonant converter, according to Equation (11), it is possible to modify the AC load resistance of the resonant circuit by regulating the rectifier phase shift angle βs to improve power efficiency in battery CV mode charging operation.

2.2.3. Inverter-to-Rectifier Phase Shift Angle Setting for Unity Power Factor Operation of DAB LCC Resonant Converter

In TPS PWM modulation signal waveform shown in Figure 2, the inverter-to-rectifier phase shift angle δ is defined as the phase angle difference between inverter voltage and rectifier voltage as expressed in (12).

$$\delta = phase\left(\frac{V_{ac\_inv}}{V_{ac\_rec}}\right) \tag{12}$$

The inverter-to-rectifier phase shift angle δ setting for unity power operation of DAB LCC resonant converter bridges can be derived from the equivalent LCC resonant circuit model with inverter ZPA constraint for compensation circuit design applied.

Figure 4 shows the equivalent LCC AC resonant circuit connected to the wireless charging transformer. The parameters for Lm, Lrp, Lrs, R1, and R2 are mutual inductance, primary winding leakage inductance, secondary winding leakage inductance, primary winding resistance, and secondary winding resistance. Zac is the resonant circuit load impedance generated by rectifier control. Zac_in is the inverter AC load impedance.

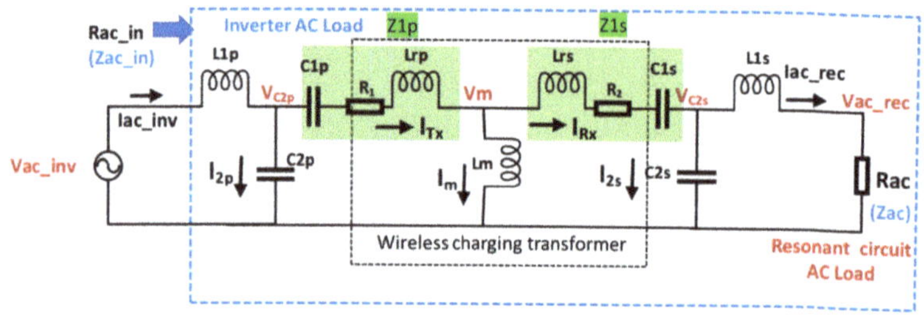

Figure 4. Equivalent LCC AC resonant circuit with detailed wireless charging transformer parameters.

In the LCC resonant circuit design, the assumption is made that unity power factor operation can be achieved in rectifier control. The rectifier voltage and current are in phase with Zac = Rac. The LCC resonant circuit parameters are designed to meet the resonant requirement (13) and (14) and inverter ZPA operation constraints (15) and (16) as explained in [2,9]. The LCC resonant circuit compensation design ensures that with the unity power factor operation of the rectifier, the LCC resonant circuit compensation parameters are designed to achieve inverter ZPA operation with Zac_in = Rac_in.

$$\omega_{sw}L_{1p} - \frac{1}{\omega_{sw}C_{2p}} = 0 \tag{13}$$

$$\omega_{sw}L_{1s} - \frac{1}{\omega_{sw}C_{2s}} = 0 \tag{14}$$

$$\omega_{sw}(L_m + L_{rs}) - \frac{1}{\omega_{sw}C_{1s}} - \frac{1}{\omega_{sw}C_{2s}} = 0 \tag{15}$$

$$\omega_{sw}(L_m + L_{rp}) - \frac{1}{\omega_{sw}C_{1p}} - \frac{1}{\omega_{sw}C_{2p}} = 0 \tag{16}$$

Assuming the wireless charging transformer parameters for the transmitter and receiver coils are identical ($L_{rp} = L_{rs} = L_r$, $R_1 = R_2 = R$). The LCC resonant compensation circuit is symmetrical with $L_{1p} = L_{1s} = L_1$, $C_{1p} = C_{1s} = C_1$, and $C_{2p} = C_{2s} = C_2$. The impedances of LCC AC resonant circuit components can be expressed as (17) to (22).

$$Z_m = j\omega_{sw}L_{\mathrm{m}} \tag{17}$$

$$Z_r = j\omega_{sw}L_{\mathrm{rp}} = j\omega_{sw}L_{\mathrm{rs}} = j\omega_{sw}L_{\mathrm{r}} \tag{18}$$

$$Z_{L1} = Z_{L1p} = Z_{L1s} = j\omega_{sw}L_{1p} = j\omega_{sw}L_{1s} = j\omega_{sw}L_1 \quad (19)$$

$$Z_{c2} = Z_{c2p} = Z_{c2s} = \frac{1}{j\omega_{sw}C_{2p}} = \frac{1}{j\omega_{sw}C_{2s}} = \frac{1}{j\omega_{sw}C_2} \quad (20)$$

$$Z_{c1} = Z_{c1p} = Z_{c1s} = \frac{1}{j\omega_{sw}C_{1p}} = \frac{1}{j\omega_{sw}C_{1s}} =, \frac{1}{j\omega_{sw}C_1} \quad (21)$$

$$Z_1 = Z_{1p} = Z_{1s} = Z_r + Z_{c1} + R \quad (22)$$

The frequency constraint (13) and (14) is re-written as (23) in component impedances. The inverter ZPA constraint (15) and (16) is re-written as (24) in component impedances.

$$Z_{L1} + Z_{c2} = 0 \quad (23)$$

$$Z_m + Z_r + Z_{c1} + Z_{c2} = 0 \quad (24)$$

Utilizing KVL and KCL rule, with constraints (23) and (24) applied, the AC currents and AC voltages of the LCC resonant circuit branches are derived as a function of rectifier AC voltage V_{ac_rec} as shown from (25) to (34).

$$V_{c2s} = \left(\frac{Z_{L1}}{Z_{ac}} + 1\right) * V_{ac\_rec} \quad (25)$$

$$I_{c2s} = \frac{1}{Z_{c2}} V_{c2s} = \frac{1}{Z_{c2}}\left(\frac{Z_{L1}}{Z_{ac}} + 1\right) * V_{ac\_rec} \quad (26)$$

$$\begin{aligned} I_{RX} &= I_{c2s} + \tfrac{V_{ac_{rec}}}{Z_{ac}} \\ &= \left(\tfrac{1}{Z_{c2}}\left(\tfrac{Z_{L1}}{Z_{ac}} + 1\right) + \tfrac{1}{Z_{ac}}\right) * V_{ac_{rec}} \\ &= \left(\tfrac{1}{Z_{c2}} + \tfrac{1}{Z_{ac}}\left(\tfrac{Z_{L1}+Z_{c2}}{Z_{c2}}\right)\right) * V_{ac_{rec}} = \tfrac{1}{Z_{c2}} * V_{ac\_rec} \text{ Apply: } Z_{L1} + Z_{c2} = 0 \end{aligned} \quad (27)$$

$$V_m = I_{RX} * Z_1 + V_{c2s} = \left(\frac{Z_1}{Z_{c2}} + \frac{Z_{L1}}{Z_{ac}} + 1\right) * V_{ac_{rec}} = \left(\frac{Z_1 + Z_{c2}}{Z_{c2}} + \frac{Z_{L1}}{Z_{ac}}\right) * V_{ac\_rec} \quad (28)$$

$$I_m = \frac{V_m}{Z_m} = \frac{1}{Z_m}\left(\frac{Z_1 + Z_{c2}}{Z_{c2}} + \frac{Z_{L1}}{Z_{ac}}\right) V_{ac\_rec} \quad (29)$$

$$\begin{aligned} I_{TX} = I_{RX} + I_m &= \left(\tfrac{1}{Z_{c2}} + \tfrac{1}{Z_m}\left(\tfrac{Z_1+Z_{c2}}{Z_{c2}} + \tfrac{Z_{L1}}{Z_{ac}}\right)\right) V_{ac\_rev} \\ &= \left(\tfrac{1}{Z_{c2}}\left(\tfrac{Z_m + Z_1 + Z_{c2}}{Z_m}\right) + \tfrac{Z_{L1}}{Z_m Z_{ac}}\right) V_{ac\_rev} \\ &= \left(\tfrac{1}{Z_{c2}}\left(\tfrac{Z_m + Z_r + Z_{c1} + R1 + Z_{c2}}{Z_m}\right) + \tfrac{Z_{L1}}{Z_m Z_{ac}}\right) V_{ac\_rev} \text{ Apply: } Z_1 = Z_r + Z_{c1} + R1 \\ &= \left(\tfrac{R1}{Z_{c2} Z_m} + \tfrac{Z_{L1}}{Z_m Z_{ac}}\right) V_{ac\_rev} \text{ Apply: } Z_m + Z_r + Z_{c1} + Z_{c2} = 0 \end{aligned} \quad (30)$$

$$\begin{aligned} V_{c2p} = I_{TX} * Z_1 + V_m &= \left(Z_1\left(\tfrac{R1}{Z_{c2}Z_m} + \tfrac{Z_{L1}}{Z_m Z_{ac}}\right) + \tfrac{Z_1+Z_{c2}}{Z_{c2}} + \tfrac{Z_{L1}}{Z_{ac}}\right) V_{ac\_rev} \\ &= \left(\tfrac{Z_1 R1 + Z_m Z_1 + Z_m Z_{c2}}{Z_{c2} Z_m} + \left(\tfrac{Z_1 + Z_m}{Z_m}\right)\tfrac{Z_{L1}}{Z_{ac}}\right) V_{ac\_rev} \end{aligned} \quad (31)$$

$$I_{c2p} = \frac{V_{c2p}}{Z_{c2}} = \left(\frac{Z_1 R1 + Z_m Z_1 + Z_m Z_{c2}}{(Z_{c2})^2 Z_m} + \left(\frac{Z_1 + Z_m}{Z_{c2} Z_m}\right)\frac{Z_{L1}}{Z_{ac}}\right) V_{ac\_rev} \quad (32)$$

$$
\begin{aligned}
I_{ac_{inv}} &= I_{TX} + I_{c2p} \\
&= \left(\frac{R1}{Z_{c2}Z_m} + \frac{Z_{L1}}{Z_m Z_{ac}} + \frac{Z_1 R1 + Z_m Z_1 + Z_m Z_{c2}}{(Z_{c2})^2 Z_m} + \left(\frac{Z_1 + Z_m}{Z_{c2}Z_m}\right)\frac{Z_{L1}}{Z_{ac}}\right) V_{ac\_rv} \\
&= \left(\frac{(R1 + Z_m)(Z_{c2} + Z_1)}{(Z_{c2})^2 Z_m} + \left(\frac{Z_{c2} + Z_r + Z_{c1} + R2 + Z_m}{Z_{c2}Z_m}\right)\frac{Z_{L1}}{Z_{ac}}\right) V_{ac_{rev}} \text{ Apply: } Z_1 = Z_r + Z_{c1} + R1 \\
&= \left(\frac{(R1 + Z_m)(Z_{c2} + Z_1)}{(Z_{c2})^2 Z_m} + \left(\frac{R2}{Z_{c2}Z_m}\right)\frac{Z_{L1}}{Z_{ac}}\right) V_{ac\_rev} \text{ Apply: } Z_m + Z_r + Z_{c1} + Z_{c2} = 0
\end{aligned}
\tag{33}
$$

$$
\begin{aligned}
V_{ac\_inv} &= I_{ac\_inv} * Z_{L1} + V_{c2p} \\
&= \left(\frac{Z_{L1}(R1 + Z_m)(Z_{c2} + Z_1)}{(Z_{c2})^2 Z_m} + \left(\frac{R2}{Z_{c2}Z_m}\right)\frac{(Z_{L1})^2}{Z_{ac}} + \frac{Z_1 R1 + Z_m Z_1 + Z_m Z_{c2}}{Z_{c2}Z_m} + \left(\frac{Z_1 + Z_m}{Z_m}\right)\frac{Z_{L1}}{Z_{ac}}\right) V_{ac\_rec} \\
&= \left(\frac{Z_{L1}(R1 + Z_m)(Z_{c2} + Z_1) + Z_{c2}(Z_1 R1 + Z_m Z_1 + Z_m Z_{c2})}{(Z_{c2})^2 Z_m} + \left(\frac{R2 * Z_{L1} + Z_{c2}(Z_1 + Z_m)}{Z_{c2}}\right)\frac{Z_{L1}}{Z_m Z_{ac}}\right) V_{ac\_rec} \\
&= \left(\frac{Z_{L1}(R1 + Z_m)(Z_{c2} + Z_1) + Z_{c2}Z_1 R1 + Z_{c2}Z_m(Z_1 + Z_{c2})}{(Z_{c2})^2 Z_m} + \left(\frac{R2 * Z_{L1} + Z_{c2}Z_1 + Z_{c2}Z_m}{Z_{c2}}\right)\frac{Z_{L1}}{Z_m Z_{ac}}\right) V_{ac\_rec} \\
&= \left(\frac{(Z_{L1}R1 + Z_m(Z_{L1} + Z_{c2}))(Z_1 + Z_{c2}) + Z_{c2}Z_1 R1}{(Z_{c2})^2 Z_m} + \left(\frac{R2 * Z_{L1} + Z_{c2}(R1 - Z_{c2})}{Z_{c2}}\right)\frac{Z_{L1}}{Z_m Z_{ac}}\right) V_{ac\_rec} \\
&= \left(\frac{Z_{L1}R1Z_1 + Z_{L1}R1Z_{c2} + Z_{c2}Z_1 R1}{(Z_{c2})^2 Z_m} - \frac{Z_{c2}Z_{L1}}{Z_m Z_{ac}}\right) V_{ac\_rec} \text{ Apply: } \begin{cases} Z_{L1} + Z_{c2} = 0 \\ Z_m + Z_r + Z_{c1} + Z_{c2} = 0 \\ R1 = R2 \end{cases} \\
&= \left(\frac{Z_{L1}R1}{Z_{c2}Z_m} - \frac{Z_{c2}Z_{L1}}{Z_m Z_{ac}}\right) V_{ac\_rec} \text{ Apply: } Z_{L1} + Z_{c2} = 0) \\
&= \left(\frac{R1}{Z_{c2}} - \frac{Z_{c2}}{Z_{ac}}\right)\frac{Z_{L1}}{Z_m} V_{ac\_rec} \\
&= \left(R1 * Z_{ac} - (Z_{c2})^2\right)\frac{Z_{L1}}{Z_m Z_{c2} Z_{ac}} V_{ac\_rec}
\end{aligned}
\tag{34}
$$

Under the condition that the rectifier operates at unity power factor with Zac = Rac, the inverter-to-rectifier phase angle δ is derived as (35).

$$
\begin{aligned}
\delta &= phase\left(\frac{V_{ac\_inv}}{V_{ac\_rec}}\right) \\
&= phase\left(\left(R1 * Z_{ac} - (Z_{c2})^2\right)\frac{Z_{L1}}{Z_{ac}Z_m Z_{c2}}\right) \\
&= phase\left(\left(R1 * R_{ac} - \left(\frac{1}{j\omega_{sw}C_{2p}}\right)^2\right)\frac{j\omega_{sw}L_{1p} * j\omega_{sw}C_{2p}}{j\omega_{sw}L_m R_{ac}}\right) \\
&= phase\left(\left(R1 * R_{ac} + \left(\frac{1}{\omega_{sw}C_{2p}}\right)^2\right)\frac{j\omega_{sw}L_{1p} * C_{2p}}{L_m R_{ac}}\right) = 90^\circ
\end{aligned}
\tag{35}
$$

Equation (35) proves that the rectifier voltage phase should lag the inverter voltage phase by 90° for DAB LCC resonant converter to operate at unity power factor.

DAB resonant circuit topology affects the inverter-to-rectifier phase shift angle δ setting for inverter and rectifier unity power factor operation. For other DAB circuit topologies, the inverter-to-rectifier phase shift angle for unity power factor operation needs to be derived from the equivalent resonant circuit model.

2.2.4. DAB LCC Resonant Converter Rectifier AC Voltage Phase Measurement from Receiver Coil Current

For LCC resonant converter, the receiver coil voltage and current are sinusoidal signals at resonant frequency, and the phase relationship between transformer winding voltage and current can be utilized to measure the phase rectifier AC voltage at resonant frequency. The resonant frequency component of rectifier AC voltage can be derived as (36) utilizing resonant condition (14), which means that the phase of rectifier AC voltage lags the receiver coil current by 90°.

$$
V_{ac\_rec} = \frac{1}{j\omega_{sw}C_{2s}}(I_{RX} - I_{ac\_rec}) - j\omega_{sw}L_{1s} * I_{ac\_rec} = \frac{1}{j\omega_{sw}C_{2s}} I_{RX} \tag{36}
$$

2.3. DAB TPS Close Loop Control Scheme for Unity Power Factor Operation of Inverter and Rectifier at Optimized Resonant Circuit AC Load Resistance

Figure 5 shows the proposed TPS control scheme to achieve unity power factor operation of the DAB converter at optimized resonant circuit AC load resistance in the wireless charging system. The battery charging voltage/current feedback control is implemented to control the inverter phase shift angle βp. The resonant circuit AC load resistance Rac feedback control is applied to regulate the rectifier phase shift angle βs. The inverter-to-rectifier phase shift angle δ is set to 90° for unity power factor operation of the DAB inverter and rectifier bridges.

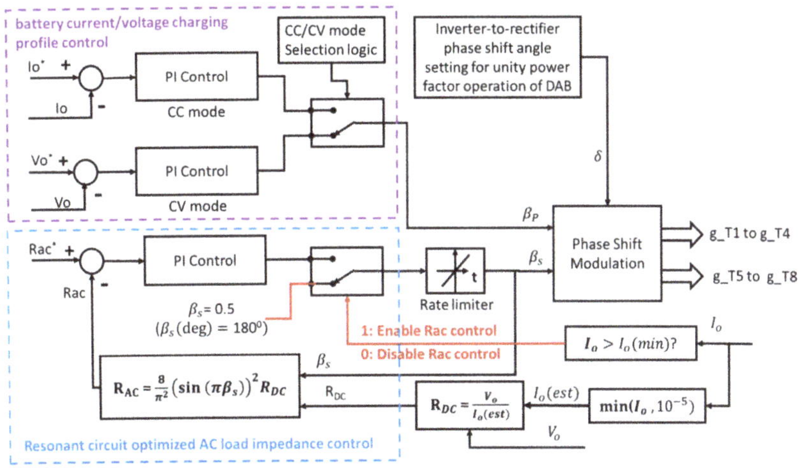

Figure 5. DAB resonant DC/DC converter TPS control scheme.

The inverter phase shift angle βp changes the inverter AC output voltage pulse-width and the input AC voltage of the resonant circuit so that the power from the inverter to the rectifier can be regulated. In CC mode control, the battery current feedback is applied to regulate the power from the inverter to the rectifier. In CV mode control, the battery voltage feedback is applied to regulate the power flow from the inverter to the rectifier.

For the inverter phase shift control, in CC mode battery charging operation, the DC/DC converter DC output current is regulated around the charging current reference by the proportional and integral (PI) control. The feedback current is the battery charging current measured after the load side filter capacitor. In CV mode charging, the DC/DC inverter output voltage is regulated around the target voltage. The feedback voltage is measured as the battery DC voltage.

The rectifier phase shift control regulates the resonant circuit AC load resistance Rac to match its optimized value. In the control algorithm implementation, the Rac feedback signal is calculated as a function of the rectifier phase shift angle βs and DC load resistance using (11), assuming the rectifier runs at unity power factor. The DC load resistance signal is estimated from the ratio of battery voltage over battery charging current. When the charging current is approaching zero, Rac feedback control is disabled because DC load resistance (R_{dc}) estimation is inaccurate. In this condition, the rectifier phase shift angle is set to its maximum value at βs = 0.5 and βs (deg) = 180. The change rate of the rectifier phase shift is limited to smooth out the dynamic transition caused by the phase shift command sudden change.

The proposed DAB resonant DC/DC converter TPS close loop control scheme can be implemented for all DAB resonant circuit topologies. However, the optimized AC load resistance value and the optimized inverter-to-rectifier phase shift angle setting to achieve unity power factor operation are to be decided by both the resonant circuit topology and the resonant circuit parameter.

2.4. DAB LCC Resonant Converter TPS Control Performance Design Analhysis

Table 3 shows the wireless charging transformer configuration and parameters. The transformer turns ratio is 1:1. The transmitter and receiver coil parameters are identical. Table 4 shows the design requirement specification. The nominal AC load is calculated from the nominal DC load resistance for the diode rectifier operation. Table 5 shows the LCC circuit parameters from the design calculation using the method given in [12].

Table 3. Transmitter (Tx) and receiver (Rx) parameters.

Symbol	Parameter	Value
$L_1 = L_2$	Self-inductance of Tx coil and Rx coil	180 μH
L_m	Mutual inductance between of Tx coil and Rx coil	41.925 μH
$L_{rp} = L_{rs}$	The leakage inductance of Tx coil and Rx coil $L_{rp} = L_1 - L_m$; $L_{rs} = L_2 - L_m$	138.075 μH
k	Coupling factor	0.221
$R_1 = R_2$	Resistance of Tx coil and Rx coil	0.5 Ω
d	Airgap between Tx and Rx	15 cm

Table 4. Design requirement for LCC resonant converter.

Symbol	Parameter	Value
V_{dc}	DC input voltage	30 V
V_o	DC output voltage	24~48 V
P_o	Maximum DC output power	50 W
f_{sw}	Resonant frequency	30 kHz
R_{dc} (norm)	Nominal DC load resistance	47 Ω
R_{ac} (norm)	Nominal AC load resistance $\text{Rac (norm)} = \frac{8}{\pi^2} R_{dc}(norm)$	38 Ω

Table 5. LCC resonant compensation network parameters.

Symbol	Parameter	Value
L1p = L1s	Compensation inductance	68.65 μH
C1p = C1s	Series compensation capacitance	248 nF
C2p = C2s	Parallel compensation capacitance	410 nF

The LCC resonant circuit design analysis assumes that the rectifier is operating at βs(deg) = 180° with unity AC power factor with battery equivalent resistance change. The method presented in [12] is applied for the LCC resonant circuit performance calculation. The switching loss and conduction loss of power electronic switches are ignored in the calculation.

Figure 6 shows the frequency sensitivity analysis of DC transconductance gain for the DAB DC/DC converter vs. the power factor angle of the DAB inverter bridge. The transconductance gain is defined as the ratio of the DC/DC converter DC output current vs. DC input voltage. It decides the DAB converter DC current output capability in CC mode charging. The benefit of the fixed AC load resistance from rectifier phase shift control is that DC/DC converter DC current output capability will not be affected by the mutual coupling factor variation in CC mode charging operation.

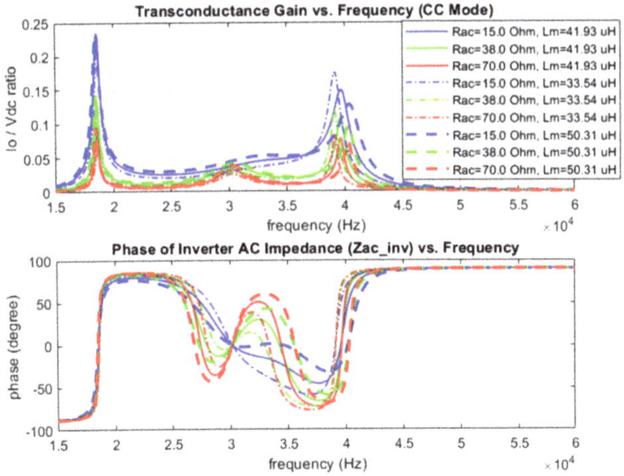

Figure 6. DC transconductance gain and inverter power factor angle with respect to DAB switching frequency change.

Figure 7 shows the frequency sensitivity analysis of DC voltage gain for the DAB DC/DC converter vs. the inverter power factor angle. The voltage gain is defined as the ratio of the DC output voltage vs. DC input voltage, and it decides the DC/DC converter DC voltage output capability. The DC voltage gain variation is significant due to the AC load resistance change and relatively minor due to the mutual inductance change. The benefit of the fixed AC load resistance from rectifier phase shift control is that the DC voltage gain variation will be significantly reduced in battery CV mode charging application.

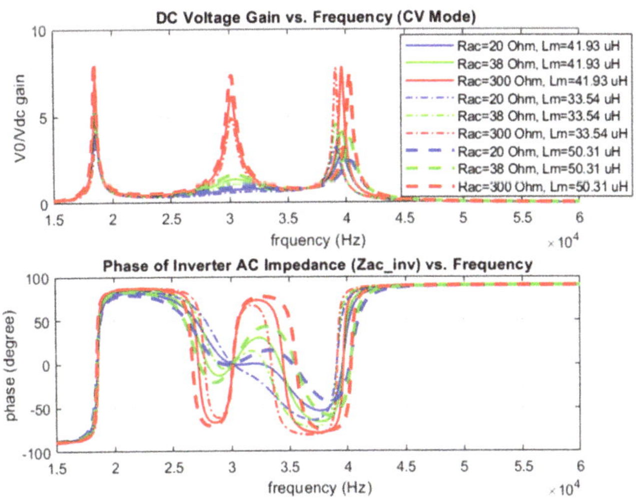

Figure 7. DC voltage gain vs. inverter power factor angle with respect to DAB switching frequency change.

Figure 8 shows DC input/output voltage/current calculated at different inverter phase shift angle βp (deg) settings vs. AC load resistance change. This result shows that DC output voltage and current magnitude at the same AC load resistance value change significantly with inverter phase shift angle βp (deg). The inverter phase shift angle can be applied to regulate the charging voltage and current in CC or CV mode charging operation. If the AC load resistance can be fixed to R_{ac} (norm) = 38 Ω using rectifier phase shift βs regulation, the maximum output voltage and maximum input and output currents are limited to smaller value ranges.

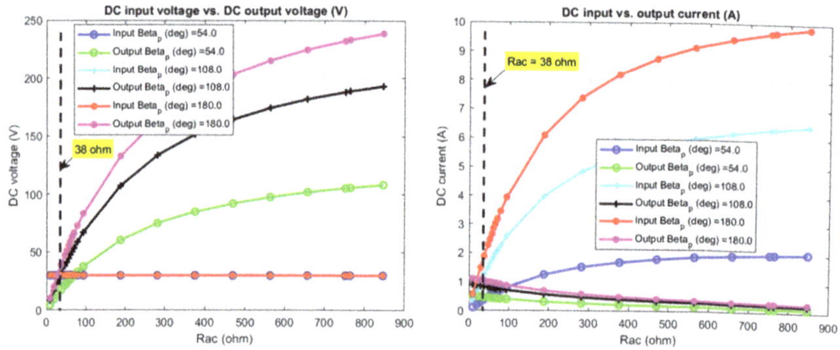

(**a**) DC input/output voltages (**b**) DC input/output currents

Figure 8. DC input/output voltages/currents vs. AC load resistance change.

Figure 9 shows DC input and DC output power and power transfer efficiency calculated at different βp (deg) under AC load resistance change conditions. This result shows that DC input and output power change with inverter phase shift angle βp (deg). However, the efficiency does not change with the inverter phase shift angle βp (deg). If R_{ac} = 38 Ω can be achieved with rectifier phase shift control, the maximum input and output power are limited to a smaller value range. Theoretically, the power efficiency can be kept above 86% for all battery operation conditions when the power switching device loss is ignored.

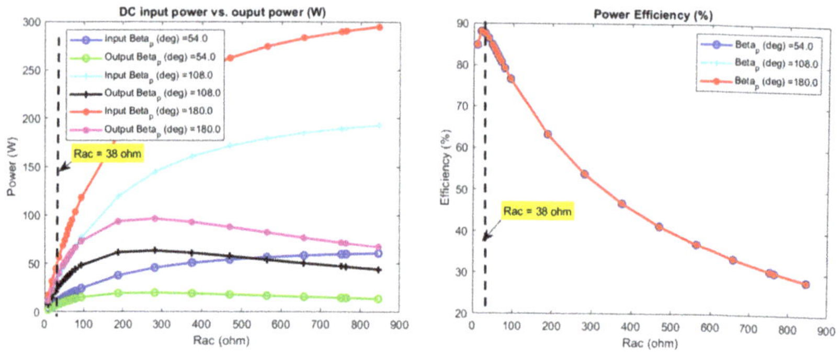

(**a**) DC input/output power (**b**) LCC circuit power efficiency

Figure 9. DC input/output power and efficiency vs. AC load resistance change.

2.5. DAB LCC Resonant Converter TPS Control Implementation in PLECS RT Box

The working principle of the phase shift PWM modulator has been explained in the previous section in the contents related to Figure 2.

Figure 10 shows the FPGA-based triple phase shift PWM modulator connection and configuration for PLECS RT box implementation. The PLECS RT BOX library PWM modulator with variable modulation index inputs, variable frequency ratio input, and variable phase shift angle inputs are used to implement this triple phase shift PWM modulation. The frequency ratio is set to “1” to operate the inverter and rectifier at a single resonant frequency. The symmetrical triangle PWM carrier signals in the range of [−1 1] are configured for the PWM modulator. The modulation index signals for all inverter and rectifier legs are set to zero to generate 50% duty cycle gating signals. The phase shift value settings for all inverter and rectifier Legs are all referenced to the phase of the PWM carrier signal of the inverter Leg 1. The phase shift of the PWM carrier signal between inverter Leg 1 and Leg 2 is Beta_p (βp). The phase shift of PWM carrier signals between rectifier Leg 1 and Leg 2 is Beta_s (βs). Rectifier Leg 1 and Leg 2 PWM carrier signal phase will shift together of delta (δ) phase angle relative to the inverter Leg 1.

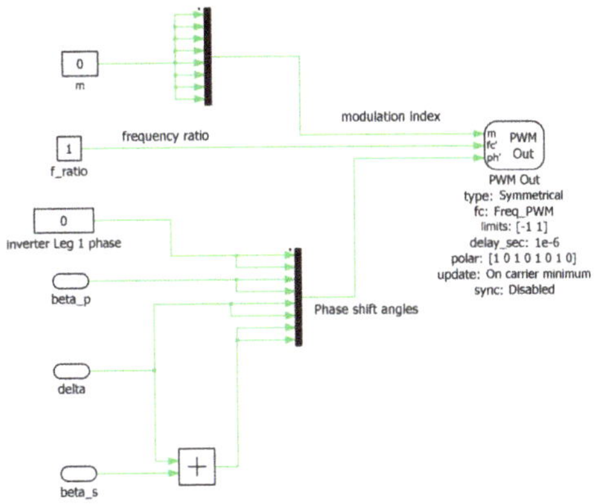

Figure 10. PLECS-RT BOX phase shift PWM modulator signal connection and configuration.

Figure 11 shows inverter phase shift angle βp generation in the PLECS RT box for experimental test purposes. In the wireless charging system's regular operation, automatic CCCV mode switching is selected to decide CC and CV mode based on comparing the measured battery voltage with the CV mode threshold and hysteresis threshold as given in (37).

$$\begin{cases} if\ Vo(k) < Vo\_min & CC\_CV\_AUTO = 0 \quad Beta\_p = Beta\_p(CC) \\ if\ Vo\ (k) > Vo\_min + \Delta Vo\_min & CC\_CV\_AUTO = 1 \quad Beta\_p = Beta\_p(CV) \end{cases} \tag{37}$$

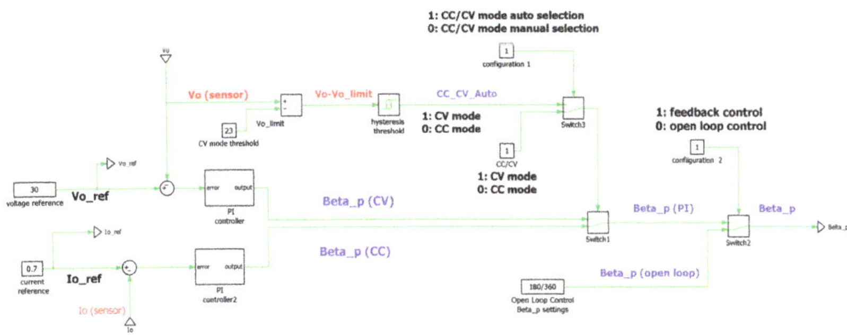

Figure 11. Inverter phase shift angle βp generation in PLECS RT Box for experimental testing purpose.

Manual CC and CV mode switching can also be configured to individually test out the dynamic performance of the CC/CV mode controller. A configuration switch is added to enable/disable the CC/CV feedback control to test DAB converter performance for inverter phase shift angle open loop control.

Figure 12 shows the rectifier phase shift angle βs generation in the PLECS RT Box real-time control for experimental test purposes. The DAB DC output voltage Vo (Vdc_rec) and current Io (Idc_rec) are measured from current/voltage sensors. The sensor measurement signals are sampled to the PLECS RT box through its high-speed individual 10 bits +/−10 V ADC convention. Before starting of TPS control, the current and voltage sensor signals are calibrated with sensor scaling and sensor offset applied. The voltage sensor ADC sampling result is calibrated for the voltage sensor scaling and offset using differential voltage probe measurement. The current sensor ADC sampling result is calibrated for the current sensor scaling and offsets using the current probe measurement.

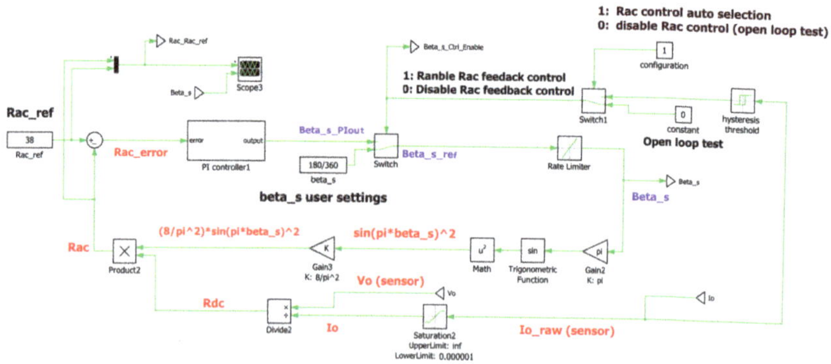

Figure 12. Rectifier phase shift angle βs generation in PLECS RT Box for experimental testing purpose.

In real-time control at "k" sample, the DC output charging current sensor signal in the unit of "A" at "k" sample is obtained as "Io_raw (k)"; the DC output battery charging voltage sensor signal in the unit of "V" is obtained as "Vo (k). A low limit saturation function is applied on the battery current measurement to avoid divided-by-zero error, as shown in (38). The estimated R_{dc} signal is constructed using (39)

$$I_O(\mathrm{k}) = \min(I_{o\_raw}(k),\ 1e-6) \tag{38}$$

$$R_{dc}(k) = \frac{V_o(k)}{I_O(\mathrm{k})} \tag{39}$$

The rectifier AC resistance signal "R_{ac}" (LCC AC load resistance) is calculated using (40), which is a digital implementation of Equation (11). In Equation (40), βs (k − 1) is the rectifier phase shift angle of the previous sample generated by the "R_{ac}" proportional-integral (PI) feedback controller.

$$\mathrm{R_{ac}}(\mathrm{k}) = \frac{8}{\pi^2}(\sin(\pi\beta_s(k-1)))^2 R_{dc}(k) \tag{40}$$

The configuration switches are added to enable/disable the "R_{ac}" feedback control to test DAB performance in open loop control manually βs setting condition. In wireless charging regular operation, R_{ac} feedback control is automatically enabled/disabled depending on the battery current hysteresis threshold settings as given in (41).

$$\begin{cases} if\ Io(k) < Io\_min & Disable\_Rac = 0, \quad \beta_{s\_ref}(k) = 0.5 \\ if\ Io\ (k) > Io\_min + \Delta Io\_min & Enable\_Rac = 1, \quad \beta_{s_{ref}}(k) = \beta_s(PI\_out,\ k) \end{cases} \tag{41}$$

The rate limiting for rectifier phase shift angle is applied to smooth out the DAB converter dynamic transition caused by sudden phase shift angle change as given in (42).

$$\begin{cases} if\beta_{s_{ref}}(k) - \beta_s(k-1) > \Delta\beta_s(nax), & then\ \beta_s(k) = \beta_s(k-1) + \Delta\beta_s(nax) \\ if\beta_{s_{ref}}(k) - \beta_s(k-1) < -\Delta\beta_s(nax), & then\beta_s(k) = \beta_s(k-1) - \Delta\beta_s(nax) \\ if - \Delta\beta_s(nax) < \beta_{s_{ref}}(k) - \beta_s(k-1) < \Delta\beta_s(nax), & then\beta_s(k) = \beta_{s\_ref}(k) \end{cases} \quad (42)$$

2.6. DAB LCC Resonant Converter Open Loop TPS Control Simulation Analysis

2.6.1. DAB LCC Resonant Converter TPS Control with Active Rectifier Control to Achieve Diode Bridge Rectify Performance

Figure 13 shows the PLECS circuit simulation results comparison TPS phase shift PWM modulation with βs (deg) = 180° vs. diode rectifier operation.

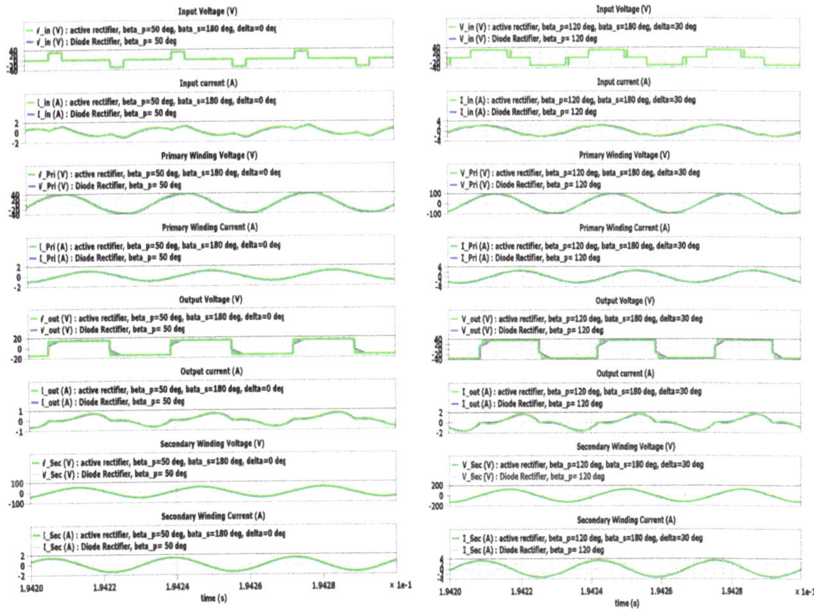

(**a**) βp (deg) = 50°, δ(deg) = 0° (**b**) βp (deg) = 120°, δ(deg) = 30°

Figure 13. Inverter/rectifier and transformer voltage and current waveforms with TPS open loop control settings to match diode bridge rectifier performance.

The above simulation results are obtained in DAB open loop TPS control with manual βp, βs, and δ settings when both invert side CC/CV feedback control and rectifier side R_{ac} feedback control are disabled. The inverter-to-rectifier phase shift angle δ settings are manually adjusted for DAB open loop TPS control to match the performance of the diode bridge rectifier operation for the given βp settings.

The simulation results show that the diode bridge rectifier's performance is much closer to the active rectifier with βs (deg) = 180° and a smaller δ (deg) angle setting. Based on the above simulation results, it is possible to estimate the initial inverter-to-rectifier phase shift angle δ (deg) from AC signal measurement in the diode bridge operation at the power-up stage. Therefore, it is possible to start the rectifier phase shift modulation smoothly from TPS control to achieve diode bridge performance when TPS control is to be implemented in separate control hardware installed in the EV car on the receiver side.

2.6.2. DAB LCC Resonant Converter TPS Control for Unity Power Factor Operation

Figure 14 shows the PLECS circuit simulation results of inverter/rectifier voltage/current and the receiver winding voltage/current waveforms obtained from TPS open loop control for the same inverter/rectifier phase shift angle βp and βs settings but different inverter-to-rectifier phase shift angle δ settings.

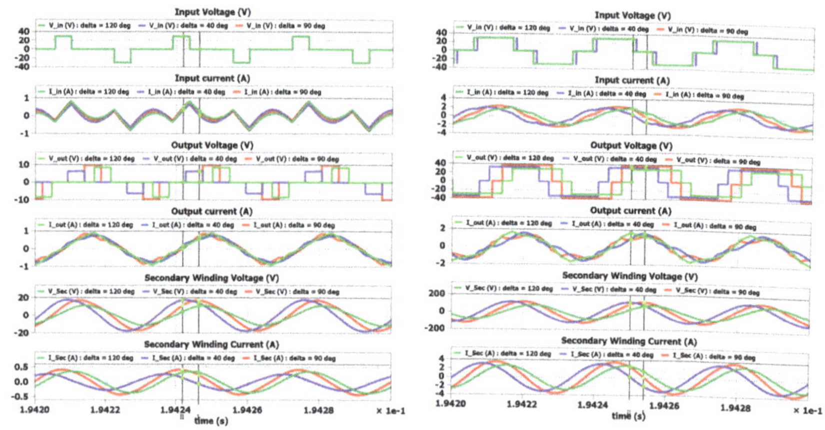

(**a**) βp (deg) = 50°, βs (deg) = 40°, δ (deg) = 40°, 90°, 120°

(**b**) βp (deg) = 120°, βs (deg) = 150°, δ (deg) = 40°, 90°, 120°

Figure 14. Inverter/rectifier AC voltage/current and receiver winding voltage/current waveforms for same βp and βs and different δ settings.

These simulation results show that the inverter-to-rectifier phase shift angle δ has a negligible effect on the rectifier AC current waveform. However, it changes the inverter and rectifier power factor angle significantly. The inverter and rectifier operate at unity power factor when the inverter-to-rectifier phase shift angle is around δ = 90°.

In TPS close loop control, the inverter-to-rectifier phase shift angle δ should be automatically adjusted to minimize inverter and rectifier reactive power flow to/from LCC resonant circuit.

2.6.3. DAB LCC Resonant Converter TPS PWM Modulation ZVS Switching Performance

In the power switch ZVS performance study, the circuit simulation needs to be conducted using the circuit parameters and settings of the experimental setup. TPS PWM modulation ZVS switching performance is evaluated in PLECS circuit simulation for SiC MOSFET C3M0120090D with on-resistance Rds = 120 mΩ, output capacitance 40 pF, and SiC MOSFET body-diode forward voltage 4.8 V.

In this study, the DAB DC output resistor is replaced with a DC voltage source from 20 V to 30 V to simulate the battery in different charging conditions. DAB DC input voltage is set to 30 V. The DC output voltage current feedback control is applied to regulate the inverter phase shift angle βp. Rectifier side R_{ac} = 38 Ω feedback control is applied to regulate the rectifier phase shift angle βs. The inverter-to-rectifier phase shift angle is manually set closer to δ = 90°. The PWM gating signal dead time is set to 200 ns.

Figure 15 shows the simulation results when DC output voltage Vo = 20 V and DC output current Io = 0.4 A with δ = 90°. This simulation emulates the condition when R_{dc} = 50 Ω when the battery is working at the changing CC mode. The simulation results show that inverter/rectifier Leg 1 and Leg 2 are all working at ZVS switching. ZVS soft-switching is achieved when Sic MOSFET is switched on with the freewheeling current flows through its body-diode. In this operation condition, adding an extra inverter-to-rectifier phase with δ = 90 + Δδ (Δδ > 0) only causes extra reactive power to flow to/from the AC resonant circuit and reduces the power efficiency.

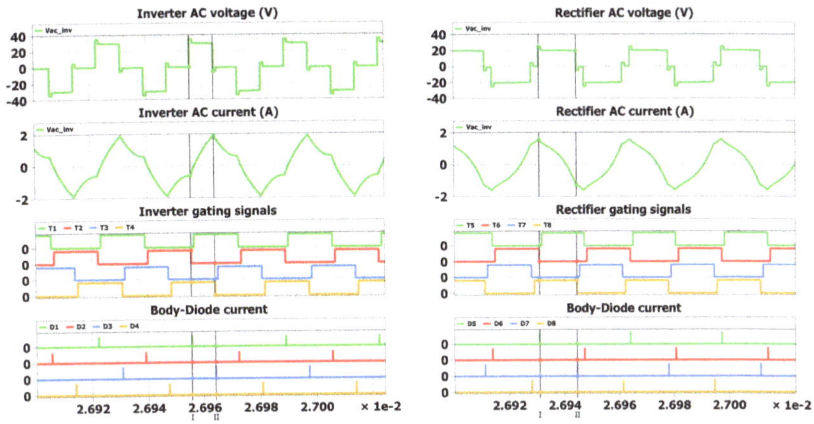

(**a**) T1&T2&T3&T4 with ZVS (**b**) T5&T6&T7&T8 with ZVS

Figure 15. Inverter/rectifier voltage/current, MOSFET gating signals, diode currents PLECS circuit simulation results for Vo = 20 V, Io = 0.4 A (R_{dc} = 50 Ω, emulate CC mode) with βp (deg) = 82.3°, βs (deg) = 151, δ (deg) = 90°.

Figure 16 shows the simulation results when DC output voltage Vo = 30 V and DC output current Io = 0.1 A with δ = 90°. This simulation case emulates R_{dc} = 300 Ω CV mode changing. The simulation results show that inverter Leg 2 and rectifier Leg 1 and Leg 2 work in ZVS switching. Inverter Leg 1 does not work in ZVS switching.

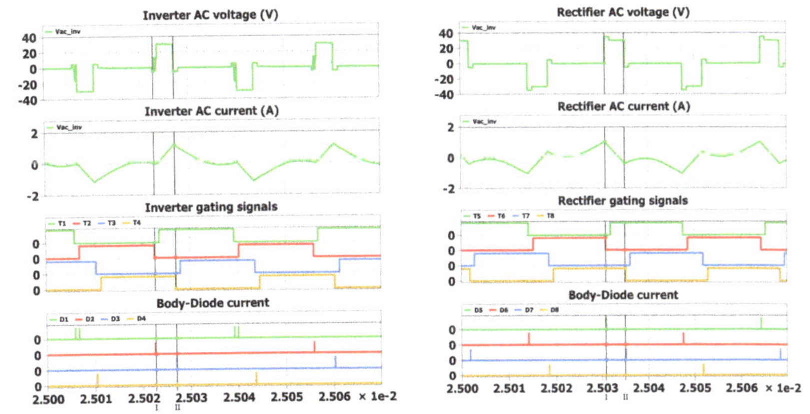

(**a**) T1&T2 no-ZVS, T3&T4 with ZVS (**b**) T5&T6&T7&T8 with ZVS

Figure 16. Inverter/rectifier voltage/current, MOSFET gating signals, diode currents PLECS circuit simulation results for Vo = 30 V, Io = 0.1 A (R_{dc} = 300 Ω, emulate CV mode) with βp (deg) = 48.3°, βs (deg) = 46.2°, δ (deg) = 90°.

Figure 17 shows the simulation results when DC output voltage Vo=30 V and DC output current Io = 0.1 A with δ = 120°. This simulation emulates R_{dc} = 300 Ω CV mode battery charging. The simulation results show that after setting δ = 90 + Δδ with Δδ = 30°

inverter Leg 2 and rectifier Leg 1 and Leg 2 are working in ZVS switching. Inverter Leg 2 is marginally working in ZVS switching.

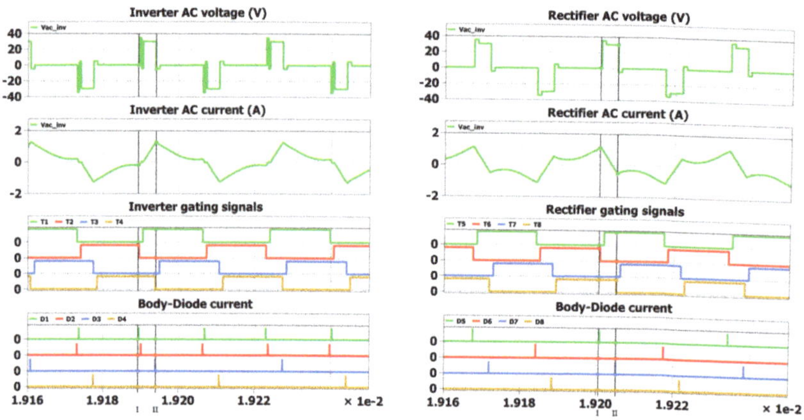

(**a**) T1&T2&T3&T4 with ZVS (**b**) T5&T6&T7&T8 with ZVS

Figure 17. Inverter/rectifier voltage/current, MOSFET gating signals, diode currents PLECS circuit simulation results for Vo = 30 V, Io = 0.1 A (R_{dc} = 300 Ω, emulate CV mode) with βp (deg) = 47.065°, βs (deg) = 46.47°, δ (deg) = 110°.

For the proposed TPS closed-loop control, the simulation study shows that when the inverter-to-rectifier phase shift angle is set to δ = 90°, both inverter and rectifier work at ZVS soft-switching mode with unity power factor in CC mode charging. However, in CV mode charging, the inverter-to-rectifier phase shift angle δ needs to increase to meet the ZVS condition. The cost paid is that the inverter and rectifier will deviate from ZPA operation with extra reactive power to/from the resonant circuit, which may increase the loss of the LCC resonant circuit. Further experimental measurement evaluation is required in the practical wireless charging system to decide whether increasing the phase shift delay δ setting is justifiable.

3. DAB LCC Resonant Converter TPS Control Validation

3.1. Experimental Hardware Setups

Figure 18 shows the experimental setup of the DAB LCC resonant converter TPS wireless charging control system. The hardware setup consists of a wireless charging transformer, DAB DC/DC converter built with 900 V, 23 A silicon carbide (SiC) MOSFET C3M0120090D inverter and rectifier bridges, a 1000 μF capacitor on the DC source side, a 2200 μF capacitor on DC load side. A variable resistance load connected to the rectifier DC output emulates the battery charging condition.

DAB LCC resonant converter operates at 30 kHz resonant frequency. SiC MOSFET C3M0120090D has on-resistance Rds = 120 mΩ, source-to-drain capacitance 40 pF, and SiC MOSFET body-diode forward voltage 4.8 V, body-diode reverse recovery charge Qrr = 115 nF, and body-diode reverse recover time 24 ns.

The LCC resonant compensation circuit consists of high-frequency ferrite core inductors and 630 V polypropylene high-frequency high voltage film capacitors. The current and voltage sensor circuit is built with a LEM LA 55-P current transducer and a LEM LV 25-P voltage transducer. PLECS RT box is used as a real-time controller for fast prototyping and experimental validation of the proposed TPS control scheme.

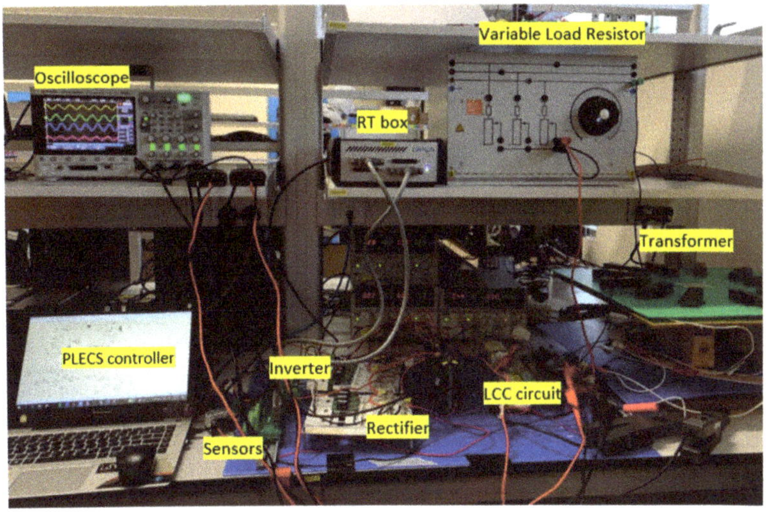

Figure 18. Experimental setup for DAB LCC resonant converter TPS control system.

In the experiment test, the input DC power supply voltage is set to 30 V. The inverter-to-rectifier phase shift angle is set to δ (deg) = 90° and rectifier phase shift βs is regulated to meet R_{ac} = 38 Ω in normal TPS close loop operation. When rectifier R_{ac} feedback close loop control is disabled, the rectifier maximum phase shift angle βs = 0.5 (βs (deg) = 180°) is applied. The measurement results are also obtained for different inverter-to-rectifier phase shift angle δ settings to validate the conclusion obtained from the TPS open loop control circuit simulation.

3.2. CC Mode and CV Mode Battery Charging Profile Feedback Control Performance

Figure 19 shows the measurement results for step response of DC output current/voltage for CC/CV mode current/voltage feedback close loop control, respectively, with R_{dc} = 54 Ω connected. These results validate that the proposed TPS close loop control is stable in the dynamic transition process for CC and CV mode current/voltage feedback control.

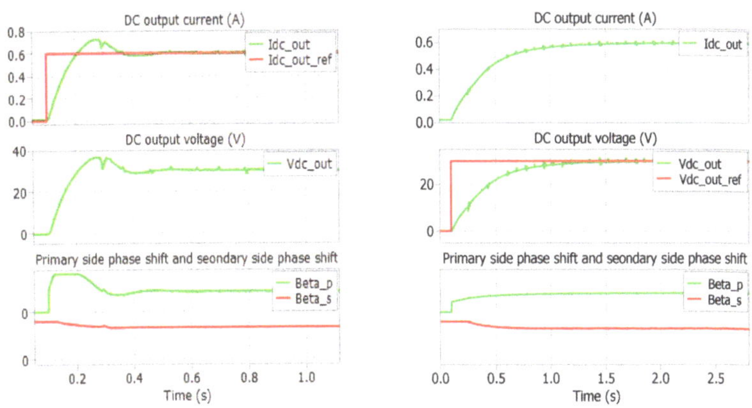

(**a**) CC mode current control (**b**) CV mode voltage control

Figure 19. DC voltage/current step response in CC/CV mode control with R_{dc} = 54 Ω.

3.3. LCC Resonant Circuit AC Voltage and Current Validation in DAB Steady State Operation

Figures 20 and 21 show the inverter/rectifier AC voltage/current waveforms experimental results vs. simulation results in the steady-state operation of the DAB LCC resonant converter. TPS close loop control is regulated at Vo = 10 V and R_{ac} = 38 Ω with R_{dc} = 56 Ω and R_{dc} = 200 Ω connected.

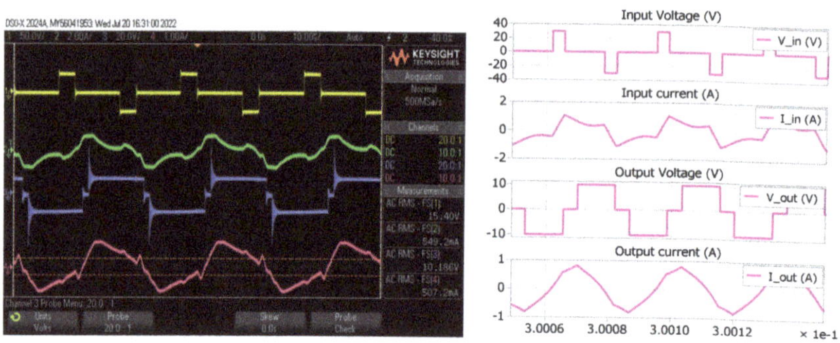

(**a**) Experiment (**b**) Simulation

Figure 20. Inverter/rectifier AC voltage and current waveforms at Vo = 10 V, R_{ac} = 38 Ω with R_{dc} = 56 Ω connected.

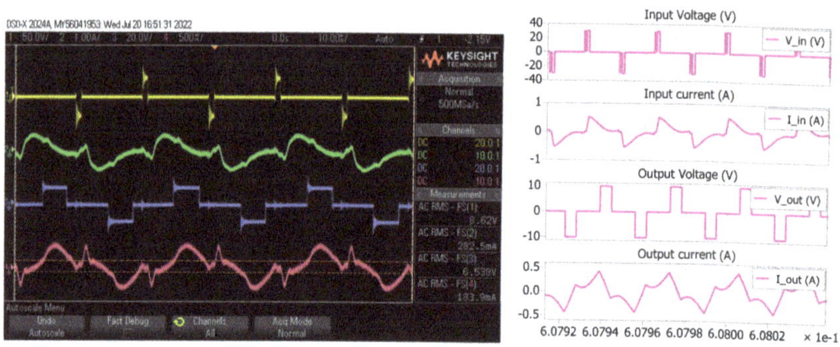

(**a**) Experiment (**b**) Simulation

Figure 21. Inverter/rectifier AC voltage and current waveforms at Vo = 10 V, R_{ac} = 38 Ω with R_{dc} = 200 Ω connected.

Table 6 compares the measured vs. simulated inverter/rectifier AC voltage and current RMS values for test cases shown in Figures 20 and 21. The RMS values of measurement results match closely with the simulation results for both R_{dc} = 56 Ω and R_{dc} = 200 Ω conditions. The phase shift angle βp and βs are greatly reduced for R_{dc} = 200 Ω, and as a result, the inverter and rectifier AC voltage and current are also reduced for R_{dc} = 200 Ω.

Figure 22 shows another test result for experimental results for inverter/rectifier AC voltage/current waveforms vs. simulation results in the steady-state operation. TPS close loop control is regulated at Vo = 30 V and R_{ac} = 38 Ω with R_{dc} = 56 Ω connected.

Table 7 compares the measured vs. simulated inverter/rectifier AC voltage and current RMS values for test cases shown in Figures 20 and 22. The RMS values of measurement results match closely with the simulation results for both Vo = 10 Ω and Vo = 30V conditions. The phase shift angle βp is greatly increased for Vo = 30 Ω. The rectifier phase shift angle βs is slightly changed for the same R_{dc} = 56 Ω connected.

Table 6. DAB Inverter/rectifier voltage/current for R_{dc} =56 Ω vs. R_{dc} = 200 Ω.

Test Cases	Vo = 10 V, R_{ac} = 38 Ω			
	R_{dc} = 56 Ω		R_{dc} = 200 Ω	
Inverter/Rectifier AC Voltage and Current	**Experiment**	**Simulation**	**Experiment**	**Simulation**
Inverter RMS voltage (V)	15.40	14.77	8.62	8.89
Inverter RMS current (A)	0.549	0.610	0.283	0.254
Rectifier RMS voltage (V)	10.19	8.27	6.54	5.66
Rectifier RMS current (A)	0.507	0.522	0.184	0.217

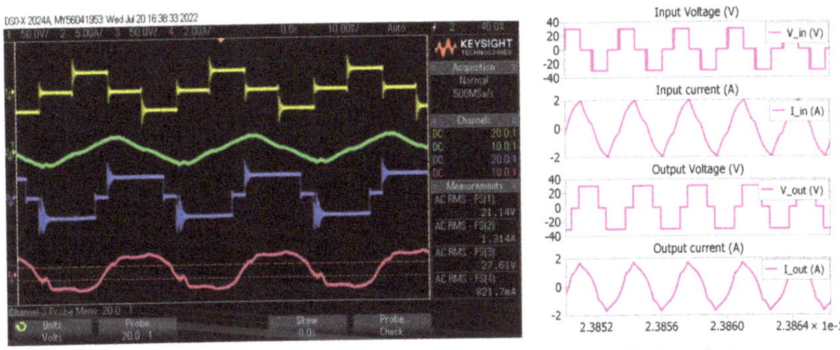

(**a**) Experiment (**b**) Simulation

Figure 22. Inverter/rectifier AC voltage and current waveforms at Vo = 30 V, R_{ac} = 38 Ω with R_{dc} = 56 Ω connected.

Table 7. DAB inverter/rectifier voltage/current for Vo = 10 V vs. Vo = 30 V.

Test Cases	R_{ac} = 38 Ω, R_{dc} = 56 Ω			
	Vo = 10 V		Vo = 30 V	
Input/Output AC Signals	**Experiment**	**Simulation**	**Experiment**	**Simulation**
Inverter RMS voltage (V)	15.40	14.77	21.14	23.91
Inverter RMS current (A)	0.549	0.610	1.31	1.32
Rectifier RMS voltage (V)	10.19	8.27	27.61	27
Rectifier RMS current (A)	0.507	0.522	0.822	1.19

3.4. Experimental Results with/without Rectifier AC Load Resistance Feedback Control

Figure 23 shows the experimental results of transmitter coil voltage (Vpri) and current (Ipr), and receiver coil voltage (Vsec) and current (Isec) for CV mode control at Vo = 10 V with R_{dc} = 150 Ω connected when rectifier R_{ac} = 38 Ω feedback control is disabled vs. enabled.

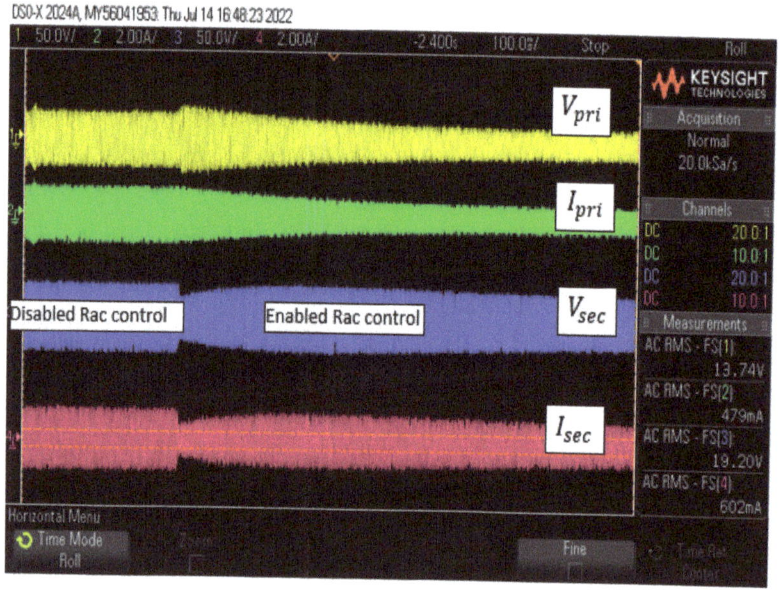

Figure 23. Transmitter/receiver current/voltage for Vo = 10 V with R_{dc} = 150 Ω connected when rectifier R_{ac} = 38 Ω feedback control is disabled vs. enabled.

Figure 24 shows the experimental results of transmitter coil voltage (Vpri) and current (Ipri) and receiver coil voltage (Vsec) and current (Isec) for CV mode control at Vo = 40 V with R_{dc} = 150 Ω connected when the rectifier R_{ac} = 38 Ω feedback control is disabled vs. enabled

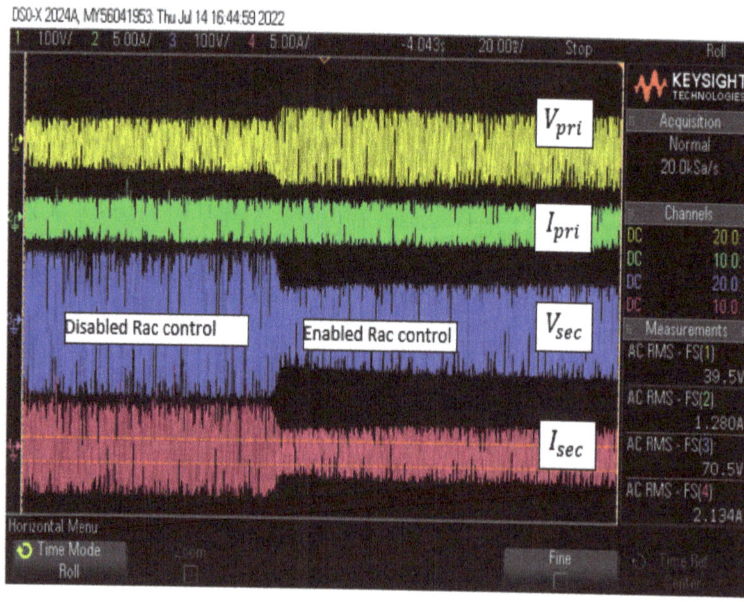

Figure 24. Transmitter/receiver coil current/voltage at Vo = 40 V with R_{dc} = 150 Ω connected when rectifier R_{ac} = 38 Ω feedback control is disabled vs. enabled.

Table 8 shows the measured transmitter/receiver winding voltage and current RMS values for R_{dc} = 150 Ω when R_{ac} = 38 Ω feedback control of rectifier phase shift angle is disabled vs. enabled. The DAB converter operates at Vo = 10 V, 20 V, 30 V, 40 V. These results show that when R_{ac} = 38 Ω feedback control is enabled, the transmitter and receiver current are greatly reduced. Therefore, the transformer winding loss is also significantly reduced.

Table 8. Transmitter/receiver RMS voltage/current with R_{dc} = 150 Ω connected.

Vo (V)	Disabled R_{ac} = 38 Ω Feedback Control βs (deg) = 180°				Enabled R_{ac} = 38 Ω Feedback Control			
	V_Pri (V)	I_Pri (A)	V_Sec (V)	I_Sec (A)	V_Pri (V)	I_Pri (A)	V_Sec (V)	I_Sec (A)
10	17.23	0.654	20.34	0.713	9.18	0.299	14.79	0.437
20	22.08	0.911	42.75	1.414	18.05	0.552	28.29	0.839
30	26.21	1.113	64.3	2.07	28.4	0.818	42.15	1.22
40	32.31	1.342	87.53	2.762	38.3	1.092	56.78	1.617

Figure 25 shows the power efficiency measurement results with R_{ac} = 38 Ω feedback control enabled (solid line) vs. with R_{ac} control disabled (βs = 0.5 and βs (deg) = 180°, dashed line) when different values of R_{dc} are connected. The measurement is conducted at different DC output voltage (Vo) levels till PI control output is saturated at a maximum phase shift angle of βp (deg) = 180°. The measurement results are plotted as output power vs. power transfer efficiency measured as the ratio of DC output power vs. DC input power.

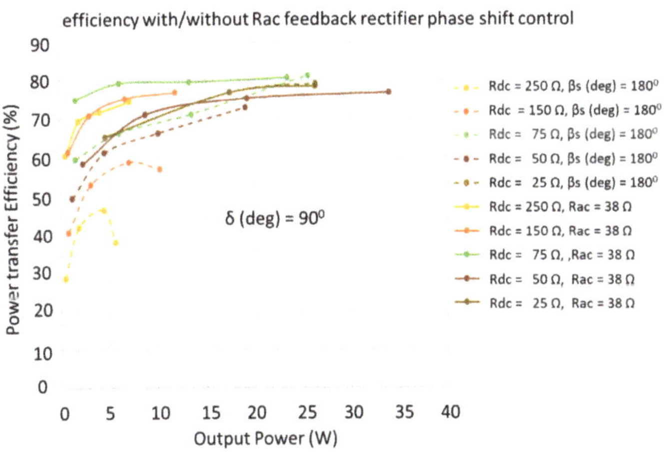

Figure 25. Power efficiency measurement with/without R_{ac} feedback control.

The measurement data shows that when R_{ac} = 38 Ω control is enabled, power efficiency is greatly improved when a larger R_{dc} is connected, especially in lower output power operation conditions of DAB. The power efficiency for R_{dc} = 250 Ω in [1 W, 5 W] output power range is improved from [30%, 40%] to [60%, 75%]. The efficiency variation range is reduced from [30%, 80%] for R_{ac} control disabled to [60%, 80%] with R_{ac} = 38 Ω control enabled.

The maximum power efficiency from measurement data around 80% is slightly lower than 86% shown in Figure 9b obtained from analytical calculation. The reason is that the SiC MOSFET switching loss and conduction loss of the DAB converter have not been included in the analytical calculation.

3.5. *Performance Evaluation for Inverter-to-Rectiiver Phase Shift Angle δ Setting Change*

In the below experimental test, TPS close loop control performance is measured for different phase shift angle δ (deg) settings to evaluate its effect on power efficiency. This test aims to validate that the DAB should work at δ (deg) = 90° to achieve unity power factor operation and maximum efficiency, as what is proved mathematically in analytical derivation and is validated in PLECS circuit simulation.

Figure 26 shows the measurement data for input DC power vs. phase shift angle δ (deg) change. The DAB converter operates at input DC voltage Vdc = 30 V and output DC voltage Vo = 10 V, 20 V, 30 V, 40 V with different DC load resistor R_{dc} connected. For the same R_{dc} and Vo, DC output power is the same; increasing input power means reduced power efficiency.

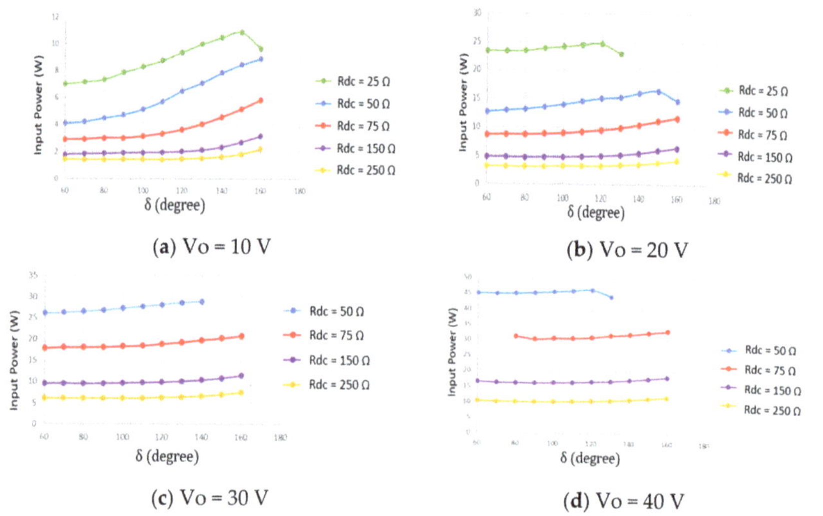

Figure 26. Input power vs. inverter-to-rectifier phase shift angle δ (deg) when DAB converter operating at Vdc = 30 V and Vo = 10 V, 20 V, 30 V. 40 V with different R_{dc} connected.

The measurement data show that the input power increases when δ (deg) > 90° for Vo = 10 V and Vo = 20 V for all R_{dc} values. Therefore, the efficiency will be reduced for a larger phase shift angle δ setting. At higher power levels for Vo = 30 V and Vo = 40 V, the input power is not sensitive to δ (deg) setting change in the range of 80° < δ (deg) < 100°. However, inverter voltage feedback control could saturate at βp = 0.5 and βp (deg) = 180° when δ (deg) > 100° and δ (deg) < 80°.

Based on the above analysis, δ (deg) = 90° is the preferred setting for DAB LCC resonant converter to operate reliability and achieve high efficiency. This conclusion tallies with the conclusion drawn from TPS open loop control simulation results for unity power factor operation of the DAB converter with δ (deg) = 90°.

4. Challenges and Possible Methods to Implement TPS Control in Decoupled Transmitter and Receiver Control Hardware

For the control scheme presented in the previous content, the TPS modulator is implemented in one PLECS RT box control hardware. Therefore, all three phase shift angles can be synchronously controlled by shifting the positive zero-crossing timing instants of the triangle PWM carrier signals of the inverter Legs and the rectifier Legs in the same PWM modulator hardware implemented in the high-timing resolution FPGA.

4.1. *Inverter and Rectifier Phase Shift PWM Modulator Implemented in Two Separate Control Hardware*

Figure 27 illustrates the inverter phase shift PWM modulators and rectifier phase shift PWM modulation signal connections when the inverter and rectifier phase shift control is to be implemented in two separate control hardware. The initial unknown random inverter-to-rectifier phase shift angle δ_init (random) in the rectifier PWM modulator is generated because of the phase timing difference of inverter Leg 1 and rectifier Leg 1 PWM carrier signals, which cannot be controlled. The inverter-rectifier phase shift angle δ is a control variable which can be adjusted using a self-tuning algorithm for TPS control. The inverter-to-rectifier phase shift angle δ is applied to remove the effect of unknown δ_init and to achieve the desired smooth power-on startup transition performance and unity power factor operation of the DAB converter in regular battery wireless charging operation.

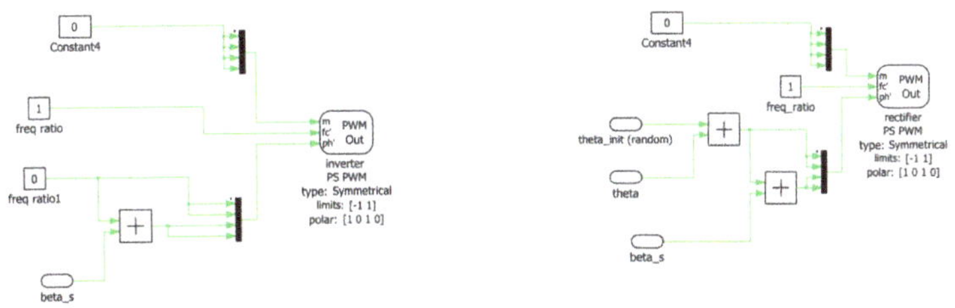

(**a**) Inverter phase shift PWM modulator (**b**) Rectifier phase shift PWM modulator

Figure 27. TPS phase shift PWM modulator signal connections when TPS control in separate inverter and rectifier control hardware.

4.2. *DAB Converter TPS Control Smooth Poweron Startup Transition with Inverter-to-Rectifier Phase Shift Angle Auto-Adjustment Scheme*

Figure 28 shows the function block diagram for the inverter-to-rectifier phase shift angle δ(deg) auto-adjustment scheme to be implemented in the same high-timing resolution rectifier FPGA of the PWM modulator with two power-on DAB converter startup transition stages added.

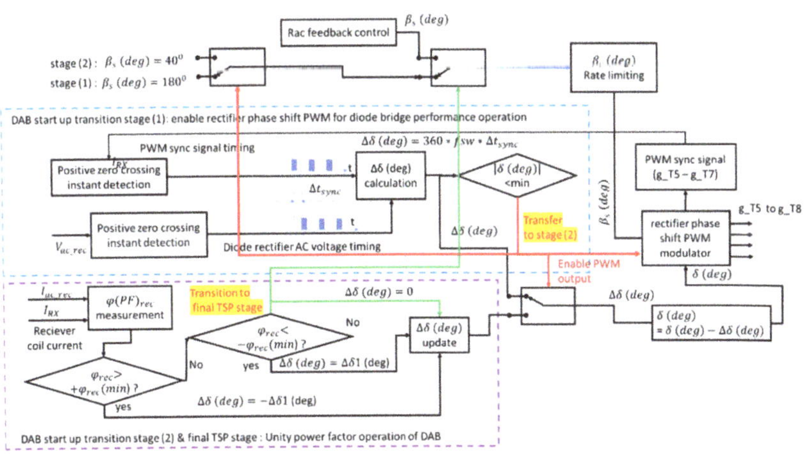

Figure 28. Function block diagram for DAB smooth startup transition from stage (1) to stage (2) and to the final TPS stage.

The transition stage (1) is to smoothly enable the rectifier phase shift PWM modulator in the DAB converter diode operation condition. The DAB converter power on startup process starts at diode rectifier operation with rectifier side PWM gating signals disabled. The inverter phase shift control is enabled for CC mode operation with a very small charging current to establish the rectifier side AC voltages for diode bridge operation. The rectifier phase shift angle is fixed to βs = 180°. The rectifier PWM carrier signal timing is auto-adjusted to match the diode rectifier timing with the inverter-to-rectifier phase shift angle change. Therefore, the rectifier phase shift control is enabled smoothly from diode bridge operation. Once the rectifier phase shift PWM modulator is enabled, the startup process transfers to the transition stage (2).

The transition stage (2) is to achieve inverter and rectifier unity power factor operation with phase shift angle δ (deg) auto-adjustment. A smaller rectifier phase shift angle is applied in an open loop condition in this stage. The inverter phase shift control operates at the same condition as the transition stage (1). The inverter-to-rectifier phase shift angle is auto-adjusted to minimize the measured rectifier power factor angle to achieve unity power factor operation. Once the DAB converter runs in unity power factor operation, the start-up transition stage (2) is completed, and TSP control is transferred to the regular power transfer operation stage.

4.3. Power on Transition Stage (1)—Smoothly Enable Rectifier Phase Shift Control in Diode Bridge Operation Condition

In the power-on transition stage (1), the rectifier phase shift PWM modulator runs internally when its output gating signals are disabled. The rectifier PWM sync signal is constructed internally in the rectifier PWM modulator FPGA by subtracting the gating signal g_T7 from g_T5. The rectifier phase shift PWM modulator starts with βs (deg) = 180°. The positive zero crossing timing difference of the PWM sync signal and diode rectifier AC voltage is used to auto-adjust δ (deg) setting until Δδ below its maximum threshold.

Figure 29 shows the operation condition when the positive zero-crossing of the rectifier PWM sync signal aligns with the diode rectifier voltage signal after successful phase shift angle δ(deg) auto-adjustment. In this condition, the rectifier phase shift PWM modulator gating signal output to the bridge power electronic switching devices is enabled. The DAB rectifier bridge phase shift PWM modulation performs similarly to the diode rectifier with smooth power on transition.

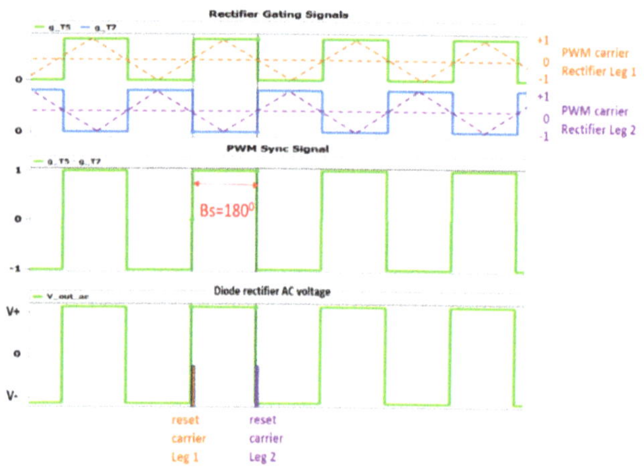

Figure 29. DAB smooth startup transition stage (1): phase shift angle δ (deg) auto-adjustment for rectifier PWM modulator smoothly enabled in diode rectifier operation condition.

4.4. Power on Transition Stage (2)—Inverter-to-Rectifier Phase Shift Angle Auto-Adjustment for DAB Unity Power Factor Operation in Rectifier Phase Shift Open Loop Control

In the power-on transition stage (2), inverter control is the same as in stage (1). The rectifier shift angle βs (deg) is smoothly reduced through the rate change limiting block to a smaller value setting (for example, βs = 40°) to improve the power factor angle auto-adjustment sensitivity. The rectifier AC power factor angle $\varphi_{rec}(PF)$ is measured. The rectifier AC power factor angle is used to adjust δ (deg) to minimize the rectifier AC power factor angle to achieve unity power factor operation of DAB bridges.

Figure 30 shows the block diagram for the rectifier AC power factor angle measurement from phase difference detection of receiver coil AC current and rectifier AC current.

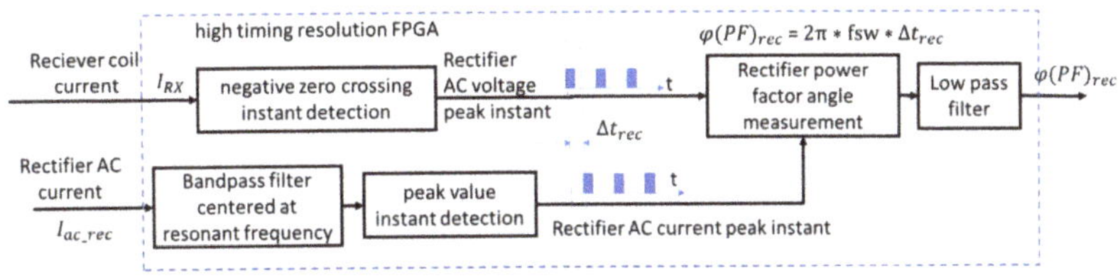

Figure 30. Rectifier AC power factor angle measurement.

The relationship of the fundamental frequency component of rectifier AC voltage and receiver coil current derived in (36) is rewritten as (43), which means that rectifier AC voltage lags the receiver coil current by 90°. Utilizing (43), the peak instant of the fundamental frequency component of rectifier AC voltage can be detected by the negative zero crossing instant of the harmonic-free receiver coil current.

$$I_{RX} = j\omega_{sw} C_{2p} V_{out\_ac} \tag{43}$$

The high order harmonics of rectifier AC current should be removed by bandpass filter centered at resonant frequency before its peak timing instant detection. The peak instant of the filtered rectifier current can be captured as the phase instant of the rectifier AC current.

The AC power factor angle of the rectifier is calculated from the measured timing error as shown in (44), where "fsw" is the DAB converter resonant and switching frequency. The low pass filtered power factor angle is used for phase shift angle δ auto-tuning adjustment purposes.

$$\varphi(PF)_{rec} = 2\pi * fsw * \Delta t_{rec} \tag{44}$$

Figure 31 illustrates the simulation results of inverter/rectifier AC voltage/current and receiver coil current for different inverter-to-rectifier phase shift angle δ (deg) settings to achieve unity power factor operation. The unity power operation is achieved when the rectifier current peak time instant aligns with the negative zero crossing of the receiver coil current.

4.5. DAB TPS Close Loop Control Stage with Inverter-to-Rectifier Phase Shift Angle Auto-adjustment for DAB Unity Power Factor Operation with Rectifier Feedback Control Enabled

In the power-on transition stage (2), when the rectifier power factor angle is below the threshold setting, the startup process is transferred from stage (2) to the normal TPS closed-loop control stage. In the final TPS control stage, R_{ac} feedback control regulation of βs angle is enabled. The inverter-to-rectifier phase shift angle δ (deg) auto-adjustment for unity power factor operation of DAB continues to run to eliminate the accumulated phase timing draft in either the inverter PWM carrier signal or rectifier PWM carrier signal. The inverter battery voltage and current control are changed to the automatic CCCV mode setting based on battery voltage measurement and SOC status estimation.

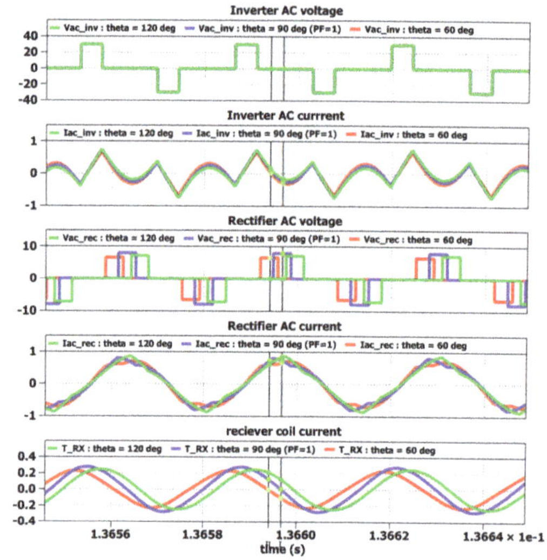

Figure 31. DAB smooth startup transition stage (2): rectifier δ (deg) auto-adjustment to achieve unity power factor control of rectifier DAB bridge.

5. Conclusions and Discussions

This paper proposes TPS closed-loop control method to realize the unity power factor operation of wireless charging DAB LCC resonant converter at optimized resonant circuit AC load resistance. The efficacy improvement for the proposed TPS control comes mainly from unity power factor operation of the DAB inverter and rectifier. ZVS in CC mode operation helps to improve the efficiency as well. The R_{ac} feedback control improves the efficiency in CV mode charging operation only. One inverter/rectifier Leg may lose ZVS switching in CV mode charging control, which may lower the efficiency at CV mode than CC mode charging.

The analytical calculation of the LCC resonant converter performance using its resonant equivalent circuit model shows that the converter output DC voltage/current and input/output power changes with inverter phase shift angle for the same resonant circuit AC load resistance level. The mathematical analysis provides the theoretical foundation to implement the battery charging current and voltage feedback control with inverter phase shift regulation. The analytical calculation result shows that the AC load resistance of the LCC resonant circuit is another crucial factor which significantly affects the voltage, current, power, and power efficiency of the resonant converter. There is an optimized AC load resistance value for maximum efficiency operation. For a unidirectional LCC resonant circuit with a diode rectifier, the significant variation of equivalent battery load resistance maps into the considerable variation of AC load resistance with a gain factor of $8/\pi^2$. The power efficiency will be degraded when the equivalent battery resistance is significantly increased in CV mode charging. The variation of AC load resistance also causes considerable DC voltage gain variation and inverter load impedance frequency curve shape changes. In DAB TPS control, the low-efficiency issue in CV mode charging can be resolved with AC load resistance feedback control around at optimal target by regulating the rectifier phase shift angle. The ratio of rectifier AC resistance and equivalent battery DC resistance is derived as a non-linear function for rectifier phase shift angle. The AC load resistance is estimated from measured DC load resistance and rectifier phase shift angle to construct the feedback control signal. Lastly, the inverter-to-rectifier phase shift angle for unity power factor operation of the DAB LCC resonant converter is mathematically proven to be $\delta = 90°$. This setting ensures minimizing the reactive power flow in LCC resonant circuit. Therefore, it maximizes the LCC resonant circuit efficiency.

The simulation analysis and experimental test are conducted to validate the performance of the proposed DAB resonant converter high-efficiency control solution. The simulation and experimental results verified that the proposed DAB control system works as designed. DC output voltage and current can be regulated stably around its reference value in either current or voltage feedback control conditions with inverter phase shift control. The unity power factor operation is achieved when the inverter-to-rectifier phase shift angle is $\delta = 90°$. The DAB converter efficiency is significantly improved when rectifier side R_{ac} feedback control is regulated around its smaller normal value when a larger DC load resistance is connected in the test. The transformer winding current magnitude is greatly reduced when R_{ac} feedback control is enabled when larger DC load resistance is connected to DC/DC converter output. The simulation and measurement results for TPS control matches well.

The ZVS performance of TPS control is conducted in the circuit simulation in the simulated CC mode control and CV mode control of battery charging condition by replacing the load resistor with a DC voltage supply to emulate the battery. The simulation results show that Sic MOSFET switches of the inverter and rectifier work at ZVS soft-switching conditions in the emulated CC mode charging with a low battery equivalent load resistance. However, in the CV mode charging with large battery equivalent load resistance, one of the inverter/rectifier Legs may lose ZVS soft switching. This explains why the measured efficiency is still around 20% lower at a smaller power condition with larger DC load resistance connected, even if the R_{ac} control is enabled. From the circuit simulation results, a slight adjustment of inverter-to-rectifier angle may not improve the ZVS switching in CV mode operation.

The proposed TPS control is validated in a small power laboratory experimental setup. The achievable efficiency is from 60% to 80% for the output power range from 1 W to 35 W. The efficiency improvement with R_{ac} feedback control enabled is around 30% at power level from 1 W to 5 W. The total efficiency measurement from this small power wireless charging DAB LCC resonant converter experimental setup is lower than the published efficiency from a large power wireless charging setup. This is because, at a low power level of around 30 W, the small SIC MOSFET switching and conduction loss contribute to a significant percentage of power efficiency reduction. Efficiency improvement from 30% to 60% at 1 W to 5 W power level with R_{ac} feedback control enabled indicates that winding loss is a significant contribution for the wireless charging system even at a very low power level. For kW range, high power EV charging application, the efficiency of the proposed TSP control scheme will be much better than the DAB control solutions published as a result of optimized R_{ac} feedback control and the unity power factor operation of DAB converter bridges resulting from a significant reduction in winding loss in the wireless charging system.

There are some engineering challenges to implementing the proposed TPS closed-loop control scheme in the wireless charging system with separate transmitter and receiver control hardware. The paper provides some basic ideas on how to implement the inverter and rectifier PWM modulation separately. The power-on startup transition stages are required to be added smoothly to enable rectifier side phase shift PWM modulation in diode rectifier operation conditions. The inverter-to-rectifier phase shift angle is to be automatically adjusted to minimize the power factor angle to achieve the unity power factor operation. The feasibility study is done with circuit simulation. However, it requires considerate engineering effort to make the idea work in a suitable wireless charging product.

Author Contributions: Conceptualization, S.C., Y.C., R.T.N. and S.L.; methodology, S.C. and Y.C.; software, S.C. and Y.C.; validation, S.C. and Y.C.; formal analysis, S.C. and Y.C.; investigation, R.T.N. and S.L.; writing—original draft preparation, S.C. and Y.C.; writing—review and editing, S.C., R.T.N. and S.L.; supervision, R.T.N. and S.L. All authors have read and agreed to the published version of the manuscript.

Funding: This research received no external funding.

Institutional Review Board Statement: Not applicable.

Informed Consent Statement: Not applicable.

Data Availability Statement: Not applicable.

Acknowledgments: The authors would like to thank Singapore Institute Technology for the Lab equipment provided for experimental validation.

Conflicts of Interest: The authors declare no conflict of interest.

Abbreviations

The following abbreviations are used in this manuscript:

DC	Direct current
AC	Alternate current
CC	Constant current
CV	Constant voltage
EV	Electric vehicle
LLC	Inductor-inductor-capacitor
LCC	Inductor-capacitor-capacitor
DAB	Dual active bridge
Rx	Receiver
Tx	Transmitter
WPT	Wireless power transfer
PLL	Phase lock loop
ZPA	Zero phase angle
ZVS	Zero voltage switching
DPS	Double phase shift
TPS	Triple phase shift
EMI	Electromagnetic interference

References

1. Ahmad, A.; Alam, M.S.; Chabaan, R. A comprehensive review of wireless charging technologies for electric vehicles. *IEEE Trans. Transp. Electrif.* **2017**, *4*, 38–63. [CrossRef]
2. Li, S.; Li, W.; Deng, J.; Nguyen, T.D.; Mi, C.C. A Double-Sided LCC Compensation Network and Its Tuning Method for Wireless Power Transfer. *IEEE Trans. Veh. Technol.* **2015**, *64*, 2261–2273. [CrossRef]
3. Galigekere, V.P.; Onar, O.; Pries, J.; Zou, S.; Wang, Z.; Chinthavali, M. Sensitivity analysis of primary-side LCC and secondary-side series compensated wireless charging system. In Proceedings of the 2018 IEEE Transportation Electrification Conference and Expo (ITEC), Long Beach, CA, USA, 13–15 June 2018; IEEE: Piscataway, NJ, USA, 2018; pp. 885–891.
4. Feng, H.; Cai, T.; Duan, S.; Zhao, J.; Zhang, X.; Chen, C. An LCC-Compensated Resonant Converter Optimized for Robust Reaction to Large Coupling Variation in Dynamic Wireless Power Transfer. *IEEE Trans. Ind. Electron.* **2016**, *63*, 6591–6601. [CrossRef]
5. Haque, M.S.; Mohammad, M.; Pries, J.L.; Choi, S. Comparison of 22 kHz and 85 kHz 50 kW wireless charging system using Si and SiC switches for electric vehicle. In Proceedings of the 2018 IEEE 6th Workshop on Wide Bandgap Power Devices and Applications (WiPDA), Atlanta, GA, USA, 31 October 2018–2 November 2018; IEEE: Piscataway, NJ, USA, 2018; pp. 192–198.
6. Nguyen, B.X.; Singh, J.; Zhang, X.; Jiang, W.; Koh, L.H.; Wang, P. Design and Control for ZVS Constant Current and Constant Voltage Wireless Charging Systems. In Proceedings of the 2018 IEEE 4th Southern Power Electronics Conference (SPEC), Singapore, 10–13 December 2018; IEEE: Piscataway, NJ, USA, 2018; pp. 1–7.
7. Zhang, X.; Cai, T.; Duan, S.; Feng, H.; Hu, H.; Niu, J.; Chen, C. A Control Strategy for Efficiency Optimization and Wide ZVS Operation Range in Bidirectional Inductive Power Transfer System. *IEEE Trans. Ind. Electron.* **2019**, *66*, 5958–5969. [CrossRef]
8. Qu, X.; Chu, H.; Wong, S.C.; Chi, K.T. An IPT battery charger with near unity power factor and load-independent constant output combating design constraints of input voltage and transformer parameters. *IEEE Trans. Power Electron.* **2019**, *34*, 7719–7727. [CrossRef]
9. Vu, V.-B.; Tran, D.-H.; Choi, W. Implementation of the Constant Current and Constant Voltage Charge of Inductive Power Transfer Systems with the Double-Sided LCC Compensation Topology for Electric Vehicle Battery Charge Applications. *IEEE Trans. Power Electron.* **2018**, *33*, 7398–7410. [CrossRef]
10. Lu, J.; Zhu, G.; Lin, D.; Zhang, Y.; Wang, H.; Mi, C.C. Realizing Constant Current and Constant Voltage Outputs and Input Zero Phase Angle of Wireless Power Transfer Systems with Minimum Component Counts. *IEEE Trans. Intell. Transp. Syst.* **2021**, *22*, 600–610. [CrossRef]

11. Lu, J.; Zhu, G.; Lin, D.; Zhang, Y.; Jiang, J.; Mi, C.C. Unified Load-Independent ZPA Analysis and Design in CC and CV Modes of Higher Order Resonant Circuits for WPT Systems. *IEEE Trans. Transp. Electrif.* **2019**, *5*, 977–987. [CrossRef]
12. Cao, S.; Naing, H.Y.Y.; Naayagi, R.T.; Lee, S.S.; Wei, F.; Tseng, K.J. Wireless charging resonant converter topology study based on analytical design computation. In Proceedings of the 2020 3rd International Conference on Energy, Power and Environment: Towards Clean Energy Technologies, Shillong, India, 5–7 March 2021; IEEE: Piscataway, NJ, USA, 2021; pp. 1–6.
13. Cao, S.; Nawawi, A.; Lim, Z.; Ang, J.; Hu, X.; Tong, C.F.; Tseng, K.J. Wireless Charging Technologies and Standardization for Electric Unmanned Crafts. In Proceedings of the 2022 International Power Electronics Conference (IPEC-Himeji 2022-ECCE Asia), Himeji, Japan, 15–19 May 2022; IEEE: Piscataway, NJ, USA, 2022; pp. 2147–2154.
14. Li, H.; Li, J.; Wang, K.; Chen, W.; Yang, X. A Maximum Efficiency Point Tracking Control Scheme for Wireless Power Transfer Systems Using Magnetic Resonant Coupling. *IEEE Trans. Power Electron.* **2015**, *30*, 3998–4008. [CrossRef]
15. Zhao, Q.; Wang, A.; Liu, J.; Wang, X. The Load Estimation and Power Tracking Integrated Control Strategy for Dual-Sides Controlled LCC Compensated Wireless Charging System. *IEEE Access* **2019**, *7*, 75749–75761. [CrossRef]
16. Li, Z.; Huang, X.; Song, K.; Jiang, J.; Zhu, C.; Du, Z. Constant current charging and the maximum system efficiency tracking for wireless charging systems employing dual-side control. In Proceedings of the 2018 International Power Electronics Conference (IPEC-Niigata 2018-ECCE Asia), Niigata, Japan, 20–24 May 2018; pp. 84–87.
17. Jiang, Y.; Wang, L.; Wang, Y.; Wu, M.; Zeng, Z.; Liu, Y.; Sun, J. Phase-Locked Loop Combined with Chained Trigger Mode Used for Impedance Matching in Wireless High Power Transfer. *IEEE Trans. Power Electron.* **2020**, *35*, 4272–4285. [CrossRef]
18. Ping, W. Modeling and Simulation of Full-Bridge Series Resonant Converter Based on Generalized State Space Averaging. *Appl. Mech. Mater.* **2013**, *347–350*, 1828.
19. Salem, M.; Jusoh, A.; Idris, N.R.N.; Alhamrouni, I. Modeling and simulation of generalized state space averaging for series resonant converter. In Proceedings of the 2014 Australasian Universities Power Engineering Conference (AUPEC), Perth, Australia, 28 September–1 October 2014; IEEE: Piscataway, NJ, USA, 2014; pp. 1–5.
20. Yang, E.X. Extended Describing Function Method for Small-Signal Modeling of Resonant and Multi-Resonant Converters. Ph.D Thesis, Virginia Tech, Blacksburg, VA, USA, 1994.

MDPI
St. Alban-Anlage 66
4052 Basel
Switzerland
www.mdpi.com

Applied Sciences Editorial Office
E-mail: applsci@mdpi.com
www.mdpi.com/journal/applsci

www.ingramcontent.com/pod-product-compliance
Lightning Source LLC
Chambersburg PA
CBHW061604220926
51850CB00005B/422
9783036593364